Nonlinear and Environmental Electromagnetics

STUDIES IN ELECTRICAL AND ELECTRONIC ENGINEERING

Vol. 1 Solar Energy Conversion: The Solar Cell (Neville)
Vol. 2 Charge Storage, Charge Transport and Electrostatics with their Applications (edited by Wada, Perlman and Kokado)
Vol. 3 Integrated Functional Blocks (Novák)
Vol. 4 Operational Amplifiers (Dostál)
Vol. 5 High-Frequency Application of Semiconductor Devices (F. Kovács)
Vol. 6 Electromagnetic Compatibility in Radio Engineering (Rotkiewicz)
Vol. 7 Design of High-Performance Negative-Feedback Amplifiers (Nordholt)
Vol. 8 Discrete Fourier Transformation and Its Applications to Power Spectra Estimation (Geçkinli and Yavuz)
Vol. 9 Transient Phenomena in Electrical Machines (P.K. Kovács)
Vol. 10 Theory of Static Converter Systems: Mathematical Analysis and Interpretation. Part A: Steady-State Processes (Slonim)
Vol. 11 Power Sources for Electric Vehicles (McNicol and Rand, Editors)
Vol. 12 Classical Electrodynamics (Ingarden and Jamiołkowski)
Vol. 13 Reliability of Analogue Electronic Systems (Klaassen)
Vol. 14 Electrets (Hilczer and Małecki)
Vol. 15 Graph Theory: Application to the Calculation of Electrical Networks (Vágó)
Vol. 16 Eddy Currents in Linear Conducting Media (Tegopoulos and Kriezis)
Vol. 17 Electrical Measurements in Engineering (Boros)
Vol. 18 Active RC Filters (Herpy and Berka)
Vol. 19 Surface-Wave Devices for Signal Processing (Morgan)
Vol. 20 Micromaching and Micropackaging of Transducers (edited by Fung, Cheung, Ko and Fleming)
Vol. 21 Nonlinear and Environmental Electromagnetics (edited by Kikuchi)
Vol. 22 Microwave Measurements of Complex Permittivity by Free Space Methods and Their Applications (Musil and Žáček)

STUDIES IN ELECTRICAL AND ELECTRONIC ENGINEERING 21

Nonlinear and Environmental Electromagnetics

Proceedings of the URSI (International Union of Radio Science) Commission E Workshop, Tokyo, October 19, 1984

Edited by

HIROSHI KIKUCHI

Nihon University, College of Science and Technology, Tokyo, Japan

ELSEVIER

Amsterdam – Oxford – New York – Tokyo 1985

ELSEVIER SCIENCE PUBLISHERS B.V.
Sara Burgerhartstraat 25
P.O. Box 211, 1000 AE Amsterdam, The Netherlands

Distributors for the United States and Canada:

ELSEVIER SCIENCE PUBLISHING INC.
52, Vanderbilt Avenue
New York, NY 10017 U.S.A.

Library of Congress Cataloging-in-Publication Data

International Union of Radio Science. Commission E.
Workshop (1984 : Tokyo, Japan)
Nonlinear and environmental electromagnetics.

(Studies in electrical and electronic engineering ; 21)
Bibliography: p.
Includes index.
1. Atmospherics--Congresses. 2. Electromagnetic waves--Transmission--Congresses. 3. Solitons--Congresses. 4. Nonlinear theories--Congresses. I. Kikuchi, H. (Hiroshi), 1926- . II. Title. III. Title: Environmental electromagnetics. IV. Series.
QC973.4.A85I58 1984 551.5'6 85-27428
ISBN 0-444-42571-3 (U.S.)

ISBN 0-444-52571-3 (Vol. 21)
ISBN 0-444-41713-3 (Series)

Printed in The Netherlands

PREFACE

Electric disturbances from lightning discharges were studied by Popov in 1895 and by Admiral Sir Henry Jackson in 1902, and the programme of the International Commission on Scientific Radio in 1913 contained the statement "Simultaneous measurements at the various receiving stations of disturbances due to atmospherics are necessary". The International Union of Radio Science, founded in 1919, immediately set up a Commission on Atmospherics (one of only four Commissions), and in 1927 Professor Kimpara joined this Commission, a collaboration that lasted several decades. He was also Chairman of the Japanese URSI Committee. His review of half a century of atmospherics is very interesting. He has included even the effect of nuclear explosions (as early as 1958). Whistlers and other VLF phenomena have been on the programme of URSI since 1950. In the next Assemblies (1952, 1954) it became clear that whistlers propagate along magnetic-field-aligned paths in the earth's outer atmosphere. For this reason it was a subject of study not only in Commission E, but also in Commission H, which studies the magnetosphere. The papers by Hayakawa, Tanaka, Sazhin and Okada, by Hayakawa, Okada, Tanaka and Ohta and by Ondoh and Nakamura are in this domain. Canadian research on atmospherics and em environments is covered by Professor Pavlasek.

Professor Marcuvitz and I were Vice-Chairmen of URSI Commission VI in the early sixties, a Commission that covered a wide range of subjects, now covered by Commissions B, C and H. His paper is an extension of the particle-wave duality (de Broglie, Schrödinger, Heisenberg), leading to an analogy between the propagation of wave packets and the trajectory of quasiparticles. Solitons were long considered as interesting mathematical solutions of nonlinear differential equations, or as rare physical phenomena. Now they may have a practical interest in optical fibre communication. Solitons propagate without change in amplitude or shape, because the effects of dispersion are balanced by the effects of nonlinearity. The peak power in single-mode fused silica fibres may amount to 1 W and the duration to 10 psec. Fibre systems based on soliton propagation appear quite practical for very high pulse rate with very little residual pulse distortion (see Unger: Trends in Optical Communication, ICC Amsterdam, 1984 pp. 153-158). In the present symposium, solitons have been treated by Kikuchi and Hirota, by Nakamura, and by Noguchi.

Professor Goto published results on undamped oscillations of electric systems as early as 1930, and he solves many problems of nonlinear oscillations in the present paper. Maginu's treatment of chaotic oscillations in a Josephson junction line complements his earlier researches in this field.

Environmental electromagnetics is the subject of three papers. Man-made noise inside a screened cable has been studied by Professor Degauque's group for several years. Reporting on lightning research activities in France, Degauque, Hamelin and Demoulin treat the effect of the lightning em pulse on a telecommunication network. Professor Struzak deals with industrial, scientific and medical rf generators, a subject on which he chaired an International CCIR Working Party. We expect agreement on limits later this year. The behaviour of animals in extremely low frequency fields was studied by Behari, Mathur and Sharma. The general subject of biological effects of em radiation is on the programme of Commissions A and E of URSI, Commission A being up to now the most active in Plenary Assemblies, combining its efforts with the Bioelectromagnetics Society and other international societies.

Ponderomotive forces in plasmas are the subject of the papers by Akama and by Sugai. The latter deals with the effect of a strong electromagnetic pulse. This forms a connection with the next four papers, in which the nuclear electromagnetic pulse is treated by a German, a Swiss, an English and a Swedish scientist. In Florence, during the last URSI Plenary Assembly, an international commission chaired by Mr. Manuel Wik reached agreement on a factual statement on the nuclear electromagnetic pulse and its associated effects. This statement was accepted unanimously by the URSI Executive Council.

I congratulate Professor Kikuchi on a very successful Symposium, and I hope that the publication of its Proceedings will benefit many interested people.

F.L.H.M. Stumpers
Chairman, URSI Commission E.

ADDRESS OF WELCOME

Mr. President, Ladies and Gentlemen, it is a great honour and delightful pleasure for me to welcome you in the name of Japanese National Committee of URSI Commission E. We are hoping that the International Workshop on Nonlinear and Environmental Electromagnetics will open the way to interesting scientific and engineering works and that this meeting will give you opportunities to tighten your friendship links. We wish you to be convinced that we will do all what we can, to make your stay in Japan the most pleasant and fruitful.

Really I made contact with URSI in 1927, the Second General Assembly at Washington, D.C., especially to Commission III - Atmospheric Disturbances, and ended the attendance at the 60th Anniversary of its General Assembly at Bruxelles in 1979.

Now I would like to deliver my speech on "Half Century Observations of Atmospherics".

Atsushi KIMPARA
Ex-President of the National Committee of URSI in Japan
Professor Emeritus, Nagoya University and Chubu University

NONLINEAR AND ENVIRONMENTAL ELECTROMAGNETICS

TABLE OF CONTENTS

V. PONDEROMOTIVE FORCE IN AN EM ENVIRONMENT

VI. ENVIRONMENTAL EFFECTS OF NEMP

SUPPLEMENT - URSI COMMISSION E ACTIVITY DURING 1982-1985

* Invited Special Paper

I. ATMOSPHERIC EMP AND VLF PHENOMENA

Nonlinear and Environmental Electromagnetics, edited by H. Kikuchi
Elsevier Science Publishers B.V., Amsterdam, 1985 – Printed in The Netherlands

HALF CENTURY OBSERVATIONS OF ATMOSPHERICS

Atsushi Kimpara
Ex-President, National Committee of URSI in Japan;
Professor Emeritus, Nagoya University and Chubu University

Half century observations of atmospherics in Japan are summarized in the following terms:

1. Thunderstorms
2. Typhoons
3. Polar fronts and jet streams
4. Sudden enhancement of atmospherics and sudden ionospheric disturbances (SEA and SID)
5. Sudden phase shift of VLF waves
6. SEA phenomena due to nuclear explosion
7. Waveform of atmospherics
8. Distribution of atmospherics in Asia.

INTRODUCTION

Electric noise due to lightning discharge, spark on pantagraph of electric train, corona discharge on high tension insulator, etc. disturbs radio communications or broadcasting, as treated in CISPR, CCIR, URSI, etc. However, atmospherics emitted from discharges in the atmosphere, inform us about the position and construction of cold front, polar front, jet stream, typhoon, hurricane and nuclear explosion in the upper atmosphere and about abnormal change in electron density of D layer due to flares in solar emissions such as ultraviolet ray and X ray. These informations are very useful for radio communications, weather forecast, national defense, etc. Hereupon these examples are described in the following, according to historical developments.

1. THUNDERSTORMS

Directional observations of atmospherics were made at six stations (Iwatsuki, Osaka, Ochiishi, Seoul, Taipei and P. Authur) from the beginning of every hour during 30 seconds per one direction, every 30 degrees clockwise starting from north, twice a month in 1928-1930 for two years. To guarantee exact simultaneity of observations, signals from Kemigawa station (36.5 kHz) and later Nauen station (16.6 kHz) were received and recorded on the receiving tape, because in those days it was impossible to rely upon standard time signals. Comparison between the position of thunderstorms observed at meteorological observatories and that of source of atmospherics determined by triangulation of two or three stations, shows the coincidence of 91.5%, and therefore the fact that thunderstorms are a powerful source of atmospherics is proved observationally.

2. TYPHOONS

Atmospherics due to typhoons were observed at the Reserach Institute of Atmospherics, Nagoya University (Toyokawa) and Ohira-Observatory (Chiba-Pref.) (distance between them is about 300 km) at 02.50, 08.50, 14.50, 17.50 and 20.50 JMT for 2 minutes at every time, with a cathode-ray oscillograph direction-finder and a photographic film, on which standard time signals were recorded: in 1950 typhoon Kezia and in 1951 typhoon Ruth. According to the study on the correlation of atmospherics with meteorological informations in the upper atmosphere, especially at a height of 3000 m, the origins of atmospherics are mainly distributed over conditional instability areas liable to develop frontal or air mass thunderstorms. The former occurs on account of instability released at troughs with ascent of air along frontal surfaces; the latter develops on account of instability within the air masses. Somw phase of the construction of typhoon were made clear by the distribution of origin of atmospherics: typhoon Kezia in September 1950 and typhoon Ruth in October 1951.

In Kezia, the origins of atmospherics are distributed most frequently on the right hand side of the center, especially in the direction perpendicular to the course, and increase slowly with distance from the centre to 400 km, where they reach a maximum and then decrease rather rapidly to 600 km, as shown in Fig. 1.

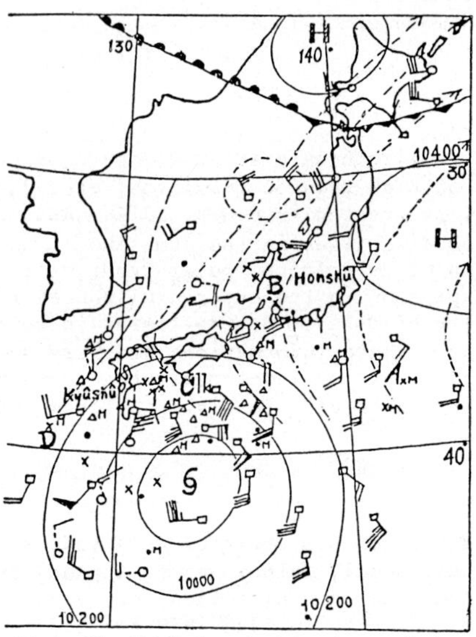

Fig. 1 a. The distribution of origins of atmospherics on September 12, 1950. (17.01 to 19.02 JMT). ---indicates surface wind, ● origins of atmospherics at 17.01, △ at 18.01, × at 19.01

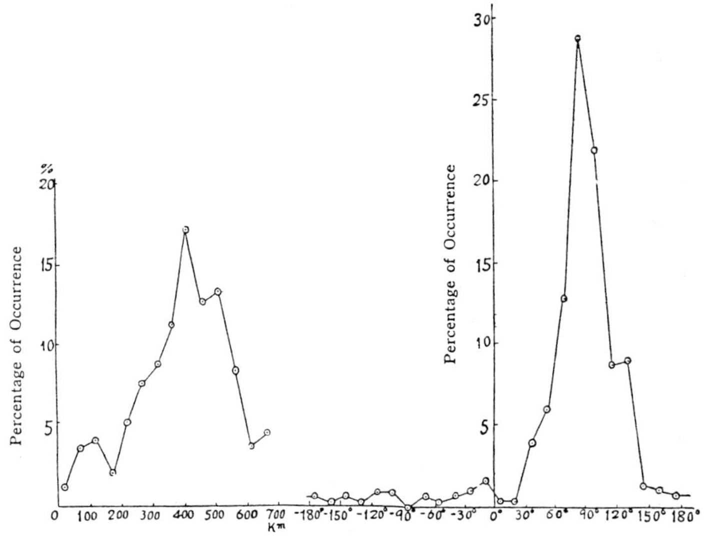

Fig. 1 b The distribution of origins of atmospherics in the neighbourhood of the Typhoon Kezia.

This maximum distribution region is considered to be the area of heavy convergence of hot and wet air mass from tropics and cold air mass from arctic regions. It seems that Kezia changed its characteristics from tropical to temperate after entering into the Sea of Japan, and origins are found to be gradually weakened on the Pacific. When it reached the middle of the Sea of Japan, new origins appeared in the Honshu area. Origins on the Pacific are considered mainly due to a convergence of hot and wet air mass from tropics and warm and wet air mass from high pressure on the Pacific, while origins on the Sea of Japan due to a convergence of cold air mass from Siberia and somewhat modified hot and wet air mass from tropics. Consequently, as the character of typhoon changed from tropical to temperate, so changed the system of origins of atmospherics from the one on the Pacific to that in the Honshu area.

Although in Kezia the origins of atmospherics were found mainly on the right hand side of the course of typhoon, origins in typhoon Ruth are found not only in the area above mentioned but also in the convergence region in front of trough of westerly waves, in 700 or 500 mb regions, situated at several hundred km before the centre, and in the similar region of easterly waves in the rear or in the centre of typhoon. There was another distribution of origins off the coast of the Japan Islands on the Pacific, sometimes prolonged over to long. 180°E, between lat. 30° and 40° N in a zonal form. This group of atmospherics makes itself quite clear in autumn and winter, while in summer it is rather vague, and it seems that this is due to a convergence in a frontal zone in the upper atmosphere as well as on the surface. In the following, four different phases of typhoon Ruth are explained.

(i) 20.50 JMT on 11 Oct., 1951. When the typhoon developed remarkably and indicated 940 mb at the centre, there appeared a trough of easterly waves in 700 mb in the southeastern sector of Ruth, and conspicuous distribution of origins was observed in front of trough. Lightning flashes and cumulonimbi in a weather map comfirm a heavy convergence over there, as shown in Fig. 2.

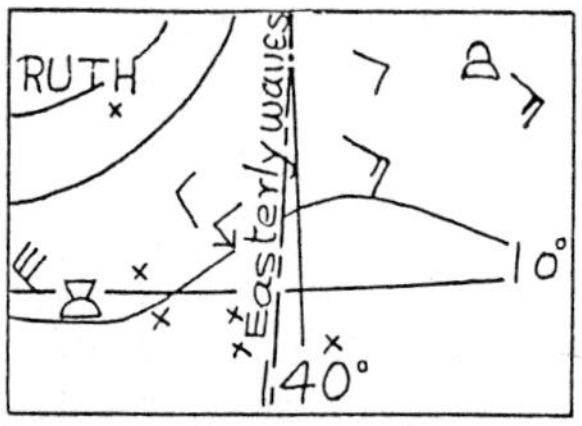

Fig. 2. Distribution of origins of atmospherics (×) at 20.50 JMT on 11 Oct., 1951.

(ii) 14.50 JMT on 9 Oct., 1951. There was a trough of westerly waves in 700 mb in front of Ruth, and an eminent distribution of origins was observed before the trough. It may be considered that convergence before the trough was strongly intensified by the wet and warm air masses due to the typhoon, as shown in Fig. 3.

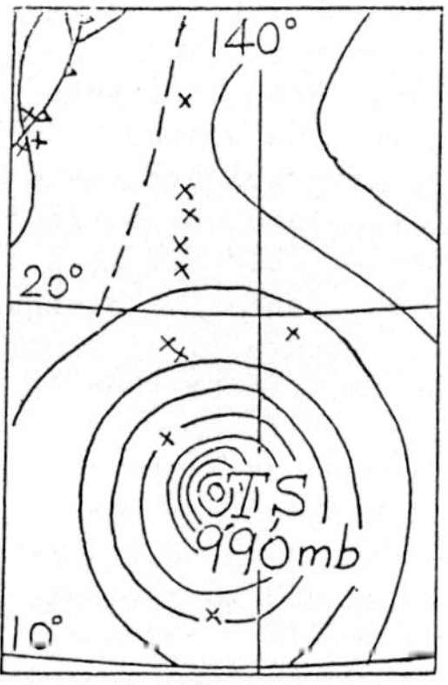

Fig. 3. Distribution of origins of atmospherics (×) at 14.50 JMT on 9 Oct., 1951.

(iii) 08.50 JMT on 10 Oct., 1951. There were troughs of westerly waves in 700 mb and 500 mb; the trough in 700 mb was found far apart before the centre, while the one in 500 mb nearly in front of the centre. Both of them extended themselves far to the north and a conspicuous distribution of origins was found before the trough in 700 mb and 500 mb respectively, as shown in Fig. 4.

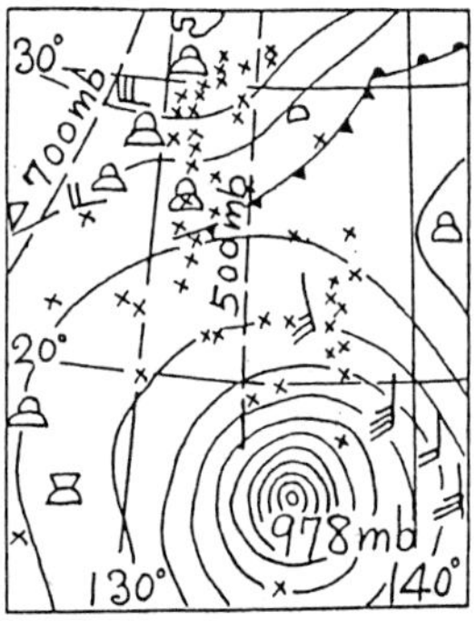

Fig. 4. Distribution of origins of atmospherics (×) at 08.50 JMT on 10 Oct., 1951.

(iv) 08.50 JMT on 11 Oct., 1951. It is easily found that the heavy convergence between the centre of Ruth and high pressure in the east, and a remarkable distribution of origins are existent there similar to the case of Kezia, as shown in Fig. 5. The difference in the distribution of origins between Ruth and Kezia may be due to that of characteristics corresponding to arriving seasons, but the properties of the trough of westerly waves determine the course of typhoon, and it should be studied further in future.

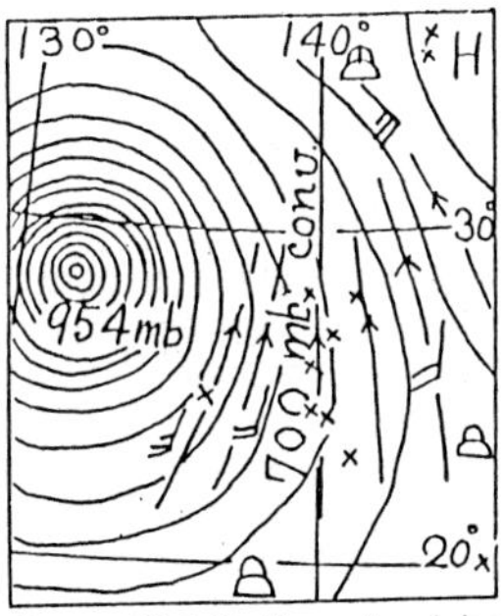

Fig. 5. Distribution of origins of atmospherics (×) at 08.50 JMT on 11 Oct., 1951.

3. POLAR FRONTS AND JET STREAMS

Remarkable distribution of origins of atmospherics are observed off the eastern coast of the Japan Islands on the Pacific, in a zonal form, in lat. 30°-40°N all the year round. According to a masking effect by strong atmospherics in the neighbourhood of Japan, it is rather difficult to find them in summer, while in spring and autumn, especially in winter, they appear explicitly in lat. 30°-40°N and long. 80°-180°E in a zonal form, and in fact they are found mainly on the Pacific, though sometimes, specifically at night, they are also detected in China, Tibet, etc.

According to meteorological informations, these atmospherics are scattered over frontal zones in the upper atmosphere and the area between these zones and fronts on the surface. Extending this idea to the middle of the Pacific, the zonal distribution of atmospherics there, is considered due to a convergence of warm air flow from the south, creeping above the cold air masses from the north; really, this kind of origins are scattered in a zonal form in lat. 30°-40°N, making a long wave which moves slowly in the north-south direction, revealing certain similarities with characteristics of jet streams in the upper atmosphere.

(i) 20.50 JMT on 11 Oct., 1951. A lot of origins are found along a front in 700 mb region, although in a high pressure area on the surface information (Fig. 6).

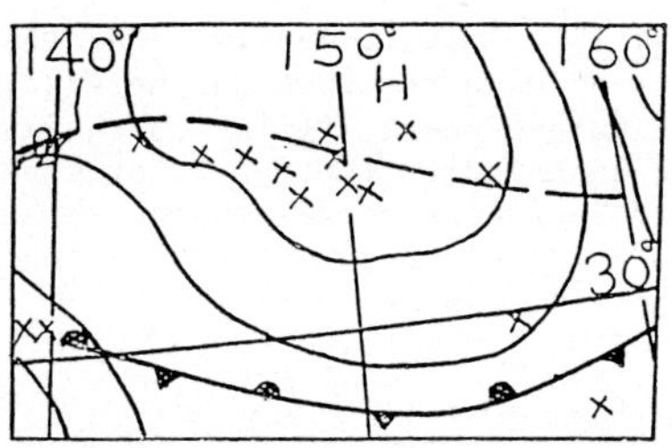

Fig. 6. Distribution of origins of atmospherics (×) at 20.50 JMT on Oct. 11, 1951.

(ii) 17.50 JMT on 9 Oct., 1951. Noticeable distributions of origins are found in the convergence area of 700 mb region between the front in the upper atmosphere and a warm front on the surface accompanied by a low pressure (Fig. 7).

(iii) 20.50 JMT on 10 Oct., 1951. Remarkable distributions of origins are scattered along the north side of a front in 700 mb region, where there is a low pressure accompanied by cold and warm fronts (Fig. 8).

(iv) 02.50 JMT on 8 Oct., 1951. A number of origins are scattered in the neighbourhood of fronts in 700 mb and 500 mb regions as well as along a cold front on the surface (Fig. 9).

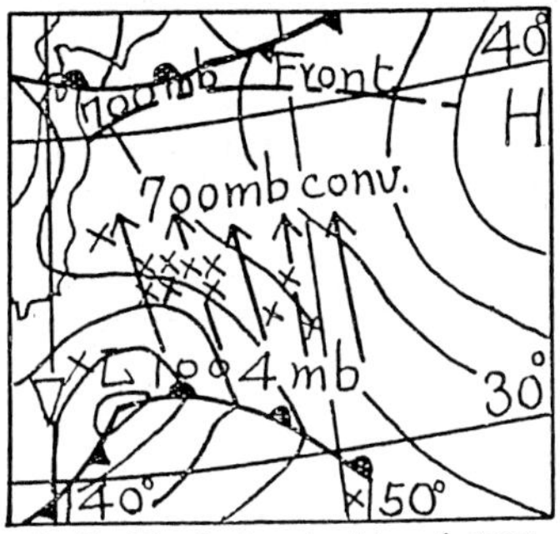

Fig.7. Distribution of origins of atmospherics (×) at 17.50 JMT on Oct. 9, 1951.

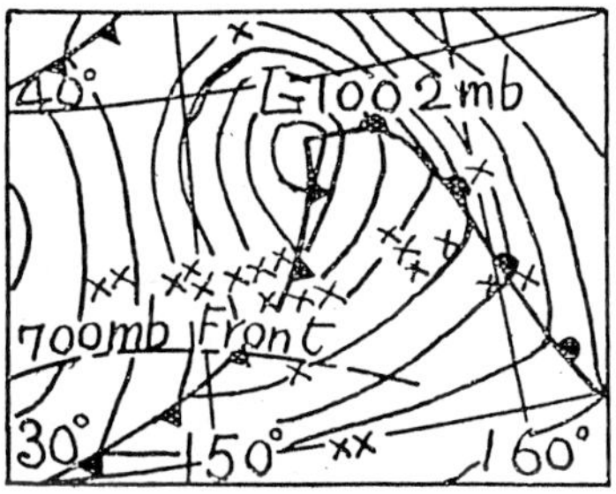

Fig.8. Distribution of origins of atmospherics (×) at 20.50 JMT on Oct. 10, 1951.

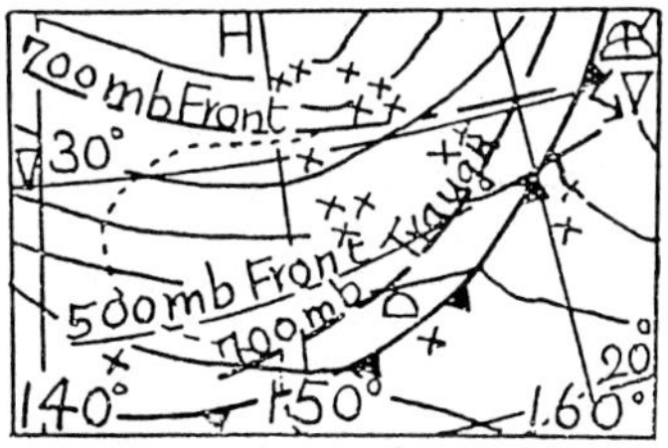

Fig. 9. Distsibution of origins of atmospherics (×) at 02.50 JMT on Oct. 8, 1951.

(v) <u>02.50 JMT on 1 Feb., 1952.</u> Distributions of atmospherics from Kamchatka to the Sea of Japan by way of Sakhalin and Hokkaido should be attributed mostly to snow showers or snow storms there; while it would be very difficult to explain those on the Pacific, scattered in lat. 35°-40°N in a zonal form without assuming the existence of convergence in the polar front in this district. At 00.00 JMT on 1 Feb. in the neighbourhood of Japan, a strong westerly air current (110-140 knots) in the upper atmosphere below 500 mb passed through from lat. 33°-34°N, long. 130°E to lat. 36°-38°N, long. 140°E. The zonal distribution of atmospherics coincides fairly well with the location of westerlies, and it seems that atmospherics exist in the region of strong westerlies, exterpolated from the above position both on the east and west side, i.e. to the continent and to the Pacific (Fig. 10).

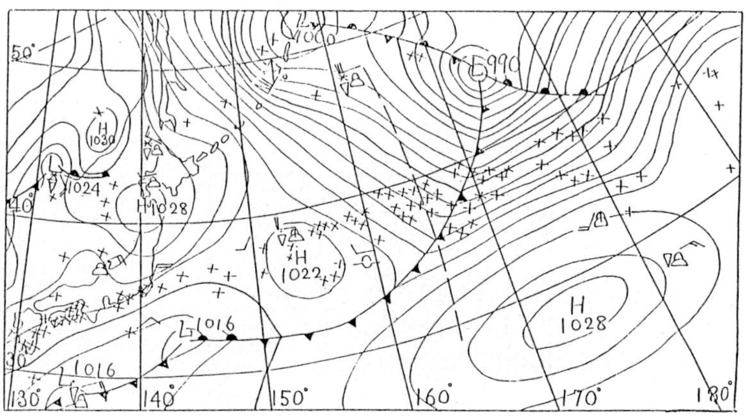

Fig.10. Distribution of origins of atmospherics (×) at 02.50 JMT on Feb. 1, 1952.

(vi) <u>28 Jan. - 3 Feb., 1951.</u> From 28 to 30 Jan., westerlies and zonal distributions of atmospherics move slowly but clearly northward, while after 30 Jan. both movements do not indicate a clear tendency; however, inspecting in detail, it is found that atmospherics return to the south on 31 Jan. and the westerlies too. On 1 Feb., atmospherics move northward again, and the westerlies too at 12.00 JMT on 31 Jan., and return to the south at 00.00 JMT on 1 Feb. On 2 and 3 Feb., movements of westerlies are quite irregualr, which seems to ascertain the branching and irregular movements of atmospherics. In fact, atmospherics move to the north again on 1 and 2 Feb., and then they split into two branches, one moving to the north and the other to the south. On 3 Feb., the north branch returns to the south, while the south one stands still. All the origins in these days are scattered in lat. 33°-43°N. Considering above results, it seems very likely that the zonal distribution of atmospherics in lat. 30°-40°N is produced in the convergence region of polar fronts under the jet streams in the westerlies of upper atmosphere (Fig. 11).

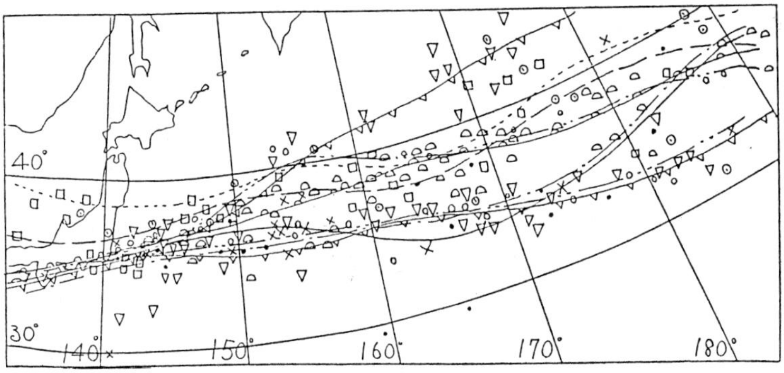

Fig. 11 Movement of zonal distribution of atmospherics from Jan. 28 to Feb. 3, 1952.

Jan. 28. 08.50 JMT ———— × Feb. 1. 02.50 JMT ⌓ ⌓ ⌓ ⌓ ⌓
,, 29. 14.50 ,, — — — □ ,, 2. 08.40 ,, ▽ ▽ ▽ ▽ ▽
,, 30. 14,50 ,, - - - - - - ⊙ ,, 3. 08.50 ,, - - - - - - ○
,, 31. 08.50 ,, —·—·— ●

4. SEA AND SID PHENOMENA

In general, solar flares of great impotance generate prominent outbursts of solar radio waves in microwave band, enhancements of atmospherics in VLF band (18-30 kHz after R. Bureau), Dellinger fadeouts in HF band radio communications, solar flare effects on the earth mangetism and fadeouts of echoes in the ionosphere measurements (SID: sudden ionospheric disturbance), with astonishing coincidence of instant of occurrence. As to atmospherics, its enhancement in the neighbourhood of maximum level corresponds fairly well with that of SID and besides the commencement of enhancement precedes SID about one hour. The reason for this is not at present physically elucidated, and it is not yet proved to be regular phenomena to make use for prediction of Dellinger fadeouts. Daily variation of intensity of atmospherics in general indicates characteristics shown in Fig. 12, which shows sudden decrease before sunrise, constancy throughout daylight hours and sudden increase near sunset.

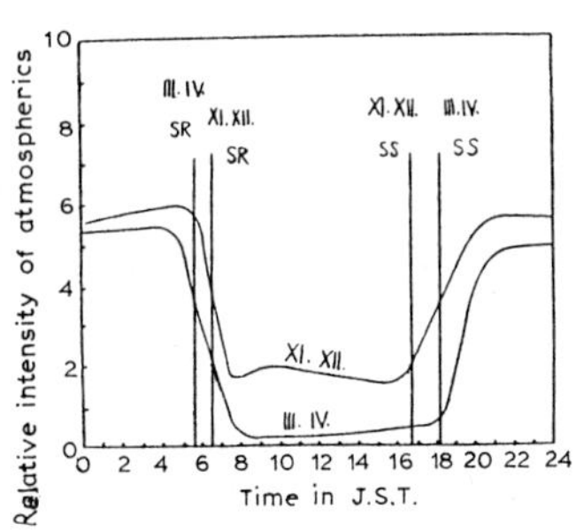

FIG. 12. Daily variation of intensity of atmospherics in spring and winter.

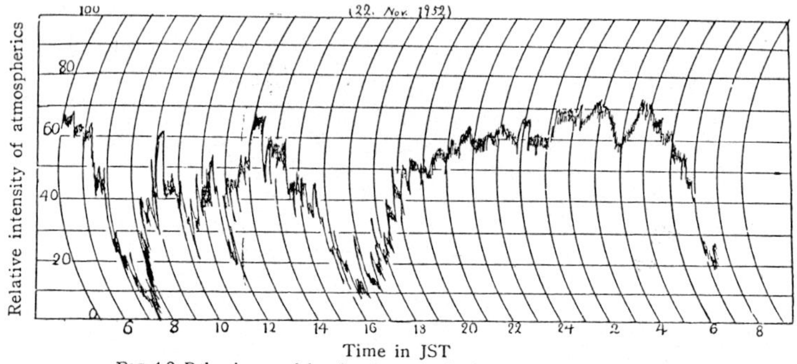

FIG. 13 .Behaviours of levels of atmospherics on 22 November, 1952.

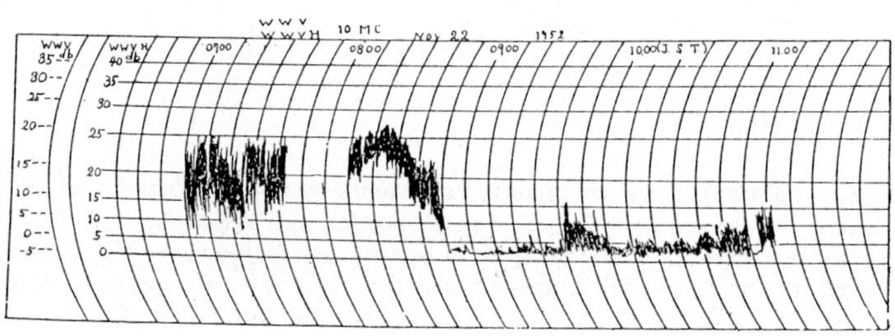

a

FIG. 14 . Behaviours of levels of short wave radio communications (WWV and WWVH at 15 and 10 Mc/s) on 22 November, 1952.

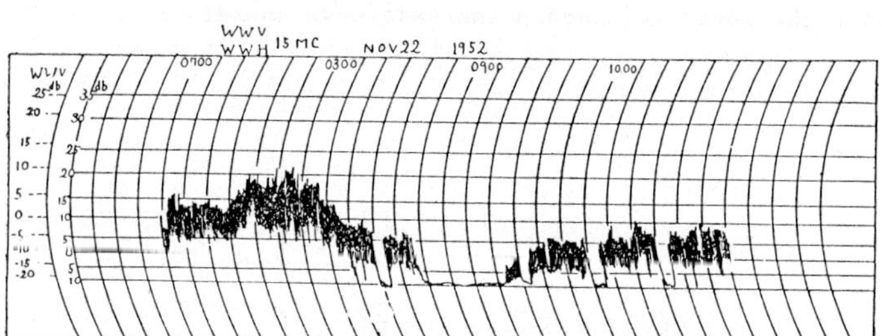

b

FIG. 14 .Behaviours of levels of short wave radio communications (WWV and WWVH at 15 and 10 Mc/s) on 22 November, 1952.

As an example of SID, those on 22 Nov., 1952 are described in the following.

(i) Atmospherics. Intensity of atmospherics at 27 kHz on 22 Nov., 1952 follows the record in Fig. 13. It begins to decrease at 05.30 JST, about one hour before sunrise (06.31 JST) at Toyokawa, and reaches a minimum at 07.30 JST as usual. At 08.25 JST it increases suddenly to the half value of night levels and at 08.50 JST it rises again rapidly, arriving at a maximum, the night levels, at 09.12 JST. At 09.18 JST it begins to decrease and at 09.50 JST it drops again quickly, reaching a minimum at 10.24 JST, and then it goes up to 11.00 JST due to meteorological disturbances unrelated to SID.

(ii) Dellinger fade-out at HF radio communications. Features of WWV and WWVH (Both of them: 15 MHz and 10 MHz) received at Hiraiso Station on 22 Nov., 1952 are shown in Fig. 14a (10 MHz) and Fig. 14b (15 MHz). That for 15 MHz, having a decreasing tendency from 08.00 JST, drops suddenly at 08.46 JST and fades out, and afterwards at 09.22 JST it resumes progressively, reaching normal levels at 09.45 JST. That for 10 MHz, falling gradually from 08.25 JST, fadeouts abruptly at 08.48 JST, and, beginning to rise again at 09.38 JST, returns to the normal value at 09.44 JST. Decrease after 09.44 JST shown usual low levels of 10 MHz in this season unconnected to SID.

(iii) Earth magnetism. The records of horizontal component, vertical component and declination, observed at Kakioka Magnetic Observatory on 22 Nov., 1952, are shown in Fig. 15. The particular crochet, known as a solar-flare effect, is clearly observed at 08.45 JST on all records.

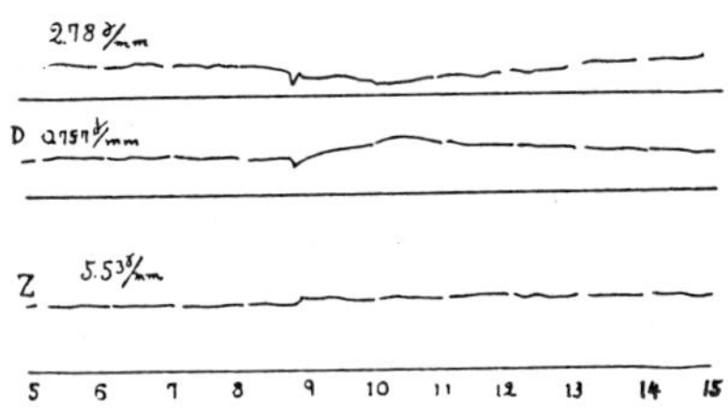

FIG.15 Behaviours of 3 components of earth magnetism on 22 November, 1952.

(iv) Solar phenomena. Results of observations at Tokyo Astronomical Observatory on 22 Nov., 1952 are shown in Table 1. Two chromospheric flares are observed at 08.45 JST in the neighbourhood of E type sunspots near the centre. The one, whose importance is 3, emits three line spectrums and the other, importance 2, two line spectrums. they began observations at 07.50 JST. Flares of importance 3 and 2 begin their activities at 08.45 JST, reach a maximum at 08.50 JST and 08.58 JST, and end at 09.23 JST and 09.27 JST respectively.

TABLE 1. Solar Flares Observed at Tokyo Astronomical Observatory on 22 November, 1952

Time of observation in JST			Approximate position		Importance
Start	Max.	End	Long.	Lat.	
08.45	08.50	09.23	4°W	8°N	3
08.45	08.58	09.27	1°E	5°N	2

These flares were prominent ones, unusually found in those period, and dark filaments were observed near flares. These dark filaments were recognized from 08.55 to 09.23 JST, and at first there were many, which emitted from the sun to the earth, reaching a maximum at 09.05 JST, while afterwards many returned to the sun.

(v) Ionosphere. Continuous measurements of (a) h'-f curves and (b) virtual heights with constant frequency were made at Kokubunji and Akita Ionosphere Observatories on 22 Nov., 1952. In (b) echoes fadeout from 08.45 to 09.30 JST. In (a) at 08.00 JST no observation was made due to troubles in equipments, at 09.00 JST no echoes, and at 10.00 JST echoes resume normal features, indicating E_1, E_2, E_2^o, E_2^x, E_p, F_1 and F_2 as shown in Fig. 16a. In Akita Observatory at 08.00 JST E_1, E_2, E_2^o, E_2^x, $2E_1$, $2E_2^o$, $2E_2^x$, F_1, F_2^o, F_2^x, $3E_1$, $3E_2$ and $2F_2$ are observed, at 09.00 JST all fadeout except E_1, E_2, F_2, and F_2^x, and at 10.00 JST echoes resume normal behaviour, showing E_1^o, E_2^o, E_2^x, $2E_2$, $3E_2$, F_1, F_2^o and F_2^x (Fig. 16b). In both Kokubunji and Akita Stations, ionospheric echoes suffer remarkable effects except F_2 echoes. The fact may be attributed to lower attenuation at higher frequencies through the abnormal D layer.

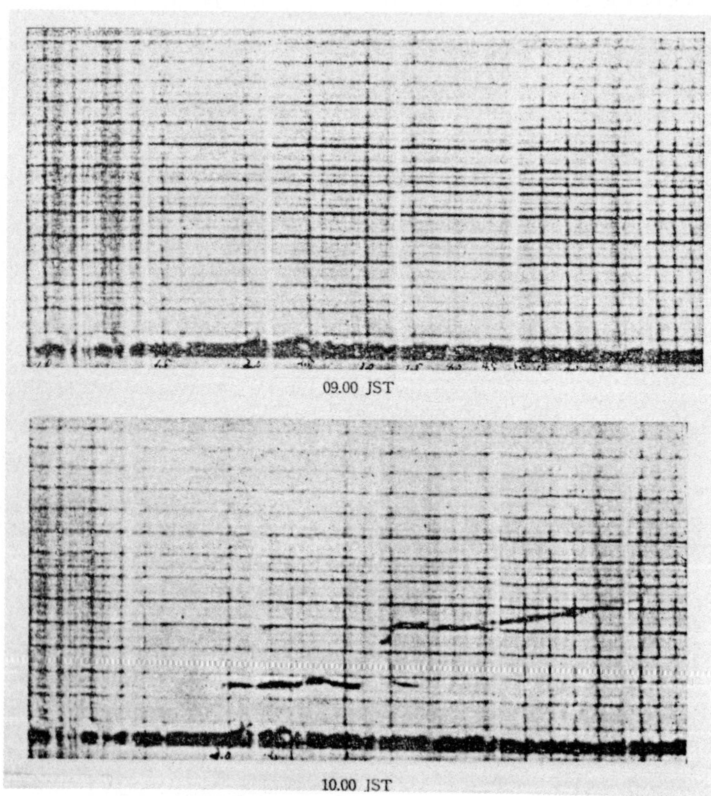

09.00 JST

10.00 JST

Fig. 16a. *h'-f* curve of Kokubunji Ionospheric Observatory on 22 November, 1952.

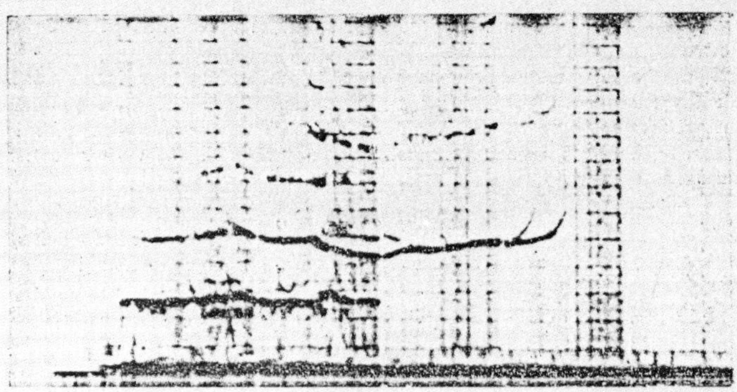

08.00 JST

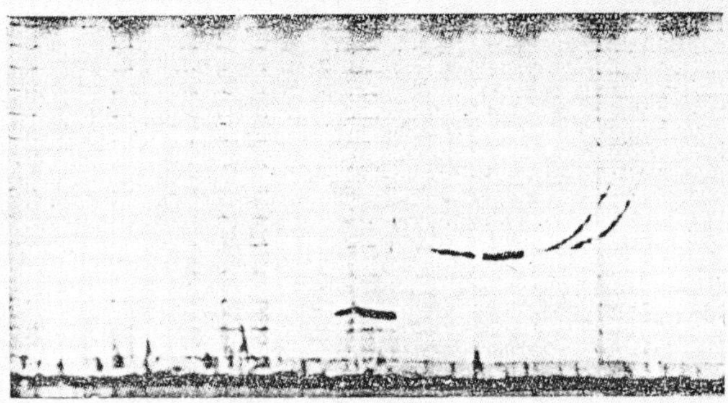

09.00 JST

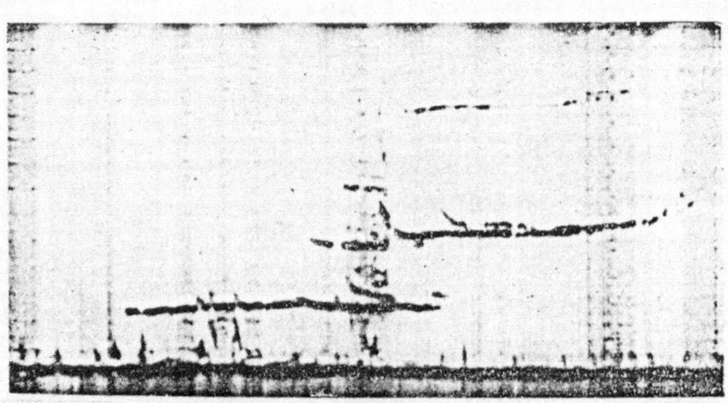

10.00 IST

Fig. 16b. *h'-f* curuve of Akita Ionospheric Observatory on 22 November, 1952.

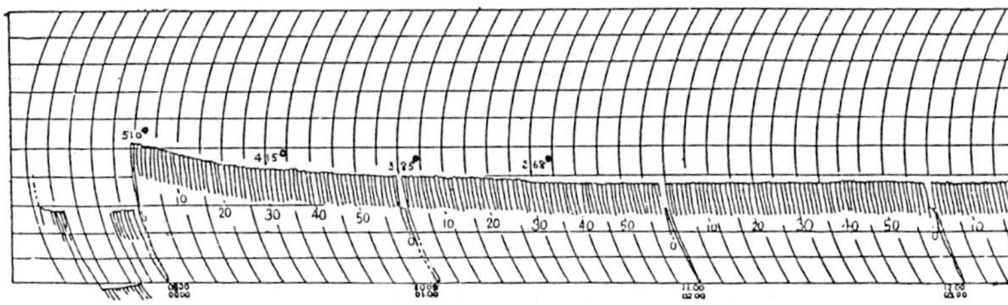

FIG. 17. Behaviours of solar radio waves at 3,750 Mc/s on 22 November, 1952.

TABLE 2a. Correlation Among Various Abnormal Phenomena on the Occasion of S.I.D. on 22 November, 1952

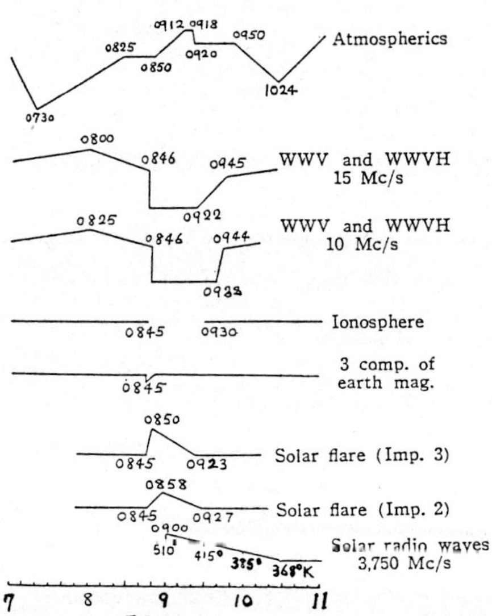

TABLE 2b. Correlation Among Solar Radio Waves, Earth Magnetism and Short Wave Radio Communications on the Occasion of S.I.D. on 22 November, 1952

Date	Solar radio waves: Max. eq. tempt.	Outbursts time in JST: Start	Outbursts time in JST: Max.	Outbursts time in JST: End	Earth magnetism: Time of solar flare effects in JST: Hor. comp.	Vert. comp.	Decli.	Clearness in correlation	WNC 37 (1.7420 kc/s): Fadeout time in JST: Start	End	Clearness in correlation	WWV. WWVH (15 Mc/s): Fadeout time in JST: Start	End	Clearness in correlation
IV/19	2,100	08.40			08.40	08.40	08.40	B	08.42		A	08.40		A
			08.48		08.48		08.48	B						
			08.50											
			08.55				08.55	B						
				09.00						09.10	A		09.24	B
IV/19	1,670	13.20			13.20	13.20		B	13.20		B			
			13.23											
			14.34		14.28	14.28		C				14.28		A
			14.46		14.46	14.46		C						
				14.50						15.05	A		15.05	A
IV/20	16,000	09.50			09.48	09.48	09.48	A A	09.50		A	09·50		A
			09.52											
				10.20						10.10	A		10.12	A
IV/21	15,000	07.25			07.25	07.25	07.25	B	07.25		A	07.25		B
			07.30		07.30	07.30	07.30	B						
				10.00						08.53	A		09.13	A
IV/23	940	07.56			07.54	07.54		C				07.58		A
			07.58											
				08.00										
IV/23	1,130	08.30			08.30	08.30		C	08.30		B	08.34		A
			08.33											
				08.35						09.00	B		08.47	A
IV/25	2,200	11.02			11.07		11.07	A				11.00		A
			11.04						11.06		A			
				11.20						11.11	B		11.16	B
IV/25	1,570	16.14												
			16.17				16.18	B	16.18	16.22	B	16.16		C
				16.19	16.24	16.24	16.24	B			B		16.18	C
V/17	1,180	08.13												
			08.14		08.15	08.15	08.15	A	08.14		C	08.18		B
				08.22									08.24	B
V/21	1,500	09.17							09.15		A	09.15		B
			09.20											
				09.24						09.34	A			
	1.100	10.49							10.50		A			
			10.52											
				10.54	10.55	10.55		A		11.02	C			
	1,240	11.04							11.04		C			
			11.09											
				11·20						11.21	A		11.42	B
V/22	1,100	08.56	08.56									↑		
	1,600		09.10											
	1,340		09.32											
				09.43								ZAN		A
	5,500	09.53			09.54		09.54	B	09.53		A			
			09.56											
				10.40						10.42		↓		
V/23	3,000	10.14			10.15			C	10.16		A	↑		
			10.20									ZAN		A
				10.40						10.37	A	↓		
V/24	2,500	13.28										13.30		A
			13.31		13.33	13.33	13.33	B					14.27	A
				13.50										

(vi) Solar radio waves at 3750 MHz. Records of solar radio waves observed at the Research Institute of Atmospherics on 22 Nov., 1952 are shown in Fig. 17. As solar activities were rather calm in this season, we used to observe from 09.00 to 12.00 JST and in consequence the outburst has already entered into the decreasing period before we could begin observations. We find therefore on the record in Fig. 17 equivalent temperatures of 510°, 415°, 385° and 386°K at 09.00, 09.30, 10.00 and 10.30 JST respectively.

In compiling above results of SID phenomena on 22 Nov., 1952, we obtain Table 2, which indicates time correlations among suden enhancement of atmospherics (SEA) at 27 kHz, Dellinger fadeout of short-wave radio communications, solar flare effects in the earth magnetism, flares on solar disc, abnormal ionospheric echo variations or disturbances (SID), and outbursts in solar radio waves at 3750 MHz, i.e. micowave band. All the abnormal phenomena, except those of atmospherics and solar radio waves, occur at the same instant of 08.45 JST with striking coincidence. As to atmospherics, detailed explanation is made at the beginning of this section. Though the beginning of outbursts of solar radio waves is unknown in this case, deducing it statistically, it can be found to coincide with 08.45 JST, from examples more than ten cases obtained in 1951. The time to reach a maximum from the beginning is also found to be 4.3 minutes (range 2 to 7 min.) statistically, and therefore one presumed that the maximum is attained at 08.49 JST. Then the commencement of outbursts on 22 Nov., 1952 is to coincide with those of solar flares and their maximum with that of the solar flare of importance 3.

5. SUDDEN PHASE SHIFT OF VLF WAVES

On the occasion of solar flare, atmospherics in VLF band (18-30 kHz after R. Bureau) and VLF signals indicate, as is well known, sudden enhancement of the field strength. On observing the phase shift of North West Cape (22.3 kHz) waves for a long time, it was found that the abnormal phase shift occurs at solar flare outbreak, as is described in the following. Distinct flares of solar X ray (2.12 Å) were observed on 5, 6, 7 and 10 March, 1970 by satellites Explorer 33 and 35: the ratios of maximum flux to that of calm sun on 5 (0416 UT), 7 (0138 UT) and 10 March (0647 UT) are proportional to remarkable phase shift of 22.3 kHz wave on 5 (0423 UT), 7 (0144 UT) and 10 (0658 UT) March, 1970: phase shifts are $136^{\circ}31$, $120^{\circ}45$, and $56^{\circ}21$ respectively as shown in Fig. 18.

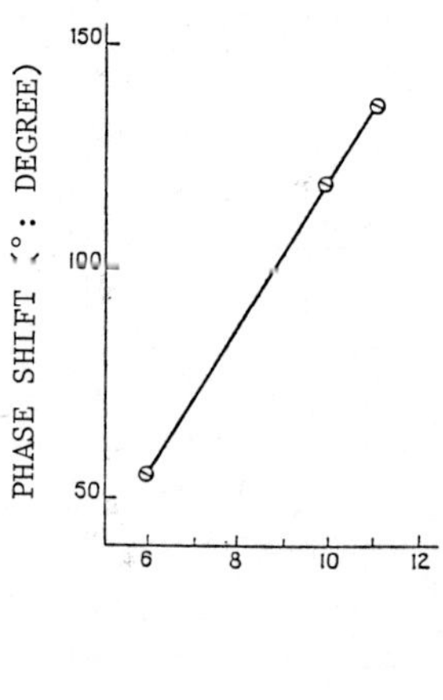

Fig. 18. Phase shift of 22.3 kHz versus the ratio of maximum flux of solar X ray to that of calm sun.

Solar flares observed on 5 (0421 UT), 7 (0141 UT), 8 (2248 UT), 9 (0534 UT) and 10 March (0657 UT) are closely correlated with abnormal phase shifts on 5, 7, 8, 9 and 10 March, 1970. These results verify quantitatively the fact that the abnormal phase shift of VLF waves has the origin in the lower ionosphere such as abnormal D layer, where nitrogen oxide NO is ionized by solar short wave X rays. Outbursts of solar radio waves at microwave band, here in our Institute at 3750 MHz, on 5 (0418 UT), 6 (0234, 0237, 0436 0540 UT), 7 (0141, 0610 UT), 8 (0103 UT), 9 (0416, 0530 UT), and 10 March (0650 UT) are found intimately correlated with these abnormal phase shift of VLF waves. Sudden enhancements of signals (SES) were observed only on 5 and 7 March, when the abnormal phase shift were particularly remarkable; therefore it seems that, as an indicator, it is better to observe the phase shift than to measure the intensity. Similar observations were made during the World Geophysical Interval (26 Jul. - 14 Aug., 1972). The comparison was made between phase variations of VLF and the flux measured by Explorer 37 (Jul., 1972) and Explorer 44 (Aug. 1972), as shown in Fig. 19. Relation between the observed phase variation of NWC (22.3 kHz) and the flux of solar X rays in band 0.5-3 Å is fairly well coincident with the flux greater than 60×10^{-5} erg/cm^2/s (curve 1 in Fig. 19). The phase variation of VLF waves is proportional to the flux of X ray in band 1-8 Å, but not linear (curve 2). The relation between the phase shift and the flux of X ray in band 8-20 Å indicates saturation (curve 3). Consequently, it is concluded that the flux X at the shortest wavelength is most effective to the ionization in D layer.

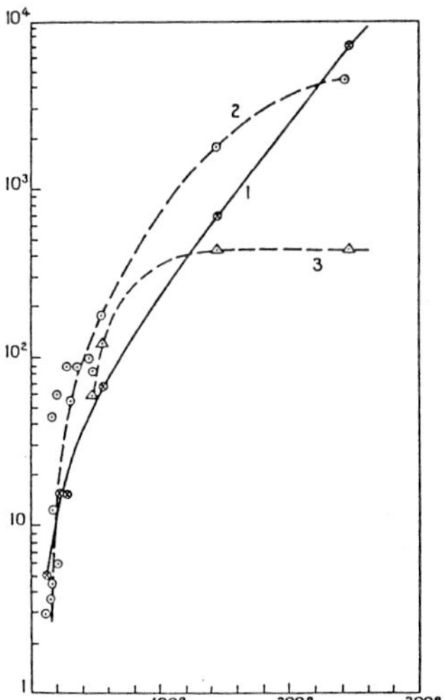

curve 1: unit 10^{-5}erg/cm^2/s; band 0.5-3 Å

curve 2: unit 10^{-4}erg/cm^2/s; band 1-8 Å

curve 3: unit 10^{-3}erg/cm^2/s; band 8-20 Å

Fig. 19. Correlation between the phase variation of 22.3 kHz and the solar X ray flux.

6. SEA PHENOMENA DUE TO NUCLEAR EXPLOSION

SEA phenomena on atmospherics at 21 and 27 kHz were observed at Toyokawa twice on 12 Aug., 1985, as shown in Table 3 and in Fig. 20. Corresponding to SEA No. 1, outbursts of solar radio waves were observed at Toyokawa (Table 4).

As to SEA No. 2, it was informed that observations were made in the Netherlands at 200, 545 and 2980 MHz, and in Czechoslovakia at 536 MHz, and besides they observed bursts only at 200 MHz, which began at 1040 UT and continued for 1 minute with intensity of $80X10^{-11}$ watt.m^{-2}.c.s^{-1}, and not at other frequencies.

In accordance with our experience, when SEA phenomena are observed at 21 and 27 kHz exclusively and not at 10 kHz, they are always deeply correlated with Dellinger fadeout, i.e. occurrence of abnormal D layers; while simultaneous increases of atmospherics at 10, 21 and 27 kHz correspond to active meteorological phenomena such as thunderstorms, typhoons, fronts, etc.

Solar flares and nuclear explosions radiate X rays, ultraviolet rays, etc. and the forms of SEA phenomena on records of atmospherics are very similar with each other, as shown in Fig. 20; although the latter has small scales and radiated energies are absorbed in the lower atmosphere, the conditions will be sometimes favourable enough to generate abnormal D layers, if the explosion is made in the upper atmosphere and has large energy.

SEA phenomena due to explosion on 1 Aug., 1958 was not found in the records of atmospherics. The neighbourhood of Japan was disturbed heavily by active meteorological phenomena at that time, and simultaneous increases in the records of atmospherics at 10, 21 and 27 kHz were observed. It seems that SEA phenomena due to nuclear explosion were masked by meteorological phenomena.

TABLE 3.

No. of observations	Frequency in kc.	Observed Time in U.T.		
		Beginning	Maximum	End
No. 1	10	non	non	non
	21	0428	0436	0506
	27	0427	0436	0500
No. 2	10	non	non	non
	21	1045	1051	1127
	27	1044	1051	1120

TABLE 4.

Frequency in Mc.	Observed Time in U.T.		Duration in min	Type	Intensity in 10^{-22} watt m^{-2} c/s^{-1}
	Beginning	Maximum			
9400	0420	0435.4	23	CD	49
3750	0420	0426.0	21	〃	72
2000	0419	0426.4	22	〃	31
1000	0422	0422.6	1.5	〃	13

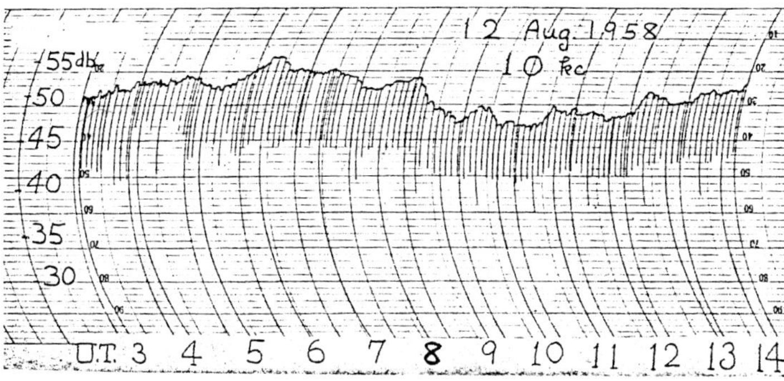

FIG.20 a

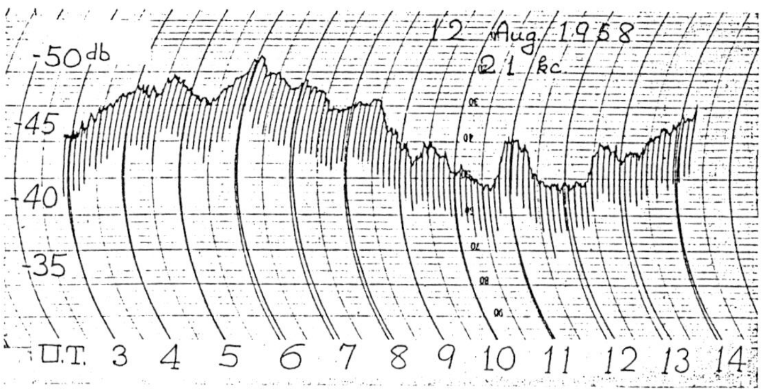

FIG20b

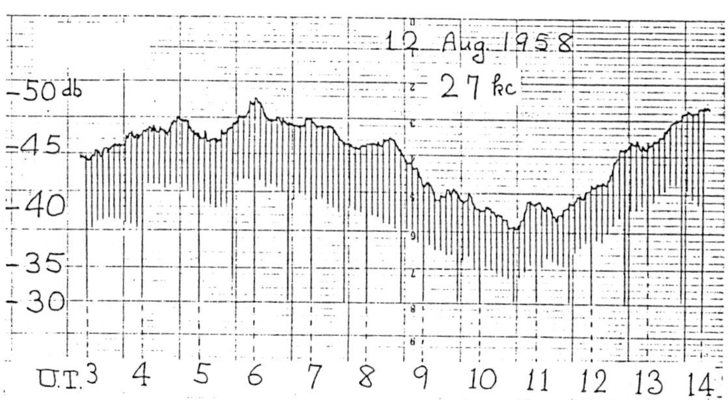

FIG.20c

7. WAVEFORM OF ATMOSPHERICS

The author observed waveforms of atmospherics for the first time, for five years from 1940 to 1944, and in 1950, improving measuring apparatuses, he started again observations of waveforms, strictly synchronized with direction finding at three points in Japan, i.e. at Toyokawa, Kikuchi and Akita It continued for seven years till 1957, getting 8,500 examples, which were analysed according to meteorological phenomena and wave propagation characteristics, as follows.

In the temperate zone it is possible to consider the waveforms arriving from a source, located by the direction finding made simultaneously with the waveform recording at three stations over Japan mentioned above. In the tropical zone, there are too many lightning discharges in the neighbourhood of observatories to discriminate one from the other, but our Institute and the other two stations, chosen for direction finding, are all in the calm district, free from lightning flashes. The waveform of atmospherics depends remarkably upon the propagation conditions, and consequently the clear difference is observed between the waveform in the daytime and the one at night.

(i) Waveform in the daytime. Main waveforms in the daytime consist of (1) stepped leader stroke type, partial discharge within cloud type and successive discharges among clouds type, (2) return streamer type, (3) multiple stroke type, while slow tail type and reflection echo type are generally found at night, few in the daytime (Fig. 25, slow tail).

(1): groups are recorded from distances less than 2,000 km along overland paths and less than 3,000 km along overseas. The main frequency range is 10 - 30 kHz, removing the pulsive superposed higher frequency components with small amplitude. Sometimes the return streamer type follows immediately after stepped leader type, and often does not; even in the latter case α and β types are generally discernible as described by Schonland, and the general features (pulse intervals, durations and so on) of three types in (1) are quite similar to those observed at short distances 10 - 100 km (Fig. 21).

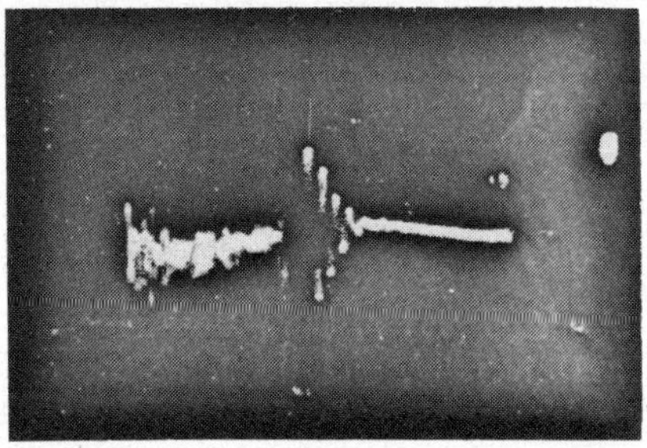

Fig. 21

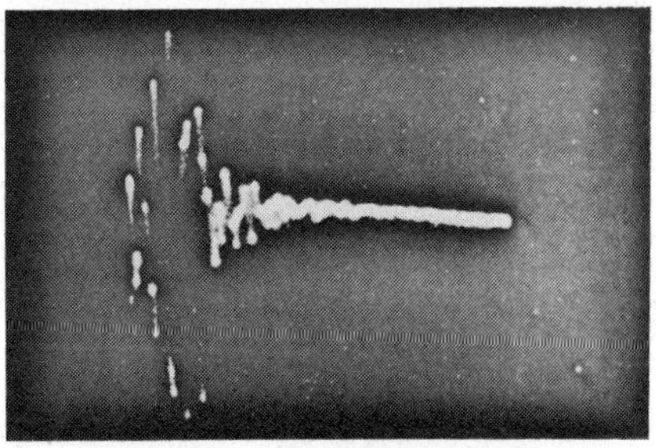

Fig. 22

(2): groups consist of (2a) waveform whose largest amplitude is found at the second peak or third and (2b) the one whose largest amplitude at the beginning of train. These are all very smooth damped waves of which (2a) is recorded from distances less than 2,000 km and (2b) from distances 1,000 - 5,000 km or more, i.e. from everywhere. Occurrence of (2b) is more numerous than that of (2a), and the frequency range of (2b) is 6 - 17 kHz on the average. Describing in more detail, 6 - 7 kHz are found at about 1,000 km distance, 7 - 11 kHz at about 3,000 km and 7 - 10 kHz at about 5,000 km (Fig. 22, 23).

(3): groups are observed in distances less than 3,000 km, and consist of more than two similar waveforms coming from the same origin, and is considered as developed from multiple strokes, comparing with the time interval and duration observed in the neighbourhood of lightning flashes electrically and optically. This type is found only in the daytime, and not observed at night due to disturbances from the interference of sky waves (Fig. 24).

(ii) Waveform at night. Main waveforms at night consist of (4) stepped leader stroke type, partial discharge within cloud type and successive discharges among clouds type, (5) return streamer type, (6) slow tail type and (7) reflection echo type.

(4): groups are observed almost in the same way as those in the daytime and frequency range is 10 - 30 kHz, but, favoured by good propagation conditions, they are recorded from long distances even more than 5,000 km along oversea paths as well as overland. Those coming from origins more than 3,000 km distance have in general the frequency range of 10 - 15 kHz (Fig. 26, 27).

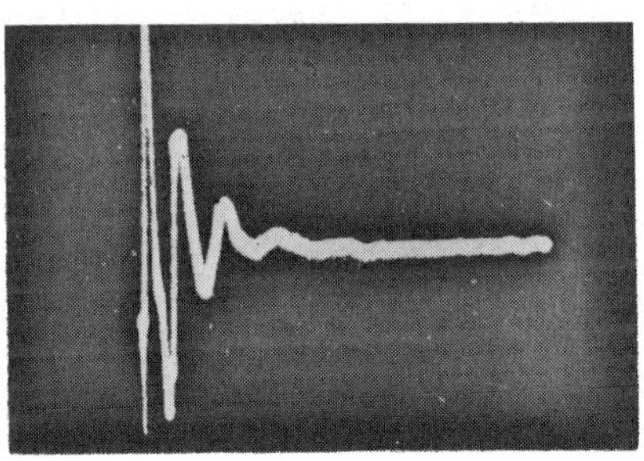

Fig. 23

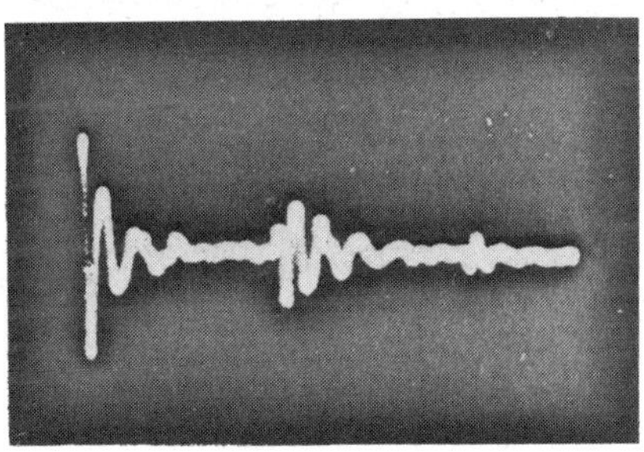

Fig. 24

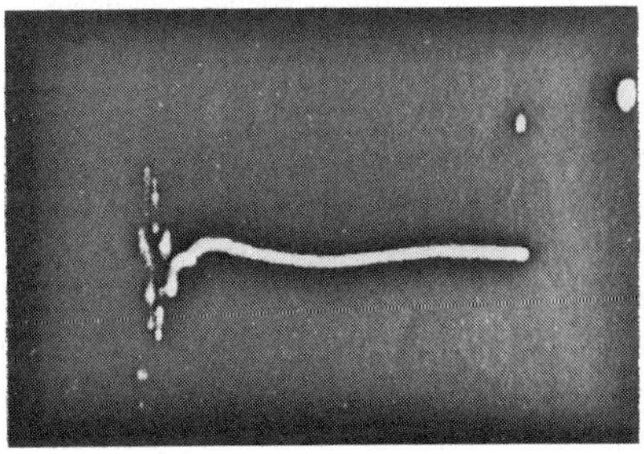

Fig. 25

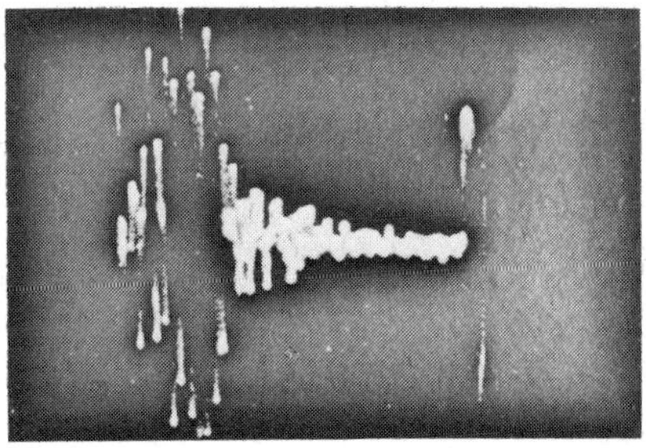

Fig. 26

(5): groups are not so smooth as in the daytime in the distance less than 3,000 km. Sometimes the peak or valley of the wave is split or superposed with pulsations of higher frequencies. Sometimes it is difficult to discriminate from reflection echo type; even in the latter case, in examining carefully, it is found to be the return streamer type. This fact is found by unreasonable value of reflection height, remarkable difference from CRDF fix, no coincidence with the calculating chart and so on. In other words, at night it is rather difficult to find a clear return streamer type within a distance less than 3,000 km; while the smooth damped waveform like that of daytime is easily found in longer distance as 3,000 -5,000 km or more. Even the latter is often taken as the reflection echo type by mistake, but, in examining with precaution, we can find it to be (5), and we may conclude that the reflection echo type is not recorded from distances more than 3,000 km at night.

(6): groups are at times observed even in the daytime (Fig. 25), but it is generally the type at night. It is recorded from everywhere or from distances 1,000 -5,000 km or more. The frequency observed is 400 - 700 Hz, average in the neighbourhood of 600 Hz, and follows very often the higher frequency types such as return streamer type or reflection echo type, but the difference of arriving time from higher frequency type is more or less indeterminate, and does not always depend on distances travelled. The amplitude of (6) is often very large compared with long travelling distances, and contains lower frequency components than Appleton observed; therefore extensive study is required before a final conclusion will be derived (Fig. 30, 31, 32 and 33).

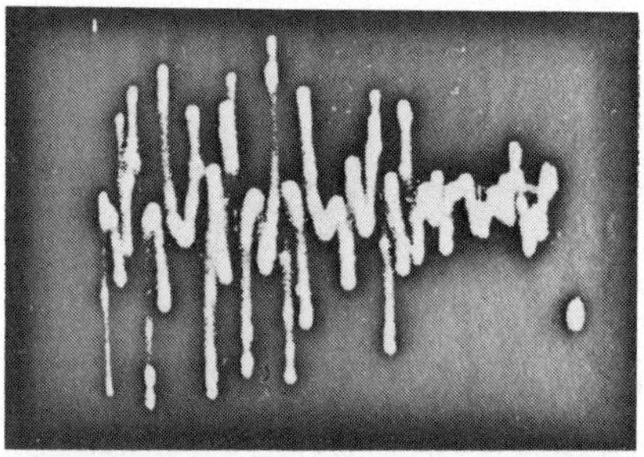

Fig. 27

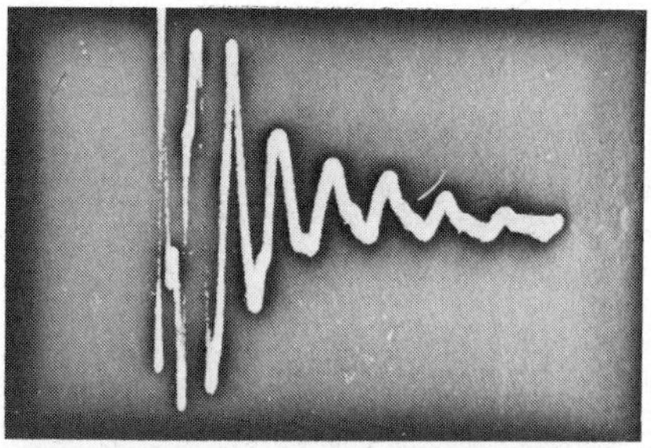

Fig. 28

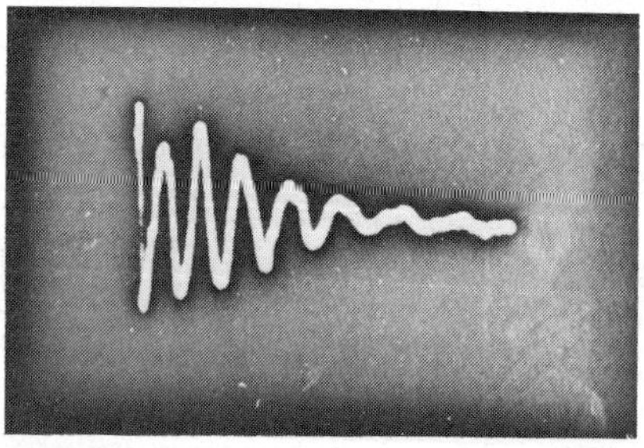

Fig. 29

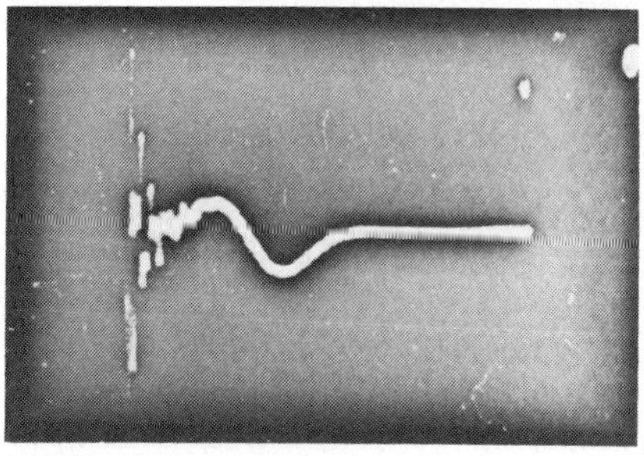

Fig. 30

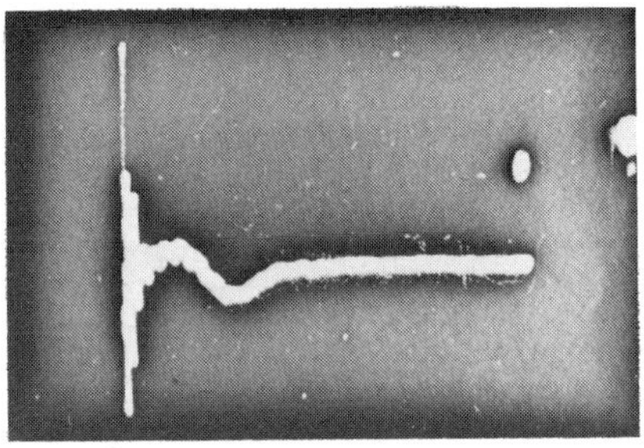

Fig. 31

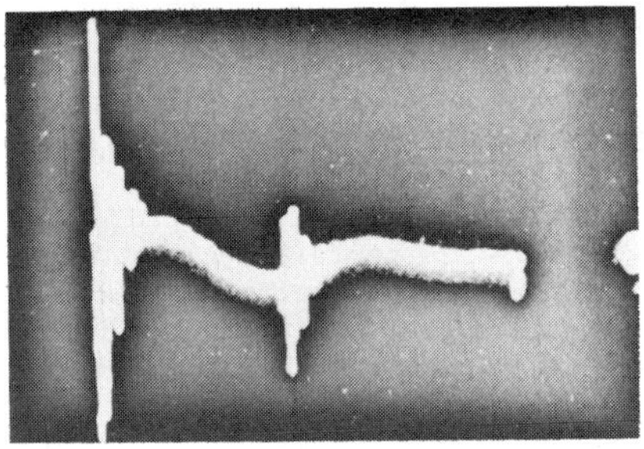

Fig. 32

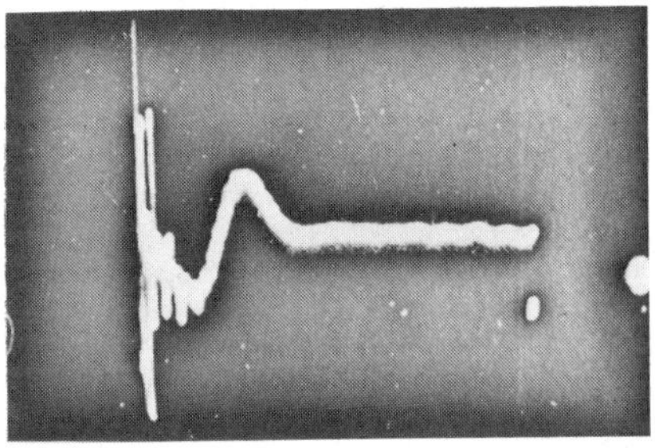

Fig. 33

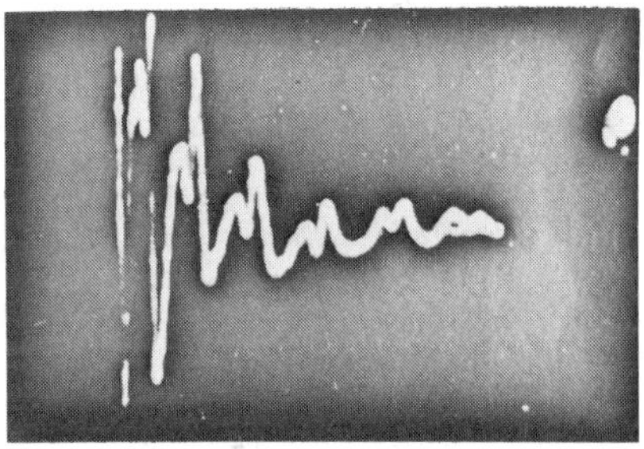

Fig. 34

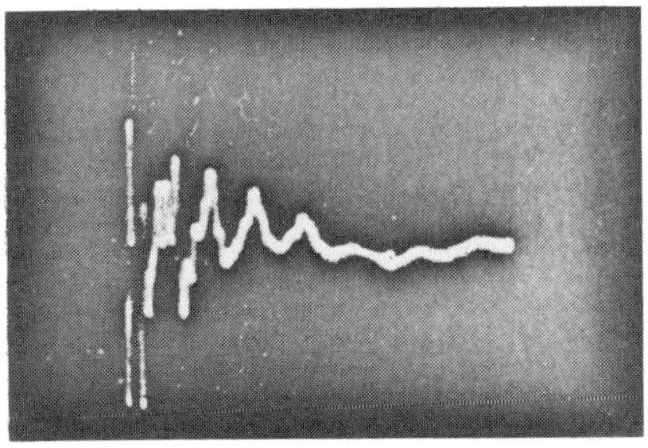

Fig. 35

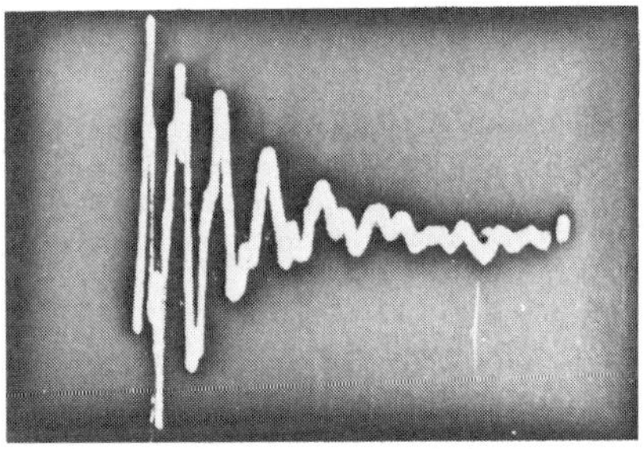

Fig. 36

(7): groups are recorded at times from distances about 1,000 km in the daytime, but in general it is observed at night in distances less than 3,000 km. Assuming the space wave components reflected between the earth and the ionosphere where the reflection is assumed to be done upon the geometric optical principle at the surface independent of incident angles, the reflection height and the distance of sources are evaluated by this type. In this paper the mean deviation between records and the calculating chart less than 10 μs and the maximum deviation 25 μs are admitted for determination of the reflection height and the distance. A perfectly clear coincidence between the records and the chart is found relatively seldom but in order to make easier the evaluation, selection was made to take positive or negative pulse series according to the clearness of waveform, and reasonable judgement was made to determine the order of sky pulses in case the initial part of wave train was complex. Thus the reflection height was found 80 - 90 km and the distance evaluated from the wavforms coincides fairly well with that of CRDF fix. Location of source at one station is expected at distances less than 3,000 km at night under favourable ionospheric and other propagation conditions (Fig. 34, 35 and 36).

8. DISTRIBUTION OF ATMOSPHERICS IN ASIA

The author made an extensive series of observations of waveform and direction finding of atmospherics, from 1943 to 1944, and gave a full account of correlation between atmospherics and weather phenomena, in addition to an explanation of seasonal variation in distribution of origin of atmospherics in the Far East. For several years after World War II, the author made fundamental researches to improve observation apparatuses. He could reach a narrow half-power beam width of 1° for cathode ray direction finder (CRDF) with unidirectional characteristics, and could also keep a strict simultaneous observation at every station relying upon JJY, the standard wave emitted by the Regulatory Bureau of Japan. Observations were made from 1952 to 1954.

(i) Correlation of atmospherics with weather phenomena. Atmospherics are radio waves emitted from discharges in the atmosphere, accompanied by thunderstorms, showers, snow storms and developed cumulonimbi; therefore origins of atmospherics coincide very well with districts occupied by these phenomena, so long as the ratio of the base line between observatories to the distance between origins and observatories is reasonable.

Cyclones also become origins provided they are deepened by fronts, troughs or other stimulating agents (Fig. 37), while in tropical storms, depressions or typhoons, origins are found in many different ways, as explained in paragraph two, i.e. in dangerous semi-circle or active regions, in converging wind line between the eye and the fronts or troughs, along the tropical front extending from the centre etc Troughs or fronts become origins when they are activated by strong insolational heating, orographic lifting, invasion of wet and warm air masses, etc.

All these phenomena are always accompanied by thunderstorms, showers, cumulonimbi, etc. when they are observed as origins.

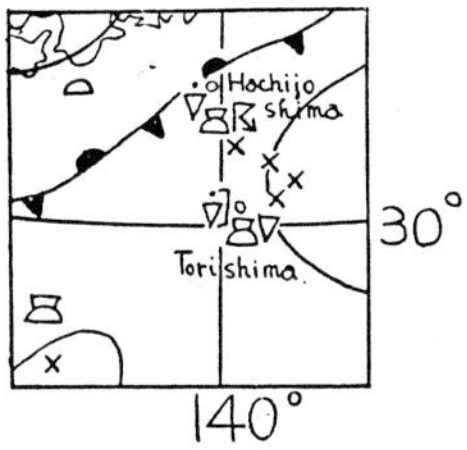

Fig. 37

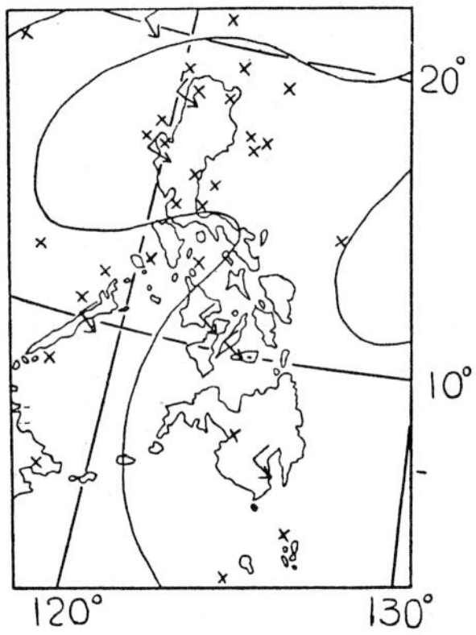

Fig. 38

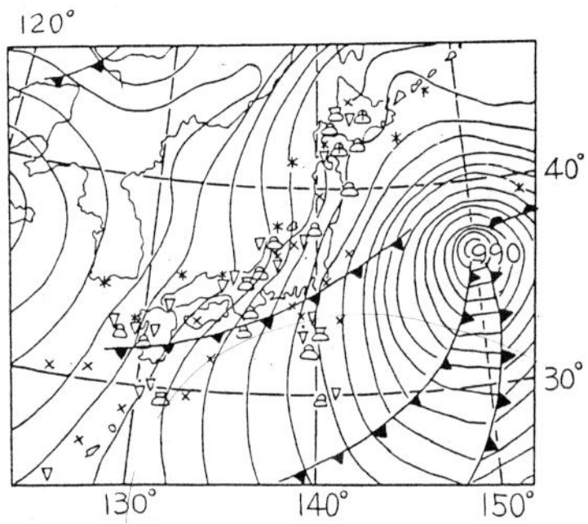

Fig. 39

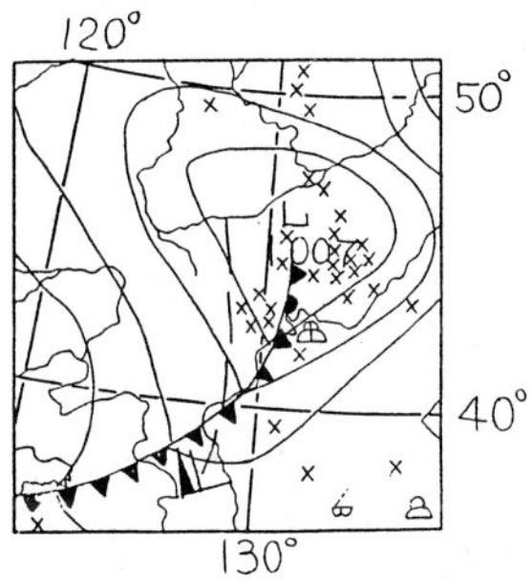

Fig. 40

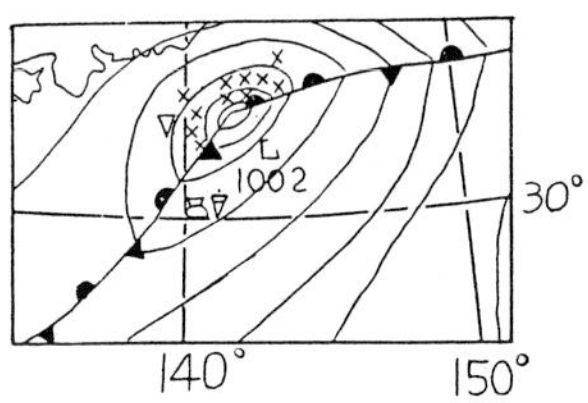

Fig. 41

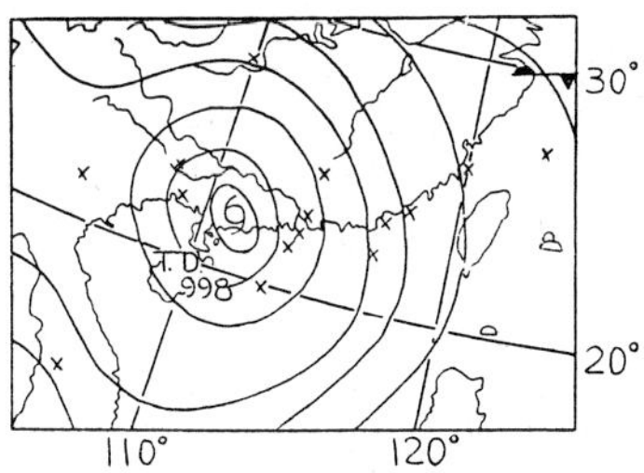

Fig. 42

In Fig. 37 we observe origins of atmospherics due to thunderstorms, showers and cumulonimbi in Hachijoshima and Torishima Islands. In Fig. 38 origins of atmospherics due to thunderstorms in the Philippines are shown. In Fig. 39 origins of atmospherics due to snow storms and cumulonimbi are distributed along the coast of the Sea of Japan and atmospherics due to showers and cumulonimbi are on the Pacific, the latter being intimately correlated with cold fronts there. In Fig. 40 an occluded front starts from a cyclone of 1,007 mb in Manchuria, and soon changes into a cold front. A trough in 700 mb passes to the west of cyclone, and there are many origins of atmospherics over convergence regions which develop around these phenomena. Fig. 41 shows origins concentrated near the centre of a cyclone of 1002 mb accompanied by two stationary fronts, one being extended to the east, the other to southeast. In Fig. 42 origins on the slope of mountains Nan-Shan and Ta-Yu-Shan, and on the coastal line of South China are due to convergence of wet and warm air masses of a tropical depression of 998 mb with the air of orographic lifting. In Fig. 43 the centre of a tropical depression is at 30°N, 157°E, from which a tropical front extends eastward and many origins are found along this front. In Fig. 44 a trough passes along Japan from Hokkaido to the Ryukyus, and origins are distributed along the trough.

(ii) Distribution of atmospherics in Asia and the Western Pacific. In summer, origins of atmospherics are distributed (1) in tropical and subtropical Asia, especially in valleys with large rivers and mountenous districts, passed by intertropical fronts (2) on the tropical convergence line of the Sea of China, i.e. the confluence region of two equatorial maritime air flows coming from the Western Pacific and the Indian Ocean, (3) in orographic lifting regions in the Indonesia Islands and Malaya, and (4) in the strong insolational heating regions in archipelagoes in the Equatorial Pacific (Fig. 45 in the daytime and Fig. 46 at night).

In winter also (1) in tropical and subtropical Asia, especially in valleys with large rivers, passed by the activated temperate front, (2) in the northern part of Australia passed by the intertropical front, (3) on the Sea of China where the cold polar continental air mass invades under the warm equatorial maritime air mass, (4) in orographic lifting regions in the Indonesia Islands including New Guinea, and (5) in the strong insolational heating regions in archipelagoes in the Equatorial Pacific (Fig. 47 in the daytime and Fig. 48 at night).

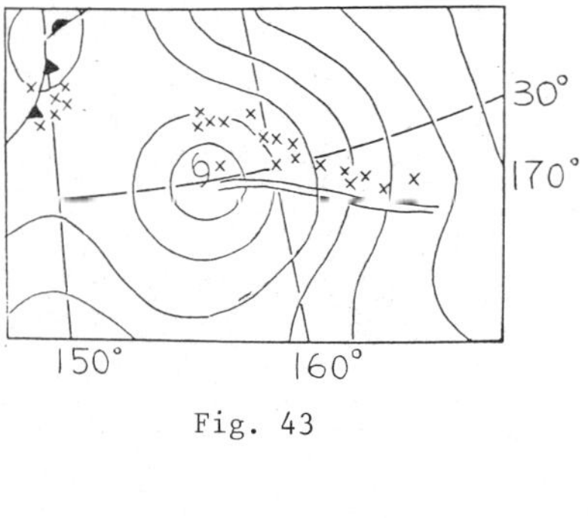

Fig. 43

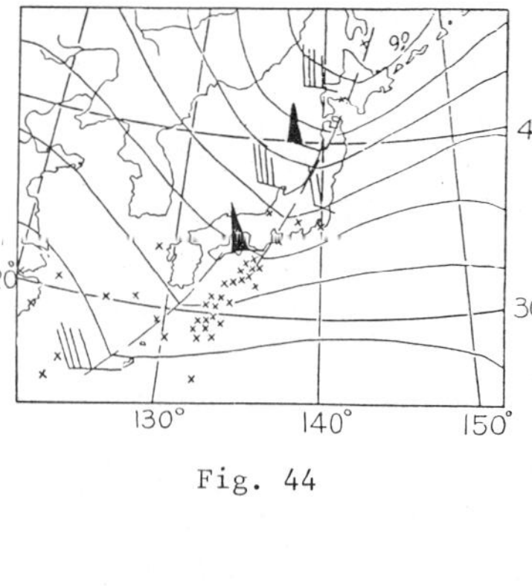

Fig. 44

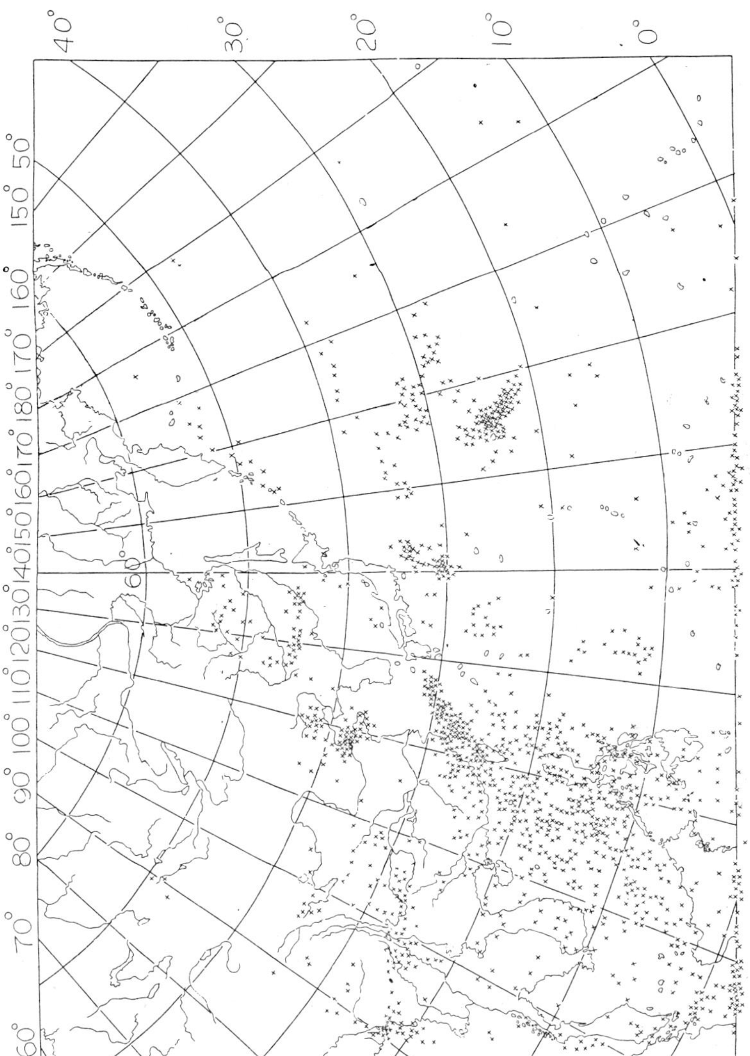

40°
30°
20°
10°
0°
60°
60°
70°
80°
90°
100
110°
120°
130°
140°
150°
160°
170°
180°
170°
160°
150°

Fig. 45

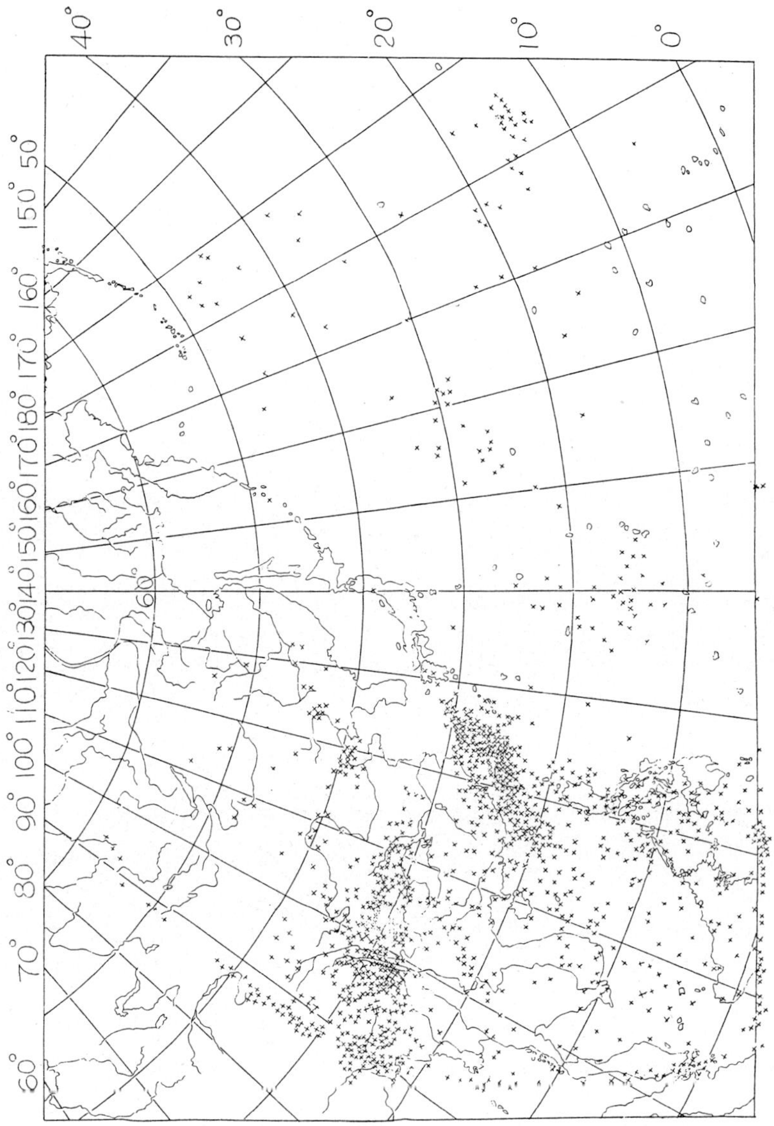
40°
30°
20°
10°
0°
60°
70°
80°
90°
100°
110°
120°
130°
140°
150°
160°
170°
180°
170°
160°
150°
50°
60

Fig. 46

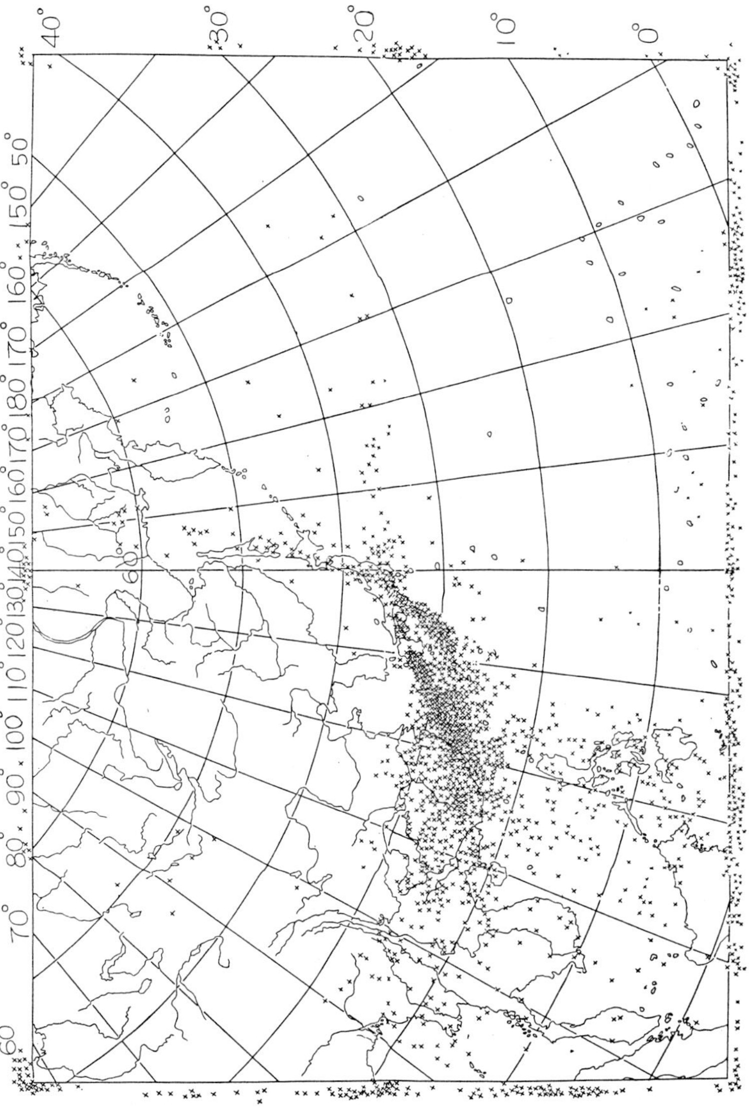
40°
30°
20°
10°
0°
60°
50°
150°
160°
170°
180°
170°
160°
150°
140°
130°
120°
110°
100°
90°
80°
70°
60°

Fig. 47

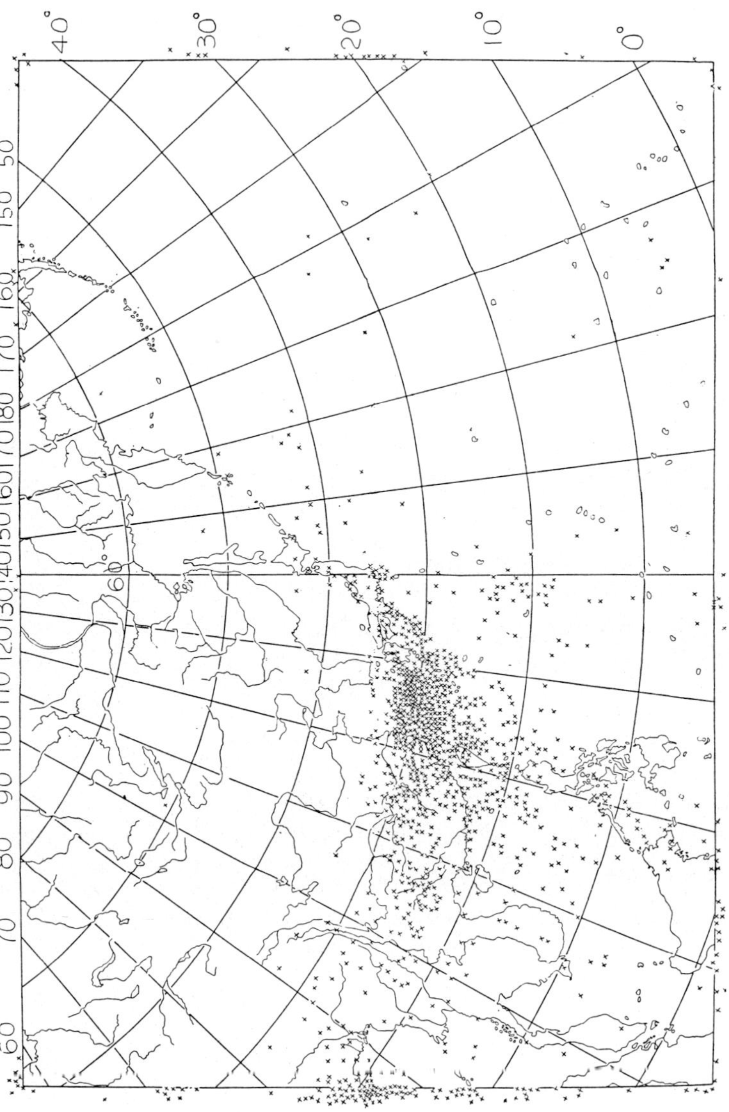

40°
30°
20°
10°
0°
50°
150°
160°
170°
180°
170°
160°
150°
140°
130°
120°
110°
100°
90°
80°
70°
60°
60

Fig. 48

REFERENCES

1. Report of Radio Research in Japan, Vol. 1, No. 1, p. 1, 1931.
2. Proceedings of Japan Academy, Vol. 27, No. 7, 1951 and Vol. 28, No. 8, 1952.
3. Journal of Geomag. Geoelec. of Japan, Vol. 5, No. 1-2, 1953.
4. Proceedings of Reseach Institute of Atmospherics, Nagoya University, Vol. 2, p. 40, 1954.
5. Comptes rendus Acad. Sc. Paris, t. 272, 28 juin 1971 and t. 280, 10 mars 1975.
6. Comptes rendus Acad. Sc. Paris, t. 248, 6 avril 1959.
7. Memoirs Fac. Eng., Nagoya Univ., Vol. 7, No. 2, Nov., 1955.
8. Memoirs Fac. Eng., Nagoya Univ., Vol. 9, No. 1, May, 1957.

Nonlinear and Environmental Electromagnetics, edited by H. Kikuchi
Elsevier Science Publishers B.V., Amsterdam, 1985 – Printed in The Netherlands

AN INTERPRETATION OF DAWN-SIDE MID-LATITUDE VLF EMISSIONS IN TERMS OF QUASI-LINEAR ELECTRON CYCLOTRON INSTABILITY

M.HAYAKAWA[1], Y.TANAKA[1], S.S.SAZHIN[2] and T.OKADA[1]

1. Research Institute of Atmospherics, Nagoya University, Toyokawa, Aichi, 442, Japan

2. pr.Kosmonavtov 18-2-44, Leningrad 196244, U.S.S.R.

ABSTRACT

Mid-latitude VLF emissions are found to take place in close association with geomagnetic substorms at two local times; one at dawn and the other in the pre-midnight. The characteristics of dawn-side VLF emissions are examined in the present paper, based on the data obtained by our VLF/ELF campaign during November,1978 to February,1979 by means of the direction finding measurements at two stations, Brorfelde in Denmark ($L\sim 3$) and Chambon-la-Foret in France ($L\sim 2$). Six remarkable events have been analyzed and the following properties have emerged; (i) the occurrence is highly peaked at dawn, L.T.$\sim$6h, (ii) the VLF emissions observed are identified as being mainly of the hiss-type, (iii) the region of maximum intensity corresponding to the latitude of generation region is estimated to be inside the plasmapause, based on the direction finding results, and (iv) the VLF emissions are found to exhibit a regular frequency drift, the frequency is increasing with local time for all the events,but one.

It is shown that all of the above features obtained can be understood within the quasi-linear model of cyclotron instability of the electrons injected in association with geomagnetic substorms. In particular, the property (iv) which is not well understood is interpreted as being related to the energy dispersion of the electrons drifting from the midnight sector of their initial injection toward the generation region of VLF emissions (dawn sector).

1. INTRODUCTION

Among different types of natural radio emissions observed on the ground as well as within the magnetosphere (see Likhter,1979; Sazhin,1982 ,1983; Anderson,1983), those which are associated with magnetic storm development attracted particular attention of the researchers (Hayashi and Kokubun,1971; Rycroft,1972; Kokubun et al.,1972; Bullough et al.,1974; Kaiser and Bullough,1975; Parady et al.,1975; Foster et al.,1976; Hayakawa et al.,1977; Hayakawa et al.,1981a,etc). The excitation of those emissions can be most easily associated with the injection of plasma sheet electrons into the inner magnetosphere and this makes it very convenient to study the corresponding wave-particle interaction processes. Moreover,these emissions provide a powerful tool for the study of the injection of electrons and the magnetospheric plasma structure. Our analysis of this paper will concern one of the most important types of these emissions, known as medium-latitude VLF emissions. In particular, we will present a detailed analysis(both from experimental and theoretical points of view) of the frequency increase of these emissions in the dawn sector. Although this phenomenon has been analyzed in several earlier publications (Vershinin,1970; Carpenter et al.,1975; Sazhin and Vershinina,1978; Vershinin et al.,1979; Park et al.,1981; Smirnova,1984;the last paper has been discussed by Sazhin et al.(1984)), its nature remains essentially unclear. We shall describe some features of this frequency increase which has not been noticed so far and present their interpretation within the quasi-linear model of mid-latitude VLF emissions suggested by Sazhin(1984).

In order to clarify the exit points of the mid-latitude VLF emissions at the ionospheric level we carried out two-stationed spaced direction finding measurements during the three northern winters between 1976 to 1979. Two stations were (1)Brorfelde in Denmark($L\sim3$) and (2) Chambon-la-Foret in France ($L\sim2$). The L as low as $L\sim3.0$ of the station roughly corresponds to storm-time plasmapause location which shifts towards lower L shells during geomagnetic disturbances (Chappell et al.,1970; Rycroft and Thomas,1970; Rycroft,1975; Hultqvist,1975). We made full use of direction findings with two different systems; (i) goniometer triangulation (Bullough and Sagredo,1973; Hayakawa et al.,1981b) and (ii)field-analysis method (Tanaka,1972; Tanaka et al.,1976; Okada et al.,1977,1981; Hayakawa et al.,1981c). Further details of the equipments are given in Hayakawa et al.(1981a).

Initial analyses of the data obtained during the last year's campaign(November 1978 to February 1979) have been briefly described in Hayakawa et al.(1981a) and the use of direction finding results has enabled them to classify the observed emissions into several categories. The most popular one among them is the "plasmapause-associated VLF emissions", which take place in close association with geomagnetic disturbances around the plasmapause and which is the subject of the present paper. Their main occurrence is concentrated at two different local times; one at dawn (L.T.$\sim$6h) and the other in the pre-midnight (L.T.$\sim$22h). The present paper is concerned with the emissions in the former L.T.'s, while the emissions in the latter L.T.'s will be reported elsewhere.

The outline of the paper is as follows. Section 2 gives the summary of the observed important characteristics of VLF emissions based on the detailed studies of the six conspicuous events. Then, in Section 3 we show that the obtained characteristics are satisfactorily interpreted with the quasi-linear model of cyclotron instability of substorm electrons.

2. SUMMARY OF THE OBSERVED CHARACTERISTICS

Six events have been analyzed in details,but we present only a few examples in the following. Three events have occurred successively on three days and we illustrate some characteristics for those events.

Fig.1 gives the temporal variations of one-hour AE and 3-hr K_p indices during the four days from 18th to 21st December 1978. The time of occurrence of VLF emissions is also indicated below the abscissa of AE index. As seen from this figure, we are in a period of relatively severe geomagnetic disturbances because the K_p index is always greater than 3. The local time of occurrence of VLF emissions is 4:20-7:50 for 19 Dec., 4:20-7:20 for 20 Dec., and 5:50-7:50 for 21 Dec. (L.T.=U.T.+1h), respectively. Taking into account the K_p curve, the emissions are found to be observed during the declining phase of K_p index. However, even during these periods, we can easily find out many particle injections as seen from the variation of AE index, which are likely to be related with the emission generation.

Fig.2 presents the temporal evolution of the frequency of emissions observed at Brorfelde on 21 Dec. 1978. The frequency interval where the intensity exceeds the prescribed threshold is indicated as a dark part and the intensity is expressed in a logarithmic (dB) scale. The frequency with the maximum intensity is indicated by a triangle and its numerical value is also given. We can see the drift rate of frequency,as well. Wide-band spectra have yielded that the observed VLF emissions are of the hiss-type.

Then, Fig.3 gives the relative position of the VLF emission region determined by the direction finding with the plasmapause location estimated by the empirical formula and our theoretical model calculation.

Based on the analyses for all the events including some presented now,we can summarize the observed characteristics of mid-latitude VLF emissions in the following.

(1) The emissions observed at Brorfelde and Chambon-la-Foret consist of two bands; one is ELF ($f \lesssim 3$kHz) and the other is VLF ($f \gtrsim 3$kHz) which is the subject of the present paper. The VLF emissions are identified as being mainly of the hiss-type.

(2) The VLF emissions were observed during the phase of decreasing geomagnetic activity as indicated by K_p index, but in a close association with geomagnetic substorms, as indicated by the peaks of AE index. The peaks of AE index which seem to be associated with the emission generation preceded the beginning of these emissions by 6-11 hours. The occurrence of the emissions is highly concentrated in the L.T. sector just around 6h.

(3) The ionospheric exit region of those VLF emissions has been determined by the direction findings at the two stations, which has yielded that the emissions are widely distributed in the ionosphere in a latitude range from the plasmapause to a lower L shell (~ 2.0) and they have penetrated the ionosphere at any point in the extended latitude range. However, the region of the maximum intensity is likely to be close to the inner edge of the plasmapause.

(4) The observed emissions exhibit a regular frequency increase for the events analyzed,but one case. The rate of frequency increase has a tendency to be

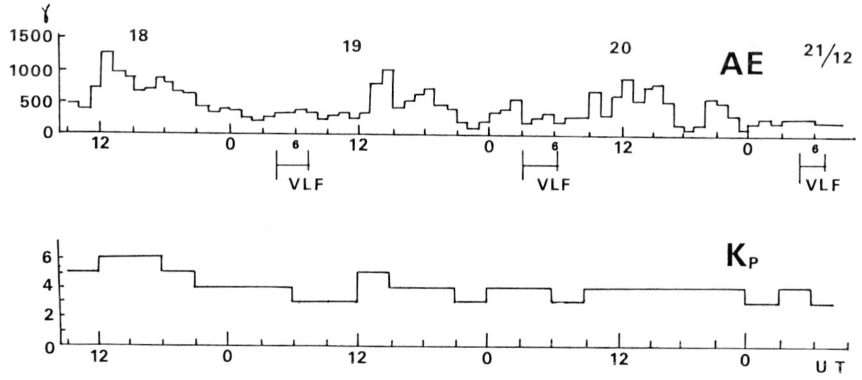

Fig.1 Temporal variation of the geomagnetic activity during the period from 18 to 21 December 1978, when we have observed dawn-side VLF emissions. The upper panel refers to the AE index and the lower the K_p index. The times of occurrence of VLF emissions are indicated below the abscissa of the AE index.

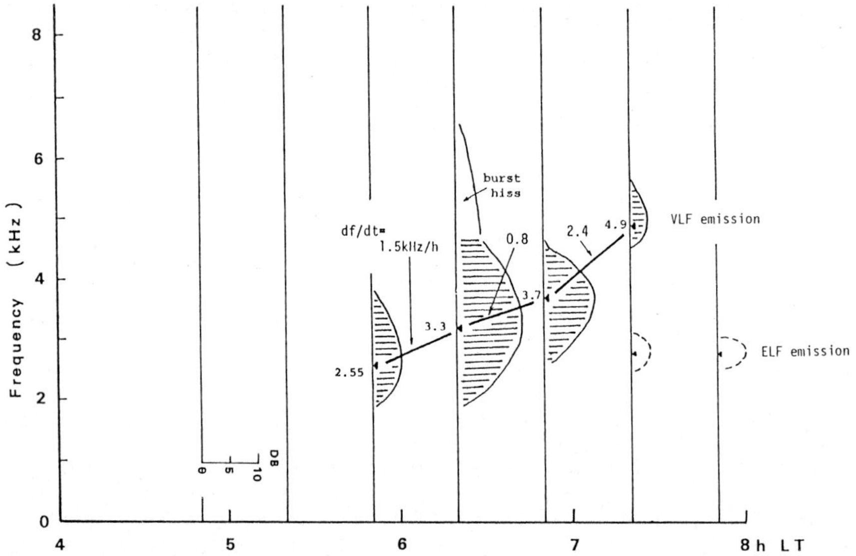

Fig.2 The frequency spectrum for the event of 21 Dec. 1978. The VLF emissions exhibiting a regular frequency drift are indicated by a dark part. While,the ELF emissions which are not the subject of the present paper with a broken envelope. The frequency of the maximum intensity for each frequency band is also indicated by a triangle,with its frequency value. The curve connecting these triangles for VLF emissions gives the frequency increasing phenomenon. The rate of frequency drift is also estimated.

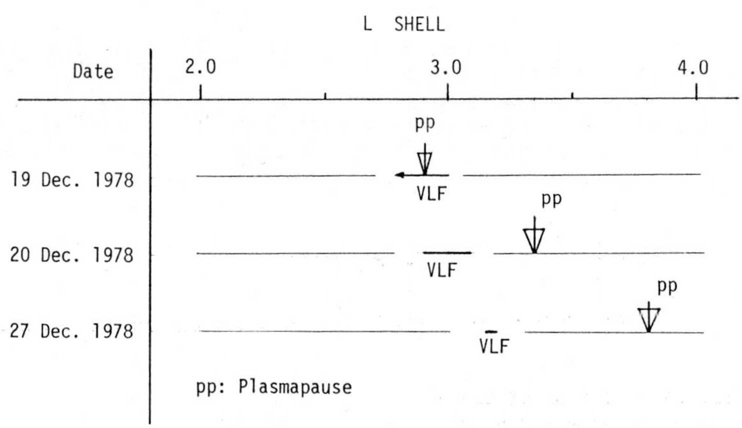

Fig.3 The relative location of VLF emission region with respect to the plasmapause position for the three events for which the direction finding results are available.

depleted at later L.T.'s in most cases, the maximum rate being as much as 2.5 kHz/hr.

Further details will be published elsewhere (Hayakawa et al.,1984).

3. INTERPRETATION OF THE OBSERVATIONAL PROPERTIES IN TERMS OF QUASI-LINEAR ELECTRON CYCLOTRON INSTABILITY

Although the emission frequency range can be in many cases satisfactorily understood in terms of the corresponding instability frequency range, one cannot, in general case, use the linear theory of the instability when interpreting the the emission amplitude. If we restrict ourselves by considering only hiss-type emissions (Point (1) in Section 2), then their quantitative model can be based on quasilinear theory. The corresponding quasilinear models have been proposed by Roux and Solomon(1970,1971), Etcheto et al.(1973), Sazhin(1977,1984) and Bespalov and Trakhtengertz(1980). Our interpretation of experimental results is based on the model of Sazhin(1984) which seems to be most compatible with the realistic magnetospheric conditions. This model being strictly speaking valid only for the hiss-type emissions, can also be used for qualitative estimates of chorus amplitude if we assume chorus spectrum as a result of a deformation of hiss spectrum (Nunn,1974; Koons,1981).

Without discussing the details of this model it is essential to mention that it is based on the supposition that all the electrons penetrating into the magnetic field tube, where the waves are excited, partly drift across this tube and partly penetrate, under the influence of the wave field, into the loss cone and further precipitate into the ionosphere, so that the electron density in this tube remains constant. (This supposition is just the same as that which was made by Etcheto et al.(1973) and Sazhin(1977).) The source function of the electrons penetrating into the considered magnetic tube is supposed not to depend on the geomagnetic latitude along the field line; total amplification of the parallel waves bouncing from one hemisphere to another is assumed to be equal to their attenuation during the reflection process at the ionosphere. A realistic distribution of magnetic induction and electron density along the field line has been taken into account.

Maximum power spectral density $\langle B_f^2 \rangle_{max}$ can be evaluated from the formula:

$$\langle B_f^2 \rangle_{max} \ (\text{Gauss}^2/\text{Hz}) = 5 \times 10^{-27} \ n_{eq} \ L^{5.5} \ W_o^{1.5} \ L_o^{4.5} \ dn_i/dt \qquad (1)$$

where n_{eq} is the equatorial electron density (in cm^{-3}) at the L shell where the waves are generated, W_o is the electron energy (in electron volt) at $L=L_o$, L_o can be arbitrarily chosen, and dn_i/dt is the velocity of the income of the electrons (in cm^{-3} sec^{-1}).

The frequency f_{max} at which $\langle B_f^2 \rangle$ is maximal is determined by the following formula:

$$f_{max} \ (\text{Hz}) = \frac{\Omega^3 \ c^2}{1.2 \times 2\pi \times 10^3 \ \Pi^2 \ v^2} \qquad (2)$$

where Ω and Π are the angular electron gyrofrequency and plasma frequency(rad/sec) at the magnetic equator where the emissions are generated, c is the velocity of the light and v is the thermal velocity of the electrons at the L shell where the waves are generated.

It would be reasonable to assume that the electrons which are responsible for wave excitation penetrate into the inner magnetosphere from the plasma sheet region near the mid-night sector and further drift to the dawn sector. In particular, the waves can be excited at any point of the electron's trajectory, however the amplitude of the excited waves is different. If we assume that the L_2shell of the drifting electrons does not change, then, as follows from Eq.(1), $\langle B_f^2 \rangle_{max} \sim n_{eq}$. One could expect that the preferable registration of the waves at L.T. ~6h is due to an increase in the ambient plasma density from the sunlit ionosphere(Point 2 in section 2). This is checked by measuring the whistler dispersion. The increase of $\langle B_f^2 \rangle_{max}$, however, seems to be not sufficient to explain the absence of the emissions at L.T. < 3h 20m, and we are to look for some other explanations. It seems reasonable for us to suppose that the observed local time dependence of the emissions is simply due to the fact that the electrons drifting to dawnside sector penetrate to lower L shells when L.T. < 6h. Hence the value of dn_i/dt at a given L shell (corresponding to Brorfelde,for example) increases with increasing L.T. in the range 0h < L.T. < 6h, which causes the corresponding increase of $\langle B_f^2 \rangle_{max}$ according to Eq.(1). Similar dawn-side increase of mid-latitude emissions has been reported in many previous papers (see for example, Thorne et al.,1974; Tsurutani and Smith,1977; Hayakawa et al.,1981a,1984).

As follows from Eq.(1) the value of $\langle B_f^2 \rangle_{max}$ being proportional to $n_{eq}\ L^{5.5}$ has a sharp peak near the inner boundary of the plasmapause. This conclusion is in good agreement with the direction finding results (Point 3 in Section 2). which indicates the region of maximum intensity of VLF emissions with respect to the plasmapause for the three events for which the direction finding data are available. The maximum intensity or main peak intensity in the ionosphere is found to be at L~3.0 which also appears to be closely related to the previous in-situ latitudinal profiles of VLF emission intensity reported by Bullough et al.(1969), Hayakawa et al.(1977) and Ondoh et al.(1980). The excited waves can be trapped near the inner edge of the plasmapause, and penetrate to lower altitudes without reflecting at the lower hybrid frequency (Inan and Bell,1977; Semenova and Trakhtengertz,1980; Sazhin and Varshavskii,1982). The wide distribution of the observed waves in the ionosphere from the plasmapause latitude to L~2.0 is probably related to the spread of the waves due to non-ducted propagation in the lower exosphere.

As follows from Eq.(2) the value of f_{max} depends on three parameters: Ω, Π and v, none of which is available from in-situ measurements. However, we can determine the average values of Ω and Π from the statistically average location of the plasmapause and electron density near its inner boundary where the emissions are believed to be excited. As seen in Fig.1 as an example, it follows that most of the emissions were observed during the periods when K_p=4-5, and so $L_{pp} \sim 3.3$(which is compatible with the result in Fig.3) and n_{eq}= $2\text{x}10^3\ cm^{-3}$(Chappell et al.,1970). Considering the dipole model for the magnetic field and substituting the corresponding values of the parameters into Eq.(2) we can obtain a useful formula relating f_{max} with the energy of the incoming electrons W (Sazhin, 1984):

$$f_{max}\ (\text{kHz}) = 20.9\ /\ W(\text{keV}) \qquad (3)$$

Formula (3) can be used not only for determining f_{max} from W, but also for determining W from f_{max}. The latter case will be of our primary interest. In particular, basing on Eq. (3), we can make a conclusion that the emissions with f_{max}=4 kHz are excited by the electrons with the energy of about 5 keV (see also Fig.3 in Sazhin's(1984) paper). In-situ observations of the electrons with close energy in the equatorial magnetosphere have been reported by many authors (see for example Rönnmark,1978).

Let us now try to make a comparison between the observations of the dawnside emissions and the electron intrusions in the nightside magnetosphere as indicated by the peaks of AE index as seen in Fig.1 ,for example. If our hypothesis that the electrons responsible for the excitation of dawnside emissions intially penetrate into the inner magnetosphere near the midnight sector is correct, then the corresponding peaks of AE index precede the times of the beginnings of the emission activity by electrons' drift time (T). This time roughly corresponds to the time interval during which the electrons are drifting from the midnight to the dawn meridian of L.T.=6h(the most probable time of registration of the considered emissions); that is, a quarter of the drift period of the electrons encircling the Earth. Thus, T(in sec) can be evaluated by the formula (see Fälthammer, 1973):

$$T\ (\mathrm{sec}) = \frac{0.7 \times 10^{9}}{0.7 + 0.3 \times \sin \alpha_e} \cdot \frac{1}{L} \cdot \frac{1}{W(\mathrm{eV})} \qquad (4)$$

where α_e, the equatorial pitch angle of the drifting electrons and W, the electron energy in eV.

When making the qualitative analysis of the process, we can neglect a weak dependence of T on α_e, which makes it possible to assume roughly $\alpha_e=\pi/4$. Supposing also that the L value of the wave generation is L$\sim$3.3 (see above) as evidenced previously by the direction findings, we can express Eq.(4) in the following form:

$$T\ (\mathrm{hours}) = 65.7\ /\ W(\mathrm{keV}) \qquad (5)$$

From Eqs.(3) and (5) we can relate the value of T with f_{max} by a very simple formula.

$$T(\mathrm{hours}) = 3\ f_{max}(\mathrm{kHz}) \qquad (6)$$

The value of f_{max} at the beginning of the events ranged from 2 to 5 kHz, which means that T ranges from 6 to 15 hours. From the inspection of AE plot presented in Fig.1 as an example, it is easy to see that the nearest peaks of AE index preceded the beginnings of the emissions by 6-11 hours (Point 2 in Section 2), which is in good agreement with the prediction of Eq.(6). Thus, our hypothesis that the electrons responsible for the emission excitation come from the midnight sector seems to be not contradictory.

Moreover, when considering AE histogram presented as in Fig.1, one can notice that the peaks of AE which are presumably responsible for emission excitation are always preceded by 4-5 hours by even greater peaks of AE. This phenomenon could be presumably associated with the changes of the magnetospheric structure during the preceding substorm (increase of a large-scale electric field, contraction of the plasmapause, increase of energetic electron density etc.) which would create necessary preconditions for the emission generation in the dawn sector. However, all these processes are to be studied more thoroughly (both theoretically and experimentally) in order that this interpretation could be more convincing.

As follows from Eq.(6) electrons responsible for higher frequency emissions (less energetic) drift longer time than the electrons responsible for lower frequency emissions (more energetic). Thus, we can expect that the electrons with different energies penetrating to the midnight magnetosphere will excite in the dawn sector the emissions with increasing frequency, which are really observed therein (Point 4 in Section 2). (This explanation of the emissions with increasing frequency has been earlier investigated by Carpenter et al.(1975), Foster et al.(1976) and Park et al.(1981); however, no quantitative model has been suggested by these authors.) Formula (6) makes it possible also to predict the value of df_{max}/dt to be as large as 3 kHz/hr. This value is larger than those which were observed in our events where df_{max}/dt has been in most cases below 1.5 kHz/hr. This discrepancy is not large if we keep in mind the roughness of our model. It can be presumably due to the fact that the model of Sazhin(1984) (as well as all the previous models) did not take into account the dependence on frequency of the reflection coefficient of whistler-mode waves from the ionosphere (r). If we take into account that r tends to decrease with increasing frequency (Hayakawa and Ohtsu,1972), then the value of f_{max} would be less than that which is predicted by Eq.(2) and this would reduce the corresponding value of df_{max}/dt. Another explanation of this discrepancy can be related to the fact that when deriving Eq.(6) it has been implicitly supposed that the emissions registered at different local times are generated at the same meridional sector, corresponding to L.T.=6h. However, if we suppose an extended region of emission generation, then the station corotating with the Earth would register the emissions excited at different L.T.'s and the corresponding values of T in Eq.(6) would increase with time. This would result in the decrease of df_{max}/dt, which has been observed during most of the events examined, although not shown. In the event of 21 Dec. in Fig.2 when this decrease of df_{max}/dt was not observed the value of df_{max}/dt appeared to be most close to that predicted by Eq.(6). Moreover, the waves at the later L.T.'s are generated in the regions corresponding to more dense plasma, and thus larger Π, as it is checked from the whistler dispersion. This would also tend to decrease the value of df_{max}/dt. In principle, the observed decrease of df_{max}/dt with increasing L.T. can indicate an extended region of emission generation.

Other explanations for the considered increase of f_{max} have been suggested by Sazhin and Vershinina(1978) and Vershinin et al.(1979). In the first model it was supposed that the frequency drift is due to the drift in the magnetosphere of the energetic electrons which resulted from the combined effects of the large-scale convection electric field and the inhomogeneity of the magnetic field. This drift would result in the increase of Ω for 0h < L.T. < 6h and its decrease for 6h < L.T.< 12h. As follows from Eq.(2), this would result in the increase of f_{max} in the first interval and its decrease in the second. However, as seen from most of our results (presented in Fig.2 as an example) the character of the frequency increase is qualitatively identical for L.T. < 6h and L.T. > 6h, which is in contradiction with this model. The only event with which it seems to be compatible is that of 30 Dec.(not shown in the present paper). Vershinin et al.(1979) have related this frequency drift with the contraction of the magnetosphere in the fixed point of the equatorial region of the magnetosphere in the vicinity of the plasmapause which take place during the period of increasing geomagnetic activity. Electron density decrease, in its turn, would lead to the decrease of Π and correspondingly to the increase of f_{max} (although Vershinin et al.(1979) based their analysis on the linear model of wave excitation, this prediction of their model is the same as that which would follow from quasi-linear model). However, the contraction of the magnetosphere which has been assumed in Vershinin et al.'s(1979) model,would take place during the periods of increasing geomagnetic activity as measured by K_p index. At the same time, all the events which were analyzed in Section 2 took place during the period when K_p index decreased and this would result, within the model of Vershinin et al.(1979), in the decrease rather than increase of f_{max}. Thus, the model of Vershinin et al.(1979) can explain in principle only the event of 30 Dec. as well as the model of Sazhin and Vershinina(1978). The process described by Vershinin et al.(1979) can also result in the decrease of the value of df_{max}/dt, when compared with that which is predicted by Eq.(6).

4. CONCLUSION

Our interpretation of the properties of the emissions is based mainly on the quasilinear model for hiss-type mid-latitude emissions which has been presented by Sazhin(1984). Source electrons are supposed to penetrate into the magnetosphere near the midnight sector at the time corresponding to the peaks in AE index and then drift toward dawn sector partly penetrating inside the plasmasphere where the waves are excited. The excited waves are further trapped near the inner edge of the plasmapause and penetrate up to the point of their registration at Brorfelde or Chambon-la-Foret. The observed frequency increase of the emissions is interpreted in terms of the energy dispersion of the drifting electrons. It is pointed out that the observed depletion of the rate of the emission frequency increase is compatible with the supposition that the generation region of these emissions is extended in longitude. The observed decrease of frequency with time (the event of 30 Dec.1978) can be understood in terms of the mechanisms considered by Sazhin and Vershinina(1978) and Vershinin et al.(1979).

ACKNOWLEDGEMENT

The VLF campaign in Europe was intially started in collaboration with Drs. L.R.O.Storey and C.Beghin of Centre de Recherches en Physique de l'Environnement, CNRS (Orleans,France) and Dr.T.S.Jorgensen of Danish Meteorological Institute(Copenhagen, Denmark), to whom we wish to express our sincere thanks for their cooperation. Thanks are also due to Japan Society of Promotion of Sciences for its financial fund for our VLF/ELF project. We are grateful to Profs. A.Iwai and J.Ohtsu of Nagoya University for their discussion and Prof.A.Kimpara, Emeritus Professor of Nagoya University for his continual encouragement.

REFERENCES

Anderson,R.R., 1983. Rev.Geophys.Space Phys., 21, 474.
Bespalov,P.A. and Trakhtengertz,V.Yu., 1980. "Problems of Plasma Theory"(Edited by Mikhailovskii,A.B.), p.88, Atomizdat, Moscow (in Russian).
Bullough,K., Hughes,A.R.W. and Kaiser,T.R., 1969. Planet.Space Sci., 17, 363.
Bullough,K. and Sagredo,J.L., 1973. Planet.Space Sci., 21, 899.
Bullough,K., Hughes,A.R.W. and Kaiser,T.R., 1974. "Magnetospheric Processes" (Edited by McCormac,B.M.), p.231, D.Reidel.
Carpenter,D.L., Foster,J.C., Rosenberg,T.J. and Lanzerotti,L.J., 1975. J.Geophys. Res., 80, 4279.
Chappell,C.R., Harris,K.K. and Sharp,G.W., 1970. J.Geophys.Res., 75, 50.
Etcheto,J., Gendrin,R., Solomon,J. and Roux,A., 1973. J.Geophys.Res., 78, 8150.
Fälthammer,C.G., 1973. "Cosmical Geophysics"(Edited by Egeland,A. et al.), p.121, Universitetsforlaget.

Foster,J.C., Rosenberg,T.J. and Lanzerotti,L.J., 1976. J.Geophys.Res., 81,2175.
Hayakawa,M. and Ohtsu,J., 1972. Planet.Space Sci., 20, 1895.
Hayakawa,M., Bullough,K. and Kaiser,T.R., 1977. Planet.Space Sci., 25, 353.
Hayakawa,M., Tanaka,Y., Iwai,A., Ohtsu,J., Storey,L.R.O., Beghin,C. and T.S. Jorgensen, 1981a. Planet.Space Sci., 29, 505.
Hayakawa,M., Tanaka,Y., Okada,T. and Iwai,A., 1981b. J.Geophys.Res., 86, 6781.
Hayakawa,M., Okada,T. and Iwai,A., 1981c. J.Geophys.Res., 86, 6939.
Hayakawa,M., Yamanaka,Y., Parrot,M. and Lefeuvre,F., 1984. J.Geophys.Res., 89, 4773.
Hayakawa,M., Tanaka,Y., Sazhin,S.S., Okada,T. and Kurita,K., 1984. Submitted.
Hayashi,K. and Kokubun,S., 1971. Rep.Ionosph.Space Res.Japan, 25, 369.
Hultqvist,B., 1975. Ann.Geophys., 31, 111.
Inan,U.S. and Bell,T.F., 1977. J.Geophys.Res., 82, 2819.
Kaiser,T.R. and Bullough,T.R., 1975. Ann.Géophys., 31, 10.
Kokubun,S., Makita,K. and Hirasawa,T., 1972. Rep.Ionosph.Space Res.Japan, 26, 138.
Koons,H.C., 1981. J.Geophys.Res., 86, 6745.
Likhter,Ya.L., 1979. "Wave instabilities in space plasmas"(Edited by Palmadesso, P.J. and Papadoulos,K.), p.3, D.Reidel.
Nunn,D., 1974. J.Plasma Phys., 11, 189.
Okada,T., Iwai,A. and Hayakawa,M., 1977. Planet.Space Sci., 25, 233.
Okada,T., Iwai,A. and Hayakawa,M., 1981. J.Atmos.Terr.Phys., 53, 679.
Ondoh,T., Nakamura,Y., Watanabe,S. and Murakami,T., 1980. J.Radio Res.Labs. Tokyo, 27, 131.
Parady,B.K., Eberlein,D.D., Marvin,J.A., Taylor,W.W.L. and Cahill,L.J.Jr., 1975. J.Geophys.Res., 80, 2183.
Park,C.G., Lin,C.S. and Parks,G.K., 1981. J.Geophys.Res., 86, 37.
Rönnmark,K., Borg,H., Christiansen,P.J., Gough,M.P. and Jones,D., 1978. Space Sci.Rev., 22, 401.
Roux,A. and Solomon,J., 1970. Ann.Géophys., 26, 279.
Roux,A. and Solomon,J., 1971. J.Atmos.Terr.Phys., 33, 1457.
Rycroft,M.J., 1972. Radio Sci., 7, 811.
Rycroft,M.J., 1975. Ann.Geophys., 31, 1.
Rycroft,M.J. and Thomas,J.O., 1970. Planet.Space Sci., 18, 65.
Sazhin,S.S., 1977. "Relationships between upper atmosphere, VLF emissions and other geophysical phenomena"(Edited by Fedyakina,N.I.), p.76 (in Russian).
Sazhin,S.S. and Vershinina,N.I., 1978. Cosmic Res., 16, 462 (in Russian).
Sazhin,S.S., 1982. Natural radio emissions in the Earth's magnetosphere, Pub. House "Nauka", Moscow (in Russian).
Sazhin,S.S., 1983. "Magnetospheric Research, No.2"(Edited by Pudovkin,M.I.), p.5, Geophysical Committee, Moscow (in Russian).
Sazhin,S.S. and Varshavskii,S.P., 1982. Geomag. $ Aeronomy, 22, 224 (in Russian).
Sazhin,S.S., 1984. Planet.Space Sci., in press.
Sazhin,S.S., Hayakawa,M., and Tanaka,Y., 1984. Planet.Space Sci., in press.
Semenova,V.I. and Trakhtengertz,V.Yu., 1980. Geomag. $ Aeronomy, 20, 1031 (in Russian).
Smirnova,N.A., 1984. Planet.Space Sci., 32, 425.
Tanaka,Y., 1972. Proc.Res.Inst.Atmospherics, Nagoya Univ., 19, 33.
Tanaka,Y., Hayakawa,M. and Nishino,M., 1976. Mem.Natl Inst.Polar Res. Tokyo, A13, p.58.
Thorne,R.M., Smith,E.J., Fiske,K.J. and Church,S.R., 1974. Geophys.Res.Lett., 1, 193.
Tsurutani,B.T. and Smith,E.J., 1977. J.Geophys.Res., 82, 5112.
Vershinin,E.F., 1970. Ann.Géophys., 26, 703.
Vershinin,E.F., Trakhtengertz,V.Yu. and Shapayev,V.I., 1979. Geomag. $ Aeronomy, 19. 362.

Nonlinear and Environmental Electromagnetics, edited by H. Kikuchi
Elsevier Science Publishers B.V., Amsterdam, 1985 – Printed in The Netherlands

DAY-NIGHT AND LATITUDINAL VARIATIONS OF WHISTLER INTENSITIES AS ESTIMATED BY DIRECTION FINDING MEASUREMENTS

M.Hayakawa, T.Okada and Y.Tanaka

Research Institute of Atmospherics,Nagoya University,
Toyokawa, Aichi, 442, Japan

K.Ohta

Department of Electronic Engineering,College of Engineering,
Chubu University, Kasugai, Aichi, 487, Japan

ABSTRACT

The field-analysis direction finding(based on the measurement of two horizontal magnetic field components and a vertical electric field component) has enabled us to estimate the absolute intensity of whistlers at medium and low latitudes. The measurements have been made at Yamaoka(geomag.lat.25°N) and Moshiri(35°N). Based on the significant amount of data at Yamaoka,it is found that the daytime intensity exhibits a maximum occurrence at 100-125 μV/m and the maximum value amounts to 225-250μV/m,while the maximum occurrence takes place at 75-100μV/m and the maximum value is 150-175μV/m at night. On the contrary,the nighttime intensity seems to be more enhanced than the daytime one at Moshiri. The occurrence histogram of absolute intensity during daytime is sharply cutoff above 500μV/m,but we can expect an occurrence of nighttime intensity up to 1mV/m. These day-night and latitudinal variations of absolute intensity of whistlers are attempted to interprete.

1. INTRODUCTION

The measurement of absolute intensity of VLF whistlers having propagated through the magnetosphere provides useful information on their propagation characteristics including the ionospheric absorption and magnetospheric propagation (ducted or nonducted)(Helliwell,1965; Hayakawa and Tanaka,1978)

The apparent field strength of a whistler can be measured in a few ways. If the whistler is very intense,it is sufficiently simple to record the amplitude as a function of time. From this record,the maximum,minimum and average values are readily obtained. Another method,less quantitative,is to match an artificially generated whistler of known amplitude aurally with the natural whistler in question(Iwai and Outsu,1958). However,the absolute intensity of whistlers cannot be accurately estimated unless both the wave normal direction and wave polarization are specified, and so the measurement of them essentially requires the adoption of the direction finding. During the last decade,several kinds of VLF direction finding methods(Tanaka,1972; Cousins,1972; Bullough and Sagredo,1973; Tsuruda and Hayashi,1975; Tanaka et al.,1976; Okada et al.,1977,1981; Leavitt et al.,1978; Ohta,1984; Ohta et al.,1984) have been developed. The results on the ionospheric exit regions have been extensively utilized to study the whistler propagation mechanism and duct properties(Matthews et al.,1979; Lester and Smith, 1980; Hayakawa et al.,1981a,b; Strangeways et al.,1983) and also the magnetospheric plasma dynamics(Sagredo and Bullough,1973; Carpenter,1980). However,the reports are as yet very lacking on the absolute intensity of whistlers,because most of the direction finding systems except those by Okada et al.(1977,1981) and Ohta et al.(1984) are intended only in finding the wave normal directions,without any interest in the wave polarization.

The plan of the paper is as follows. Section 2 describes the procedure how to determine the absolute intensity of whistlers after the wave normal direction and polarization are specified by means of our field-analysis direction finding methods(Okada et al.,1977,1981; Okada and Iwai,1980; Ohta,1984; Ohta et al.,1984). The measurements have been made at two different geomagnetic latitudes;Yamaoka (geomag.lat. 25°N) and Moshiri(35°N), and the absolute intensities at the two stations have been presented in Sections 3 and 4,respectively. Section 5 describes the day-night and latitudinal variations and the physical implications are presented in terms of the joint influence of the source intensity,the ionospheric absorption and magnetospheric propagation characteristics.

2. ESTIMATION OF ABSOLUTE INTENSITY BY MEANS OF DIRECTION FINDING MEASUREMENTS

An estimation of the field intensity of whistlers relies on the primary determination of their direction of arrival and wave polarization. So,we here give a brief description of the field-analysis direction finding system. See Okada et al.(1977,1981) for further details. Fig.1 illustrates the situation of

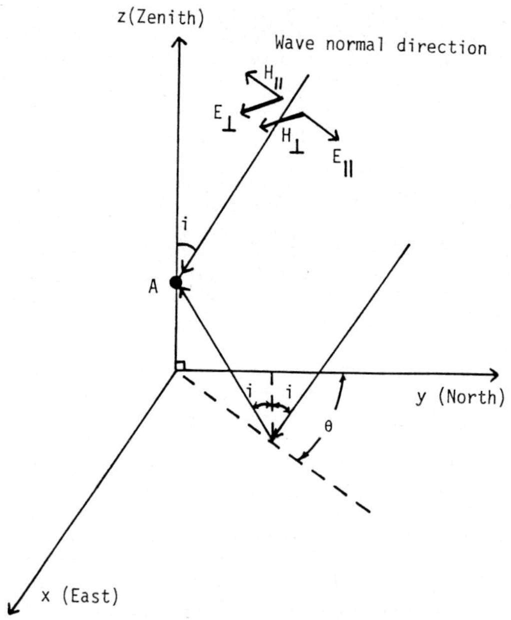

Fig.1 The configuration of the problem. The downgoing wave is incident on the direction finder located at the point A,with an incident angle i and with an azimuthal angle θ. The elliptically polarized wave is decomposed into the two components; transverse electric(TE) and transverse magnetic(TM) modes,as shown in the figure. The electromagnetic field at A is expressed by a superposition of the field of the direct ray and that of the ray reflected from the ground. See Tanaka(1972),Tanaka et al.(1976) or Okada et al.(1977,1981) for the detailed expressions of the electromagnetic fields and the approximate formulas by assuming the perfect conductor as the ground.

the problem in which the downgoing wave is incident on the direction finder with an incident angle i and with an azimuthal angle θ. The incident wave is assumed to be,generally,elliptically polarized and the components of the wave magnetic field perpendicular(corresponding to the TM mode) and parallel(the TE mode) to the incident plane are expressed by,

$$H_{\perp} = A_{\perp} \exp(j\omega t)$$
$$H_{\parallel} = A_{\parallel} \exp(j(\omega t - Q))$$

where $A_{\perp}$ and $A_{\parallel}$ are the amplitudes of respective components and Q indicates the phase difference of the two components. Then,the polarization parameter,p in the wavefront is defined by,

$$p = \frac{H_{\parallel}}{H_{\perp}} = u - j\, v$$

The field-analysis direction finding determines the arrival direction (i,θ) and wave polarization (u,v) either by taking the correlations of the electric and magnetic field components(Okada et al.,1981) or by directly measuring the amplitude ratios and phase differences between the magnetic fields and electric field(Okada et al.,1977; Ohta,1984). When those four parameters(i,θ,u,v) are decided,the maximum magnetic field intensity of a whistler lying in the wavefront is given by,

$$H_{max}^2 = (H_{\perp}^2 + H_{\parallel}^2)_{max}$$
$$= \left[(A_{\perp} \cos\omega t)^2 + \left\{ A_{\perp} (u \cos\omega t + v \sin\omega t)^2 \right\}^2 \right]_{max}$$

The corresponding value of the electric field E_{max} is given by,

$$E_{max} = Z_0 H_{max}$$

where Z_0 is the wave impedance of free space, and E_{max} is the absolute intensity treated in the following sections.

3. WHISTLER INTENSITY AT YAMAOKA(GEOMAG.LAT. 25°)

The direction finding measurement at Yamaoka is intended to be fully automatic at a specific frequency of 5 kHz(Ohta et al.,1984) and the automatic measurement has been continued since December,1981. Nine intense events are selected for daytime whistlers and four events for nighttime whistlers. Based on those events,the occurrence histograms of the absolute intensity are presented in Fig. 2(a) for daytime and in Fig.2(b) at night,respectively. The local time of the daytime events falls in the interval from ~15h to ~19h L.T.(L.T.=U.T.+1h). The duration of all the events is not larger than 2 hours,those kind of occurrence bursts being considered to be attributed to the life of a single duct(Hayakawa et al.,1983). The detailed study of the propagation characteristics on ducted propagation based on the direction finding results on the two intense events have been published elsewhere in Ohta et al.(1984),but we here have to mention ,at least,that the ionospheric exit points of all the daytime whistlers we studied are identified as being very close to the observing station. The corresponding local time of nighttime events ranges from 3h30m to ~9h L.T.,with each duration being nearly of the same magnitude as that of the daytime whistlers. The ionospheric exit regions of nighttime whistlers are again very close to the station. A comparison between Figs.2(a) and 2(b) clearly indicates that more

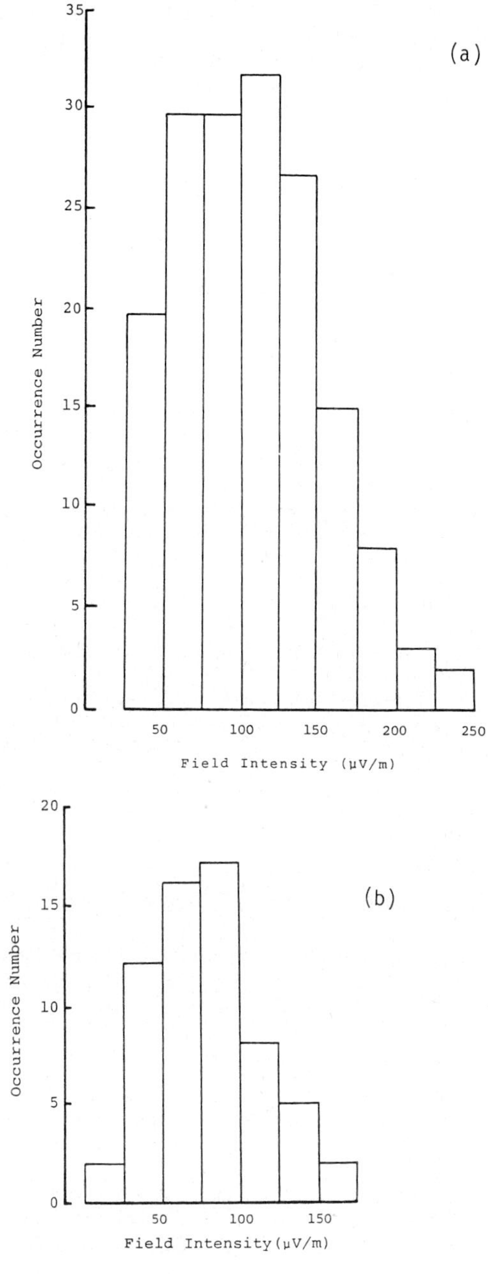

Fig.2 The occurrence histogram of the absolute intensity of whistlers observed at Yamaoka(geomag.lat.25°N). (a) Daytime and (b) nighttime. The daytime figure is based on the nine active events,while Fig.(b) is based on the four events. The measuring frequency is 5 kHz.

intense whistlers are observed during daytime than at night. The daytime intensity in Fig.2(a) exhibits a relatively flat distribution in a range from 50 to 150 μV/m with a maximum occurrence at 100-125 μV/m. Then,the occurrence distribution of the intensity shows a decrease above 150 μV/m, and the maximum value we have recorded is 225-250 μV/m. While,the intensity at night in Fig.2(b) indicates a rather well-defined occurrence peak at 50-100 μV/m, nearly about one half that of the daytime whistlers, and the maximum value detected is 150-175 μV/m,being 75 μV/m smaller than that of the daytime ones. The decrease of the occurrence against intensity,or the gradient, is nearly the same at day and night.

Previously,Okada et al.(1977) carried out the field-analysis direction finding at 4.5kHz at Takayama(25°) very close to Yamaoka,but the observation period was too short to obtain a statistical distribution as in Fig.2 in the present paper, and they succeeded in measuring the absolute intensity for a few intense whistlers during the most active period over a few days' observation period. Their peak measured intensity only at day was about 200 μV/m,which lends further support to the present statistical results of Fig.2.

4. WHISTLER INTENSITY AT MOSHIRI(GEOMAG.LAT.35°)

The timing of the direction finding at Moshiri has been made by monitoring the auxilliary real-time whistler analyzer(Okada et al.,1981; Hayakawa et al., 1983) in order to acquire as abundant useful direction finding data as possible.

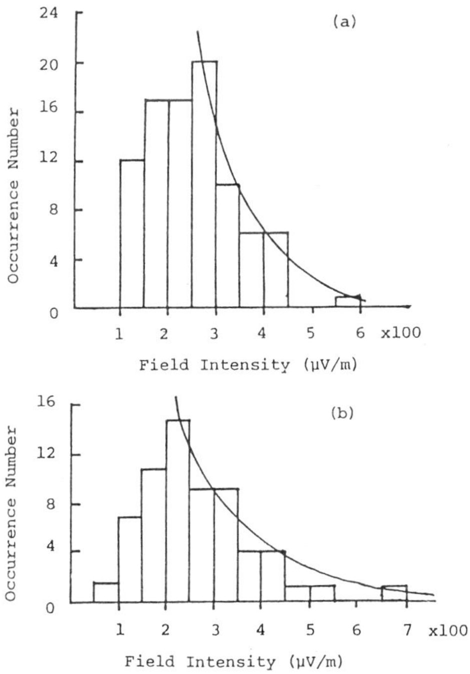

Fig.3 The occurrence histogram of the absolute intensity of whistlers observed at Moshiri(geomag.lat. 35°). (a)Daytime and (b)nighttime.

The observation was made for about two months,January and February in 1978. Then, 22 January 1978(15:10-18:07 h L.T.) was analyzed as the most active daytime event and,correspondingly, 20 February,1978(05:00-07:40 h L.T.) was chosen as the most active nighttime event. The observation frequency was 5.6 kHz,being slightly higher than that at Yamaoka. Both daytime and nighttime whistlers have been identified as having penetrated the ionosphere close to the station(Hayakawa et al.,1981a). The results on the absolute intensity are presented in Fig. 3; Fig.3(a) refers to daytime and Fig.3(b) nighttime. Although the amount of data at Moshiri is quite limited,the occurrence distribution in Fig.3 provides us with useful information,because the most active events are selected for analysis during the two months' observational period. A comparison of the figures indicates that the maximum occurrence takes place nearly at the same value such as 250-300 μV/m at day and at 200-250 μV/m at night,respectively. A concentration of occurrence in a range from 100 to 450 μV/m is recognizable at day. While,the occurrence distribution of the nighttime intensity is found to be widely distributed, and hence the higher intensity such as ~700 μV/m is expected at night. The curve in each figure is drawn by the eye-fit approximation of the distribution and it indicates a much sharper cutoff at higher intensities at day than at night. Hence,the whistler intensity of the order of ~1 mV/m would be expectable only at night.

5. DAY-NIGHT AND LATITUDINAL VARIATIONS OF WHISTLER INTENSITY AND DISCUSSION

Fig.4 illustrates a different form of presentation of the previous figures

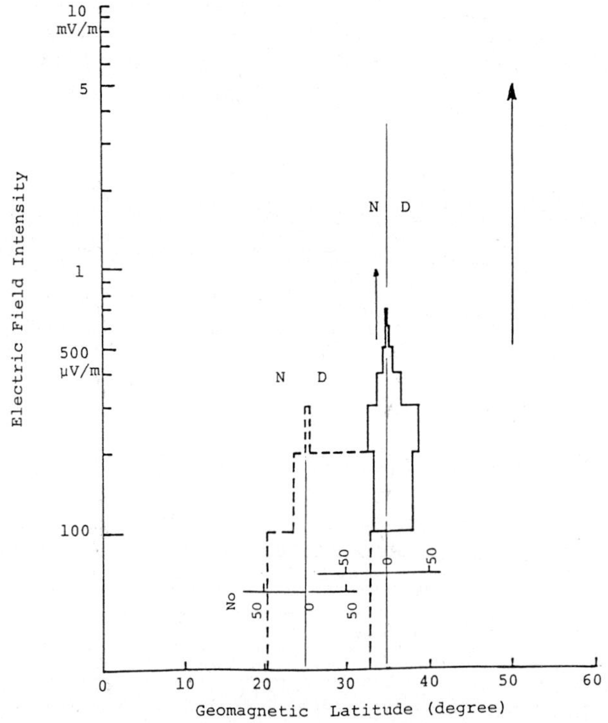

Fig.4 The latitudinal variation of the field intensity as constructed by the results obtained in the present paper and those at higher latitudes(Helliwell,1965)(shown by a solid line with an arrow at the latitude of 50°). N and D refer to daytime and nighttime,respectively

(Figs.2 and 3) as a function of geomagnetic latitude. The most important factors during the course of the propagation from one hemisphere to another include,(i) the source intensity, (ii) the ionospheric transmission loss in passing through the lower ionosphere and (iii)the propagation loss in the magnetosphere. As far as the third point is concerned,the ground whistlers even at the latitudes we are now dealing with,have been found to be attributed to the propagation trapped in field-aligned ducts and to suffer negliginle propagation loss(Hayakawa and Tanaka,1978; Hayakawa et al.,1981a,b; Ohta et al.,1984), and furthermore the whistlers at each station are found to have transmitted the ionosphere very close to each station. The second factor of the ionospheric transmission(reflection and absorption) loss has been theoretically calculated by means of the full-wave computation by Pitteway and Jespersen(1966). The electron density profiles for the typical daytime and nighttime conditions used for the computations are shown in Fig.5. These profiles are assumed to be independent of geomagnetic latitude, but the gyrofrequency is varied with geomagnetic latitude according to the dipole model. The collision frequency profile is also illustrated in Fig.5. Fig.6 gives the latitudinal dependence of the ionospheric transmission loss on the assumption of the vertical incidence just for one transit through the ionosphere. Hence,the total loss will be twice the value in Fig.6. The shaded region corresponds to the frequency band between 3 and 10 kHz where we expect the enhanced whistler energy. The observing frequencies of 5 and 5.6kHz in Sections 3 and 4 fall within this shaded region.

Based on Fig.6,we first look at the day-night variation at Moshiri as shown in Figs. 3 and 4. The theoretical curves of 5.6 kHz(calculated but not indicated in the figure) suggests that as the day-night difference in the absorption for one transverse through the ionosphere at the relevant latitude is about 7 dB,the total loss at night is 7 dB x 2 = 14 dB less than that at day,indicating about 5 times higher intensity being anticipated during nighttime. This theoretical expectation based on the equal intensity at the duct entrance at day and night seems to be in partial contradiction with the observation in Fig.3,because the maximum intensity as the most informative quantity,seems to amount to about 1 mV/m at night according to the anticipated curve in Fig.3(b),while the daytime

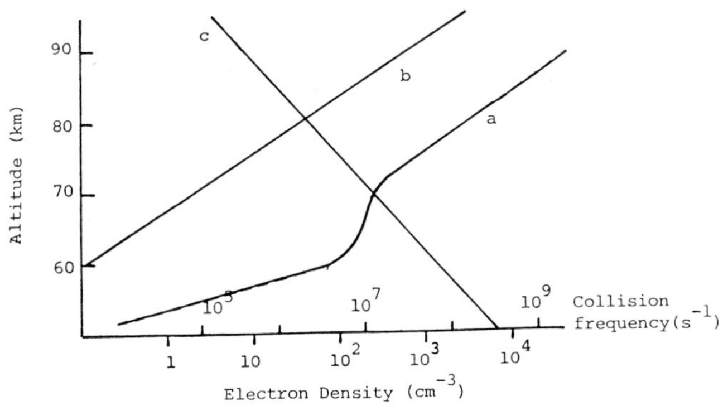

Fig.5 The electron density profiles of daytime(a) and nighttime(b) conditions and the collision frequency profile(c) used in the computation of the ionospheric transmission loss. These profiles are assumed to be independent of latitude,but the gyrofrequency is varied based on the dipole model.

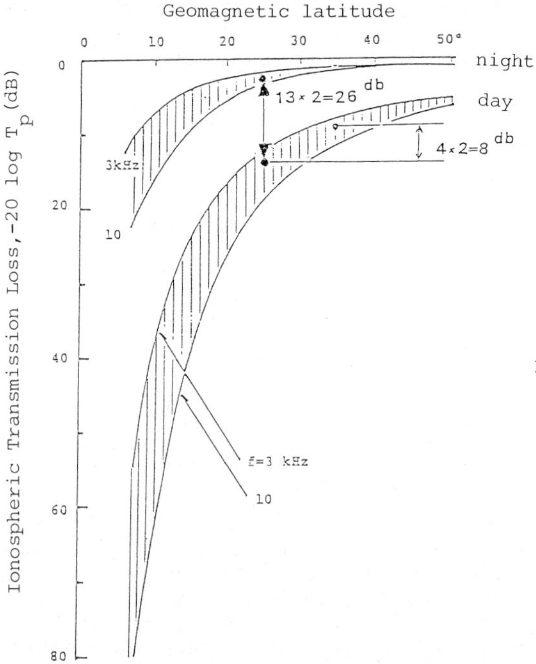

Fig.6 The latitudinal dependence of the ionospheric transmission (reflection plus absorption) loss for one transit, as deduced from -20 log T_p(dB) where T_p is the transmission coefficient of the penetrating mode. The vertical incidence is assumed.

intensity is sharply cutoff around 600 μV/m. The cloud distributions taken by the meteorological satellite are not available for these two events,and we follow the statistical study by Tanaka and Hayakawa(1980). According to them,the day-night difference in the atmospherics intensity in the conjugate region of Moshiri is about a few dB stronger at day than at night, and this might be,in part,responsible for the discrepancy. The latitudinal experimental variation of daytime intensity between Moshiri and Yamaoka will be shown to be satisfactorily interpreted in terms of the latitudinal variation of the ionospheric transmission loss for the vertical incidence, and this might lead us to speculate that the duct entrance region of daytime whistlers observed at Moshiri seems to be just inside the active thunderstorm region as in the case for Yamaoka and so some kinds of mechanisms are operative to reduce the nighttime intensity,such as the relative distance between the active source region and the duct entrance point and/or the enhanced attenuation due to scattering at the duct entrance and/or exit point at night as suggested by Hasegawa et al.(1978) and Hayakawa and Tanaka(1978). This speculation requires future investigation invloving the conjugate thunderstorm activity.

Then,the day-night asymmetry of intensity at Yamaoka(25°) is treated. The theoretical curve in Fig.6 suggests that there exists a strong day-night asymmetry in the ionospheric absorption of about 13 x 2 =26 dB; in other words,the nighttime intensity should be about 20 times that of daytime one on the assumption of the equal source amplitude at the duct entrance. This kind of a large difference is quite in contradiction with the observation which has yielded that the daytime whistlers are, rather, about twice more intense during daytime than at night, as in Figs.2 and 4. Tanaka and Hayakawa

(1980) have tried to interprete this phenomenon and they have concluded that the most important factor responsible for this discrepancy is that the active source regions are well separated at day and night. Namely,the daytime whistlers are intense because the thunderstom active region at day is located just around the conjugate region of Yamaoka. While,at night its conjugate region is well away from the active thunderstorm region and the wave is incident onto the duct after the Earth-ionosphere waveguide propagation,resulting in small amounts of energy transmission and hence weaker nighttime intensity.

The direction finding results on the ionospheric exit regions(not shown) have revealed that all of the whistlers observed both at day and night are found to be attributed to the propagation trapped in field-aligned ducts existing very close to the station of Yamaoka. So,we have examined the cloud distributions taken by the Japanese meteorological satellite for all the events analyzed in the present paper in order to know the thunderstorm activity near the duct entrance region. As the result,the entrance points of daytime ducts are found to be located just within the active region,probably an origin of lightning discharges. While,at night the conjugate region of Yamaoka is away from the nearest active cloud region by not smaller than one thousand kilometers. Now we try to interprete the observational fact or to compare the experimental results with the theoretical prediction. We take the source intensity at the duct entrance point during daytime as the reference(0 dB). Although the nighttime duct entrance points are found to be far away from the active thunderstorm region,the cloud picture provides no information on the intensity at the nighttime duct entrance. Hence, we again follow Tanaka and Hayakawa(1980), and the intensity at night is taken to be -5 dB at the conjugate point of Yamaoka by adopting the average source intensity from the CCIR data. The difference in the ionospheric absorption loss has been estimated to be -13 dB x 2 = -26 dB from Fig.6; that is, the daytime intensity should be 26 dB weaker than the nighttime one. While, the nighttime whistlers have suffered from the oblique incidence($i > 80°$) after the Earth-ionosphere waveguide propagation, and so the nighttime waves would be,additionally, more attenuated due to the oblique incidence. For example,the transmission coefficient at oblique incidence($i=80°$) is found to be about one fifth that for the vertical incidence(Hayakawa,1974) and so 14 dB more loss at night. However,when we consider more oblique incidence such as $i > 80°$(as is normal for the waveguide propagation),we expect much more loss. Hence,these theoretical predictions do not seem to be inconsistent with the experimental fact that the daytime intensity is about a few dB stronger than the nighttime one.

Since the conjugate region of daytime whistlers observed at Yamaoka and Moshiri are considered to be just within the active thunderstorm region,then a comparison of the daytime intensity is made at the two latitudes. The observation has yielded that the intensity at Moshiri is roughly limited by the maximum value of $\sim$600 μV/m ,while the maximum value at Yamaoka is $\sim$350 μV/m, being about one half. This seems to be in good agreement with the theoretical prediction on the latitudinal dependence of about 3 dB x 2 = 6 dB; or the intensity ratio being about twice on the assumption of equal amplitude in the atmospheric intensity at each conjugate region. The atmospherics intensity at the duct entrance can be,actually,found by Tanaka and Hayakawa(1980) to be of the same magnitude at the conjugate points of Moshiri and Yamaoka, and then the intensity ratio obtained between these two stations is reasonably interpreted simply in terms of the latitudinal effect of the ionospheric absorption loss.

Finally we would like to point out an interesting need of accurate intensity measurement in the study of whistler amplification and the associated particle precipitation due to wave-particle interactions in the magnetosphere(Carpenter and LaBelle,1982; Doolittle and Carpenter,1983). The simultaneous measurement of intensity of a causative atmospheric and the associated amplified whistler would provide useful additional information to the wave-particle interaction studies.

ACKNOWLEDGEMENT

The authors wish to express their sincere thanks to Profs.J.Ohtsu and A.Iwai of Nagoya University and to Prof.H.Eguchi of Chubu University for their useful discussion. Thanks are also due to Prof.A.Kimpara for his continued encouragement.

REFERENCES

Bullough,K. and Sagredo,J.L.,1973. Planet.Space Sci., 21, 899.
Carpenter,D.L., 1980. J.Geophys.Res., 85, 4157.
Carpenter,D.L. and LaBelle,J.W., 1982. J.Geophys.Res., 87, 4427.
Cousins,M., 1972. Tech.Rep. SEL-72-013, Radioscience Lab. Stanford Univ.
Doolittle,J.H. and Carpenter,D.L., 1983. Geophys.Res.Lett., 10, 611.
Hasegawa,M., Hayakawa,M. and Ohtsu,J., 1978. Ann.Geophys., 34, 317.
Hayakawa,M., 1974. Doctoral Thesis, Nagoya University, Nagoya,Japan.
Hayakawa,M. and Tanaka,Y., 1978. Rev.Geophys.Space Phys., 16, 111.
Hayakawa,M., Okada,T. and Iwai,A., 1981a. J.Geophys.Res., 86, 6939.
Hayakawa,M., Tanaka,Y., Okada,T. and Iwai,A., 1981b. J.Geophys.Res., 86, 6781.
Hayakawa,M., Tanaka,Y., Okada,T. and Ohtsu,J., 1983. Ann.Geophysicae, 1, 515.
Helliwell,R.A., 1965. Whistlers and Related Ionospheric Phenomena, Stanford Univ. Press, Stanford,Calif.
Iwai,A. and Outsu,J., 1958. Proc.Res.Inst.Atmospherics, Nagoya Univ., 5, 50.
Leavitt,M.K., Carpenter,D.L., Seely,N.T., Padden,R.R. and Doolittle,J.H., 1978. J.Geophys.Res., 83, 1601.
Lester,M. and Smith,A.J., 1980. Planet. Space Sci., 28, 645.
Matthews,J.P., Smith,A.J. and Smith,I.D., 1979. Planet.Space Sci., 27, 1391.
Ohta,K., 1984. Trans.Inst.Electr.Comm.Engrs.Japan, J67-B, 869(in Japanese).
Ohta,K., Hayakawa,M. and Tanaka,Y., 1984. J.Geophys.Res., 89, 7557.
Okada,T., Iwai,A. and Hayakawa,M., 1977. Planet.Space Sci., 25, 233.
Okada,T. and Iwai,A., 1980. Trans.Inst.Electr.Engrs.Japan, 100,21(in Japanese).
Okada,T., Iwai,A. and Hayakawa,M., 1981. J.Atmos.Terr.Phys., 43, 679.
Pitteway,M.L.V. and Jespersen,J.L., 1966. J.Atmos.Terr.Phys., 28, 17.
Sagredo,J.L. and Bullough,K., 1973. Planet.Space Sci., 21, 913.
Strangeways,H.J., Madden,R.A. and Rycroft,M.J., 1983. J.Atmos.Terr.Phys., 45,387.
Tanaka,Y., 1972. Proc.Res.Inst.Atmospherics, Nagoya Univ., 19, 33.
Tanaka,Y., Nishino,M. and Hayakawa,M., 1976. Mem.Natl Inst.Polar Res., Tokyo, A13(Aeronomy), pp.58.
Tanaka,Y. and Hayakawa,M., 1980. Ann.Geophys., 36, 577.
Tsuruda,K. and Hayashi,K., 1975. J.Atmos.Terr.Phys., 37, 1193.

Nonlinear and Environmental Electromagnetics, edited by H. Kikuchi
Elsevier Science Publishers B.V., Amsterdam, 1985 – Printed in The Netherlands

CHARACTERISTICS OF RISER EMISSIONS OBSERVED BY ISIS SATELLITES

T. Ondoh

Radio Research Laboratories, Tokyo, 184, Japan

ABSTRACT

Giant riser emissions and VLF hisses with fine rising frequency structures in frequency range between 8 kHz and 20 kHz are observed at mid latitudes by ISIS satellites. The giant riser emissions, of which frequency rises from 9 kHz to 18 kHz in 2.5 seconds, occur often at invariant latitudes from 47° to 60° in the recovery phase of magnetospheric substorms.

Groups of the normal riser emissions, of which frequency rises from about 5 kHz to about 10 kHz in 1.5 seconds, were repeatedly observed in a period of 26 - 27 seconds at invariant latitudes from 56.5° to 61.9° by the ISIS-2 on March 10, 1981 of the geomagnetically quiet day. Geomagnetic pulsations of Pc 3 were also observed simultaneously with the periodic appearance of the normal riser emissions at Memambetsu observatory(geomag. lat. 34.1°N).

The periodic occurrence of the normal riser emissions is explained by the periodic modulation of riser-emission growth rate which results from the particle precipitations modulated by the geomagnetic pulsations of Pc 3.

The giant and normal riser emissions occur in mid latitudes which are lower than the occurrence latitudes of the chorus emissions. The rise times of giant and normal riser emissions are considerably longer than the chorus rise times below 0.5 seconds. The frequency rise and frequency range of giant and normal riser emissions are also larger than those of chorus emissions observed around the auroral zone. The low-energy protons in the inner radiation zone seem to generate the giant and normal riser emissions in the plasmasphere.

Numerical results of model calculations on the frequency-time spectra of the riser emissions indicate that the transverse proton cyclotron instabilities in the plasmasphere are favourable for explaining the giant and normal riser emissions observed in mid latitudes by ISIS satellites.

1. INTRODUCTION

The riser emissions was first described together with other types of VLF emissions by Gallet(1959). Since then, most studies of mid-latitude VLF emissions have been mainly concerning VLF phenomena at frequencies below 10 kHz. But there has been no report on the giant riser emissions. Recently we have found the giant riser emissions appearing between 8 kHz and 20 kHz from VLF electric field data (50 Hz - 30 kHz) of the ISIS satellites received at

Kashima station(geomag. lat. 25.8°N. long. 206.4°E). In this paper we report characteristics of the giant riser emissions and periodic occurrence of the normal riser emissions observed by the ISIS satellites in the topside ionosphere of mid-latitudes. And also we will discuss on the generation mechanism of these phenomena in mid latitudes.

The apogee, perigee and inclination of the ISIS-1 orbit as of October, 1979 are 3511 km, 574 km and 88.4° respectively, and those of the ISIS-2 orbit as of October 1979 are 1423 km, 1354 km and 88.2° respectively.

2. GIANT RISER EMISSIONS OBSERVED IN MID LATITUDES

Riser emissions observed in the topside ionosphere of mid-latitude are classified into two types ; the giant riser emissions of which frequency rises from about 8 kHz to 18 kHz or more in a few seconds and the normal riser emissions of which frequency rises from about 5 kHz to about 10 kHz within two seconds. The frequency range and the frequency rise of giant riser emissions are larger than those of the normal riser emissions, and the typical rise time of 2.5 seconds for the giant riser emissions is longer than that of 1.5 seconds for the normal riser emissions, while the frequency of chorus emissions which occur mainly around the auroral zone rises typically from about 0.5 kHz to 3 kHz or less in the rise time below 0.5 seconds (Helliwell, 1965). The chorus emissions and artificially stimulated emissions are interpreted as the coherent emissions due to the phase bunching of electrons in the electron cyclotron resonance, that is, the transverse electron cyclotron instabilities around the equatorial plane on the auroral field line of L = 4 (Helliwell,1967). So the rise time and frequency rise of observed emissions are very important parameters in relation to the phase bunching time (about 100 msec) and interaction length (about 1000 km) of the wave-particle interaction along the geomagnetic field line.

Fig. 1 shows a frequency-time spectrum of the giant riser emission which was first observed at 2356 UT(0849 LT) on Nov. 27, 1972 by ISIS-2. At this time, the ISIS-2 lay at invariant latitude of 53.9° (L= 2.88) and at height of 1368 km. The frequency of the giant riser emission increases from 9 kHz to 18 kHz in 2.5 seconds. So, the time change rate of the giant riser frequency is 3.5 kHz/s which is comparable with 2.0 kHz/s of the Auroral Kilometric Radiation and it is larger than the time change rate of the chorus frequency.

An examination of time variation of AE indices indicates that the giant riser emissions in Fig. 1 occurred in the recovery phase of geomagnetic substorm.

Fig. 2 shows an example of VLF hiss with fine rising frequency structures observed by Alouette-2 on Feb. 25, 1971. Using AE indices and satellite VLF data (AL-2, ISIS-1 and ISIS-2) received at Kashima station from Feb. 1971 to Sept. 1973, we have statistically examined whether the giant riser emission or VLF hisses with fine rising frequency structures occurs or not for the geomagnetic substorm.

Table 1 Substorm dependency of giant riser emissions and VLF hisses with fine rising frequency structures observed between 8 kHz and 20 kHz in mid-latitude topside ionosphere.

	Quiet time	Expansion phase	Recovery phase
Riser emissions	2	0	12
VLF hiss with fine rising frequency structures	4	4	10
Total	6	4	22

Table 1 indicates that giant riser emissions or VLF hisses with fine rising frequency structures appearing between 8 kHz and 20 kHz occur mainly in the recovery phase of geomagnetic substorms. The giant riser emissions or VLF hisses with fine rising frequency structures were observed for 07 - 12 hours and for 15 - 19 hours in local time by Alouette-2 and ISIS satellites received at Kashima station during the above data period. Thus the giant riser emission is also the daytime phenomenon like the chorus emission in the auroral zone. Fig. 3 shows a latitudinal distribution of occurrence rate of the giant riser emission or the VLF hiss with fine rising frequency structures per one degree in geomagnetically invariant latitude. It is clearly seen that the giant riser emissions or VLF hisses with fine rising frequency structures between 8 kHz and 20 kHz occur mainly at invariant latitudes between 47° and 60° in the topside ionosphere. The observed facts described in this section suggest that the giant riser emissions are generated by precipitating protons of the ring current in the recovery phase of geomagnetic substorms.

3. DISCUSSION ON THE GENERATION MECHANISM OF GIANT RISER EMISSIONS OBSERVED IN THE TOPSIDE IONOSPHERE OF MID-LATITUDES

It is generally required to explain the observed f-t spectra of discrete emissions that the particles, which can generate the observed emission frequencies at different points in space and time along the same field line, have approximately the same energy. We discuss on the generation mechanism of the giant riser emissions by demonstrating numerical possibilities for the observed f-t spectra. Fig. 1 shows the f-t spectra of the giant riser emissions observed by ISIS-2 at invariant latitude of 53.9°(L = 2.88). The electron gyrofrequency and electron plasma frequency are f_{He}= 39.4 kHz and f_p= 284.6 kHz(N = 1000/cm^3, Chappell, 1972) respectively at the top of the geomagnetic field line passing through invariant latitude of 54° (L = 2.89). Helliwell(1967) has interpreted the chorus emissions and artificially stimulated emissions observed in the auroral zone in terms of a feedback model of cyclotron interaction between whistler-mode waves and energetic electrons along the auroral field lines. The frequency rise and rise time of the chorus emission are typically 2 kHz and 0.5 seconds. We first discuss a possibility of the electron cyclotron radiations emitted backward from energetic electrons as an origin of the giant riser emission along the geomagnetic field line of L = 2.89.

The backward Doppler-shifted frequency of the electron cyclotron radiations is given by $f = f_{He}/(1 + n.v_{\parallel}/c)$, where $v_{\parallel}= v.\cos\alpha$, α is the particle pitch angle, v is the electron velocity, c is the light velocity and n the refractive index of whistler mode waves. The energy of electrons with velocity of v is expressed by $E_e= 250(v/c)^2$ keV.

In the geomagnetically equatorial plane on the geomagnetic field line of L = 2.89, the backward Doppler-shifted electron cyclotron waves of 9 kHz can be generated from electrons of 12.9 keV for n = 15.1, where the equatorial loss cone on the field line of L = 2.89 is α = 9.6° for geomagnetic field of B_e = 0.0141 gauss on the equatorial plane and B_m = 0.5055 gauss at mirror altitude of 300 km on the field line of L = 2.89. The particle's pitch angle is calculated by an expression of $\alpha = \sin^{-1}(B/B_m)^{1/2}$ on the same geomagnetic field line. We use here the centered dipole approximation for the geomagnetic field , $r = r_e\cos^2\theta$ and $B = B_o(r_E/r)^3.(1 + 3\sin^2\theta)^{1/2}$ where r is the geocentric distance, θ the geomagnetic latitude, r_E= 6370 km and B_o= 0.34 gauss.

The parallel velocity of electrons which emit the backward Doppler shifted electron cyclotron waves of 9 kHz in the equatorial plane is estimated at $v_{\parallel} = (f_{He}/f - 1).c/n = 6.711 \times 10^4$ km/s. The frequency of the giant riser emissions in Fig. 1 increases from 9 kHz to 18 kHz in 2.5 seconds. A travel time of the electrons with 12.9 keV during the rise time of 2.5 seconds bocomes 1.678×10^5 km along the geomagnetic field line of L = 2.89. The length of geomagnetic field line of L = 2.89 from the equatorial plane to the ground can be calculated as 1.553×10^4 km from the expression given by Ondoh(1961). The travel distance of electrons emitting the backward Doppler-shifted cyclotron waves of 9 kHz in 2.5 seconds is much longer than the length of L = 2.89 field line from the equatorial plane to the ground. Hence the backward Doppler-shifted electron cyclotron radiations is not favourable for explaining the f-t spectra of giant riser emissions observed around L = 3 in the topside ionosphere. The interaction distance along the auroral field line estimated by Helliwell(1967) is of the order of 1000 - 2000 km which are comparable with the phase bunching distance in the electron cyclotron resonance. The length of the auroral field line (L = 4) on which the chorus emissions and ASE are explained by the electron cyclotron resonance (Helliwell, 1967) is much longer than that of the geomagnetic field line around L = 3. Thus, the rise time and frequency rise of giant riser emissions are considerably large so that they can not be explained by the backward Doppler shifted electron cyclotron radiations.

Then we consider a possibility of the electron cherenkov radiation as an origin of the giant riser emissions along the geomagnetic field line of L = 2.89. The frequency of the cherenkov radiation from an electron with parallel velocity of v is given by $f = (v/c)^2.(f_p/f_H)^2\cos^2\emptyset$ for the whistler mode waves where $\emptyset$ is the wave normal angle to the geomagnetic field line. Substituting f = 9 kHz, f_p= 284.6 kHz and f_{He} = 39.4 kHz in the equatorial plane on the field line of L = 2.89 and also f = 18 kHz, f_p= 636 kHz and f_{He}= 245 kHz at altitude of 5000 km on the field line of L = 2.89 into the above expression respectively, we have $v/c \geqq 6.62 \times 10^{-2}$ and E_e= 1.1 keV for generating the 9 kHz waves in the equatorial plane and also $v/c \geqq 1.04 \times 10^{-1}$ and E_e= 2.7 keV for generating the 18 kHz waves at altitude of 5000 km. The transit distances of electrons with 1.1 keV and 2.7 keV during the rise time of 2.5 seconds become 4.96×10^4 km and 7.83×10^4 km respectively. The transit distances of the electrons along the geomagnetic field line of L = 2.89 are longer than the field line length of 1.55×10^4 km between the equatorial plane and the ground. Thus, the electron cherenkov radiation is not also favorable for the f-t spectra of the giant riser emissions, since the electrons of approximately same energy moving along the same field line are required to explain the observed frequency rise of the discrete emissions during the observed rise time of 2.5 seconds.

The Auroral Kilometric Radiation appearing for 50 kHz to 500 kHz is explained by the moving source excitation mechanism, df/dt = df/ds.ds/dt where ds is the line element along the auroral field line and ds/dt is the velocity of the ion sound waves of about 10 km/s (Gurnett et al.,1979). In this mechanism, the ionsound waves propagating down along the auroral field line excite electron cyclotron waves or electron plasma waves in the ambient plasma. But the moving-source excitation mechanism can not explain the giant riser emission in mid latitudes since f_{He} and f_p along the mid-latitude field lines are greater than the frequencies of the giant riser emissions.

The IPDP with rising frequency structures in ULF range is explained by a travel time difference between different energetic particles from the particle injection region near the midnight meridian to the generation region in the geomagnetically equatorial plane (Lanzerotti and Fukunishi,1974). But we can not apply this mechanism to the giant riser emissions since the different emission frequencies are generated in the different meridians on the equatorial plane and since the travel time difference in the particle's drift motion is expected to be above one hour which is much longer than the observed rise time of giant riser emissions, 2.5 seconds.

The OGO-3 observations of protons and electrons with energy of

200 eV $\leq$ E_p $\leq$ 50 keV in the equatorial plane show that the low-energy protons of the ring current are distributed for L = 1.5 - 4.0 R_E with a maximum intensity around L = 3.0 and that the low-energy electrons are distributed for L = 2.0 - 6.0 R_E with a maximum intensity around L = 4.0 R_E for a geomagnetic storm (Frank, 1967). The proton and electron radiation zones correspond to the inner zone of energetic protons and the outer zone of energetic electrons respectively (Van Allen, 1968). So, it is expected that the low-energy protons in the inner zone produce plasma wave phenomena around the geomagnetic field line of L = 3.0 in the plasmasphere.

Next we discuss the proton cyclotron radiation as an origin of the giant riser emission. For simplicity, we consider the parallel propagation of the forward Doppler-shifted proton cyclotron radiations to the geomagnetic field line, so called the anomalous Doppler-effect when the parallel proton velocity, $v_{\parallel}$ is greater than the parallel wave phase velocity, c/n(Ginzburg,1964).

In this case, the rotation direction of emitted waves is the same as that of protons around the geomagnetic field line in the coordinate moving at the proton velocity which is larger than the phase velocity of emitted waves, although the rotation direction of emitted waves is different from that of protons in the rest coordinate. In the anomalous proton cyclotron resonance, the expression of $f(n.v_{\parallel}/c - 1) = f_{Hp}$ also satisfies the conditions for the coherent radiations and for the transverse cyclotron instabilities. So we can expect the energy transfer from protons to waves in the proton cyclotron resonance, that is, the transverse proton cyclotron instabilities. The anomalous Doppler -shifted frequency of proton cyclotron radiations is given by $f = f_{Hp}/(n.v_{\parallel}/c - 1)$, where n is the refractive index of R-mode proton cyclotron waves. We estimate approximately the refractive index by substituting $f^* = f_{Hp}$ into $n^2 = 1 + f_p^2/(f^* + f_{Hp}).(f_{He} - f^*)$ given by Spitzer (1982). The proton gyrofrequencies, f_{Hp} in the equatorial plane (altitute of 12070 km) and at altitude of 9500 km on the geomagnetic field line of L = 2.89 are 21.5 Hz and 40 Hz respectively. Table 2 gives the plasma parameters in the equatorial plane and at altitude of 9500 km on the geomagnetic field line of L = 2.89.

Table 2 Typical plasma parameters in the plasmasphere along the geomagnetic field line of L = 2.89 where $f_{Hp} = f_{He}/1836$ and f_{He}=2800.B kHz.

Altitude	Geomag. lat.	f_{He} kHz	f_{Hp} Hz	B gauss	N/cm^3	f_{Hp} kHz	n
Equatorial plane	0°	39.4	21.5	0.0141	1000	284.6	224
9500 km	21.9°	73.3	40	0.0262	3000	493	209

The computed refractive indices are respectively n = 224 in the equatorial plane and n = 209 at altitude of 9500 km on the field line of L = 2.89. The forward Doppler-shifted proton cyclotron frequencies generated on the field line of L = 2.89 are assumed to be 9 kHz in the equatorial plane and 18 kHz at altitude of 9500 km respectively. We have $v_{\parallel}/c = (f_{Hp}/f + 1)/n = 4.475 \times 10^{-3}$ for f = 9 kHz and f_{Hp}= 21.5 Hz in the equatorial plane and $v_{\parallel}/c = 4.795 \times 10^{-3}$ for f = 18 kHz and f_{Hp}= 40 Hz at 9500 km on the L = 2.89 field line. Thus, the derivation of parallel proton velocity is independent of the proton pitch angle. Proton pitch angles in the equatorial plane and at 9500 km on the L = 2.89 field line are calculated as α = 9.6° and α = 13.2° respectively from $\alpha = \mathrm{Arcsin}(B/B_m)^{1/2}$, where B_m= 0.5055 gauss is the geomagnetic field at the mirror point of 300 km. Using $v/c = v_{\parallel}/(c.\cos\alpha)$ and the proton energy of $E_p = 459 \times 10^3.(v/c)^2$ keV, we have E_p= 9.5 keV for generating the 9 kHz waves in the equatorial plane and E_p = 11.1 keV for generating the 18 kHz waves at 9500 km on the L = 2.89 field line. The field line length of L = 2.89 between the equatorial plane and altitute of 9500 km is approximately equal to the cross product of the parallel proton velocity, v = 1340 km and the observed rise time of 2.5 seconds. Thus the f-t spectra of giant riser emissions observed at invariant latitude of 54° by ISIS-2 are explained by the anomalous Doppler-shifted cyclotron radiations emitted forward from protons of about 10 keV in the plasmasphere.

4. PERIODIC OCCURRENCE OF RISER EMISSION GROUPS OBSERVED BY ISIS SATELLITE

Kimura (1974) has reviewed observational and theoretical works on the interrelation between quasi-periodic VLF emissions and geomagnetic pulsations in high latitudes. Recently many authors have also investigated short-period geomagnetic pulsations associated with repetitive VLF emissions (Harker et al., 1974 ; Fraser-Smith and Helliwell,1980 ; Sato and Kokubun,1981 ; Sato, 1984). However, these phenomena were almost observed at high latitudes above the auroral zone.

Fig. 4 shows narrow-band field-intensity increases at 8 kHz observed by ISIS-2 between 0257 and 0301 UT on March 10, 1981. This event occurred around geomagnetic local time of 11 hours over the northern Siberian district at geomagnetically invariant latitudes from 56° to 63° or for meridional distance of about 700 km. The narrow-band data at six frequencies are processed from magnetic tapes of ISIS VLF wide-band data by narrow-band DC amplifiers with the minimum reading circuit. The charging and discharging time constants of the minimum reading circuit are 10 seconds and 10 milli-seconds respectively. Detailed f-t spectra of this event are given in Fig. 5, where a bottom interval between white dots on each plate is one second and the full scale of the ordinate is 10 kHz. Groups of normal riser emissions occurred repeatedly in a period of 26 - 27 seconds. The normal riser emissions, of which frequency rises from 5 kHz to 9 kHz in 1.5 seconds, seem to be triggered from the steady narrow-band hiss around 5 kHz.

The geomagnetic pulsations with period from 20 seconds to 27 seconds, so called Pc 3 occurred simultaneously with the periodic appearance of riser emission groups for 0257 - 0301 UT on March 10, 1981 at Memambetsu, Japan as shown by Fig. 6. The high- and low-latitude ends of the periodic occurrence of riser emissions are effectively 61.9° and 56.5° in invariant latitude. The length of geomagnetic field lines between the equatorial plane and the ground are computed as 3.316×10^4 km at 61.9° and 2.234×10^4 km at 56.5° from the expression given by Ondoh (1961). The ratios of field line lengths to the occurrence period, 26 seconds of the riser emission become 5100 km/s at 61.9° and 2230 km/s at 56.5°. These are of the order of the propagation velocity of hydromagnetic waves along the geomagnetic field lines. It seems likely that the periodic occurrence of normal-riser emission groups is caused through the modulation of riser emission source by geomagnetic pulsations of Pc 3. The geomagnetic pulsations of Pc 3 occur normally on the dayside in mid-latitudes between the auroral zone and the plasmapause latitude(Lanzerotti and Southwood, 1979). Consequently, it seems likely that the geomagnetic pulsations of Pc 3 occurred simultaneously with the periodic appearance of riser emission groups in a wide range of mid latitudes below the auroral zone on March 10, 1981.

Coroniti and Kennel (1970) and Kimura(1974) have discussed theoretically the periodic modulation of VLF emissions by hydromagnetic waves, so called geomagnetic pulsations in the magnetosphere. In the small relaxation limit, the periodic modulation of VLF emission growth rate results from the large amplitude particle precipitations modulated by geomagnetic pulsations.

On the other hand, the geomagnetic pulsation gradually modulates the equilibrium electron distribution in the magnetosphere in the quasi-static limit. The modulation of the electron diffusion coefficient and VLF-emission energy are linearly proportional to the amplitude of geomagnetic pulsation. In this case, the modulation of VLF wave intensity is inphase with the geomagnetic pulsation. In either limit, the periodic modulation of VLF emission is caused through modulation in the source or generation process of VLF emission by geomagnetic pulsations. The recurrent period of riser emission groups observed by ISIS-1 is 26 - 27 seconds which are very close to the period of PC 3 observed at Memambetsu 9geomag. lat. 34.1°N, long. 208.8°E), Japan. Therefore, the periodic occurrence of normal-riser emission groups observed from 56° to 63° in invariant latitude seems to be caused by the geomagnetic pulsations of Pc 3 in the plasmasphere. The phase relation between the riser emission groups and geomagnetic pulsations of Pc 3 is not clear since the time resolution of rapid-run magnetogram at Memambetsu is poor.

Then we discuss possible generation mechanism of the normal riser emissions as shown in Fig. 5 along the geomagnetic field line of L = 3.28 (invariant latitude of 56.5°) since the regularly periodic appearance of normal-riser emission groups occurs between 56.5° and 62° in invariant latitude. The normal riser emissions have the frequency rise from 5 kHz to 9 kHz in 1.5 seconds. So the frequency range of normal riser emissions is larger than that (0.5 - 3 kHz) of chorus emissions and the rise time of normal riser emissions is also longer than that (below 0.5 seconds) of chorus emissions. The occurrence latitude of normal riser emissions are lower than those of chorus emissions. Thus it is difficult to explain the occurrence characteristics of normal riser emissions by the backward Doppler-shifted electron cyclotron radiations as discussed on the giant riser emissions previously.

We demonstrate this difficulty by a numerical example. At the top of geomagnetic field line of L = 3.28, the electron gyrofrequency is 26.9 kHz and we assume the electron density of $500/cm^3$ (f_p= 201.2 kHz) there. So the refractive index of the whistler mode becomes n = 17.3. We have the parallel electron velocity of $v_{\parallel}$= 76800 km/s from the expression of $v_{\parallel}/c = (f_{He}/f - 1)/n$. The travel distance of the electrons during the rise time of 1.5 seconds is 114300 km. This is much longer than the length of the geomagnetic field line of L = 3.28 from the equatorial plane to the ground, 2.234×10^4 km. So the backward Doppler-shifted cyclotron radiations from the electrons moving away from the observer is difficult to explain the f-t spectra of normal riser emissions.

Next we show a numerical example of the anomalous Doppler-shifted proton cyclotron radiations which explain the f-t spectra of normal riser emissions. Refractive indices of R-mode proton cyclotron waves are computed as n = 232 in the equatorial plane and n = 236 at altitude of 13000 km on the geomagnetic field line of L = 3.28 for the plasma parameters given in Table 3.

Table 3 Typical plasma parameters in the plasmasphere along the geomagnetic field line of L = 3.28.

Altitude	Geomag. lat.	f_{He}	f_{Hp}	N	f_p	n
14540 km	0°	26.9 kHz	14.7 kHz	$500/cm^3$	201.2 kHz	232
13000 km	15.7°	37.4 kHz	20.4 kHz	$1000/cm^3$	284.6 kHz	236

When the forward Doppler-shifted proton cyclotron frequencies of 5 kHz and 9 kHz are generated respectively in the equatorial plane and at altitude of 13000 km on the field line of L = 3.28, we have $v_{\parallel}/c = 4.32 \times 10^{-3}$ for 5 kHz waves in the equatorial plane and $v_{\parallel}/c = 4.25 \times 10^{-3}$ for 9 kHz waves at 13000 km from an expression of $v_{\parallel}/c = (f_{Hp}/f + 1)/n$.

Thus the parallel proton velocity for the Doppler-shifted cyclotron radiations is obtained only from the plasma parameters at the generation region but not from the proton's pitch angle.

The proton pitch angles in the equatorial plane and at 13000 km on the field line of L = 3.28 are estimated respectively at α= 7.8° and α = 9.3° for the geomagnetic field of 0.5165 gauss at the mirror altitude of 300 km. So we can calculate the proton energies of E_p= 8.7 keV for generating the 5 kHz waves in the equatorial plane and E_p= 8.5 keV for generating the 9 kHz waves at altitude of 13000 km on the field line of L = 3.28 by using the expressions of E_p= 459 x $10^3(v/c)^2$ keV and $v/c = v_{\parallel}/ (c.\cos\alpha)$. The travel distance of protons with 8.6 keV during the observed rise time of 1.5 seconds is approximately equal to the length of geomagnetic field line of L = 3.28 between the equatorial plane and altitude of 13000 km. Hence the f-t spectra of normal riser emissions observed at invariant latitude of 56.5° by ISIS-1 are explained by the anomalous Doppler-shifted cyclotron radiations emitted forward from protons of 8.6 keV in the outer plasmasphere.

5. CONCLUSION

Giant riser emissions and VLF hisses with multiple rising-frequency structures appearing between 8 kHz and 20 kHz have been observed mainly at invariant latitudes from 47° to 60° in the recovery phase of geomagnetic substorms by Alouette-2, ISIS-1 and ISIS-2. The frequency rise, rise time of 2.5 seconds and frequency range of giant riser emissions are much larger than those of the chorus emissions which occur mainly around the auroral zone.

Groups of normal riser emissions of which frequency rises from 5 kHz to 9 kHz in 1.5 seconds were observed repeatedly with period of 26 - 27 seconds between invariant latitudes of 56.5° and 61.9° on March 10, 1981 of the geomagnetically quiet day by ISIS-1. Geomagnetic pulsations of Pc 3 were also observed simultaneously with the periodic occurrence of normal riser emission groups at Memambetsu (geomag. lat. 34.1°N),Japan. The periodic modulation of riser emission growth rate seems to result from the proton precipitations modulated by the geomagnetic pulsations of Pc 3.

The frequency rise, rise time of 1.5 seconds and frequency range of normal riser emissions are also larger than those of chorus emissions in the auroral zone. The Doppler-shifted cyclotron radiations emitted backward from energetic electrons moving away from the observer are difficult to explain the observed frequency rises during the observed long rise-times of giant and normal riser emissions in mid latitudes.

The results of numerical calculations show that the anomalous Doppler-shifted cyclotron radiations emitted forward from energetic protons, that is, the transverse proton cyclotron instabilities in the plasmasphere are favourable for explaining the f-t spectra of giant and normal riser emissions observed in the topside ionosphere of mid-latitude.

ACKNOWLEDGEMENTS

I wish to thank members of the Satellite Control Section, Kashima Branch, RRL for their collaboration in data acquisition from the satellites.

REFERENCES

Chappell, C. R. : 1972, Rev. Geophys. Space Science, 10, pp. 951 - 979.

Coroniti, F. V. and Kennel, C. F. : 1970, J. Geophys. Res.,75,pp. 1279 - 1289.

Frank, L. A. : 1967, J. Geophys. Res.,72,pp. 185 - 196.

Fraser-Smith, A. C. and Helliwell, R. A. : 1980, Geophys. Res. Lett.,7 , pp. 851 - 853.

Gallet, R. M. : 1959, Proc. IRE, 47, pp. 211 - 231.

Ginzburg, V. L. : 1964, The Propagation of Electromagnetic Waves in Plasma, Pergamon Press, London.

Gurnett, D. A., Anderson, R. R., Scarf, F. L., Fredricks, R. W. and Smith,E. J. : 1979, Space Sci. Rev., 23, pp. 103 - 122.

Harker, K. J., Crawford, F. W. and Fraser-Smith, A. C. : 1974, J. Geophys. Res., 79, pp. 1836 - 1843.

Helliwell, R. A. : 1965, Whistler and Related Ionospheric Phenomena, Stanford Univ. Press.

Helliwell, R. A. : 1967, J. Geophys. Res., 73, pp. 4773 - 4790.

Kimura, I. : 1974, Space Sci. Rev., 16, pp. 389 - 411.

Lanzerotti, L. J. and Fukunishi, H. : 1974, Rev. Geophys. Space Phys., 12, pp. 724 - 729.

Lanzerotti, L. J. and Southwood, D. J. : 1979, Solar System Plasma Physics,III, ed. by Kennel, C. F., Lanzerotti, L. J. and Parker, E. N., North-Holland, Amsterdam, p. 111.

Ondoh, T. : 1961, J. Atmos. Terr. Phys., 21, pp. 284 -286.

Sato, N. and Kokubun, S. : 1981, J. Geophys. Res., 86, pp. 9 - 18.

Sato, N. : 1984, J. Geophys. Res., 89, pp. 2781 - 2787.

Spitzer, Jr., L. : 1962, Physics of Fully Ionized Gases, 2nd Edition, Interscience Publ. Inc., New York, p. 77.

Van Allen, J. A. : 1968, Physics of the Magnetosphere, ed., by Carovillano, R. L., McClay, J. F. and Radoski, H. R., Springer-Verlag, New York Inc., New York and D. Reidel Publ. Co., pp. 147 - 217.

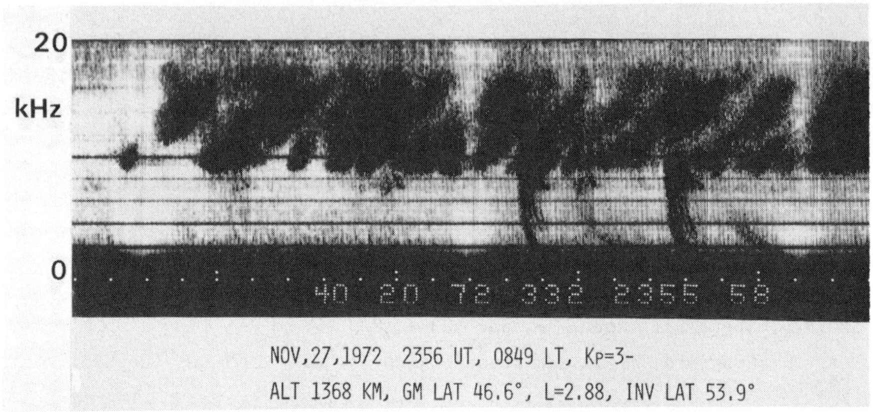

Fig. 1 Giant riser emissions observed at 2356 UT on Nov. 27, 1972 by ISIS-2 (Geomag. lat. 46.6°N, long. 162.8°E ; L = 2.88 ; Altitute 1368 km ; Kp = 3-).

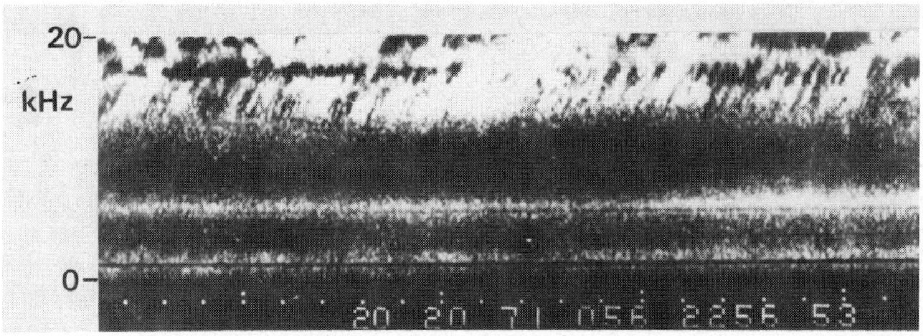

Fig. 2 VLF hiss with fine rising structures observed at 2257 UT on Feb. 25, by Alouette-2 (Geomag. lat. 43.6°N, long. 144.9°E ; L = 2.88, Altitute 2915 km ; Kp = 5-), where the full scale of the ordinate is 20 kHz in frequency.

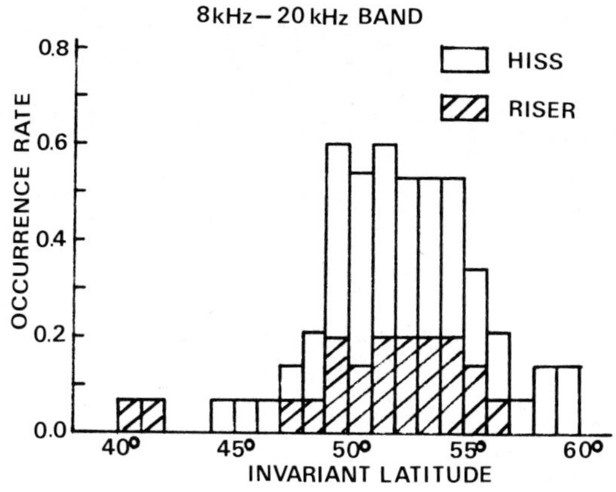

Fig. 3 Latitudinal distribution of occurrence rate of giant riser emissions and VLF hisses with fine rising structures appearing between 8 kHz and 20 kHz observed by AL-2, ISIS-1 and ISIS-2 at Kashima, Japan.

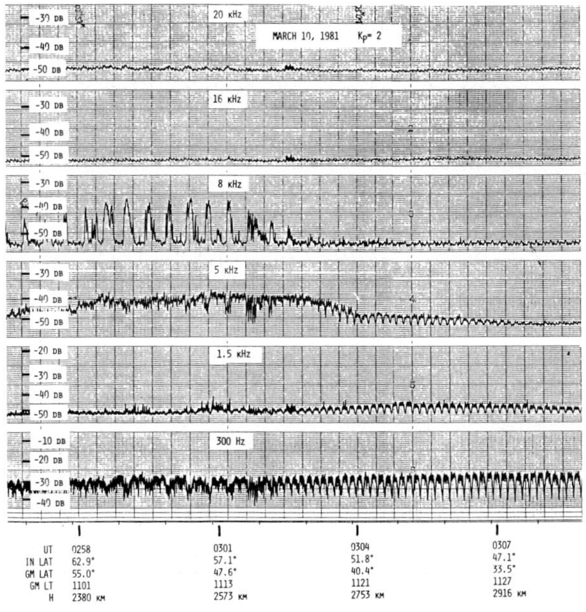

Fig. 4 Narrow-band intensity data at six frequencies observed by ISIS-2 on March 10, 1981. The periodic occurrence of field intensity at 8 kHz is clearly seen between 0258 and 0301 UT.

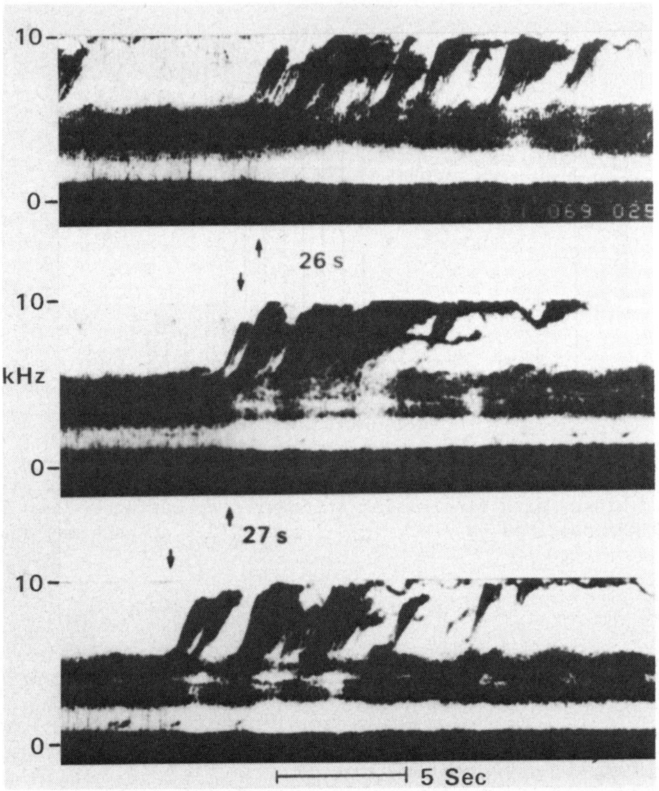

Fig. 5 Detailed f - t spectra of normal riser emissions observed at 0259 UT on March 10, 1981 by ISIS-1 (Geomag. lat. 52.5°N, long. 196.2°E; L = 4.24, Altitute 2446 km). The times of 26 and 27 seconds are the period between riser emission groups in each plate.

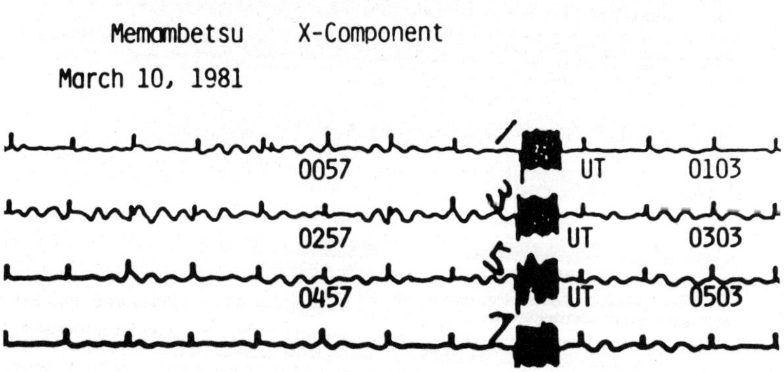

Fig. 6 Rapid-run magnetogram of geomagnetic X-component observed on March 10, 1981 at Memambetsu, Japan. The geomagnetic pulsations of Pc 3 are seen between 0050 and 0504 UT.

Nonlinear and Environmental Electromagnetics, edited by H. Kikuchi
Elsevier Science Publishers B.V., Amsterdam, 1985 – Printed in The Netherlands

REVIEW OF CANADIAN STUDIES OF ATMOSPHERIC E.M.P. PHENOMENA AND OF E.M. ENVIRONMENT MEASUREMENTS

T.J.F. Pavlásek
Faculty of Engineering, McGill University

Canadian research in atmospheric E.M.P. is motivated by concern for effects on major natural resources (e.g. naturally caused forest fires), on energy distribution systems (e.g. stability of large scale, geographically extended power grids) and on the massively proliferating microcircuit based equipment and systems in the communications, computer and instrumentation areas.

Investigations in this area cover a broad range of activities, including studies of:

- the fundamental behaviour of naturally occuring sources of E.M.P.
- particular forms of humanly contrived sources (other than N.E.M.P.)
- detection, measurement and monitoring
- development of sensors, monitoring systems, data management evaluation and interpretation
- study of effect on and response of "victims" of E.M.P.
- analytical/numerical, data processing techniques for analysis and evaluation

Activities are also reported on "real life" and laboratory measurement, monitoring, simulation, analysis and prediction of the ambient E.M. environment in high density metropolitan core areas, both outdoors and indoors.

Those engaged in the above research activities include researchers in government establishments, industrial and utilities' laboratories, universities and regulatory and standardization bodies.

(Refer to the "SUMMARY" at the end)

II. NONLINEAR WAVES, QUASI-PARTICLES AND SOLITONS

Nonlinear and Environmental Electromagnetics, edited by H. Kikuchi
Elsevier Science Publishers B.V., Amsterdam, 1985 – Printed in The Netherlands

Quasiparticle View of Wave Propagation

NATHAN MARCUVITZ, FELLOW, IEEE

Invited Paper

Abstract—A mixed analytical and computational method of treating the propagation of single mode-type wavepackets or beams in inhomogeneous, time-varying, linear, nonlinear, or turbulent media is described. On averaging out fast space–time variations, one views a wavetrain kinetically as a system of "quasiparticles" whose distribution in position and momentum space permits the calculation of space–time dependent wavetrain properties such as envelope amplitude, phase, etc. Individual quasiparticles, which have an energy (Hamiltonian) descriptive of the mode-type, move along "ray" paths characteristic of the medium. The overall system of quasiparticles evolves in a nonsingular (caustic free) manner dependent on initial configuration and on effects of both trapped and untrapped quasiparticles. Starting from a kinetic description of the distribution of quasiparticles, one derives a "fluid dynamic" description that is equivalent to the Whitham and other averaging procedures for analyzing wavetrain propagation; a "macroparticle" description akin to known "centroid" analyses of localized wavetrains is also derivable. The particle-wave duality implied in the above analysis is discussed both from a quantum mechanical and classical standpoint. A few examples of the use of the quasiparticle method are presented.

I. Introduction

As often noted, similar wave phenomena occur in fields as varied as electromagnetic, acoustic, hydrodynamic, etc., and in media that range from free space, semiconductors, ocean, ionosphere, etc. Waves frequently have the form of finite wavepackets or beams and a central question is how such wavetrains are distorted on propagation through linear, nonlinear, or turbulent media. As is known, there are computational difficulties associated with conventional modal or ray treatments of wavetrain propagation in the above media, particularly when many modes or rays are excited in open inhomogeneous waveguiding structures. A commonly used technique in such computations is to average out the fast-space–time wavetrain behavior and determine only slow space–time properties characteristic of the transport of power, phase, etc. that are of primary interest in wave propagation. In the following we shall describe how a wavepacket or beam may be viewed as a finitely extended, moving and deforming, system of point "quasiparticles" whose dynamics permits one both to visualize and approximate the average space–time propagation properties of a wavepacket or beam in diverse media.

Quasiparticles are distinguished by their "energy" Hamiltonian, which is a function of their position and momentum but is generally of a different form than that for real particles. Individual quasiparticles constantly change their position and momentum with respect to a time or a guiding spatial coordinate; most of the discussion will be confined to the former (wavepacket) case with the understanding that equivalent remarks apply to the latter (beam) case. A system or cloud of quasiparticles, viewed in position and momentum phase space, assumes overall configurations that are representative of the instantaneous state of a wavepacket or beam; these distributions may or may not change with time. For a linear medium the stationary distributions are indicative of excited eigenstates that represent either oscillatory or guided wave modes supported by the relevant medium; for a nonlinear medium they represent solitary waves or solitons. Time dependent quasiparticle distributions can be described in terms of variables that are either: kinetic (momentum and space dependent), fluid-dynamic (space dependent); or those of a localized macroparticle (just time dependent), with increasing coarseness of description. The kinetic or phase space description utilizes the Wigner distribution function [1] known from quantum mechanics. The fluid dynamic description is equivalent to Whitham's elegant variational procedure [2] for determining average characteristics of wavetrain propagation, as well as to other similar averaging techniques [3], [4]. The macroparticle description is akin to procedures defining the centroid properties of propagating wavetrains [5], [6]. All are derivable from the kinetic description, which moreover possesses certain advantages in analytic and numerical computations vis-à-vis the moment or Whitham procedures.

Individual quasiparticles will be shown to follow trajectories that at least for linear media are identical to those of optic type rays. A system of quasiparticles, representative of a finite wavetrain, will however evolve along different types of trajectories indicative of the direction of average power flow at any point in the wavetrain. These latter non-ray-trajectories, which track "transport" properties of the total field, are determined from moments of the kinetic description in a manner that avoids the usual singularity (caustic, etc.) difficulties of ray-optics, a significant advantage of the quasiparticle method.

One recalls that distributed physical systems support oscillatory or guided waves characteristic of distinctive eigen or quasi-mode types [7], each distinguished by a dispersion relation. Different mode types are represented by different types of quasiparticles, each distinguished by its own energy (Hamiltonian). For homogeneous and stationary media an oscillatory mode type is distinguished by a dispersion relation $\omega(k)$ with ω denoting the temporal frequency and k the spatial wavenumber. For a guided mode type the dispersion relation is of the form $\kappa(k_t, \omega)$ with κ denoting the spatial wavenumber along the guiding direction, k_t the wavenumber transverse thereto, and ω the temporal frequency. In the presence of sources the excited modes may take the form of oscillatory wavepackets or guided wave beams that are localizable in

Manuscript received January 25, 1980; revised July 21, 1980. This work was supported by the Office of Naval Research under Contract N00014-77-C-0169.
The author is with the Polytechnic Institute of New York, MRI, Farmingdale, NY 11735.

space–time. For example, at one instant of time a wavepacket may be described by the spatial form

$$\psi(x) = \sqrt{2}a \exp[-(x/b)^2/2] \cos(k_0 x + \phi) \quad \text{(I.1a)}$$

comprising oscillatory modes of wavenumber k centered about $\pm k_0$ (if $k_0 >> 1/b$); at a subsequent time this spatial form will be modified in a manner depending on the dispersion relation $\omega(k)$ characteristic of the type of excited oscillatory modes. Similarly, at some incident plane, a simple steady state beam may have a transverse cross-sectional dependence

$$\psi(x) = a \exp[-(x/b)^2] \exp(ik_0 x) \quad \text{(I.1b)}$$

composed of guided modes with transverse wavenumber k_t centered about $k_t = k_0$ (if $k_0 >> 1/b$); at some exit plane this beam will display an altered transverse form dependent on the dispersion relation $\kappa(k_t, \omega)$ characteristic of the type of excited guided modes. The wavepacket in (1a) will be viewed at a given time as two distinct systems of identical quasiparticles, one with "momenta" in the approximate range $k_0 \pm 1/b$ and the other in the range $-k_0 \pm 1/b$, both systems located at $x = 0$ within the neighborhood of $\pm b$. At a later time both systems will have altered momenta and locations depending on their "energy" $\omega(k)$. Similarly, at a given plane, the beam in (1b) can be regarded as a single system of identical quasiparticles with "momenta" proximate to the range $k_0 \pm 1/b$ and located about $x = 0$ within the neighborhood of $\pm b$. At a different plane the quasiparticle x-momenta and positions will be altered in a manner dependent on their momentum $\kappa(k_t, \omega)$ along the guiding direction.

The simple oscillatory or guided wavetrains indicated above, composed of waves of a single-mode type, are usually characterized by fast carrier and by slow modulational space–time scales. The quasiparticle description of the propagation of such wavetrains is based on averaging out the fast behavior and describing only the slow or averaged space–time dependence of primary interest. To illustrate the averaging for the wavepacket $\psi(x)$ defined in (1a) we introduce the 2-point product

$$\psi(x_1)\psi^*(x_2) = a^2 \exp[-(x/b)^2 - (\xi/2b)^2] \cdot [\cos 2(k_0 x + \phi) + \cos k_0 \xi]$$

where $\xi = x_1 - x_2$ and $x = (x_1 + x_2)/2$ define the "fast" variable ξ and slow variable x. The phase dependent fast variability of the $\cos 2(k_0 x + \phi)$ term may be averaged out by integration over a suitable period of x or ϕ to yield the 2-point average

$$\langle \psi(x_1)\psi^*(x_2)\rangle = a^2 \exp[-(x/b)^2 - (\xi/2b)^2] \cos k_0 \xi$$

which displays only a slow variability with x; for turbulent fields one forms an ensemble average. For $\xi = 0$ one obtains the usual mean-square average

$$\langle |\psi(x)|^2\rangle = a^2 \exp[-(x/b)^2].$$

Such averages are of importance in applications. They are readily amenable to computer analysis and they form the basis for a kinetic quasiparticle picture of wave processes. To provide this basis one seeks from the defining equations for $\psi(x)$ to infer corresponding equations for averaged variables that permit one to calculate wavetrain characteristics such as: average speed, frequency, wavenumber, envelope amplitude, wavenumber width, etc. It is because of the use of averaged variables for analysis of wavepacket or beam dynamics that one is led to a simple physical interpretation in terms of the dynamics of quasiparticles. Such an interpretation is indicative of a wave–"particle" duality that is well known in quantum dynamics and is not completely unfamiliar in classical wave dynamics where at least at linear levels it is often expressed in "ray-optic" terms. However, as noted, it is equally applicable for nonlinear and turbulent wave propagation. In Section II quasiparticle descriptions of a single mode type wave equation are derived; in Section II-A at a kinetic level, in Section II-B at a fluid dynamic level, and in Section II-C at a macroparticle level. The underlying wave–particle duality constituting the basis for a quasiparticle description is discussed in Section III; particle–wave duality characteristic of quantum mechanics is reviewed in Section III-A and a similar classical wave–quasiparticle duality is developed in Section III-B. Examples of quasiparticle applications to wavepacket propagation in a number of media are presented in Section IV: an idealized quadratically variable medium in Section IV-A, a localized exponential medium in Section IV-B, a time dependent quadratic medium in Section IV-C, and a nonlinear medium in Section IV-D.

II. The Quasiparticle Approximation

A wavepacket formed of oscillatory modes of a single type will be viewed as a system of point quasiparticles, each described by its position $x(t)$, momentum $k(t)$, and energy $\omega(k, x, t)$. The individual quasiparticle motion is determined by the functional form of ω and differs with each mode type. A kinetic description of the overall quasiparticle system in k, x space at time t is given in terms of a distribution function $F(k, x, t)$ which provides a physical picture of how the wavepacket evolves in space–time. A fluid description is obtained from the k moments of $F(k, x, t)$ and a macroquasiparticle description is recovered from the k and x moments of $F(k, x, t)$. Our initial task is to derive a defining equation for $F(k, x, t)$ from the equations descriptive of the wavepacket field.

A. *Kinetic Description*

Consider a single-mode type of wave problem described locally by a complex-wave function $\psi(x, t)$ satisfying

$$i \frac{\partial \psi}{\partial t} = H\psi \quad \text{(II.1)}$$

with the linear or nonlinear operator (assumed Hermitian for simplicity)

$$H = H\left(\frac{1}{i}\frac{\partial}{\partial x}, x, t\right)$$

or

$$H = H\left(\frac{1}{i}\frac{\partial}{\partial x}, x, t, |\psi|\right)$$

and with ψ subject to an initial condition $\psi(x, 0)$. Such problems arise for example when one considers the "forward scattered" wave descriptive of propagation in a nonuniform, nonstationary and/or nonlinear medium. Although the analysis is applicable to three dimensions, for simplicity of presentation only one spatial dimension x will be considered; the analysis is thus descriptive of a generalized time dependent transmission line. To identify the quasiparticle descriptors as well as the kinetic distribution function F from the defining equation for ψ, one introduces the double Fourier representation

$$\psi_1 \psi_2^* = \iint \phi_1 \phi_2^* \exp\left[i(k_1 x_1 - k_2 x_2)\right] \frac{dk_1\, dk_2}{(2\pi)^2} \quad \text{(II.2)}$$

where $\phi_1 = \phi(k_1, t)$ and $\phi_2 = \phi(k_2, t)$ are, respectively, the Fourier amplitudes of $\psi_1 = \psi(x_1, t)$ and $\psi_2 = \psi(x_2, t)$. The product $\psi_1 \psi_2^*$ displays both a fast and slow spatial behavior and it is essential for the quasiparticle interpretation to average out the very fast behavior. If $\psi(x, t)$ and hence $\phi(k, t)$ are stochastic, the averaging is explicit with (II.2) being expressed in terms of ensemble averages as

$$\langle \psi_1 \psi_2^* \rangle = \iint \langle \phi_1 \phi_2^* \rangle \exp\left[i(k_1 x_1 - k_2 x_2)\right] \frac{dk_1\, dk_2}{(2\pi)^2} \quad \text{(II.2a)}$$

and the average symbol $\langle\ \rangle$ should be retained but for simplicity will be omitted throughout the following. If the ψ and ϕ are deterministic, this symbol should be understood as an average over a suitably fast spatial period or phase. After the averaging, "slow" (x or κ) and "fast" (ξ or k) spatial or wave-number variables are defined by

$$x = (x_1 + x_2)/2$$
$$\kappa = k_1 - k_2 \quad \text{(II.3)}$$

and

$$\xi = x_1 - x_2$$
$$k = (k_1 + k_2)/2$$

whence since

$$k_1 x_1 - k_2 x_2 = k\xi + \kappa x$$
$$dk_1\, dk_2 = dk\, d\kappa.$$

Equation (II.2) can be rewritten as

$$\psi_1 \psi_2^* = \int F(k, x, t) \exp(ik\xi) \frac{dk}{2\pi} \quad \text{(II.3a)}$$

where

$$F(k, x, t) = \int \phi_1 \phi_2^* \exp(i\kappa x) \frac{d\kappa}{2\pi}. \quad \text{(II.3b)}$$

The reality property $F(k, x, t) = F^*(k, x, t)$ of the fast (correlation) spectrum of $\psi_1 \psi_2^*$ follows readily from (II.3); however, positive definiteness is not in general assured. Fourier transformations inverse to (II.3) are manifestly

$$F(k, x, t) = \int \psi_1 \psi_2^* \exp(-ik\xi)\, d\xi \quad \text{(II.4a)}$$

$$\phi_1 \phi_2^* = \int F(k, x, t) \exp(-i\kappa x)\, dx. \quad \text{(II.4b)}$$

For $\xi = 0$ and $\kappa = 0$, equations (II.3a) and (II.4b) yield

$$|\psi(x, t)|^2 = \int F(k, x, t) \frac{dk}{2\pi}$$

$$|\phi(k, t)|^2 = \int F(k, x, t)\, dx$$

whence one infers that

$$\int |\psi(x, t)|^2\, dx = \int |\phi(k, t)|^2 \frac{dk}{2\pi}$$
$$= \int F(k, x, t) \frac{dk\, dx}{2\pi}. \quad \text{(II.5)}$$

Even though only reality but not positive definiteness of $F(k, x, t)$ is assured, equation (II.5) suggests the identification of $F(k, x, t)$ as a number density of quasiparticles in k, x phase space. The consequent interpretation of $|\psi|^2$ as a quasiparticle number density in real space and of $|\phi|^2$ as a number density in momentum space are in conformity with a similar concept in quantum mechanics (cf. Section III). In the latter instance because of a different normalization $|\psi|^2$ and $|\phi|^2$ can be given a probabilistic interpretation.

The defining equation for $F(k, x, t)$ can be deduced from (II.1) if one infers therefrom that

$$i \frac{\partial}{\partial t} (\psi_1 \psi_2^*) = (H_1 - H_2^*)(\psi_1 \psi_2^*) \quad \text{(II.6)}$$

where

$$H_1 = H\left(\frac{1}{i} \frac{\partial}{\partial x_1}, x_1, t\right)$$
$$H_2^* = H\left(-\frac{1}{i} \frac{\partial}{\partial x_2}, x_2, t\right).$$

Noting the definition (II.3) and forming the inverse transform (II4a) of (II.6), one obtains

$$i \frac{\partial F}{\partial t} = \int \exp(-ik\xi) \left[H\left(\frac{1}{i} \partial_\xi + \frac{1}{2i} \partial_x, x + \frac{\xi}{2}, t\right) - H\left(\frac{1}{i} \partial_\xi - \frac{1}{2i} \partial_x, x - \frac{\xi}{2}, t\right)\right] \psi_1 \psi_2^*\, d\xi.$$

After commuting the operator H so that it acts only on $\exp(ik\xi)$ with a consequent change in sign of ∂_ξ, one finds as the formally exact defining equation for F [1], [8]:

$$i \frac{\partial F}{\partial t} = \left[H\left(k + \frac{1}{2i} \partial_x, x - \frac{1}{2i} \partial_k, t\right) - H\left(k - \frac{1}{2i} \partial_x, x + \frac{1}{2i} \partial_k, t\right)\right] F. \quad \text{(II.7a)}$$

Equation (II.7a) can be rewritten as

$$\frac{\partial F}{\partial t} = -2 \sin \frac{1}{2} \left(\frac{\partial}{\partial k} \frac{\vec{\partial}}{\partial x} - \frac{\partial}{\partial x} \frac{\vec{\partial}}{\partial k}\right) H(k, x, t) F \quad \text{(II.7b)}$$

where the arrow implies operation on F rather than on H. Setting $\omega = H(k, x, t)$ and assuming a weak dependence of F and H on x, k, one expands H in first order to obtain as the "quasiparticle approximation" to (II.7):

$$\frac{\partial F}{\partial t} + \frac{\partial \omega}{\partial k} \frac{\partial F}{\partial x} - \frac{\partial \omega}{\partial x} \frac{\partial F}{\partial k} = 0. \quad \text{(8)}$$

Equation (II.8) has the form of a collisionless "kinetic" equation indicative of a distribution of "particles" in k, x phase space at time t. For non-Hermitian H (complex ω) or for multimode descriptions, a collision term will appear on the right-hand side of (II.8). The distribution function $F(k, x, t)$ defines the number density of such "particles" per unit volume of the phase space and the kinetic equation indicates the "characteristic" trajectories along which the "particles" move. The characteristic equations of the partial differential (II.8), namely:

$$\frac{dx}{dt} = \frac{\partial \omega}{\partial k}$$

$$\frac{dk}{dt} = -\frac{\partial \omega}{\partial x} \qquad \text{(II.9)}$$

identify the trajectories of quasiparticles of position $x(t)$, momentum $k(t)$ and energy $\omega(k, x, t)$. On a quasiparticle trajectory, one observes in the quasiparticle approximation that for the mode type (II.1)

$$\frac{dF}{dt} = 0$$

$$\frac{d\omega}{dt} = \frac{\partial \omega}{\partial t} \qquad \text{(II.10)}$$

The implied constancy of F on a quasiparticle trajectory permits the time dependent solution of (II.8), which evolves from a given initial state $F(k, x, 0)$, to be written conventionally as

$$F(k, x, t) = F(k_0(k, x, t), x_0(k, x, t), 0) \qquad \text{(II.11)}$$

where $k_0(k, x, t), x_0(k, x, t)$ define coordinates at time $t = 0$ of a quasiparticle trajectory ending at k, x at time t (cf. Appendix A). Note that in the quasiparticle approximation if $F(k, x, 0)$ is positive definite, so also is $F(k, x, t)$.

"Fluid dynamic" properties of a quasiparticle system can be inferred from the k-moments of the kinetic distribution function F. Such properties are representative of macroscopic characteristics of wavepacket solutions of (II.1) in regions wherein wavepackets form. If a typical wavepacket solution is represented as

$$\psi(x, t) = a(x, t) \exp [i\theta(x, t)] \qquad \text{(II.12)}$$

then as noted in (II.5) the local amplitude $a(x, t)$ follows from $F(k, x, t)$ via

$$a^2(x, t) = \int F(k, x, t) \frac{dk}{2\pi} \qquad \text{(II.13)}$$

One observes that the amplitude squared $a^2(x, t)$ of the wavepacket, or the zeroth moment of F, is indicative of the number density of quasiparticles. The local phase $\theta(x, t)$ of the wavepacket can be determined from a local wavenumber and frequency, defined as:

$$K(x, t) = \frac{\partial \theta}{\partial x}$$

$$\Omega(x, t) = -\frac{\partial \theta}{\partial t} \qquad \text{(II.14a)}$$

by noting that

$$\theta(x, t) = \theta(x_0, 0) + \int (K\, dx - \Omega\, dt) \qquad \text{(II.14b)}$$

where the integration in x, t space is over an appropriate path extending from (arbitrary) $x_0, 0$ to x, t. To find the local wavenumber and frequency one first observes, in view of (II.14a), that

$$K(x, t) = -i \left. \left(\psi_2^* \frac{\partial}{\partial x_1} \psi_1 - \psi_1 \frac{\partial}{\partial x_2} \psi_2^* \right) \Big/ 2a^2 \right]_{\xi=0}$$

$$\Omega(x, t) = i \left. \left(\psi_2^* \frac{\partial}{\partial t} \psi_1 - \psi_1 \frac{\partial}{\partial t} \psi_2^* \right) \Big/ 2a^2 \right]_{\xi=0} \qquad \text{(II.14c)}$$

whence, using (II.3a) and (II.1), one derives from (II.14c) in the limit $\xi = 0$ the relations

$$K(x, t) = \frac{1}{a^2} \int k F(k, x, t) \frac{dk}{2\pi}. \qquad \text{(II.15a)}$$

$$\Omega(x, t) = \frac{1}{a^2} \int \left[\omega(k, x, t) - \frac{1}{8} \frac{\partial^2 \omega}{\partial k^2} \frac{\partial^2}{\partial x^2} - \frac{1}{8} \frac{\partial^2 \omega}{\partial x^2} \frac{\partial^2}{\partial k^2} + \frac{1}{4} \frac{\partial^2 \omega}{\partial k \partial x} \frac{\partial^2}{\partial k \partial x} \cdots \right] F \frac{dk}{2\pi} \qquad \text{(II.15b)}$$

which identify the local space–time dependent wavenumber K and frequency Ω as averages or moments of the quasiparticle distribution function F. Thus via (II.3), (II.8), (II.13), and (II.14), the space–time dependent amplitude and phase characteristics of a propagating wavepacket can be found directly from a kinetic description starting from initial values $F(k, x, 0)$ and $\theta(x, 0)$, which are derivable from $\psi(x, 0)$. In particular, equation (II.15b) provides a space–time dependent "dispersion relation" relating $\Omega(x, t)$ and $K(x, t)$, cf. equation (II.23) below.

B. Fluid Dynamic Description

"Fluid dynamic" equations descriptive of a propagating wavepacket can be inferred from moments of the collisionless quasiparticle kinetic equation (II.8). This "fluid" description is closely related to the conservation equations obtained by Whitham [2] from his averaged variational formalism for propagating wavepackets. In the quasiparticle system representative of the wavepacket one defines a local "velocity" v and "force" f by means of the F moments:

$$a^2 v(x, t) = \int \frac{\partial \omega}{\partial k}(k, x, t)\, F(k, x, t) \frac{dk}{2\pi}$$

$$a^2 f(x, t) = -\int \frac{\partial^2 \omega}{\partial k^2} \frac{\partial \omega}{\partial x}(k, x, t)\, F(k, x, t) \frac{dk}{2\pi}. \qquad \text{(II.16)}$$

Then from the quasiparticle kinetic equation (II.8), written in the form:

$$\frac{\partial F}{\partial t} + \frac{\partial}{\partial x}\left(\frac{\partial \omega}{\partial k} F\right) - \frac{\partial}{\partial k}\left(\frac{\partial \omega}{\partial x} F\right) = 0 \qquad \text{(II.17)}$$

one obtains using (II.13) and (II.16), by forming in a conventional manner the zeroth and velocity moments of F, the "fluid" equations:

$$\frac{\partial a^2}{\partial t} + \frac{\partial}{\partial x}(a^2 v) = 0 \qquad \text{(II.18a)}$$

$$\frac{\partial}{\partial t}(a^2 v) + \frac{\partial}{\partial x}(a^2 \overline{v^2}) - a^2 f = 0 \qquad \text{(II.18b)}$$

where for simplicity it has been assumed that ω is time independent and separably dependent on k and x (i.e., $\partial_k \omega$ is independent of x). Using (II.18), one can rewrite (II.18b) in the more conventional "fluid" form

$$\frac{\partial v}{\partial t} + v\frac{\partial v}{\partial x} + \frac{1}{a^2}\frac{\partial P}{\partial x} = f \qquad \text{(II.18c)}$$

if, as in (II.15), one defines a mean-square velocity $\overline{v^2}$ by

$$a^2\overline{v^2} = \int \left(\frac{\partial \omega}{\partial k}\right)^2 F \frac{dk}{2\pi} \qquad \text{(II.19a)}$$

and introduces a local quasiparticle "pressure" $P(x, t)$ via

$$a^2\overline{v^2} = a^2 v^2 + P. \qquad \text{(II.19b)}$$

Equations (II.18) constitute two equations for four unknowns and hence additional information on P and f is required for closure. Determination of the requisite information is perhaps best clarified by a particular example.

For the case where the quasiparticle "energy" is specified by a separable dispersion relation

$$\omega(k, x, t) = \frac{k^2}{2} + V(x, t) \qquad \text{(II.20a)}$$

corresponding to (II.1) with

$$H = -\frac{1}{2}\frac{\partial^2}{\partial x^2} + V(x, t) \qquad \text{(II.20b)}$$

one infers from (II.16) and (II.15a) that

$$v = K(x, t)$$

$$f = -\frac{\partial V}{\partial x}. \qquad \text{(II.21a)}$$

To determine the local "pressure" P one relates K^2 to $\overline{k^2}$ by observing that:

$$-\left(\psi_2^* \frac{\partial^2}{\partial x_1^2}\psi_1 + \psi_1 \frac{\partial^2}{\partial x_2^2}\psi_2^*\right)_{\xi=0}$$

$$= \int \left[\left(k + \frac{1}{2i}\frac{\partial}{\partial x}\right)^2 + \left(k - \frac{1}{2i}\frac{\partial}{\partial x}\right)^2\right] F \frac{dk}{2\pi}$$

Evaluating the right-hand side, inserting (II.12) and (II.14a) into the left side and equating, one finds

$$K^2 - \frac{1}{a}\frac{\partial^2 a}{\partial x^2} = \overline{k^2} - \frac{1}{4a^2}\frac{\partial^2}{\partial x^2}a^2. \qquad \text{(II.21b)}$$

Hence, since $v = K$ and $\overline{v^2} = \overline{k^2}$, from (II.19b)

$$P = \frac{1}{4}\frac{\partial^2}{\partial x^2}a^2 - a\frac{\partial^2}{\partial x^2}a \qquad \text{(II.21c)}$$

and therefore for all media (independently of the potential V), the quasiparticle "pressure" P exerts a force given by

$$\frac{1}{a^2}\frac{\partial P}{\partial x} = -\frac{\partial}{\partial x}\left(\frac{1}{2a}\frac{\partial^2 a}{\partial x^2}\right). \qquad \text{(II.21d)}$$

On use of the identities (II.21), equations (II.18a) and (II.17b) become a closed set of "fluid" equations for the local quasiparticle density a^2 and velocity K, viz.

$$\frac{\partial a^2}{\partial t} + \frac{\partial}{\partial x}(a^2 K) = 0 \qquad \text{(II.22a)}$$

$$\frac{\partial K}{\partial t} + K\frac{\partial K}{\partial x} - \frac{\partial}{\partial x}\left(\frac{1}{2a}\frac{\partial^2 a}{\partial x^2}\right) = -\frac{\partial V}{\partial x}. \qquad \text{(II.22b)}$$

One observes from (II.22b) that the "pressure" force (II.21d) driving the quasiparticles can either be enhanced or compensated by the force arising from the potential V. In the context of the Whitham formalism for the case (II.20), it is of interest to note from (II.20a), (II.15b), and (II.21b) that the local frequency Ω takes the form

$$\Omega = \frac{K^2}{2} - \frac{1}{2a}\frac{\partial^2}{\partial x^2}a + V \qquad \text{(II.23)}$$

and thereby identifies (II.23) as the averaged space–time dependent "dispersion relation" in the Whitham formalism; a result that also follows from (II.15b) and should be distinguished from the quasiparticle "energy" $\omega = k^2/2 + V$. The relation

$$\frac{\partial K}{\partial t} + \frac{\partial \Omega}{\partial x} = 0$$

implied by (II.14a) and termed the "wave conservation equation" by Whitham then becomes with the aid of (II.23) the "fluid momentum" equation (II.22b). It is to be observed that $a(x, t)$ and $K(x, t)$ can be obtained either by solution of the fluid equations (II.22) or via (II.13) and (II.16) directly from the quasiparticle distribution $F(k, x, t)$, as obtained from solution of the kinetic equation (II.8). In many instances the latter procedure is computationally simpler.

C. Macro-Particle Description

Another, albeit coarser description of the dynamics of a quasiparticle system, views the overall system as a single macroparticle. The average position, momentum, size, etc., of the macroparticle are dependent only on time and represent the average location, wavenumber, width, etc., of the corresponding wavepacket. The macroparticle description may be derived from the kinetic equation (II.17) by taking suitable k and x moments. Thus on multiplication of (II.17) first by x and then by k, and on integration over both k and x, one obtains dynamical equations for the average position $\overline{x}(t)$ and momentum $\overline{k}(t)$ of the macroparticle, namely,

$$\frac{d}{dt}\overline{x} = \overline{\frac{\partial}{\partial k}\omega}$$

$$\frac{d}{dt}\overline{k} = -\overline{\frac{\partial}{\partial x}\omega} \qquad \text{(II.24)}$$

where the average quasiparticle coordinates are defined in terms of the distribution function F by

$$N\overline{x}(t) = \iint x\, F(k, x, t)\frac{dk\, dx}{2\pi}$$

$$N\overline{k}(t) = \iint k\, F(k, x, t)\frac{dk\, dx}{2\pi} \qquad \text{(II.25a)}$$

and the time dependent averages

$$N\overline{\frac{\partial \omega}{\partial k}} = \iint \frac{\partial \omega}{\partial k} F(k, x, t) \frac{dk\, dx}{2\pi}$$

$$N\overline{\frac{\partial \omega}{\partial x}} = \iint \frac{\partial \omega}{\partial x} F(k, x, t) \frac{dk\, dx}{2\pi}$$

$$N = \iint F(k, x, t) \frac{dk\, dx}{2\pi} \qquad \text{(II.25b)}$$

where the above integrals are taken over the entire k, x phase space. For the case in which the dispersion relation defining the quasiparticle energy $\omega(k, x)$ is real and separable, as in (II.20), the number of quasiparticles N is constant and the dynamical equations (II.24) become:

$$\frac{d}{dt}\overline{x} = \overline{k}$$

$$\frac{d}{dt}\overline{k} = -\overline{\frac{\partial V(x)}{\partial x}} \qquad \text{(II.26)}$$

Similarly, from (II.17) the higher moment averages

$$\overline{x^2} = \overline{x(t)^2} = \iint x^2 F(k, x, t) \frac{dk\, dx}{2\pi N}$$

$$\overline{k^2} = \overline{k(t)^2} = \iint k^2 F(k, x, t) \frac{dk\, dx}{2\pi N}$$

$$\overline{xk} = \overline{x(t)\, k(t)} = \iint xk\, F(k, x, t) \frac{dk\, dx}{2\pi N} \qquad \text{(II.27)}$$

are defined by the indicated moments of F via

$$\frac{d}{dt}\overline{x^2} = 2\overline{xk}$$

$$\frac{d}{dt}\overline{k^2} = -2\overline{k\frac{\partial V(x)}{\partial x}}$$

$$\frac{d}{dt}\overline{xk} = \overline{k^2} - \overline{x\frac{\partial V(x)}{\partial x}}. \qquad \text{(II.28)}$$

For potentials $V(x)$ of higher than the second degree in x, or for potentials nonlinearly dependent on F, equations (II.26) and (II.28) are not closed in that the first-order moments $\overline{x}, \overline{k}$ are dependent on higher moments $\overline{x^2}, \overline{k^2}$, etc. One notes that the mean-square width of the macroparticle is $\overline{x(t)^2} - \overline{x(t)}^2$, the mean-square wavenumber if $\overline{k(t)^2} - \overline{k(t)}^2$, etc., which yield corresponding information about the wavepacket being represented.

It should be observed that it may be simpler to obtain the above macroparticle averages directly from the kinetic distribution function, if the latter is known, via (II.25) and (II.27). However, the dynamical equations (II.26) and (II.28) are useful for revealing invariant as well as time-dependent characteristics of the macroparticle motion. Another procedure for determining a number of macroparticle averages follows from another kinetic distribution function (related to the "radar ambiguity" function) defined by

$$S(\kappa, \xi, t) = \int \psi_1 \psi_2^* \exp(-i\kappa x)\, dx$$

where the κ dependence is now indicative of the slow x-dependence of $\psi_1 \psi_2^*$. The S distribution can be found either directly from its defining equation or by transform of the F distribution, whence the desired macroparticle averages follow as

$$N\overline{x(t)^n} = \left(i\frac{\partial}{\partial \kappa}\right)^n S(0, 0, t) = \int x^n |\psi|^2\, dx$$

$$N = \int |\psi|^2\, dx = S(0, 0, t).$$

The classical wave–macroparticle duality described herein is the converse of the particle–wave duality encountered in quantum mechanical applications. Since the converse derivation of a quantum wave dynamics is of interest in its own right and since it illuminates from a different angle the duality just discussed, it will be briefly described in the following section. The derivation of the basic quantum wave equation from real particle dynamics will be nonpostulational in contrast to many conventional presentations; in its converse classical counterpart in Section III-B it furthermore provides a somewhat different derivation of the macroquasiparticle equations.

III. Particle-Wave Duality

A. *Quantum Case*

At quantum levels, fluctuation effects must be considered and hence the classical motion of a single real particle may be viewed as a statistical average of a microscopic wave dynamics. More specifically, if a real point particle of unit mass, with a classical position q and momentum p at time t, is described by a time independent energy Hamiltonian

$$H(p, q) = \frac{p^2}{2} + V(q) \qquad \text{(III.1)}$$

then its classical motion, defined by the trajectory equations,

$$\frac{dq}{dt} = \frac{\partial H}{\partial p}$$

$$\frac{dp}{dt} = -\frac{\partial H}{\partial q}$$

$$\frac{dH}{dt} = 0 \qquad \text{(III.2)}$$

must be viewed in a statistical sense. Thus one defines a probability density $|\psi(q, t)|^2$ of finding the particle at position q at time t, with the complex-wave function satisfying the probability normalization requirement

$$1 = \int_{-\infty}^{+\infty} \psi^*(q, t)\, \psi(q, t)\, dq \quad \text{or} \quad 1 = (\psi, \psi) \qquad \text{(III.3)}$$

and hence the average particle position at time t is given by

$$\overline{q}(t) = \int_{-\infty}^{+\infty} \psi^*(q, t)\, q\, \psi(q, t)\, dq \quad \text{or} \quad \overline{q} = (\psi, Q\psi). \qquad \text{(III.4)}$$

The right-hand expressions are conventional Hermitian inner product representations of those on the left; Q is a time independent "position operator" with representative q in the ψ

basis. Similarly, one defines a probability density $|\phi(p,t)|^2$ for the particle to have momentum p at time t, with the complex-wave function $\phi(p,t)$ also satisfying a probability normalization requirement

$$1 = \int_{-\infty}^{+\infty} \phi^*(p,t)\,\phi(p,t)\,dp \quad \text{or} \quad 1 = \{\phi,\phi\} \qquad \text{(III.5)}$$

and hence with the average particle momentum at time t given by

$$\overline{p}(t) = \int_{-\infty}^{+\infty} \phi^*(p,t)\,p\,\phi(p,t)\,dp \quad \text{or} \quad \overline{p} = \{\phi, P\phi\} \qquad \text{(III.6)}$$

with P a "momentum operator" whose representative in the ϕ basis is p. The bracketed expressions on the right of equations (III.5) and (III.6) are evidently Hermitian inner product representations of those on the left, but in p space rather than the q space products in (III.3) and (III.4).

From (III.3) and (III.5), using the Fischer–Riesz theorem [5], one infers the existence in q space of an orthonormal basis $\Psi_p(q)$ with respect to which the wave function $\phi(p,t)$ is the transform of $\psi(q,t)$, viz.

$$\psi(q,t) = \int_{-\infty}^{+\infty} \phi(p,t)\,\Psi_p(q)\,dp \qquad \text{(III.6a)}$$

with

$$\delta(p-p') = \int_{-\infty}^{+\infty} \Psi_p^*(q)\,\Psi_{p'}(q)\,dq \quad \text{or} \quad \delta(p-p') = (\Psi_p, \Psi_{p'}) \qquad \text{(III.6b)}$$

representative of the orthonormality properties of the $\Psi_p(q)$. One observes that (III.6a) and (III.6b) imply the normalization requirements

$$(\psi,\psi) = \{\phi,\phi\} = 1. \qquad \text{(III.6c)}$$

Since $\phi(p,t)$ is the transform of $\psi(q,t)$, then $p\phi(p,t)$ must be the transform of some q-space wavefunction that can be identified as

$$P\psi(q,t) = \int_{-\infty}^{+\infty} p\,\phi(p,t)\,\Psi_p(q)\,dp. \qquad \text{(III.7)}$$

This identification in terms of the momentum operator P follows from the relation, deducible from (III.6) and (III.7),

$$(\psi, P\psi) = \{\phi, p\phi\} = \overline{p}(t) \qquad \text{(III.8)}$$

where the as yet unknown representative of the operator P in the ψ basis is to be determined from its known representative p in the ϕ basis. Incidentally, since from (III.6a) and (III.7),

$$\int \phi(p,t)\,P\Psi_p(q)\,dp = \int \phi(p,t)\,p\Psi_p(q)\,dp, \qquad \text{for every } \phi(p,t)$$

one notes that the Ψ_p are q-space eigenfunctions of the operator P with eigenvalues p, viz.

$$P\Psi_p(q) = p\Psi_p(q).$$

Equations (III.4) and (III.8) not only identify the classical position $\overline{q}(t)$ and momentum $\overline{p}(t)$ variables as probabilistic averages of the ψ wave field, but also lend significance to the usual quantum mechanic postulational association of time independent operators and classical variables. A further observation, following from the reality of the classical variables is that these operators are Hermitian, i.e.,

$$\overline{q}(t) = (\psi, Q\psi) = (Q\psi, \psi)$$

$$\overline{p}(t) = (\psi, P\psi) = (P\psi, \psi). \qquad \text{(III.9a)}$$

More generally, one associates the energy operator $H(P,Q)$ with the average of the classical Hamiltonian, via

$$\overline{H(p,q)} = (\psi, H(P,Q)\,\psi) \equiv (\psi, H\psi) = (H\psi, \psi) \qquad \text{(III.9b)}$$

where for a classical deterministic particle it should be noted that $\overline{H(p,q)} = H(\overline{p},\overline{q})$ since classically $p(t)$ and $q(t)$ have zero fluctuations from the average $\overline{p}(t), \overline{q}(t)$. A defining equation for ψ may now be determined by imposition of the requirement that the classical equations (III.2) are "average" equations, whence from (III.9b) and the constancy of the average Hamiltonian:

$$\frac{d}{dt}\overline{H(p,q)} = 0 = \left(\frac{\partial\psi}{\partial t}, H\psi\right) + \left(\psi, H\frac{\partial\psi}{\partial t}\right)$$

$$0 = \left(\frac{\partial\psi}{\partial t}, H\psi\right) + \left(H\psi, \frac{\partial\psi}{\partial t}\right) \qquad \text{(III.10)}$$

Equation (III.10) then implies that (i = imaginary unit)

$$H\psi = ci\frac{\partial\psi}{\partial t}$$

is a possible (but not unique) defining equation for ψ. In the absence of fluctuations the real proportionality constant c approaches zero but in quantum mechanics the proportionality constant is the Planck constant $\hbar$, and hence

$$H(P,Q)\,\psi = \hbar i\frac{\partial\psi}{\partial t} \qquad \text{(III.11)}$$

becomes the Schroedinger equation defining the wave function $\psi(q,t)$; because of the arbitrariness of the proportionality constant, it should be noted that quantum mechanics is but one possible example of a particle–wave duality.

The operator P is as yet undefined in the ψ basis. We observe from the interpretation of (III.2) and (III.1) as averages that

$$\frac{d}{dt}\overline{q}(t) = \overline{\frac{\partial H(p,q)}{\partial p}}$$

$$= \overline{p}(t)$$

whence from (9a) and (11)

$$\left(\frac{\partial\psi}{\partial t}, Q\psi\right) + \left(\psi, Q\frac{\partial\psi}{\partial t}\right) = (\psi, P\psi)$$

or

$$\left(\psi, \frac{HQ - QH}{\hbar/i}\,\psi\right) = (\psi, P\psi).$$

Using the operator expression for the Hamiltonian H, following from (III.1), one has

$$\frac{i}{2\hbar}(P^2Q - QP^2) = P$$

which is satisfied if

$$PQ - QP = \frac{\hbar}{i}$$

and hence in the ψ basis one infers the desired P representation:

$$P \to \frac{\hbar}{i}\frac{\partial}{\partial q} \quad \text{and} \quad Q \to q. \tag{III.12}$$

Thus the Schroedinger equation (III.11) takes the more explicit form

$$H\left(\frac{\hbar}{i}\frac{\partial}{\partial q}, q\right)\psi = \hbar i \frac{\partial \psi}{\partial t} \tag{III.13a}$$

or for H defined as in (III.1),

$$\left[\frac{\hbar^2}{2}\frac{\partial^2}{\partial q^2} + V(q)\right]\psi = \hbar i \frac{\partial \psi}{\partial t} \tag{III.13b}$$

defines the wave function ψ explicitly.

From the above derivation of (III.13) for ψ, it is manifest that the wave field ψ has been defined so that the averages (III.9a) reproduce, in the limit $\hbar \to 0$, the real classical motion defined by (III.2). Since for $\hbar \neq 0$, the ψ field contains more information than the classical equations one can deduce from ψ other (quantum) properties of the particle motion such as indeterminacy relations for the mean square fluctuations of particle position and momentum, etc. Also, for the subsequent discussion, it is of interest to note that if one introduces in the ψ basis time independent operators $\dot{Q}$, $\dot{P}$ via

$$\frac{d\overline{q}(t)}{dt} = (\psi, \dot{Q}\psi)$$

$$\frac{d\overline{p}(t)}{dt} = (\psi, \dot{P}\psi)$$

then from (III.11) *et seq*., one infers dynamical (operator) equations of motion

$$\dot{Q} = \frac{i}{\hbar}[H, Q]$$

$$\dot{P} = \frac{i}{\hbar}[H, P] \tag{III.14}$$

where $[H, Q]$ is the commutator HQ-QH, etc. Equations (III.14) are abstract equations of interest not only in the time dependent ψ basis (Schroedinger) but also in the time independent Heisenberg basis.

B. Classical Case

Converse to the discussion in Section III-A, average properties of a localized single-mode classical wave field can be described in terms of the dynamics of a single "macroquasiparticle." As already discussed in Section II-C, a macroparticle is representative of a localized space–time wave structure whose dynamics is characterized by position, "momentum," and "energy" variables similar to, but distinct from those descriptive of a real particle. From the viewpoint of Section III-A we demonstrate now that if a wavefield is given, a macroparticle description is unique. In contrast to the nonuniqueness of probabilistic wave descriptions of a given real particle dynamics, we thereby again derive the macroparticle dynamics of Section II-C. As above, an overall wave field is decomposed into normalized local mode equations of the form

$$i\frac{\partial \psi}{\partial t} = H\left(\frac{1}{i}\frac{\partial}{\partial x}, x, t, |\psi|\right)\psi = H\psi \tag{III.15}$$

where $\psi = \psi(x, t)$ is a complex wave function and H is in general an x, t dependent nonlinear hermitean operator indicative of the "energy" or frequency of the mode type in question.

For H an Hermitian operator in x-space, local mode equations of the type (III.15) imply that the wave function ψ possesses the property

$$\frac{\partial}{\partial t}(\psi, \psi) = 0 \tag{III.16}$$

where, as in Section III-A, the Hermitian inner product (ψ, ψ) is normalized to unity, with $\psi \to 0$ as $|x| \to \infty$. A time dependent average property $\overline{u}(t)$ of the wave function ψ is obtained from a classical Hermitian operator U via:

$$\overline{u}(t) = (\psi, U\psi). \tag{III.17}$$

If the operator U is both x and t dependent, the time derivative of $\overline{u}$ follows from (III.17) as

$$\frac{d}{dt}\overline{u}(t) = \left(\psi, \frac{HU - UH}{-i}\psi\right) + \left(\psi, \frac{\partial U}{\partial t}\psi\right)$$

whence in view of (III.15) and (III.17)

$$\frac{d}{dt}\overline{u}(t) = i\overline{[H, U]} + \overline{\frac{\partial U}{\partial t}}. \tag{III.18}$$

For prescribed H, equation (III.18) leads to "equations of motion" for the average properties of a localized wavepacket whose exact space–time evolution is described by (III.15).

For example, let $\psi(x, t)$ be described by the linear wave equation

$$i\frac{\partial \psi}{\partial t} = \left(\frac{-1}{2}\frac{\partial^2}{\partial x^2} + V(x)\right)\psi = H\psi \tag{III.19}$$

with the initial condition $\psi = \psi(x, 0)$. Define $k = -i(\partial/\partial x)$ as a wave momentum operator and

$$\overline{x}(t) = (\psi, x\psi)$$

$$\overline{k}(t) = \left(\psi, -i\frac{\partial}{\partial x}\psi\right) \tag{III.20}$$

as the average position and average "momentum" of a macroparticle descriptive of the wavepacket $\psi(x, t)$. The identification of $\overline{k}(t)$ as the average momentum of the wavepacket follows from writing $\psi = a(x, t)\exp[i\theta(x, t)]$, whence $\overline{k}(t) = (\psi, (\partial\theta/\partial x)\psi)$ since $a \to 0$ as $|x| \to \infty$. The trajectory equations and other dynamical properties of the "macroparticle" can be inferred from the following commutator relations between the k operator and an arbitrary function $w = w(x)$ of the operator x, viz.

$$i[k, w] = w'(x) \tag{III.21a}$$

$$i[k^2, w] = kw'(x) + w'(x)k \tag{III.21b}$$

$$i[k^2, kw + wk] = k^2w'(x) + w'(x)k^2 + 2kw'(x)k \tag{III.21c}$$

$$i[kx + xk, w] = 2xw'(x). \tag{III.21d}$$

$w'(x)$ is the x derivative of $w(x)$. Thus from (III.21b) and

(III.21a) with $w = x$ and $w = V(x)$, respectively, one obtains from (III.18), with $H = k^2/2 + V(x)$, the trajectory equations:

$$\frac{d\overline{x}}{dt} = \overline{k}$$

$$\frac{d\overline{k}}{dt} = -\overline{\frac{dV(x)}{dx}} \qquad \text{(III.22)}$$

subject to initial conditions $\overline{x}(0)$, $\overline{k}(0)$ derivable from $\psi(x, 0)$ and (III.20). Similarly, for the higher moment averages

$$\overline{x(t)^2} = (\psi, x^2\psi)$$

$$\overline{k(t)^2} = (\psi, k^2\psi)$$

$$\overline{kx + xk} = (\psi, (kx + xk)\psi)$$

one infers from (III.18) and (III.21) the defining equations

$$\frac{d\overline{x^2}}{dt} = \overline{kx + xk}$$

$$\frac{d\overline{k^2}}{dt} = -\overline{(kV'(x) + V'(x)k)}$$

$$\frac{d}{dt}(\overline{kx + xk}) = 2\overline{k^2} - 2\overline{xV'(x)} \qquad \text{(III.23)}$$

just as in the previous discussion in Section II-C.

As a simple example, let $V(x) = b^2x^2/2$, whence (III.22) becomes

$$\frac{d\overline{x}}{dt} = \overline{k}$$

$$\frac{d\overline{k}}{dt} = -b^2\overline{x}. \qquad \text{(III.24)}$$

IV. Examples

To illustrate the applicability of the quasiparticle view we consider a class of pulsewave problems in uniform linear and nonlinear waveguiding media having transversely varying indices of refraction $n(x, t) = 1 + \delta n(x, t)$ that vary weakly with x and t. The relevant (2-mode) wave equation is chosen as:

$$\left[\partial_z^2 + \partial_x^2 - \left(\frac{n}{c}\right)^2 \partial_t^2\right] \Psi(z, x, t) = 0 \qquad \text{(IV.1)}$$

With a pulsed carrier of wavenumber k_0 and frequency ω_0 the wavefunction Ψ can be expressed as

$$\Psi = \exp\left[i(k_0 z - \omega_0 t)\right] \psi(z, x, t) \qquad \text{(IV.2)}$$

where the complex modulation amplitude ψ is weakly dependent on z, x, t. On substitution of (IV.2) into (IV.1), assuming that $\partial_t\psi \ll -i\omega_0\psi$, $\partial_z\psi \ll ik_0\psi$, $\delta n \ll 1$, and introducing new coordinates via the normalizations

$$\frac{ct}{k_0 d^2} \to t$$

$$\frac{z}{k_0 d^2} \to z$$

$$\frac{x}{d} \to x$$

the equation defining $\psi(z, x, t)$ becomes

$$i(\partial_t + \partial_z)\psi + \tfrac{1}{2}\partial_x^2\psi + k_0^2 d^2 \delta n(x, t)\psi = 0$$

where d characterizes the scale of the field variations in the transverse x-direction. Finally, on shifting by means of the coordinate change $z - t \to z$ to a reference frame moving along z with unit velocity, one obtains the z-independent single mode type, "forward scattering", equation

$$i\,\partial_t\psi = \left[-\tfrac{1}{2}\partial_x^2 - k_0^2 d^2 \delta n(x, t)\right]\psi. \qquad \text{(IV.3a)}$$

As an initial condition we choose for analytical purposes a pulse of the simple form

$$\psi(z, x, 0) = \exp(-x^2/2)\, s(z) \qquad \text{(IV.3b)}$$

and observe that the x-dependence can be chosen quite arbitrarily in numerical computations; the quasiparticle technique becomes particularly useful when there is a relatively rapidly varying x-dependent phase. For linear media with z-independent refractive index the arbitrary pulse shape $s(z)$ assures a similar field variation along z and hence this z dependence will be elided. For nonlinear media wherein the index depends nonlinearly on the amplitude $|\psi|$, the discussion will be restricted to the case $|s(z)| = 1$. As previously observed, on appropriate replacement of t by z the pulse problem in (IV.3) becomes identical to that for a steady state "beam" propagating in a medium of refractive index $\delta n(x, z)$ (whence $s(z) \to s(t) \to \exp(-i\omega t)$), with the initial beam shape given by (IV.3b).

It is evident that (IV.3a) is of the type (II.1) with the operator

$$H = -\tfrac{1}{2}\partial_x^2 + V(x, t, |\psi|)$$

$$V(x, t) = -k_0^2 d^2 \delta n(x, t)$$

Four related cases with progressively increasing analytical difficulty will be discussed, viz.

$$V(x, t) = b^2x^2/2 \qquad \text{(IV.4a)}$$

$$= -(bB)^2 \exp\left[-(x/B)^2\right]/2 \qquad \text{(IV.4b)}$$

$$= b^2x^2\frac{(1 + wt)}{2} \qquad \text{(IV.4c)}$$

$$= -b^2|\psi|^2/2 \qquad \text{(IV.4d)}$$

where in (IV.4d) the nonlinear potential $V(x, t)$ should be written as a functional $V\{|\psi|\}$. The first two t-independent cases a) and b) are described by quasiparticles of "energy" $\omega(k, x) = (k^2/2 + V(x))$ whose position $x = x(t)$ and momentum $k = k(t)$ oscillate with time in the potential well $V(x)$. For $b = 2$ the potentials are depicted in curves a) and b) of Figs. 1(a) and (b) for both $B = 1$ and 2; it should be noted that constant differences in potential do not affect the quasiparticle dynamics. The time dependent nonlinear potential in (IV.4d) coincides with the potential (IV.4b) only at $t = 0$, as follows from IV.3b). As a preliminary observation, it is seen that the potential (IV.4a), which incidentally is physically realizable only for small x, traps all of the quasiparticles, whereas only some are trapped in the potentials (IV.4b) and (IV.4d). For sufficiently small x the potentials in cases (IV.4a) and (IV.4b) become identical and hence imply the same quasiparticle dynamics for small x.

Case a) $\omega = (k^2 + b^2x^2)/2$

This well known case of a "particle" in a quadratic well potential can be readily solved analytically since the quasiparticle trajectory equations (II.9) become

$$\frac{dx}{dt} = k$$

$$\frac{dk}{dt} = -b^2x \qquad \text{(IV.5)}$$

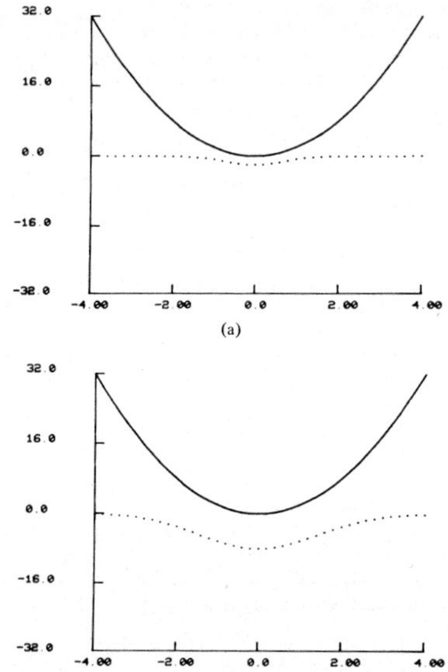

Fig. 1. Plots of potentials $V(x) = (bx)^2/2$ (solid line) and $V(x) = -(b/B)^2/2 \exp[-(x/B)^2]$ (dotted line). (a) $b = 2, B = 1$. (b) $b = 2, B = 2$.

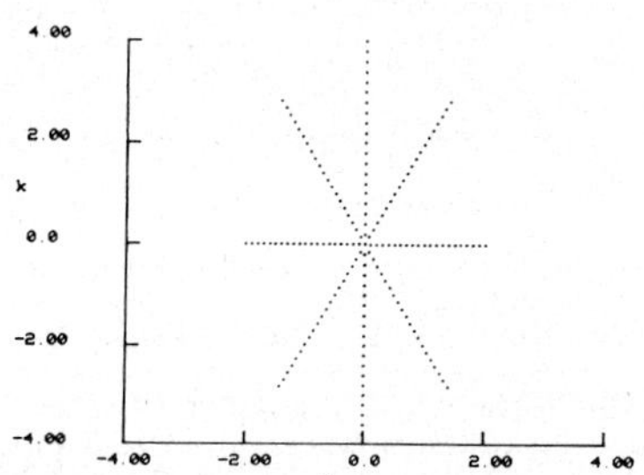

Fig. 2. Quasiparticle trajectory in k, x phase space $\omega = (k^2 + 4^2x^2)/2$. $0 < t < 1.57$.

and yield the analytic solutions:

$$x(t) = x_0 \cos bt + \frac{k_0}{b} \sin bt$$

$$k(t) = k_0 \cos bt - x_0 b \sin bt \qquad \text{(IV.6a)}$$

corresponding to the initial quasiparticle coordinates $x(0) = x_0$ and $k(0) = k_0$. For $b = 2$, Fig. 2 depicts the locations of a group of quasiparticles initially located at $x_0 = 0$ with momenta k_0 evenly distributed along the interval $-4 < k < +4$ at times $bt = m(\pi/4)$ $(m = 1, 2, 3, 4)$. The periodicity with time period $bt = \pi$ of these locations implies "trapping" of all quasiparticles by the square well potential; the individual quasiparticle trajectories are seen to be elliptical and they imply a periodic spreading and contraction of the quasiparticle positions and momenta in

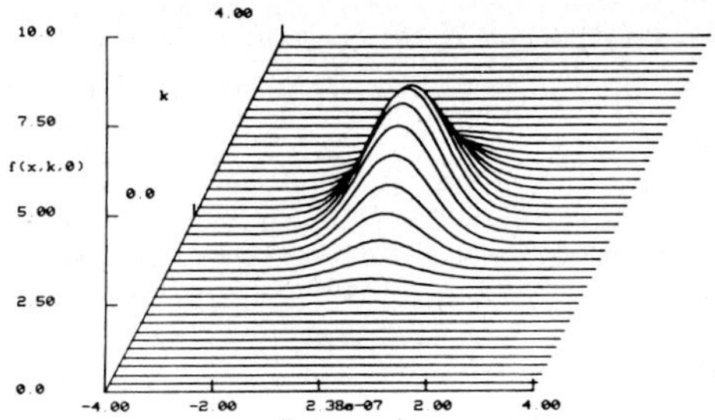

(a)

Fig. 3. (a) Initial quasiparticle space density $F(k, x, 0) = \sqrt{4\pi} \exp[-(x^2 + k^2)]$.

phase space. On solving for x_0 and k_0, one finds from (IV.6a), or directly from (IV.5) that

$$x_0 = x \cos bt - \frac{k}{b} \sin bt$$

$$k_0 = k \cos bt + x\, b \sin bt \qquad \text{(IV.6b)}$$

define the position $x_0(k, x, t)$ and momentum $k_0(k, x, t)$ at time $t = 0$ of a quasiparticle which at t has the phase space coordinates x, k; as noted in (II.11) this information is required for the updating of the quasiparticle distribution function. In view of the initial condition (IV.3b), the quasiparticle distribution function at $t = 0$ follows from the transform relation (II.4a) as

$$F(k, x, 0) = \sqrt{(4\pi)} \exp[-(x^2 + k^2)] \qquad \text{(IV.7a)}$$

and is displayed in the k, x phase space plot of Fig. 3(a). At a subsequent time t the distribution function follows from (II.11) and (IV.6b) as

$$F(k, x, t) = \sqrt{(4\pi)} \exp\left\{-\left[\left(x \cos bt - \frac{k}{b} \sin bt\right)^2 + (k \cos bt + xb \sin bt)^2\right]\right\} \qquad \text{(IV.7b)}$$

The overall distribution in phase space evidently does not vary with time if $b = 1$; for this special case,

$$F(k, x, t) = F(k, x, 0) \qquad \text{(IV.7c)}$$

which implies that when $b \neq 1$ the initial state $\psi(x, 0) = \exp(-x^2/2)$ is an oscillatory eigenmode of the medium. More generally any other initial exponential type distribution $F(k, x, 0)$ that is a function of $x^2 + k^2$ will not change with time and hence will be representative of a higher eigenmode. To illustrate the time evolution of $F(k, x, t)$ in (IV.7b) for $b \neq 2$, plots of $F(k, x, t)$ in the k, x phase space are shown in Figs. 3(b)–(e) at times $bt = (\pi/4, \pi/2, 3\pi/4), \pi$. The cycle period $bt = \pi$ is apparent. The time dependent shape in k, x phase space of the overall system of quasiparticles, although evidently following from the individual quasiparticle trajectories depicted in Fig. 2, displays macroscopic features that are distinctly different from the microdynamics of the individual quasiparticles. Alternatively and more generally stated, overall transport properties (amplitude, phase, etc.) of a wave complex are related to but are not readily inferred from the individual rays that compose the complex. Ordinary ray calculations of

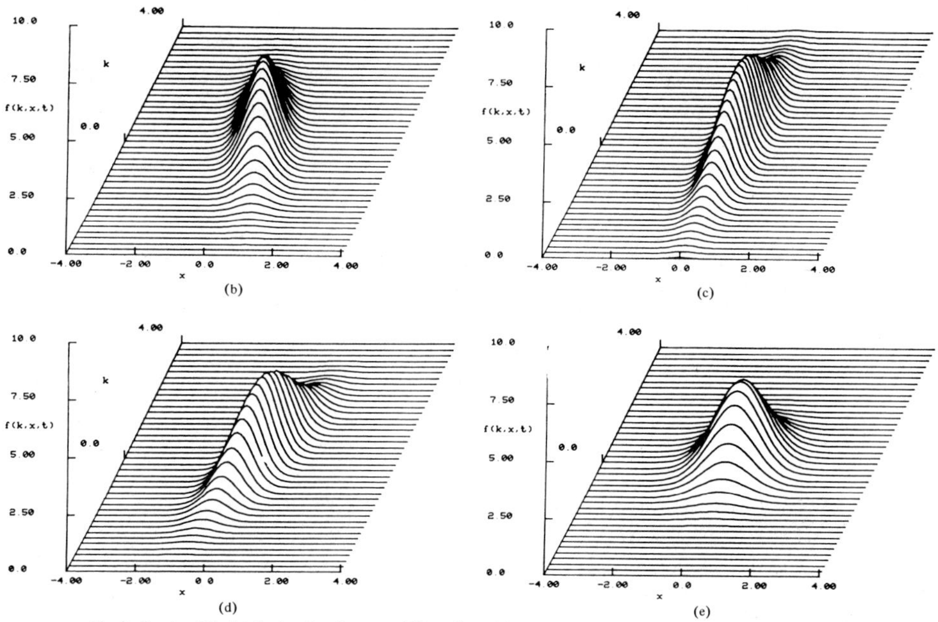

Fig. 3. Quasiparticle distribution function. $\omega = (k^2 + 4x^2)/2$. (b) $t = 0.393$. (c) $t = 0.785$. (d) $t = 1.18$. (e) $t = 1.57$.

transport properties encounter difficulties at ray turning points (caustics), a difficulty not present in the quasiparticle description. With the knowledge of the kinetic distribution $F(k, x, t)$ from (IV.7b) one can calculate, via (II.13) and (II.15), "fluid dynamic" wave properties which represent on the average the space–time dependent amplitude $a(x, t)$ and phase $\theta(x, t)$ of the propagating wavepacket. One thereby obtains (note from (II.14) that $\theta = \int (Kdx - \Omega dt)$)

$$a^2(x,t) = \frac{1}{D}\exp\left[-\left(\frac{x}{D}\right)^2\right]$$

$$K(x,t) = -\left(\frac{x}{2D^2}\right)\left(b - \frac{1}{b}\right)\sin 2bt$$

$$\Omega(x,t) = 0.5\left\{\left(\frac{x}{2D^2}\right)^2\left[\left(b - \frac{1}{b}\right)^2 \sin^2 2bt - 4\right] + b^2x^2 + D^{-2}\right\} \qquad \text{(IV.8a)}$$

where

$$D^2 = 0.5\left[\left(1 + \frac{1}{b^2}\right) + \left(1 - \frac{1}{b^2}\right)\cos 2bt\right].$$

The square law dependence of the phase on x^2 implies periodic "focusing and defocusing" of the wavepacket with time. This behavior is illustrated in the amplitude $a(x, t)$ and in the $a^2K(x, t)$ plots in Fig. 4 which are calculated from $F(k, x, t)$ of (IV.7b) by means of the relations (II.13) and (II.15). In the limiting case $b = 0$, corresponding to free-space propagation in the "forward scattering" paraxial approximation, (IV.8a) reduce to

$$D^2 = 1 + t^2$$

$$K(x,t) = xt/D^2$$

$$\Omega(x,t) = [x^2(t^2 - 1) + D^2]/2D^4. \qquad \text{(IV.8b)}$$

For the eigenmode case $b = 1$, (IV.8) yield the familar results.

$$a^2(x,t) = \exp(-x^2)$$

$$K(x,t) = 0$$

$$\Omega(x,t) = \tfrac{1}{2} \qquad \text{(IV.8c)}$$

the constancy of Ω in this case identifies $\frac{1}{2}$ as the lowest oscillatory eigenvalue of the medium. Incidentally, it should be noted in this case that since the operator H is only of the second degree in both ∂_x and x, the quasiparticle approximation and hence (IV.8) are exact solutions of (IV.3a). As an added comment, it should be observed that one can employ the "fluid" equations (II.22) and (II.23) to evaluate the above "fluid dynamics" properties of the wavepacket; however, the kinetic evaluation via (II.13) and (II.15) is more direct and usually the simpler procedure, particularly when as in this case $F(k, x, t)$ is known explicity. The macroparticle description of Section II-C characterizes the wavepacket evolution in terms of averages that can be obtained either directly from the kinetic distribution function F by integration over k, x phase space as in (II.25) and (II.27), or equivalently from the distribution F as given in (IV.7b). There follows from (II.25)

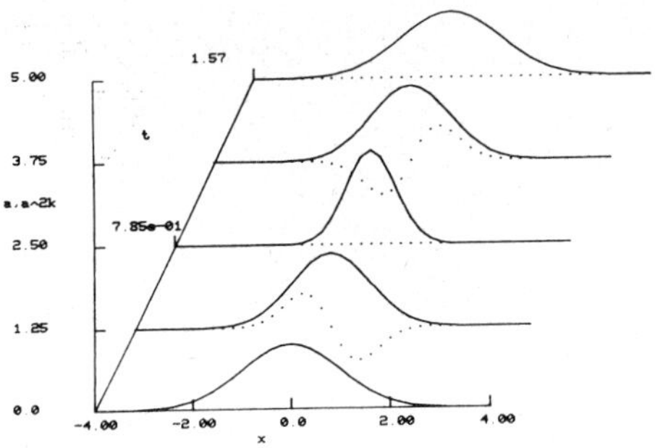

Fig. 4. Wave amplitude and wavenumber versus x for $0 < t < 1.57$. $\omega = (k^2 + 4x^2)/2$.

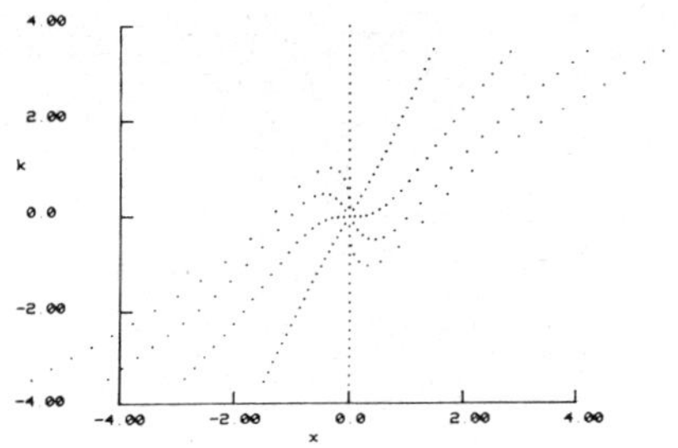

Fig. 5. Quasiparticle trajectory in k, x phase space $\omega = (k^2 - 4x^2 \exp(-x^2))/2$, $0 < t < 1.57$.

and (II.27):

$$\overline{x}(t) = 0$$

$$\overline{k}(t) = 0$$

$$\overline{x^2(t)} = [1 + (1/b^2) + (1 - (1/b^2)) \cos 2bt]/4$$

$$\overline{k^2(t)} = [1 + b^2 + (1 - b^2) \cos 2bt]/4 \qquad \text{(IV.9)}$$

for the average position and mean square width of the macroparticle, as well as for its average momentum and mean square momentum width, all as a function of time. The evident periodic variability of the mean-square macroparticle width provides quick rough measures of the propagating wavepacket; the special case $b = 1$ yields the values $\overline{x^2(t)} = \overline{k^2(t)} = 0.5$ for the eigenmode, while the values $\overline{x^2(t)} = 0.5\ (1 + t^2)$ and $\overline{k^2(t)} = 0.5$ for $b = 0$ distinguish propagation in free space.

Case b) $\omega = [k^2 - (bB)^2 \exp - (x/B)^2]/2$

This example corresponds to propagation of a wavepacket in a physically realizable uniform medium with an exponential transversely tapered refractive index corresponding to the potential plot in curves b) of Fig. 1. The quasiparticle trajectory equations (II.9) in this case become

$$\frac{dx}{dt} = k$$

$$\frac{dk}{dt} = -b^2 x \exp\left[-\left(\frac{x}{B}\right)^2\right]$$

and cannot be solved simply by analytical means. However a numerical solution is readily obtained. For the $b = 2$ and $B = 1$ case, Fig. 5 displays at time intervals $bt = m(\pi/4)$ with $m = 1, 2, 3, 4$ the successive locations of a group of quasiparticles initially at $x = 0$ with momenta evenly distributed on the interval $-4 < k < +4$. One observes in conformity with the potential plot in curves b) of Fig. 1 that only quasiparticles with small momenta $k < bB$ (negative "energy") are trapped in the well, while those with positive "energy," i.e., momenta $k > bB$, escape. Knowing the individual quasiparticle motion and the quasiparticle distribution (IV.7a) in phase space at $t = 0$, one can numerically update the quasiparticle distribution $F(k, x, t)$ at subsequent times via (II.11). Sequential plots of numerically calculated distributions $F(k, x, t)$ at successive times $bt = m\ \pi/4$ are shown in Figs. 6(a)–(d). Because of the presence of both untrapped and trapped quasiparticles there is no overall periodicity in the time evolution of the distributions. The spreading in phase space of the untrapped quasiparticles with large momenta is apparent; the cyclical contraction and expansion of the distribution in the $x = 0 = k$ vicinity arises from the motion of the quasiparticles with small momenta and evidently resembles similar plots in Figs. 3(a)–(d) of trapped quasiparticle effects. For a deeper potential well with say $B = 2$ (cf. Fig. 1(b)) and the same initial distribution as before, most of the quasiparticles would be trapped in the vicinity of $x = 0 = k$; hence the time evolution of $F(k, x, t)$ would be almost identical to that for the square well potential of case a).

The amplitude $a(x, t)$ and average wavenumber $K(x, t)$ of the wavepacket represented by the above distribution of quasiparticles can be inferred from appropriate numerically evaluated k-integrals of $F(k, x, t)$ in accordance with (II.13) and (II.15a). The solid curves in Fig. 7 display the so-obtained amplitude as a function of x for times t corresponding to the associated $F(k, x, t)$ plots in Figs. 6(a)–(d). The lack of overall periodicity, the spreading amplitude wings, and the cyclical contraction and expansion of the amplitude center originate from the presence of both untrapped and trapped quasiparticles and are in contrast with the plots in Fig. 4 wherein all quasiparticles are trapped. The dotted curves in Fig. 7 depict the quasiparticle current density $a^2(x, t)\ K(x, t)$ as a function of x for the same times t as above. The initial flow of quasiparticle current toward the $x = 0$ center and the subsequent outward flow is cyclical and elucidates the origin of the central amplitude contraction and expansion as due to trapped particle effects. Although the wavenumber $K(x, t)$ is usually a more important characteristic of wavepacket evolution than the current density $a^2(x, t)\ K(x, t)$, it is computationally simpler to calculate the latter directly and the former indirectly.

Case c) $\omega = [k^2 + b^2 x^2 (1 + wt)]/2$

This example of a simple time varying quadratic potential is intended to illustrate the effects of both spatial inhomogeneity and temporal nonstationarity as a preliminary to the consideration of nonlinear wave propagation, which is also characterized by an inhomogeneous and nonstationary potential. The trajectory equations for quasiparticles of the indicated energy $\omega(k, x, t)$ are defined via (II.9) by

$$\frac{dx}{dt} = k$$

$$\frac{dk}{dt} = -b^2 x(1 + wt).$$

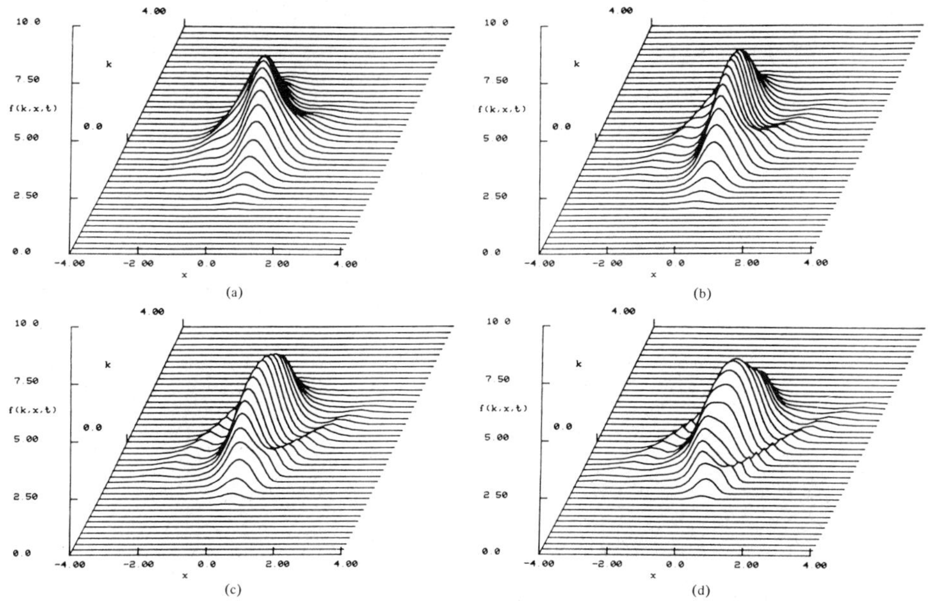

Fig. 6. Quasiparticle phase space density $\omega = (k^2 - 4x^2 \exp(-x^2))/2$. (a) $t = 0.393$. (b) $t = 0.785$. (c) 1.18. (d) 1.57.

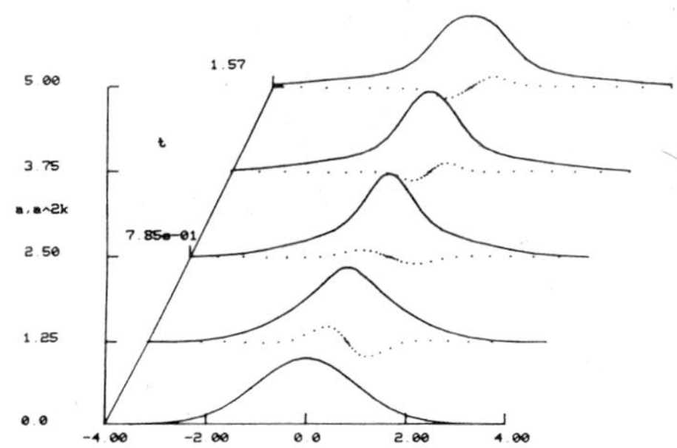

Fig. 7. Wave amplitude and wavenumber versus x for $0 < t < 1.57$. $\omega = (k^2 - 4x^2 \exp(-x^2))/2$.

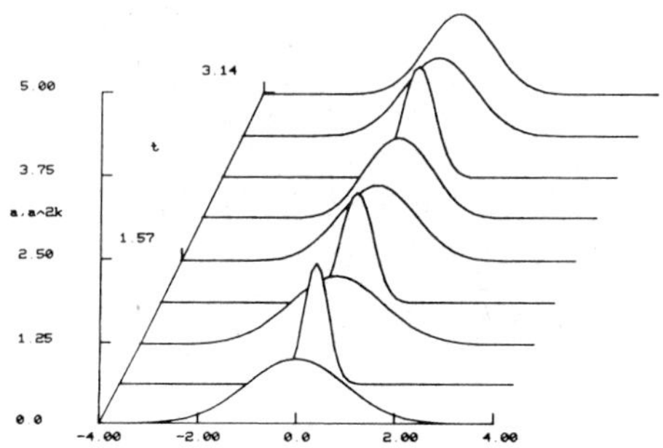

Fig. 8. Wave amplitude and wavenumber versus x for $0 < t < 3.14$. $\omega = [k^2 + (1 + t)\, 4x^2]/2$.

They are readily solved numerically; they permit updating of the kinetic distribution function $F(k, x, t)$ from the initial distribution (IV.7a); concomittantly they determine the space-time evolution of the amplitude and phase characteristics of the propagating wavepacket. Despite the added analytical complexity due to the medium nonstationarity, numerical calculations are performed with the same overall computational program as in the previous examples.

The resulting data for this case of wavepacket propagation in a medium characterized by a time varying quadratic well potential are not significantly different than that for a stationary quadratic well. Within the "forward scattering" constraint that the differential refractive index $\delta n(x, t) \ll 1$, the linear temporal increase of the well depth destroys the periodicity of the plots in Figs. 2–4, and implies focusing and defocusing effects appropriate to an increasing equivalent b^2. These effects are noted in the time evolution of the x-dependent amplitude $a(x, t)$ displayed in the "hidden line" plots of Fig. 8.

Case d) $\omega = [k^2 - b^2 |\psi(x, t)|^2]/2$

This example, descriptive of wavepacket propagation in a dispersive nonlinear medium, has received considerable attention in the literature both numerically [9] and analytically [10] under the heading of the "nonlinear parabolic or Schroedinger equation." The relevant potential is manifestly both inhomogeneous and nonstationary. In the $t = 0$ vicinity the potential follows from the initial condition on $\psi(x, 0)$

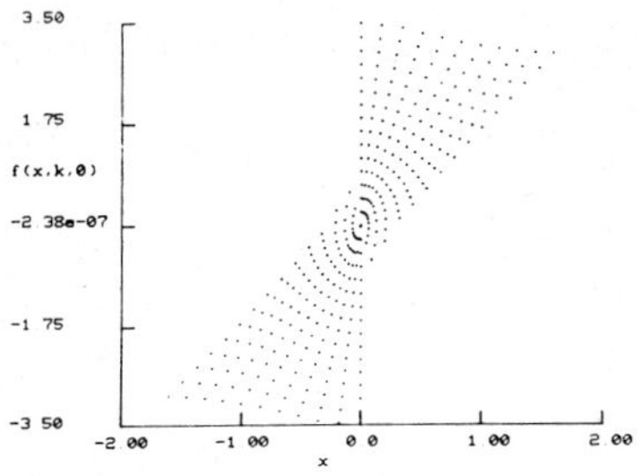

Fig. 9. Quasiparticle trajectory in k, x phase space $\omega = [k^2 - 4a^2(x, t)]/2$, $0 < t < 0.500$.

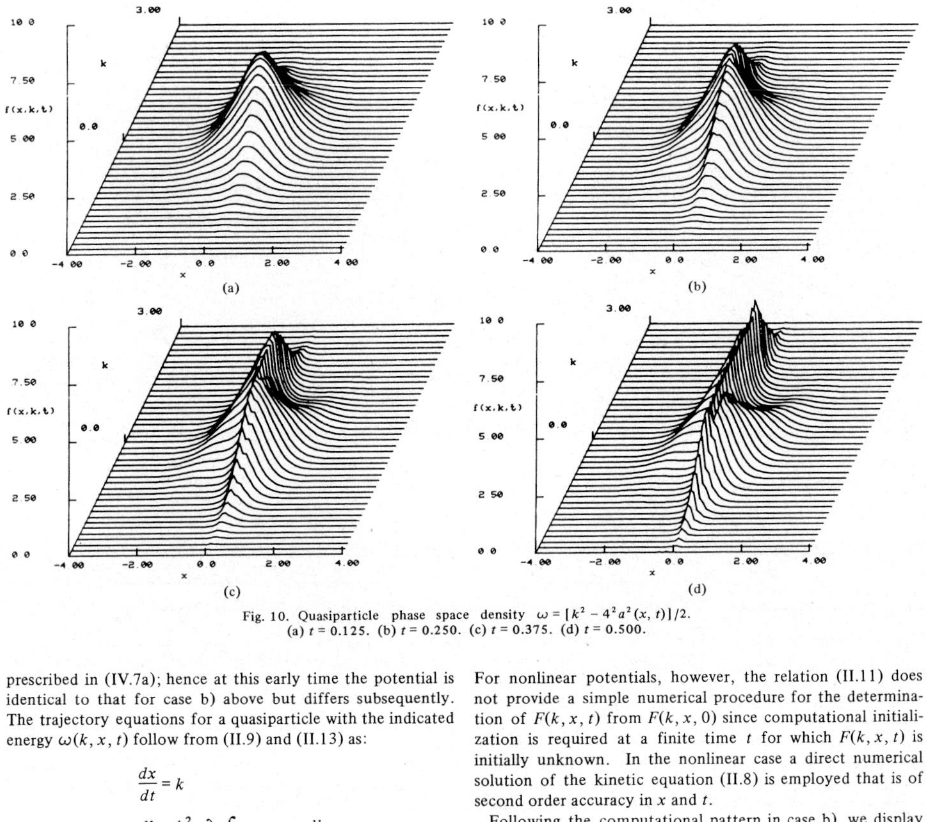

Fig. 10. Quasiparticle phase space density $\omega = [k^2 - 4^2a^2(x, t)]/2$.
(a) $t = 0.125$. (b) $t = 0.250$. (c) $t = 0.375$. (d) $t = 0.500$.

prescribed in (IV.7a); hence at this early time the potential is identical to that for case b) above but differs subsequently. The trajectory equations for a quasiparticle with the indicated energy $\omega(k, x, t)$ follow from (II.9) and (II.13) as:

$$\frac{dx}{dt} = k$$

$$\frac{dk}{dt} = \frac{b^2}{2}\frac{\partial}{\partial x}\int F(k, x, t)\frac{dk}{2\pi}$$

The dependence on $F(k, x, t)$ implies that these equations must be solved self-consistently with the solution of the kinetic equation (II.8). These trajectory and kinetic equations may be solved by numerical computation if $t = 0$ is the initial time. For nonlinear potentials, however, the relation (II.11) does not provide a simple numerical procedure for the determination of $F(k, x, t)$ from $F(k, x, 0)$ since computational initialization is required at a finite time t for which $F(k, x, t)$ is initially unknown. In the nonlinear case a direct numerical solution of the kinetic equation (II.8) is employed that is of second order accuracy in x and t.

Following the computational pattern in case b), we display in Fig. 9 for $b = 2$, at equal spaced time steps ranging from $t = 0$ to $t = 0.5$, the successive locations of quasiparticles initially at $x = 0$ with momenta evenly distributed within the interval $-2 < k < 2$. In contrast to the corresponding trajectory plots in Fig. 5 for the stationary linear case b), The potential, which in this case is time varying and contracting, traps a

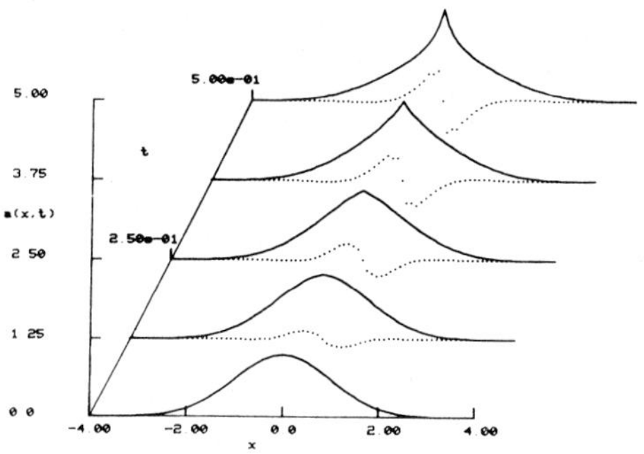

Fig. 11. Wave amplitude and wavenumber for $0 < t < 0.5$. $\omega = [h^2 - 4a^2(x,t)]/2$.

relatively small number of quasiparticles in the $x = 0$ vicinity and permits a large number to escape. Updating of the initial quasiparticle distribution (IV.7a) is effected by direct numerical solution of the kinetic (II.8). Results of such numerical calculations of $F(k,x,t)$ at equally spaced time steps between $t = 0.125$ and $t = 0.5$ are displayed in Figs. 10(a)–(d) for the case $b = 2$. The overall plotting time is shorter than in the corresponding plots of Fig. 6 because of the existence at about $t = 0.5$ of an analytical singularity that limits the validity of the solution to this range. The singularity is apparent in the formation of cliff like structures in the quasiparticle distribution at about $t = 0.5$; they arise because the nonlinear potential gives rise to a negative central "pressure" that drives quasiparticles toward $x = 0$ with the result that negative momentum quasiparticles pile up on the positive side of $x = 0$ while positive momentum quasiparticles pile up on the negative $x = 0$ side. The different contributions of trapped and untrapped quasiparticles to this piling up process can also be inferred from careful following of individual quasiparticle paths in the Fig. 9 trajectory plots. Numerical k-integration of $F(k,x,t)$ in accordance with (II.13) and (II.15) then provide the data for the amplitude $a(x,t)$ (solid lines) and the average current density $a^2(x,t)\,K(x,t)$ (dotted) plots in Fig. 11; the latter illustrate the effects of the piling up at $x = 0$ of the trapped quasiparticles and also the spreading effects due to the untrapped particles.

V. Summary

Analysis of wavepacket propagation through rather arbitrary media has been effected by averaging out fast space-time dependence and concentrating on relatively slowly varying macroscopic variables such as the average amplitude, wavenumber, frequency, etc. The wavepacket is viewed as a system of quasiparticles, each of which moves in phase space along "ray" trajectories characteristic of the mode-type and the medium; the overall configuration of the quasiparticle system evolves in accordance with a kinetic equation whose solution depends on the initial shape of the wavepacket and on the quasiparticle dynamics. The desired macroscopic wavepacket properties are evaluated from appropriate moments of the distribution function. Knowledge of the latter permits both a "fluid dynamic" and a "macroparticle" description of the average properties of a propagating wavepacket. For simplicity the above analysis has been limited to one spatial dimension and to one propagating mode type; the extension to three dimensions is straightforward but mode coupling requires a somewhat greater effort. Examples relating to wave propagation in turbulent media have been omitted because they require the use of renormalization methods; they will be reported elsewhere. It should also be noted that steady state beam propagation can be treated by similar quasiparticle methods on replacement of the time coordinate by the coordinate in the direction of beam propagation.

Appendix A

A "creation" green's function g describing the k, x phase space distribution at time t of a single quasiparticle of energy $\omega(k,x,t)$ created at k', x' at time t' is defined by

$$(\partial_t + \partial_k\omega\,\partial_x - \partial_x\omega\,\partial_k)\,g(k,x,t;k',x',t') = \delta(k-k')\,\delta(x-x')\,\delta(t-t') \qquad (A.1)$$

and subject to the condition that

$$g = 0, \qquad \text{for } t \leqslant t'.$$

At time $t = t'+$ one then infers from (A.1) that

$$g(k,x,t'+;k',x',t') = \delta(k-k')\,\delta(x-x').$$

At any subsequent time $t > t'$ one also observes from (A.1) that on a quasiparticle trajectory which starts at k', x' at time t', the quasiparticle coordinates

$$x = \hat{x}(t;k',x',t')$$
$$k = \hat{k}(t;k',x',t')$$

are defined by

$$\frac{d\hat{x}}{dt} = \partial_k\omega(\hat{k},\hat{x})$$

$$\frac{d\hat{k}}{dt} = -\partial_x\omega(\hat{k},\hat{x}). \qquad (A.2)$$

On this trajectory the green's function is constant, i.e.,

$$\frac{d}{dt}g = 0$$

and hence one can represent g as

$$g(k,x,t;k',x',t') = \delta(k - \hat{k}(t;k',x',t'))\,\delta(x - \hat{x}(t;k',x',t'))\,U(t-t')$$

where $U(t)$ is the unit step function which vanishes for $t < 0$.

An "annihilation" (adjoint) Green's function g^+ similarly describes the k', x' phase space distribution at time t' of a single quasiparticle of energy $\omega(k',x',t')$ annihilated at the point k, x at time t and is defined by

$$-(\partial_{t'} + \partial_{k'}\omega\,\partial_{x'} - \partial_{x'}\omega\,\partial_{k'})\,g^+(k',x',t';k,x,t) = \delta(k'-k)\,\delta(x'-x)\,\delta(t'-t) \qquad (A.3)$$

and subject to $g^+ = 0$ for $t' \geqslant t$, whence one derives that at $t' = t-$

$$g^+(k',x',t-;k,x,t) = \delta(k'-k)\,\delta(x'-x).$$

From (A.1) and (A.3) one deduces, on multiplication of (A.1) by g^+ and (A.3) by g and on subtraction and integration over an appropriate k, x, t space, the adjointness property

$$g^+(k',x',t';k,x,t) = g(k,x,t;k',x',t'). \qquad (A.4)$$

On a trajectory whose coordinates

$$x' = \hat{x}(t'; k, x, t)$$

$$k' = \hat{k}(t'; k, x, t)$$

satisfy the same quasiparticle equations (A.2), but with t replaced by t' and with the requirement that the trajectory ends at

$$\hat{x}(t; k, x, t) = x$$

$$\hat{k}(t; k, x, t) = k$$

one infers from (A.3) that

$$\frac{dg^+}{dt'} = 0.$$

It then follows that

$$g^+(k', x', t'; k, x, t)$$
$$= \delta(k' - \hat{k}(t'; k, x, t))\, \delta(x' - \hat{x}(t'; k, x, t))\, U(t - t') \qquad (A.5)$$

The solution $F(k, x, t)$ of the kinetic equation (II.8) follows from the initial condition $F(k, x, 0)$ and (A.1) as

$$F(k, x, t) = \iiint g(k, x, t; k', x', 0)\, F(k', x', 0)\, dk'\, dx'$$

since the initial condition can be regarded as an impulsive excitation at $t' = 0$. Hence, on use of the adjointness relation (A.4) and (A.5), one obtains

$$F(k, x, t) = F(\hat{k}(0; k, x, t), \hat{x}(0; k, x, t), 0) \qquad (A6)$$

which represents the desired quasiparticle distribution function in terms of the initial distribution and the phase space coordinates at $t' = 0$ of a quasiparticle which at time t has the coordinates k, x. In stationary linear media it is to be observed that the desired quasiparticle coordinates can be represented as

$$\hat{k}(0; k, x, t) = \hat{k}(-t; k, x, 0) \equiv k_0(k, x, t) \qquad (A.7a)$$

$$\hat{x}(0; k, x, t) = \hat{x}(-t; k, x, 0) \equiv x_0(k, x, t) \qquad (A.7b)$$

which represent the coordinates of a quasiparticle at time $-t$ which passes through the point k, x at zero time; and are relatively simple to obtain numerically. Equations (A.6) and (A.7) then lead to the result quoted in (II.11).

Acknowledgment

The author expresses his appreciation to Dr. S. Barone and to David Ming-Whei Wu for their criticism and help in preparation of this manuscript; the careful reading and comments of an unknown reviewer should also be noted.

References

[1] E. Wigner, "On the quantum correction for thermodynamic equilibrium," *Phys. Rev.*, vol. 40, June 1932.

[2] G. B. Whitham, *Linear and Nonlinear Waves.* New York: Wiley-Interscience, 1974.

[3] V. I. Karpman and E. M. Krushkal, "Modulated waves in nonlinear dispersive media," *Sov. Phys.–JETP*, vol. 28, no. 2, Feb. 1969.

[4] B. B. Kadomtsev and V. I. Karpman, "Nonlinear waves," *Sov. Phys. Uspekhi*, vol. 14, no. 1, July 1971.
V. N. Tsytovitch, *Nonlinear Effects in Plasma.* New York: Plenum Press, 1970.
V. I. Karpman, *Nonlinear Waves in Dispersive Media.* London, England: Pergamon Press, 1975.

[5] S. N. Vlasov, V. A. Petrishchev, and V. I. Talanov, "Averaged description of wave beams · · ·," *Radiofiz.*, vol. 14, no. 9, Sept. 1971.

[6] M. C. Newstein and D. Ramakrishnan, "The averaged description of the propagation of optical beams," Polytech. Inst. New York, Tech. Rep.

[7] L. B. Felsen and N. Marcuvitz, *Radiation and Scattering of Waves.* Englewood Cliffs, NJ: Prentice-Hall, 1973.

[8] I. M. Besieris, "The wave-kinetic method · · ·," Virginia Polytechnic Institute, rep. VPI-E-79-5, 1979.
H. Bremmer, "General remarks concerning theories · · · in random media," *Radio Sci.*, vol. 8, no. 6, pp. 511–534, June 1973.
M. J. Bastiaans, "Transport equation for the Wigner distribution function," *Opt. Acta*, vol. 26, no. 10, pp. 1265–1272, 1979.

[9] V. E. Zakharov, V. V. Sobolev, and V. S. Synakh, "Behavior of light beams in nonlinear media," *Sov. Phys.–JETP*, vol. 33, no. 1, July 1971.
V. E. Zakharov and V. S. Synakh, "The nature of the self-focusing singularity," *Sov. Phys.–JETP*, vol. 41, no. 3, 1975.

[10] V. E. Zakharov and A. B. Shabat, "Exact theory · · · of nonlinear waves," *Sov. Phys.–JETP*, vol. 34, pp. 62–69, 1972.

Nonlinear and Environmental Electromagnetics, edited by H. Kikuchi
Elsevier Science Publishers B.V., Amsterdam, 1985 – Printed in The Netherlands

NONLINEAR ELECTROMAGNETICS IN TERMS OF QUASI-PARTICLES AND SOLITONS AND ITS APPLICATION TO NONLINEAR DISPERSIVE AND DISSIPATIVE MEDIA

H. Kikuchi and M. Hirota
Nihon University, College of Science and Technology, Tokyo 101, Japan

As a unified approach to nonlinear electromagnetics in a broad sense, covering nonlinear wave packets, pulses, trains and/or charged particles, a new quasi-particle description is introduced in terms of its position, momentum, energy and phase space density. The first three variables are governed by the dynamical equations analogous to Hamilton's canonical equations of analytical dynamics and the phase space density is described by the kinetic equation. This has been originated by Professor Marcuvitz for wave propagation in very general media, inhomogeneous, time-varying, linear, nonlinear or turbulent. This concept makes possible a unified particle view of nonlinear waves in the Lagrangian system of coordinates and provides them with a new physical insight of the particle-wave duality. The method is applied to nonlinear, dispersive and weakly dissipative media with several examples described by the Korteweg-de Vries-Burgers and nonlinear Schrödinger equations with reference to charged-particle beams and space charge waves, distributed parameter lines, electrodynamics in the earth's plasmasphere, modulated waves, etc. While both stationary and nonstationary solutions are considered, temporal and spatial evolution of waves are investigated numerically for the latter case as an initial value problem by using the Fourier analysis and a digital computer and are interpreted in terms of quasi-particles. These numerical results ensure analytical but approximate new solutions obtained otherwise in fairly good agreement.

1. INTRODUCTION

Recently there has been a renewed interest in nonlinear phenomena in many different branches of physics and engineering, e.g. in applied mathematics, acoustics, hydrodynamics, solid state physics, plasma physics, radio physics, optics, etc., although each of these fields has its own specific problems, methods and results. While a number of different approaches were devised in the past for individual nonlinear phenomena, attempts to form a unified or clear view of nonlinear waves and oscillations have rather recently been made only for the last one and a half decades. One systemstic approach along this line is a generalization of so-called "simple waves" to a dispersive or dissipative medium, leading to the Korteweg-de Vries (KdV) or Burgers equation, respectively (Kadomtsev and Karpman, 1971; Karpman, 1975). Another approach is the so-called adiabatic theory of nonlinear waves as formulated in the most general form by Whitham (1974), leading to the nonlinear parabolic equation (often referred to as the nonlinear Schrödinger equation) when the wave amplitude and phase vary slowly in space and time. The purpose of the present paper is to introduce a new quasi-particle concept as a unified approach to nonlinear electromagnetics and plasmas with particular reference to the field of environmental electromagnetics.

Conventional electromagnetics usually restricts itself to wave propagation in linear media on the basis of Maxwell's equations. For a large amplitude wave or pulse with short duration, propagation media or transmission lines often exhibit nonlinear characters leading to a variety of novel phenomena. When charged particles are involved, as usually referred to as electrodynamics or plasma dynamics, the Lorentz force term inherently introduces nonlinear effects. In order to deal with these nonlinear effects systematically, covering complex wave packets, pulses, wavetrains and/or charged-particle beams, a quasi-particle description familiar in quantum statistics is introduced in terms of its position, momentum, energy and phase space density. The first three variables are governed by the dynamical equations analogous to Hamilton's canonical equations of analytical dynamics and the phase space density is described by the kinetic equation. This has been initiated by Professor Marcuvitz (1970, 1980, this issue) for wave propagation in very general media, inhomogeneous, time-varying, linear, nonlinear or turbulent. This concept makes possible a unified view of nonlinear waves with a new physical insight of the particle-wave duality.

To effect this, the Lagrangian system of coordinates, which follow the zero-order trajectory of the particles, is employed for wave-packets accompanying soliton generation in nonlinear, dispersive and weakly dissipative media described by the Korteweg-de Vries-Burgers and nonlinear Schrödinger equations with numerical and analytical solutions. Then, the dispersive and dissipative terms are considered as the perturbation to the nonlinear background composed of real or equivalent particles. Both stationary and nonstationary solutions are considered with several examples, i.e. charged-particle beams and space charge waves, distributed parameter lines, electrodynamics in the earth's plasmasphere, modulated waves, etc. For the latter case, temporal and spatial evolution of waves are investigated numerically as an initial value problem by using the Fourier analysis and a digital computer. The results are interpreted in terms of quasi-particles, and are compared with analytical but approximate solutions.

For the K-dV-Burgers equation, the initial condition is chosen to be sinusoidal. Then, its temporal evolution is decomposed into three stages, i.e. initial, transition and quasi-stationary. In the initial stage, the nonlinear term is dominant with negligible dispersion and dissipation, forming a so-called simple wave and causing wave steepening and breaking. In the transition period, dispersion is growing rapidly, producing quasi-particles or solitons for sufficiently small dissipation. For a dissipation-dominated case, however, a soliton-like structure rapidly damps out after wave breaking and tends to a shock-like structure. The third stage represents the quasi-stationary state where solitons or shock formation has been completed, since the nonlinear term nearly balances the dispersion or dissipation term. Such a temporal evolution is reversed after a half of the recurrence time and the waveform is going back to the initial state through almost completely inverse processes. Then, the recurrence time decreases with increasing dispersion. The number, amplitudes and recurrence time of solitons in a non-dissipative system are obtained approximately from the eigenvalues of the Schrödinger equation with a potential well equal to the initial value of the K-dV equation. These analytical solutions are in fairly good agreement with those observed from numerical results.

For the nonlinear Schrödinger equation, the initial condition is chosen to be a modulated wave. In the strongly nonlinear regime where the nonlinear coefficient exceeds a certain critical value and where waves are modulationally unstable, modulation is growing until a half of the recurrence time, forming an envelope soliton with conversion of amplitude into phase modulation and then the envelope soliton is eventually going back to the initial state through inverse processes. The recurrence time increases with decreasing nonlinearities, tending to infinity in the marginal state. In the weakly nonlinear regime below the critical value, the wave amplitude tends to decrease with time until a half of the recurrence time, while the recurrence time decreases with decreasing nonlinearities. Except for very strong nonlinearities, it is generally sufficient to consider the carrier and the first side-band waves in the spectrum, making an analytical treatment feasible. The temporal development of the side-band wave amplitudes and the recurrence time are expressed approximately in terms of Jacobi's elliptic function and elliptic integral of the first kind, respectively. The agreement between analytical and numerical solutions is very good except for the immediate vicinity of the marginal state.

2. QUASI-PARTICLE DESCRIPTION OF NONLINEAR DISPERSIVE AND DISSIPATIVE MEDIA

We apply a quasi-particle description to nonlinear dispersive and disspative media specifically to elucidate the complex behaviour of temporal and spatial evolution of nonlinear waves. In general, a nonlinear dispersive system is decomposed into the zero-order structure of a nonlinear background and the first order of dispersive perturbation. Introducing a unified particle view, the former is characterized by real or equivalent particles and the latter is described in terms of wave-packets or quasi-particles. When a dissipation is involved in the system, this is regarded as a result of collisions between real and quasi-particles and among themselves. Then, a whole system can be described in terms of mutual interactions between real and quasi-particles. Figure 1 represents its conceptual diagram. With the introduction of a method of analytical dynam-

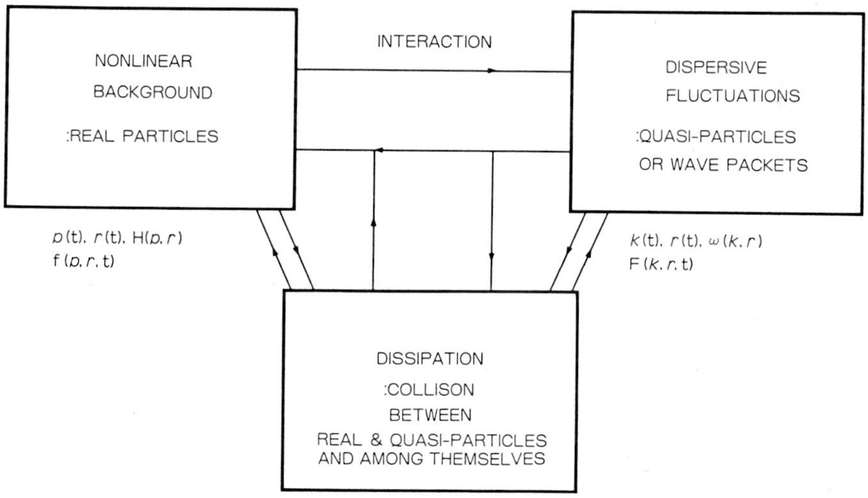

Fig.1. Decomposition of Nonlinear Dipersive and Dissipative System

ics, the problem of nonlinear waves in a dispersive and dissipative medium can be reduced to the determination of the trajectories and densities of real and quasi-particles in appropriate phase spaces. Since their trajectories and densities are described by Hamilton's canonical equations and the kinetic equations respectively, our task is to solve these equations. Thus, partial differential equations composed of nonlinear, dispersive and dissipative terms are interpreted in physical terms as the interaction accompanied by collisions between particles and quasi-particles and among themselves. Particularly, the production and annihilation of quasi-particles or solitons are governed by the kinetic equation for their distribution in phase space (Marcuvitz, 1970, 1980, this issue).

The nonlinear background and dispersive perturbation or fluctuations are expressed by a dynamical system described in terms of the momenta, positions, hamiltonians and phase space densities of real and quasi-particles at time t, $\underline{P}(t)$; $\underline{k}(t)$, $\underline{r}(t)$, $H(\underline{p},\underline{r})$; $\omega(\underline{k},\underline{r})$, $f(\underline{p},\underline{r},t)$; $F(\underline{k},\underline{r},t)$. Then, the trajectory equations, kinetic equations, total particle numbers, energies, and momenta are written respectively for real particles and for quasi-particles as follows:

Real Particles (Nonlinear Background):

$$\frac{d\underline{r}}{dt} = \nabla_p H(\underline{p},\underline{r}), \qquad \frac{d\underline{p}}{dt} = -\nabla H(\underline{p},\underline{r}) \tag{1}$$

$$\left(\frac{\partial}{\partial t} + \nabla_p H\cdot\nabla - \nabla H\cdot\nabla_p\right) f = \frac{\delta f}{\delta t} \tag{2}$$

$$\int f(\underline{p},\underline{r},t)\,d\underline{p}\,d\underline{r}, \quad \int H(\underline{p},\underline{r}) f(\underline{p},\underline{r},t)\,d\underline{p}\,d\underline{r}, \quad \int \underline{p} f(\underline{p},\underline{r},t)\,d\underline{p}\,d\underline{r}, \tag{3}$$

Quasi-Particles (Dispersive Fluctuations):

$$\frac{d\underline{r}}{dt} = \nabla_{\underline{k}}\omega(\underline{k},\underline{r}), \qquad \frac{d\underline{k}}{dt} = -\nabla\omega(\underline{k},\underline{r}) \tag{4}$$

$$\left(\frac{\partial}{\partial t} + \nabla_k\omega\cdot\nabla - \nabla\omega\cdot\nabla_k\right) F = \frac{\delta F}{\delta t} \tag{5}$$

$$\frac{1}{(2\pi)^3}\int F(\underline{k},\underline{r},t)\,d\underline{k}\,d\underline{r}, \quad \frac{1}{(2\pi)^3}\int \omega F(\underline{k},\underline{r},t)\,d\underline{k}\,d\underline{r}, \quad \frac{1}{(2\pi)^3}\int \underline{k} F(\underline{k},\underline{r},t)\,d\underline{k}\,d\underline{r}, \tag{6}$$

where $\delta f/\delta t$ is a "collision" term, representative of the coupling of background particles among themselves and to quasi-particles, while $\delta F/\delta t$ is a collision term, representative of the coupling of quasi-particles among themselves and to background particles and represents the production and annihilation of quasi-particles. In a collisionless case, the background or quasi-particle energy, H or ω is kept constant on a collisionless trajectory.

We are now ready for application of this method to nonlinear dispersive and dissipative media as described by the Korteweg-de Vries-Burgers and nonlinear Schrödinger equations.

3. SIMPLE WAVES - CHARGED-PARTICLE BEAMS AND SPACE CHARGE WAVES

Consider a beam of noninteracting charged particles whose velocity v satisfies the relation

$$\frac{dv}{dt} = \frac{\partial v}{\partial t} + v\frac{\partial v}{\partial z} = 0 \tag{7}$$

where z is the coordinate in the rest frame of reference. Introducing the Lagrangian system of coordinates x which move with a constant zero-order velocity v_0 of the beam, i.e.

$$x = z - v_0 t \tag{8}$$

and decomposing the beam velocity v into the d.c. velocity v_0 and the perturbation u around it, we write

$$v(z,t) = v_0 + u(x,t) = v_0 + u(z-v_0t,t). \tag{9}$$

Substituting this into Eq.(7), we again arrive at the same relation for the velocity perturbation

$$\frac{du}{dt} = \frac{\partial u}{\partial t} + u\frac{\partial u}{\partial x} = 0 \tag{10}$$

This represents the simplest form of nonlinear waves as elucidated below and is termed "simple waves" (Kadomtsev and Karpman, 1971).

Under the initial condition

$$u(x,0) = u_0(x), \tag{11}$$

the solution of Eq.(10) can be written as

$$u(x(t),t) = u(x(0),0) = u(x_0,0) = u_0(x_0). \tag{12}$$

Then, the trajectory x = x(t) of each particle in the Lagrangian system of coordinates satisfies

$$\frac{dx}{dt} = u = u_0(x_0) \tag{13}$$

Integrating this, the trajectory equation becomes

$$x = u_0(x_0)t + x_0 \tag{14}$$

and then Eq.(12) can be rewritten as

$$\begin{aligned} u(x(t),t) &= u_0(x_0) = u_0(x - u_0(x_0)t) \\ &= u_0(x - u[x(t),t]t) \end{aligned} \tag{15}$$

This indicates that the initial perturbation moves with a velocity equal to its amplitude; the greater its amplitude is, the faster its speed is. The waveform is thus distorted—the particles with $u > 0$ run ahead, and those with $u < 0$ lag behind (Fig. 2a). From Eq.(15), we have

$$u_x = \frac{u_0'}{1 + tu_0'} , \qquad u_t = -\frac{u_0 u_0'}{1 + tu_0'} . \tag{16}$$

Thus, the region of positive velocity gradient is flattened with time but that of negative gradient is steepened (Fig. 2a), reaching $u_x = \mp \infty$ at the breaking time t_B and coordinate x_{0B} which are obtained from the denominator of Eqs.(16) = 0 and $(-u_0')_{max}$, (Fig. 3a), i.e.

$$u_0''(x_{0B}) = 0, \qquad t_B = -\frac{1}{u_0'(x_{0B})} . \tag{17}$$

For example, consider

$$u_0(x_0) = \cos \pi x_0 \tag{18}$$

Then, substituting the derivatives

$$u_0' = -\pi \sin \pi x_0 , \qquad u_0'' = -\pi^2 \cos \pi x_0 \tag{19}$$

into Eqs.(17), we have

$$x_{0B} = \frac{1}{2} , \qquad t_B = \frac{1}{\pi} . \tag{20}$$

We apply these results to electron beams. Then, the charge and current densities ρ, i are written as

$$\rho = -Ne , \qquad i = \rho v = -Ne(v_0 + u) , \tag{21}$$

where N is the electron density. Substituting Eqs.(21) into the equation of continuity

$$\frac{\partial i}{\partial z} + \frac{\partial \rho}{\partial t} = 0 \tag{22}$$

and referring to Eq.(8), we have

$$\frac{\partial N}{\partial t} + \frac{\partial}{\partial z}(Nv) = \frac{\partial N}{\partial t} + \frac{\partial}{\partial x}(Nu) = \frac{\partial N}{\partial t} + u\frac{\partial N}{\partial x} + N\frac{\partial u}{\partial x} = 0 . \tag{23}$$

To solve Eq.(23) with respect to N, we recall the relation, referring to Eqs. (12), (13) and (16),

$$\frac{dN}{dt} = \frac{\partial N}{\partial t} + \frac{dx}{dt}\frac{\partial N}{\partial x} = \frac{\partial N}{\partial t} + u\frac{\partial N}{\partial x} = -N\frac{\partial u}{\partial x} = -\frac{Nu_0'(x_0)}{1 + tu_0'(x_0)} . \tag{24}$$

Integrating this, we obtain finally

$$N = \frac{N_0}{1 + tu_0'} . \tag{25}$$

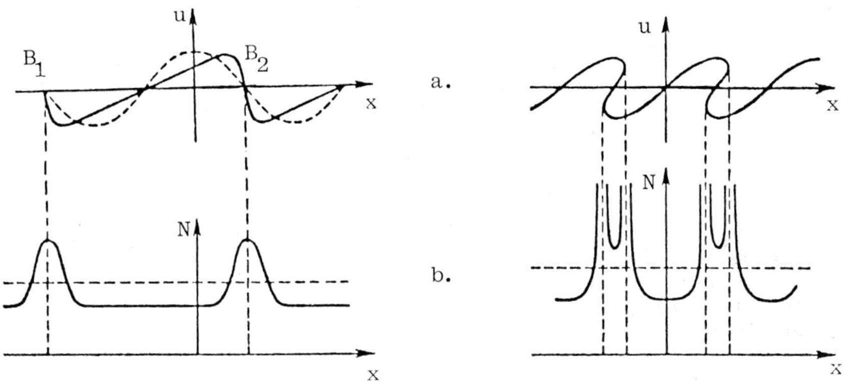

a. beam-velocity perturbation. b. particle density.

Fig. 2. The temporal evolution of simple waves ($0 < t < t_B$).

Fig. 3. The temporal evolution of simple waves ($t > t_B$).

where $N_0(x_0) = N(x(0),0)$ = initial electron density.

Figures 2 and 3 represent the temporal evolution of the velocity perturbation (a) and particle density of the modulated electron beam (b), respectively (Kadomtsev and Karpman, 1971); dashed lines refer to the initial distribution and solid lines refer to the waveform at time $t < t_B$ (Fig. 2) or $t > t_B$ (Fig. 3). The beam-velocity modulation and its nonlinear coupling to the particle density cause the increase or decrease in the density in the region of negative or positive velocity gradient respectively, thus leading to the so-called "bunching" of the particles and exhibiting the maximum of the particle density at the point of a maximum in negative velocity gradient (B_1 and B_2 in Fig. 2). This is just the bunching used for generation of microwaves in klystrons. The increase of the particle density continues until the velocity gradient and itself tend to infinity at the points B_1 and B_2 where "wave breaking" occurs (Fig. 3a) and double singularities appear in the density (Fig. 3b). This leads to generation of harmonics and also to amplification of small density oscillations.

4. A NONLINEAR DISTRIBUTED PARAMETER LINE WITH DISPERSION AND DISSIPATION

Consider the distributed parameter line represented schematically in Figure 4. The series impedance along the line is composed of the series inductance L and resistance R, and the capacitance C' parallel to it. The shunt admittance of the line is composed of a nonlinear shunt capacitance C. The distribution of transverse voltage and current along the L-R line is governed by the relation

$$V(z + \Delta z) - V(z) = -L\Delta z \cdot \frac{\partial}{\partial t} I(z) - R\Delta z \cdot I(z) \ . \qquad (26)$$

On the other hand, the equation of continuity or conservation of charge at the position z is written as

$$\frac{\partial}{\partial t} Q(z)\Delta z = -I(z) + I(z - \Delta z)$$

$$+ \frac{C'}{\Delta z} [\frac{\partial}{\partial t} \{v(z + \Delta z) - v(z)\} + \frac{\partial}{\partial t} \{v(z - \Delta z) - v(z)\}, \tag{27}$$

where Q is the charge accumulated in the capacitance C and the right-hand side represents the currents flowing into C through L-R and C' from both sides of z. Making $\Delta z \to 0$ in Eqs.(26) and (27), one obtains the fundamental equations for nonlinear distributed parameter transmission lines

$$\frac{\partial V}{\partial z} = -(R + L \frac{\partial}{\partial t})I , \tag{28}$$

$$\frac{\partial I}{\partial z} = - \frac{\partial}{\partial t} (Q - C'\frac{\partial^2 V}{\partial z^2}) . \tag{29}$$

Differentiating Eqs.(28) and (29) partially with respect to z and t, respectively and eliminating I, it follows after some rearrangement that

$$L \frac{\partial^2 Q}{\partial t^2} + R \frac{\partial Q}{\partial t} = \frac{\partial^2 V}{\partial z^2} + LC' \frac{\partial^4 V}{\partial^2 t \partial z^2} + RC' \frac{\partial^3 V}{\partial t \partial z^2} . \tag{30}$$

Recalling that Q is a function of V and expanding Q into a Taylor series around the d.c. bias voltage V , we have, retaining terms up to the second order,

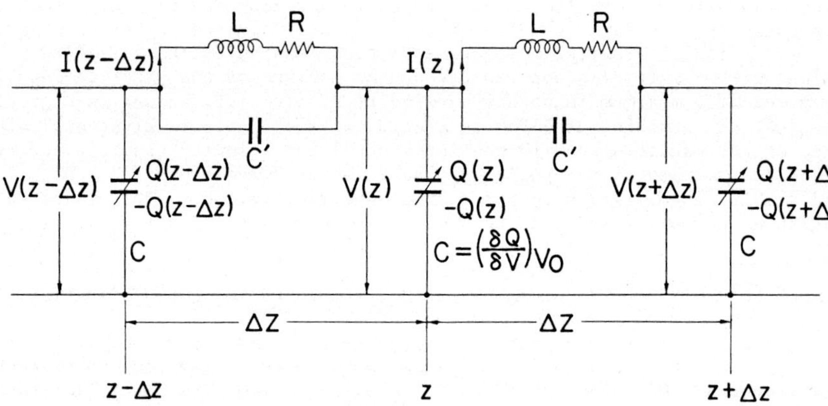

Fig. 4. Schematic representation of a nonlinear distributed parameter line.

$$Q(V) = Q(V_0) + \left(\frac{\partial Q}{\partial V}\right)_{V_0}(V - V_0) + \frac{1}{2}\left(\frac{\partial^2 Q}{\partial V^2}\right)_{V_0}(V - V_0)^2$$

$$= Q(V_0) + V_0 C u\left(1 - \frac{1}{2}\varepsilon u\right) , \tag{31}$$

where $u = (V - V_0)/V_0$ is the normalized signal voltage, $C = (\partial Q/\partial V)_{V_0}$, the linear shunt capacitance, and $\varepsilon = -(V_0/C)(\partial^2 Q/\partial V^2)$ is a parameter representing weak nonlinear saturation. Consider first the case when the shunt capacitance C is linear and let $\varepsilon = 0$. Further, making $C' \to 0$, $R \to 0$, the line reduces to a distributed LC transmission line along which the propagation velocity of current and voltage becomes, as is well known,

$$v_0 = \frac{1}{\sqrt{LC}} . \tag{32}$$

Hereupon, introducing a coordinate system moving with the velocity v and stretching the space and time coordinates as

$$x = \varepsilon^{1/2}(z - v_0 t), \quad \tau = \varepsilon^{3/2} t , \tag{33}$$

the partial differential operators with respect to the new space and time coordinates are written as

$$\frac{\partial}{\partial z} = \varepsilon^{1/2}\frac{\partial}{\partial x} , \quad \frac{\partial}{\partial t} = \varepsilon^{1/2}\left(-v_0\frac{\partial}{\partial x} + \varepsilon\frac{\partial}{\partial \tau}\right) . \tag{34}$$

If one rewrite Eq.(30) in terms of $u(x,\tau)$, using Eqs.(31), (32) and (34), we have, retaining terms up to $O(\varepsilon^2)$,

$$\frac{2}{v_0}\cdot\frac{\partial^2 u}{\partial x\partial\tau}\left(u\frac{\partial u}{\partial x}\right) + \frac{\varepsilon^{3/2}}{v_0}\cdot\frac{R}{L}\left[\frac{\partial u}{\partial x} - \varepsilon\left(\frac{\partial u}{v_0\partial\tau} + u\frac{\partial u}{\partial x}\right)\right]$$
$$+ \frac{C'}{C}\cdot\frac{\partial^4 u}{\partial x^4} - \frac{\varepsilon^{-1/2}}{v_0}\cdot\frac{R}{L}\cdot\frac{C'}{C}\cdot\frac{\partial^3 u}{\partial x^3} = 0 , \tag{35}$$

where it is seen in the bracket [ ] that the second term $O(\varepsilon)$ can be neglected, compared to the first leading term. Then, integrating Eq.(35) once with respect to x, we obtain finally

$$\frac{\partial u}{\partial\tau} + \frac{v_0}{2}\left(u\frac{\partial u}{\partial x} + \frac{C'}{C}\frac{\partial^3 u}{\partial x^3}\right) + \frac{\varepsilon^{-3/2}}{2}\cdot\frac{R}{L}\left(u - \varepsilon\frac{C'}{C}\frac{\partial^2 u}{\partial x^2}\right) = 0 . \tag{36}$$

This is just a generalized form of the Korteweg-de Vries-Burgers equation. The second and third terms on the left-hand side represent nonlinear and dispersive effects, respectively; the latter is produced by the capacitance C' parallel to L-R. The fourth and fifth terms represent dissipation; the former is the loss due to the resistance R alone and the latter represents the increase in loss due to the capacitance C'.

The particular case R = 0 leads to the so-called "Korteweg-de Vries equation" for a nonlinear lossless distributed parameter line

$$\frac{\partial u}{\partial \tau} + \frac{v_0}{2}\left(u\frac{\partial u}{\partial x} + \frac{C'}{C}\frac{\partial^3 u}{\partial x^3}\right) = 0 \ . \tag{37}$$

5. STATIONARY SOLUTIONS OF THE KORTEWEG-DE VRIES-BURGERS EQUATION

Putting $t = v_0\tau/2$ in Eq.(36), one can write a generalized Korteweg-de Vries-Burgers equation in the form

$$u_t + uu_x + \alpha u_{xxx} + \lambda u - \mu u_{xx} = 0 \qquad (\alpha,\ \lambda,\ \mu > 0) \ . \tag{38}$$

When $\lambda = 0$, Eq.(38) is termed the Korteweg-de Vries-Burgers equation and is often used as a basic model for nonlinear dispersive and dissipative systems. Consider first its stationary solutions, putting

$$u(x,\ t) = q(\theta), \quad \theta = \kappa x - \nu t, \qquad c = \frac{\nu}{\kappa} \ , \tag{39}$$

where κ is the wavenumber, ν, the frequency, and c is the velocity. Then Eq. (38) ($\lambda = 0$) becomes an ordinary differential equation written as

$$-\nu\dot{q} + \kappa q\dot{q} + \alpha\kappa^3\dddot{q} - \mu\kappa^2\ddot{q} = 0 \ . \tag{40}$$

Integrating this with respect to θ and making $q \to 0$, $\dot{q} \to 0$, $\ddot{q} \to 0$ as $\theta \to \infty$, we have

$$-cq + \frac{1}{2}q^2 + \alpha\kappa^2 - \mu\kappa\dot{q} = 0 \ . \tag{41}$$

This can be regarded as the equation of motion for a nonlinear harmonic oscillator with friction within a potential well described as

$$V(q) = \frac{1}{6}q^2(q - 3c) \ , \tag{42}$$

where θ and q represent time and the position of an equivalent particle, respectively, its momentum p, Lagrangian L and Hamiltonian H being written as

$$p = \alpha\kappa^2\dot{q} \ , \tag{43}$$

$$L = T(q,\dot{q}) - V(q) = \frac{p^2}{2\alpha\kappa^2} - \frac{1}{6}q^2(q - 3c) = \frac{\alpha\kappa^2}{2}\dot{q}^2 - \frac{1}{6}q^2(q - 3c) \ , \tag{44}$$

$$H = \dot{q}\frac{\partial L}{\partial \dot{q}} - L = p\dot{q} - L = \frac{p^2}{2\alpha\kappa^2} + V(q) = \frac{\alpha\kappa^2}{2}\dot{q}^2 + \frac{1}{6}q^2(q - 3c) \ . \tag{45}$$

Then Eq.(41) is rewritten simply as

$$\frac{dH}{dq} = \mu\kappa\dot{q} \ . \tag{46}$$

For the case of no dissipation, $\mu = 0$, H = constant, and the total energy of the system is conserved. If $\mu \neq 0$, there is dissipation of energy for $\dot{q} \neq 0$.

We now consider three cases.

(i) Case $\alpha = 0$. Eq.(38) ($\lambda = 0$) becomes the so-called Burgers equation

$$u_t + uu_x - \mu u_{xx} = 0. \tag{47}$$

Then Eq.(41) becomes

$$\dot{q} = \frac{q}{2\mu\kappa}(q - 2c) \tag{48}$$

and the trajectory in the phase space is schematically shown in Figure 5a. Integrating Eq.(48), we obtain

$$q = c(1 - \tan h \frac{c}{2\mu\kappa}\theta) . \tag{49}$$

This represents a monotonic shock as shown in Figure 5b and the half-width of "shock thickness" is written as

$$\delta = \frac{2\mu\kappa}{c} . \tag{50}$$

(ii) Case $\mu = 0$. Eq.(38) ($\lambda = 0$) leads to the so-called K-dV equation

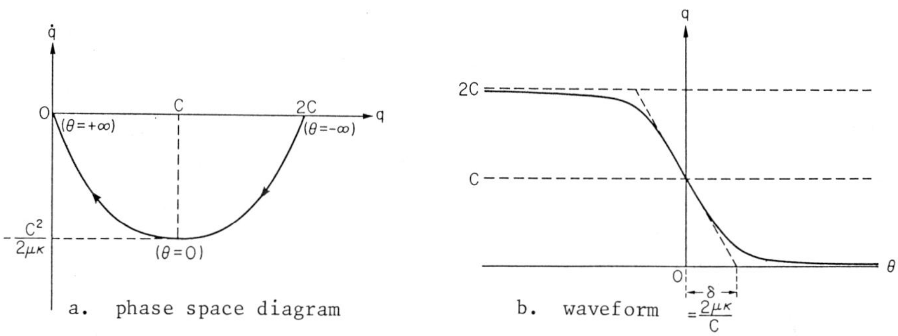

Fig. 5. Schematic diagram of a monotonic shock ($\mu^2 > 4\alpha c$).

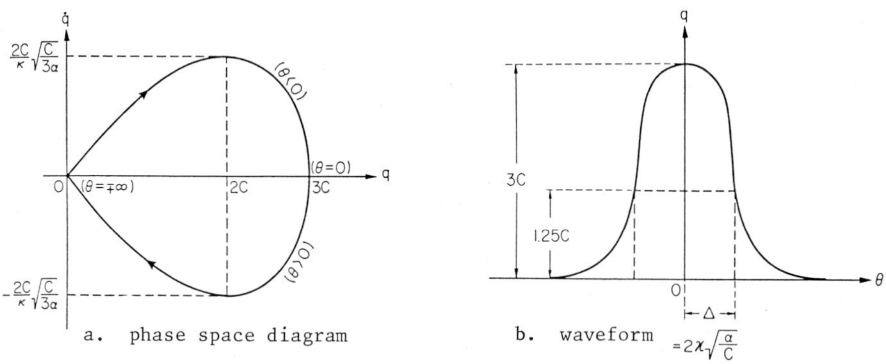

Fig. 6. Schematic diagram of a soliton.

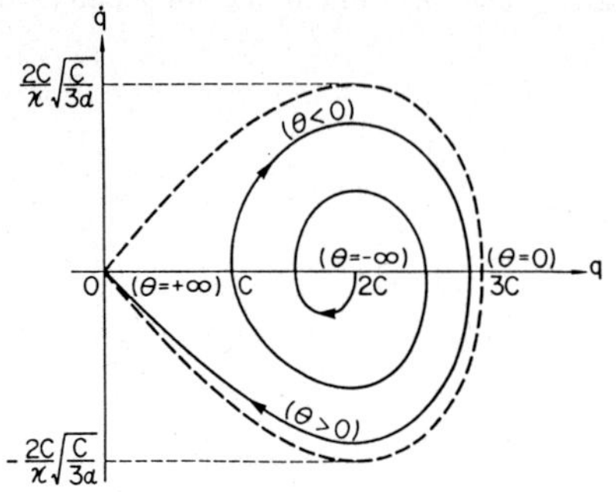

a. phase space diagram

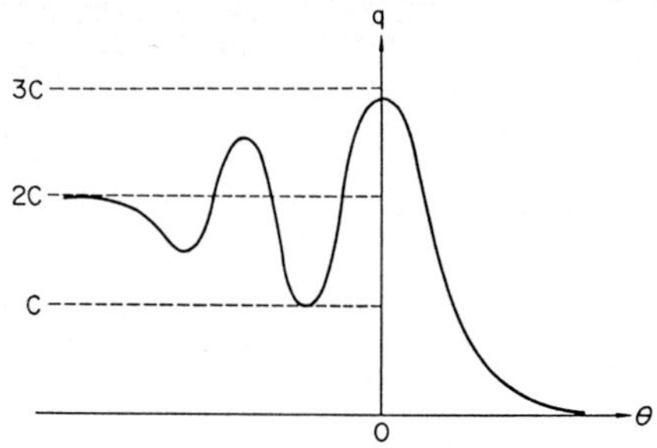

b. waveform

Fig. 7. Schematic diagram of a dispersive shock with oscillatory structure ($\mu^2 < 4\alpha c$).

$$u_t + uu_x + \alpha u_{xxx} = 0 \,. \tag{51}$$

Then Eq.(41) becomes

$$\dot{q}^2 = -\frac{q^2}{3\alpha\kappa^2}(q - 3c) \tag{52}$$

and the trajectory in phase space is shown in Figure 6a. The integration of Eq.(52) yields the solution

$$q = 3c\ \mathrm{sec\,h}^2\left(\sqrt{\frac{c}{\alpha}}\ \frac{\theta}{2\kappa}\right) \,. \tag{53}$$

This represents a solitary wave or so-called "soliton" as shown in Figure 6b and the half-width of its pulse is expressed as

$$\Delta = 2\kappa\sqrt{\frac{\alpha}{c}} \,. \tag{54}$$

(iii) Case $\alpha \neq 0$, $\mu \neq 0$. Eq.(41) is written as

$$\ddot{q} = \frac{\mu}{\alpha\kappa}\dot{q} - \frac{q}{2\alpha\kappa^2}(q - 2c) \tag{55}$$

and the trajectory in phase space varies depending upon the relative importance of dispersion and dissipation. For a dissipation-dominant case, i.e. $\mu^2 > 4\alpha c$, the trajectory is seen to be that in Fig. 5a with a nodal point at $q = 2c$, thus forming a monotonic shock, as seen in Fig. 5b. Conversely, for a dispersion-dominant case, i.e. $\mu^2 < 4\alpha c$, the trajectory in phase space is shown to be spiral with a spiral point at $q = 2c$ in Figure 7a, thus exhibiting a distinct oscillatory structure behind the shock front, as illustrated in Figure 7b. With further increasing dispersion and/or decreasing dissipation, the first waves of the structure behind the shock tend to solitons but with oscillating tail. For a limit $\mu \to 0$, the K-dV-Burgers equation tends to the K-dV equation and we eventually have some periodic solitons with a maximum amplitude 3c and a half-width $\Delta = 2\kappa\sqrt{\alpha/c}$, one of which is shown in Fig. 6b.

6. NONSTATIONARY SOLUTIONS OF THE KORTEWEG-DE VRIES-BURGERS EQUATION

We now consider nonstationary solutions of the Korteweg-de Vries-Burgers equation

$$u_t + uu_x + \alpha u_{xxx} - \mu u_{xx} = 0 \quad (\alpha > 0,\ \mu > 0) \tag{56}$$

as an initial value problem with the initial condition

$$u(x, 0) = u_0(x) \ . \tag{57}$$

Then, its temporal evolution can usually be decomposed into three stages, i.e. initial, transition, and quasi-stationary. In the initial stage ($0 < t < t_B$), the nonlinear term is dominant with negligible dispersion and dissipation, causing wave steepening and breaking. In the second stage, i.e. the transition period ($t \gtrsim t_B$), dispersion is growing rapidly, producing quasi-particles or solitons for sufficiently small dissipation. For a dissipation-dominant case, however, a soliton-like structure rapidly damps out after wave breaking and tends to a shock-like structure. The third stage represents the quasi-stationary state where solitons or shock formation has been completed, since the nonlinear term nearly balances the dispersion and/or the dissipation term ($t >> t_B$).

For the case of no dissipation or sufficiently small dissipation, the temporal evolution of a periodic initial state such as $u_0 = \cos \pi x$ exhibits a recurrence phenomenon. Beyond a half of the recurrence time t_R, the development follows the course inverse to the processes from $t = 0$ to $t = t_R/2$. As $t \to t_R - t_B$, dispersion decreases rapidly and quasi-particles or solitons die out, but being converted into real particles constituting a nonlinear background. The phenomenon is eventually going back to the initial state at $t = t_R$. We now discuss each stage in detail.

(i) Initial stage: $0 < t < t_B$. In Eq.(56), the first two terms are dominant and other terms are negligible, thus leading to the equation of nonlinear background

$$u_t + uu_x \simeq 0 \tag{58}$$

which is nearly the same as Eq.(10). Therefore, the solution described in Section 3 applies approximately to the present period.

(ii) Transition period: $t \gtrsim t_B$. In Eq.(56), dispersion (the third term) and dissipation (the fourth term) increase and become dominant rapidly. Therefore, decomposing u into a nonlinear background $u^{(0)}$ and the perturbation $u^{(1)}$ as

$$u = u^{(0)} + u^{(1)}, \quad |u^{(1)}| << |u^{(0)}| \tag{59}$$

and substituting this in Eq.(58), a quasi-linear approximation results in

$$u_t^{(0)} + u^{(0)}u_x^{(0)} + \alpha u_{xxx}^{(0)} - \mu u_{xx}^{(0)} \simeq 0 \ , \tag{60}$$

$$u_t^{(1)} + u^{(0)}u_x^{(1)} + u_x^{(0)}u^{(1)} + \alpha u_{xxx}^{(1)} - \mu u_{xx}^{(1)} = 0 \,. \tag{61}$$

Putting $u^{(1)} \sim \exp(-i\omega t + ikx)$ with Fourier expansion and referring to

$$\frac{\partial}{\partial t} = -i\omega \,, \quad \frac{\partial}{\partial x} = ik \,, \quad \frac{\partial^2}{\partial x^2} = -k^2 \,, \quad \frac{\partial^3}{\partial x^3} = -ik^3 \,, \tag{62}$$

we obtain from Eq.(61) the complex dispersion relation

$$\omega = \omega_r + i\omega_i = u^{(0)}k - \alpha k^3 - i(u_x^{(0)} + \mu k^2) \,, \tag{63}$$

where the real part ω_r is the wave angular frequency and the imaginary part ω_i represents wave growth or damping when it is positive or negative, i.e. for $u^{(0)} + \mu k^2 <$ or > 0, respectively. For no dissipation $\mu = 0$, the wave damps in the region of positive slope $u^{(0)} > 0$ but grows in the region of negative slope $u^{(0)} < 0$.

The motion of a centroid or guiding centre of wave-packets with a frequency width $\omega_i(k)$ around a centre angular frequency $\omega_r(k)$ can be regarded as the motion of quasi-particles of momentum k and energy ω and their trajectory equatios are written from Eqs.(4) and (63) as

$$\dot{x} = \frac{\partial}{\partial k}\omega_r(k,x) = u^{(0)} - 3\alpha k^2 \,, \tag{64}$$

$$\dot{k} = -\frac{\partial}{\partial x}\omega_r(k,x) = -ku_x^{(0)} \,. \tag{65}$$

Then, the kinetic equation with respect to the phase space density of quasi-particles is written from Eq.(5) in the form

$$\frac{dF}{dt} = 2\omega_i F = -2(u_x^{(0)} + \mu k^2)F \,, \tag{66}$$

where

$$\frac{d}{dt} = \frac{\partial}{\partial t} + \frac{\partial\omega_r}{\partial k}\cdot\frac{\partial}{\partial x} - \frac{\partial\omega_r}{\partial x}\cdot\frac{\partial}{\partial k} \,. \tag{67}$$

Eq.(66) represents the ratio of soliton generation $(u^{(0)} + \mu k^2 < 0)$ or annihilation $(u^{(0)} + \mu k^2 > 0)$.

(iii) Quasi-stationary state ($t >> t_B$). In this state, solitons or shock formation has already been completed far beyond the breaking time and the motion of wave-packets can be regarded to be quasi-stationary. For no dissipation, when all the solitons are present and each soliton comes to run individually with its own velocity, the temporal evolution of the envelope of the maximum amplitudes of those solitons are expressed, referring to Eq.(57), as

$$\frac{du_0}{dt} = \frac{\partial u_0}{\partial t} + \{u_\infty + (u_0 - u_\infty)/3\}\frac{\partial u_0}{\partial x} \,, \tag{68}$$

where $u_\infty = [u(x)]_{x=\pm\infty}$. This indicates that the envelope of a group of solitons forms a simple wave with the temporal development described in Section 3. With

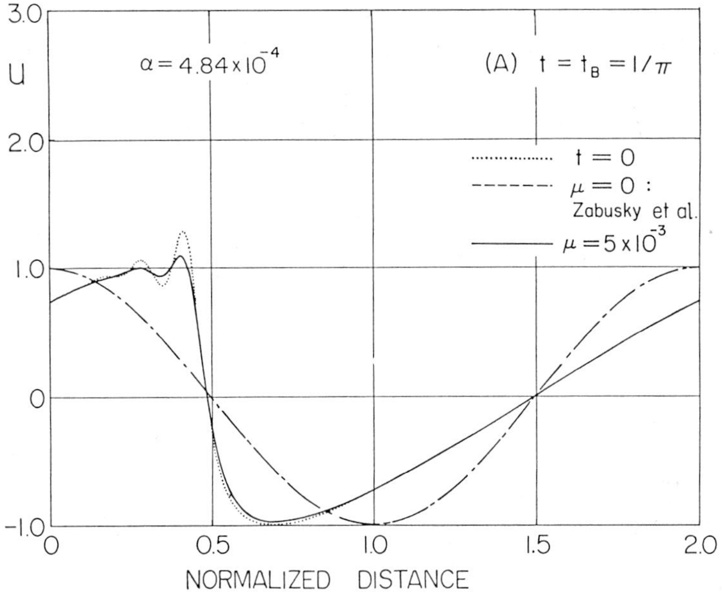

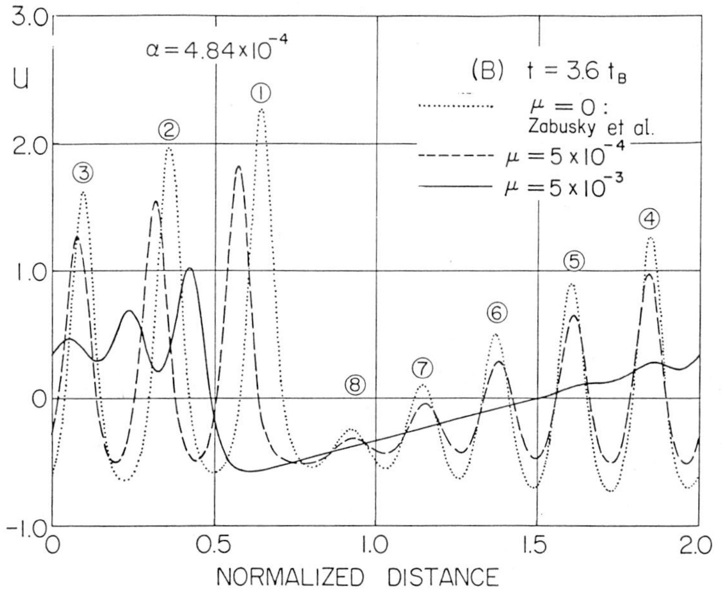

Fig. 8. The temporal evolution of the waveform described by the Korteweg-de Vries-Burgers equation (A, B).

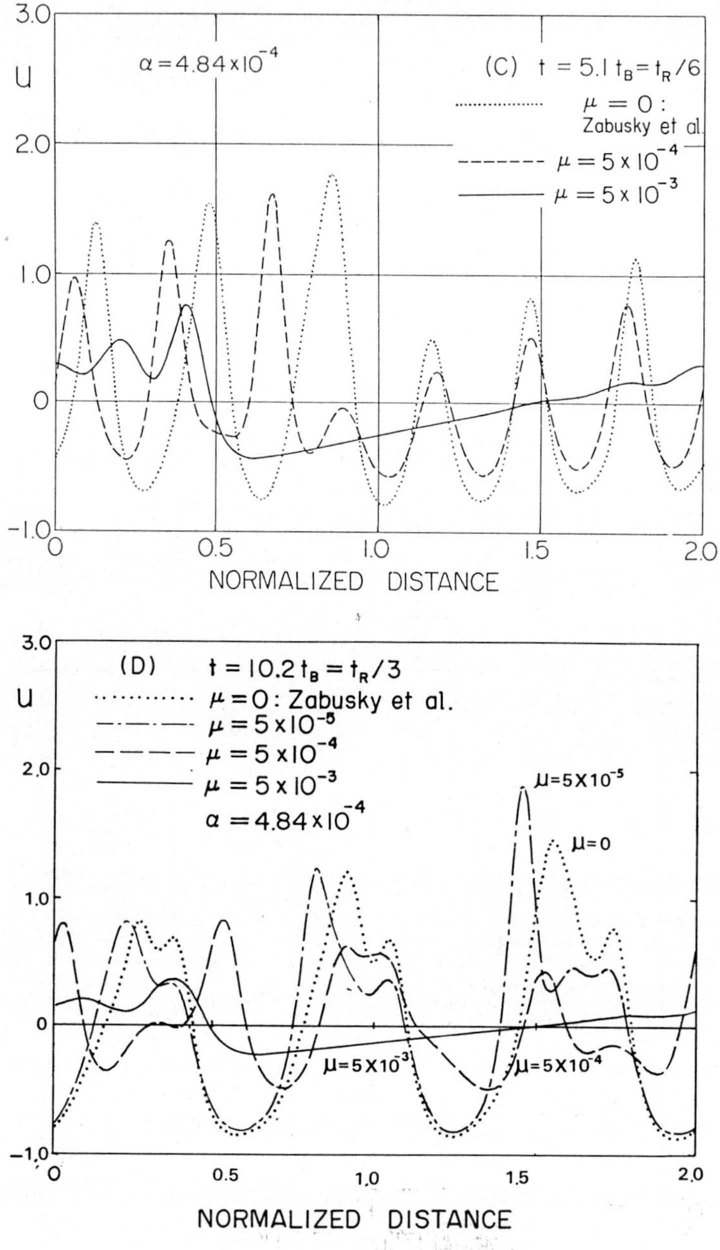

Fig. 8. The temporal evolution of the waveform described by the Korteweg-de Vries-Burgers equation (C, D).

increasing dissipation, the amplitude of the envelope of a group of solitons decreases and tends to a shock wave with oscillatory structure just behind the front but with decreasing amplitude and velocity. For sufficiently large dissipation the shock remains almost at rest.

Figures 8 to 10 represent some examples of nonstationary numerical solutions or temporal evolution of the K-dV-Burgers equation under the initial condition (57). In particular, Fig. 8 shows how the introduction of dissipation modifies the KdV solution obtained by Zabusky and Kruskal (1965) for no dissipation. As seen at $t = t_B$ in (A), the effects of dispersion and dissipation are so small as yet that the waveforms for three cases with different dissipation are all identical except at and immediately behind the shock front where small oscillatory structure appears slightly, indicating the applicability of a simple wave model described in Section 3 during the initial state for $0 \leq t < t_B$ at least as an approximate solution. However, this small structure forms a nucleus for quasi-particles or solitons and grows rapidly, thus creating them for $t \gtrsim t_B$. It is seen at $t = 3.6\ t_B$ in (B) that a train of eight solitons has been formed for no or small dissipation but for the latter case with decreasing amplitude. Since the velocity of a soliton is proportional to its amplitude, it is already seen that the soliton no. 1 for $\mu = 0$ precedes that for $\mu = 5.0 \times 10^{-4}$ and that the waveform for $\mu = 5.6 \times 10^{-3}$ forms a shock with stationary oscillatory structure just behind its front almost at rest, rather than generation of a train of solitons. As time goes further to $t = t_R/6 = 5.1\ t_B$ (t_R: recurrence time), solitons nos. 7 and 8 moving backwards tend to be captured into soliton no. 1 for no or weak dissipation and the shock for larger dissipation (solid curve) almost remains at rest with decreasing amplitude. Fig.(D) represents the waveforms during soliton collisions at $t = t_R/3$ for the above three and one additional cases of different dissipation. For no dissipation, three pairs of double or composite solitons are formed and remaining two solitons have been captured into them. In addition, one can see some traces of another soliton for two cases of weaker dissipation (chain and dashed curves). For larger dissipation (solid curve), one of structures behind the shock front almost disappears and the shock itself also becomes degraded.

Fig. 9 represents an example of the temporal evolution of the waveform (a) and its Fourier components (b) for relatively strong dispersion and dissipation. In (a) only two solitons are seen to be generated because of relatively large dispersion and the recurrence of waveform is rather well at $t = 1.65$, indicating that proper dissipation may rather improve its recurrence as a result of suppression of its harmonic components as seen in (b) where the recurrence time is also indicated by arrows at time of a maximum of its fundamental component. Incompleteness of the recurrence is explained in terms of small harmonic components (second and third) still remaining at the recurrence time where they become a minimum. It is also seen in (b) that inclusion of dissipation tends to extend the recurrence time.

Fig. 10 shows another example of the temporal evolution of the waveform (a, b, c) and its Fourier components (d) for somewhat smaller dispersion and dissipation than for Fig. 9. In (a) the formation of three solitons can be seen at $t = 2.36\ t_B$ and the waveform (solid curve) has no appreciable deviation from that for no dissipation (dashed curve). Fig.(b) is the waveform at the first recurrence time with more incomplete recurrence than for Fig. 9, since the second and third harmonic components in (d), particularly the latter become larger than for Fig. 9. Fig.(c) shows the waveform recurrent to (a) where three solitons have again been created in fairly good recurrence for no dissipation (dashed curve) but with appreciable decrease in soliton amplitude in the presence of dissipa-

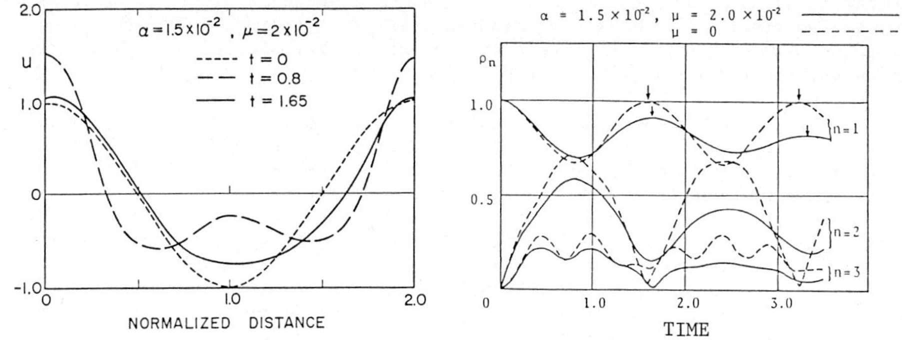

Fig. 9. The temporal evolution of the waveform and its Fourier components for relatively strong dispersion and dissipation.

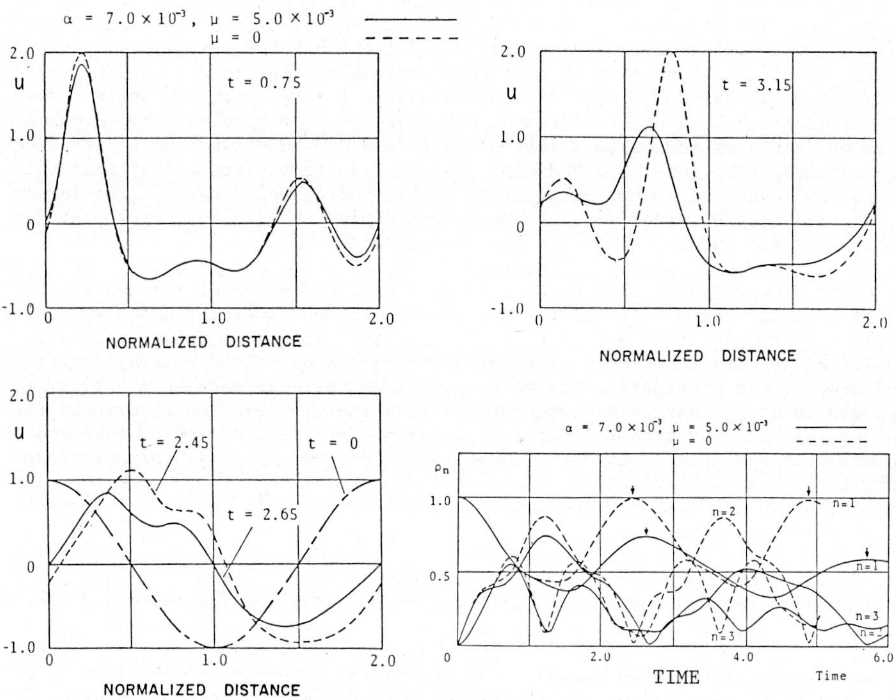

Fig. 10. The temporal evolution of the waveform and its Fourier components for somewhat smaller dispersion and dissipation.

tion (solid curve). It is seen in (d) that the second recurrence time becomes longer than the first one due to dissipation and that both recurrence times extend with increasing dissipation.

Numerical solutions mentioned above have been obtained by means of two methods, Fourier analysis and a finite difference method. For relatively weak nonlinear cases with relatively large dispersion and/or dissipation, Fourier analysis is more useful, as might be expected, for studying the small nonlinear interactions of Fourier components. The interactions transfer energy between different components and, through nonlinear terms, generate new components from existing ones when transferring Eq.(56) into a finite difference equation. For fully nonlinear cases with relatively small dispersion and/or dissipation, however, generations and distributions of Fourier components become so broad and complicated as not to allow us computer calculations. According to our experience, it is advantageous to work directly in the time-space domain rather than in the frequency-wavenumber domain for such a fully nonlinear case. As a result, we employed both methods of difference, depending upon degrees of nonlinearity in Eq.(56).

We now restrict ourselves to the KdV equation without dissipation ($\mu = 0$ in Eq.(56)) under the initial condition (57) and proceed to an analytical method to solve it. Consider the Schrödinger equation

$$\Psi_{xx} + \frac{1}{6\alpha}\,[\lambda_n + u_0(x)]\,\Psi = 0, \tag{69}$$

where the potential $-u_0(x) = -u(x, 0)$, the eigenvalues λ_n and the wave functions $\Psi(x)$. According to Gardner et al. (1967), the discrete eigenvalues of Eq. (69) completely determines the amplitudes A_n and velocities c_n ($n = 1, 2, \cdots, N$) of all the solitons produced from the initial perturbation by the relations

$$A_n = 3c_n = -2u_0\lambda_n \tag{70}$$

and the number of solitons is equal to the number of the discrete eigenvalues. For the case when the soliton solution $u(x, t)$ is asymptotically constant at infinity ($u \to u_\infty$ as $x \to \pm\infty$), Eqs.(70) should be replaced by

$$A_n = 3(c_n + 2u_\infty\lambda_n)\lambda_n \tag{71}$$

We now apply this theory to the present problem and make an approximation for the potential $-u_0(x)$ ($x \geq 0$)

$$\begin{aligned} u_0(x) &= \cos \pi x \simeq 1 - \frac{\pi^2}{2}x^2 \qquad &&\text{for } x \geq 0 \\ &= 0 &&\text{for } x < 0 \end{aligned} \tag{72}$$

Then the eigenvalues of Eq.(70) can be written as

$$\lambda_n = (2n + 1)\pi\sqrt{3\alpha} - 1 \leq 0 \tag{73}$$

Figure 11 shows the number of solitons versus a nonlinearity parameter $(6\alpha)^{-\frac{1}{2}}$. Horizontal bars and solid circles refer to analytical and observational results, respectively, based on the waveforms obtained from numerical calculations. Both results seem to be in reasonable agreement, taking into account an approximation for the potential in Eq.(72). Figure 12 represents the dependence of the recurrence time on dispersion. Solid circles and a connecting solid line

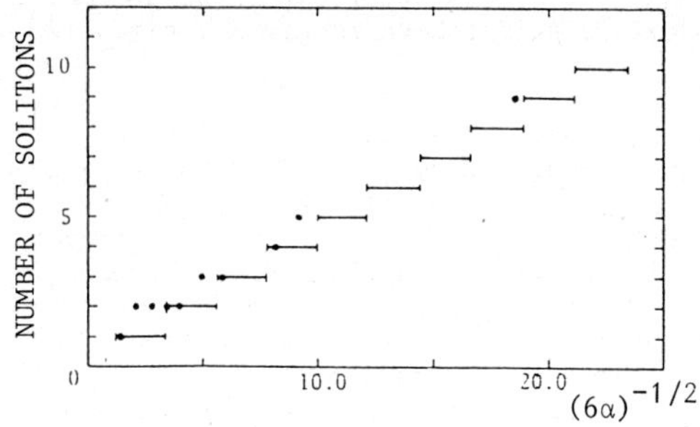

Fig. 11. Dependence of the number of solitons on nonlinearity.

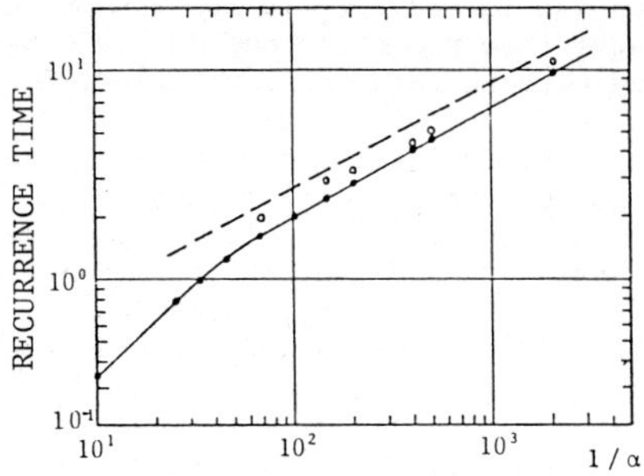

Fig. 12. The effect of dispersion on the recurrence time.

refer to numerical solutions, while a dashed line refers to analytical results obtained from Eq.(70). The difference between them is thought to be due partly to a finite value of u_∞ (asymptotic value of u at infinity) for each soliton. In particular, a larger deviation for large dispersion is thought to be due to an approximation for the potential in Eq.(72). Further, open circles refer to the results of corrections by introducing u_∞ obtained from numerical calculations into Eq.(71) and indicate better agreement with observational results.

7. SHOCKS, SOLITONS AND THE PLASMAPAUSE - APPLICATION OF NONLINEAR ELECTROMAGNETICS TO THE PLASMASPHERE

As an example of applications of the Korteweg-de Vries-Burgers equation, we consider the problem of the formation of the earth's plasmasphere and plasmapause. According to our model (Kikuchi, 1976), the onset of the plasmapause formation is assumed to begin near a midnight equatorial region, where electrostatic shocks are thought to be formed by the intersection of the solar wind induced convective flow and the earth with magnetic fields. These plasma discontinuities or electrostatic shocks tend to be developed and linked to the dayside as well as polewards along a geomagnetic field line with the aid of both the earth's corotational flow and the convective flow controlled geomagnetically, thus forming a global torus-like plasmasphere.

For the initial stage of the local formation of the nightside equatorial plasmapause, we can construct a rather simple but plausible model so as to make a mathematical treatment feasible. Based upon the fluid equations for a cold ion plasma and the Boltzmann equation for a hot electron plasma together with Poisson's equation, it turns out that the model can eventually be described approximately in terms of the K-dV-Burgers equation with respect to the ion density N

$$\frac{\partial N}{\partial t} + N\frac{\partial N}{\partial x} + \alpha\frac{\partial^3 N}{\partial x^3} = \mu\frac{\partial^2 N}{\partial x^2}\ , \qquad (74)$$

where α and μ are the dispersion and dissipation coefficients, respectively, dependent on geomagnetic conditions. Since the ion density is assumed to be stationary for a steady magnetic condition as well as other plasmapause parameters,

TABLE 1: Typical Plasmapause Parameters

C_o = ion sound speed = 20.3 km/sec for T- = 2.5 x 10^{-4} °K or KT- = 2.32 eV

$2\tilde{\delta}$ = shock or plasmapause thickness = 0.1 L ≈ 640 km for 2δ = 1.

Geomagnetic Phase / Parameter (normalized)	Prestorm Quiet Time	Storm Main Phase	Post -Storm Recovery
C	5	10	5
M	1.5	2	1.5
μ	1.25	0.63	0.1
α	< 0.078	0.01	0.2
$D = \mu^2 - 4\alpha C$	> 0	0	- 3.99
δ	0.5	0.13	
Δ			0.4
Shock Profile	Monotonic Shock	Shock in Steepening Transition from Monotonic to Soliton-like Shock	Shock with Wave-like Structure Soliton-like Shock
Primary Dissipation Mechanism	Trapped Electrons	Particle Reflection	Landau Damping

C = coefficient of supersonic component

M = Mach number of supersonic flow, μ = dissipative coefficient,

α = dispersive coefficient, D = determinant,

$\delta = \frac{2\mu}{C}$ = half of shock or plasmapause thickness

$\Delta = 2\sqrt{\frac{\alpha}{C}}$ = half of soliton width

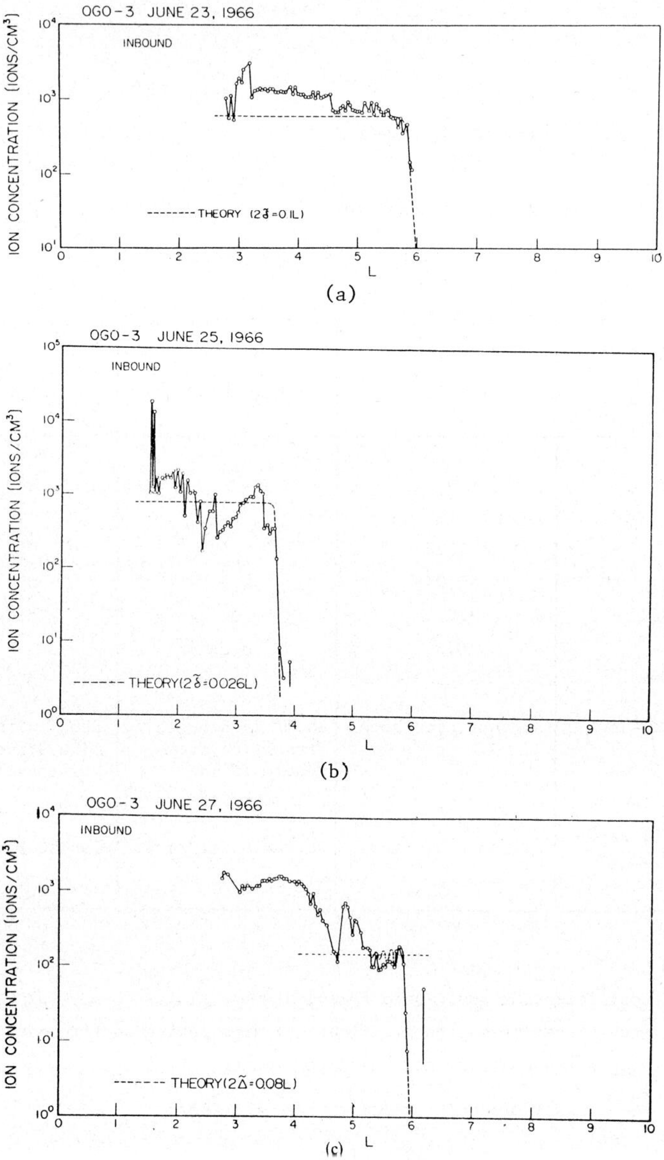

Fig. 13. Comparison of the present model with OGO satellite data for the 1966 June storm event. The dashed curve refers to the present theory. (a) A prestorm midnight profile of thermal ion plasma and a monotonic shock. (b) A main phase midnight profile and a critical shock in the transition from a monotonic to an oscillating structure. (c) A recovery midnight profile and a soliton-like shock with oscillating structure.

the theory in Section 5 applies to the present problem. Table 1 represents various typical plasmapause parameters together with a summary of results thus obtained.

Figure 13 represents a comparison of theoretical results (dashed curves) with OGO satellite data (solid curves) for a typical magnetic storm event with a sequence of prestorm, main phase and poststorm recovery. Fig.(a) is a prestorm, midnight ion plasma profile showing a typical undisturbed feature with no significant structure besides the plasmapause and corresponds to a monotonic shock with a thickness of $2\tilde{\delta} = 0.1$ L (L: McIlwain's invariant shell parameter): dissipation-dominant case for $\mu^2 \gg 4\alpha c$ (Fig. 5). Fig.(b) shows a main phase midnight profile and a critical shock in the transition from a monotonic to an oscillatory structure: critical case when nonlinear, dispersive and dissipative terms balance one another for $\mu^2 = 4\alpha c$. The shock front or the plasmapause is steepened considerably, compared with the prestorm profile, and its thickness is decreased as small as $2\tilde{\delta} = 0.026L$ because of increasing convective flow velocity. The observed structured plasma in the plasmasphere indicates a growing phase of the storm rapidly varying in a nonstationary state, because the quasi-stationary main phase profile obtained from the same OGO-3 during the July 1966, storm event does not exhibit structured profile, although the plasmapause is very much compressed and steepened. Fig.(c) represents a recovery midnight profile and a corresponding soliton-like shock with oscillatory wave-like structure just inside the plasmapause, whose scale sizes of small irregularities are similar and are observed to be 0.1 ~ 0.2L, indicating a good agreement between theory and experiment: dispersion-dominant case for $\mu^2 < 4\alpha c$ (Fig. 7). Thus the formation mechanisms of the plasmapause and associated fine structure irregularities are well described in terms of the K-dV-Burgers equation.

8. NONLINEAR SCHRÖDINGER EQUATION, QUASIPARTICLE AND FLUID DYNAMIC DESCRIPTIONS

Consider a single-mode type of wavepackets in a nonlinear medium described by (Marcuvitz, 1970, 1980, this issue)

$$i\frac{\partial\Psi}{\partial t} - \bar{\omega}(\underline{k};|\Psi|^2)\Psi = 0\,, \qquad \bar{\omega}(\underline{k};|\Psi|^2) = \omega(\underline{k}) - \beta|\Psi|^2. \tag{75}$$

Assuming that the wavepacket energy of a carrier wave is concentrated in the vicinity of its centre angular frequency $\omega_0 = \omega(\underline{k}_0)$ and wavenumber $\underline{k}_0 = k_0\underline{x}$ and setting

$$\Psi(\underline{k},t) = e^{-i\omega_0 t}\psi(\underline{k},t)\,, \tag{76}$$

the angular frequency ω can be expanded about its centre wavenumber $\underline{k}_0$ as

$$\omega(\underline{k}) = \omega(\underline{k}_0) + v_{g0}(\underline{k} - \underline{k}_0) + \frac{1}{2}v'_{g0}(\underline{k} - \underline{k}_0)^2 = \omega_0 - \frac{1}{2}v'_{g0}\frac{\partial^2}{\partial x^2}, \tag{77}$$

where v_{g0} is the group velocity. Take a coordinate system moving with its group velocity. Substituting Eqs.(76) and (77) into Eq.(75) and referring to

$$\underline{k} - \underline{k}_0 = -i\frac{\partial}{\partial x}\,, \qquad v'_{g0} = \frac{\partial^2}{\partial k^2}\omega(\underline{k}_0)\,,$$

we attain to the nonlinear Schrödinger equation

$$i\frac{\partial\psi}{\partial t} + \frac{v'_{g0}}{2}\frac{\partial^2\psi}{\partial x^2} + \beta|\psi|^2\psi = 0\,, \tag{78}$$

where the second and third terms on the left-hand side represent dispersion and nonlinearity, respectively. In particular, for the case of nondispersion, Eq. (78) reduces to the equation of nonlinear background alone, i.e.

$$i\frac{\partial \Psi}{\partial t} + \beta|\Psi|^2\Psi = 0 \tag{79}$$

and putting

$$\Psi = \psi_0 e^{i\phi(t)}, \qquad \phi(t) = \omega t + \phi_0 , \tag{80}$$

its solution is written as

$$\omega = \beta|\Psi_0|^2. \tag{81}$$

To solve Eq.(78), let

$$\Psi = a(x,t)e^{i\phi(x,t)} . \tag{82}$$

Then we have the relations between the amplitude a and phase ϕ

$$\frac{\partial \phi}{\partial t} + \frac{v'_{g0}}{2}\left(\frac{\partial \phi}{\partial x}\right)^2 - \beta a^2 - \frac{v'_{g0}}{2a}\frac{\partial^2 a}{\partial x^2} = 0 , \tag{83}$$

$$\frac{\partial a^2}{\partial t} + \frac{\partial}{\partial x}\left(a^2 v'_{g0}\frac{\partial \phi}{\partial x}\right) = 0 . \tag{84}$$

Further, putting

$$\phi(x,t) = kx - \hat{\omega}t, \tag{85}$$

k and $\hat{\omega}$ exhibit weak time-space depedence for quasi-stationary cases and can be written as

$$k = \frac{\partial \phi}{\partial x}, \qquad \hat{\omega} = -\frac{\partial \phi}{\partial t} \tag{86}$$

Substituting Eqs.(85) and (86) into Eq.(83), we obtain the dispersion relation

$$\hat{\omega} = \frac{v'_{g0}}{2}k^2 - \beta a^2 - \frac{v'_{g0}}{2a}\cdot\frac{\partial^2 a}{\partial x^2} . \tag{87}$$

This accompanies generation of wave-packets or quasi-particles and represents the relation between their momentum k and energy ω. Referring to

$$v'_{g0}\frac{\partial \phi}{\partial x} = kv'_{g0} = \frac{\partial \hat{\omega}}{\partial k} = v_g , \qquad v_{g0} = 0 ,$$

Eq.(84) can be rewritten as

$$\frac{\partial a^2}{\partial t} + \frac{\partial}{\partial x}(v_g a^2) = 0 \tag{88}$$

which represents the wave kinetic equation with respect to the quasi-particle density of wave-packets akin to the equation of continuity. Then the trajectory equations of quasi-particles in phase space become

$$\frac{dx}{dt} = \frac{\partial\hat{\omega}}{\partial k} = v_g = kv'_{g0} \ , \tag{89}$$

$$\frac{dk}{dt} = -\frac{\partial\hat{\omega}}{\partial x} = \beta\frac{\partial a^2}{\partial x} + \frac{v'_{g0}}{2}\frac{\partial}{\partial x}\left(\frac{1}{a}\frac{\partial^2 a}{\partial x^2}\right) \tag{90}$$

and the collisionless kinetic equation for quasi-particles is written in the Lagrangian microscopic description as

$$\frac{D}{Dt}F = 0, \quad \frac{D}{Dt} = \frac{\partial}{\partial t} + \frac{\partial\hat{\omega}}{\partial k}\frac{\partial}{\partial x} - \frac{\partial\hat{\omega}}{\partial x}\frac{\partial}{\partial k} \tag{91}$$

or

$$\frac{\partial F}{\partial t} + \frac{\partial}{\partial x}\left(\frac{\partial\hat{\omega}}{\partial k}F\right) - \frac{\partial}{\partial k}\left(\frac{\partial\hat{\omega}}{\partial x}F\right) = 0 \ . \tag{92}$$

Then the macroscopic fluid density N and velocity v are expressed respectively as the zeroth and velocity moments of F in the form

$$N(x,t) = \int_{-\infty}^{\infty} F(k,x,t)\frac{dk}{2\pi} \equiv |\Psi(x,t)|^2 = a^2 \tag{93}$$

and

$$v(x,t) = \frac{1}{N}\int_{-\infty}^{\infty}\frac{\partial\hat{\omega}(k,x)}{\partial k}F(k,x,t)\,\frac{dk}{2\pi} \equiv v_g \ . \tag{94}$$

We now proceed to the fluid-dynamic description that are derived from moments of Eq.(91) or (92) in the Eulerian or macroscopic form and obtain the equation of continuity

$$\frac{\partial N}{\partial t} + \frac{\partial}{\partial x}(Nv) = 0 \tag{95}$$

and the equation of motion of fluids

$$\frac{\partial v}{\partial t} + v\frac{\partial v}{\partial x} + v'_{g0}\frac{\partial\hat{\omega}}{\partial x} = 0 \ . \tag{96}$$

Eq.(95) can also be written as

$$\frac{\partial N}{\partial t} + v\frac{\partial N}{\partial x} + N\frac{\partial v}{\partial x} = 0 \ . \tag{97}$$

Substituting Eq.(90) into Eq.(96) and recalling the equations of gas dynamics and state, a neglect of higher order terms leads, on referring to Eq.(93), to

$$\frac{\partial v}{\partial t} + v\frac{\partial v}{\partial x} - \beta v'_{g0}\frac{\partial N}{\partial x} = \frac{dv}{dt} + \frac{1}{Nm}\frac{\partial P}{\partial x} = \frac{dv}{dt} + \frac{c^2}{N}\frac{\partial N}{\partial x} = 0 \ , \tag{98}$$

where

$$c^2 = \frac{1}{m}\frac{dp}{dN} = -\beta v'_{g0}N = -\beta v'_{g0}a^2 \ ,$$

$$= c_0^2 \left(\frac{N}{N^0}\right)^{\gamma-1} = c_0^2 \left(\frac{a}{a_0}\right)^{2(\gamma-1)} = c_0^2 \left(1 + \frac{a-a_0}{a_0}\right)^{2(\gamma-1)} , \tag{99}$$

c and c_0 are the velocity of nonlinear acoustic waves and its linear approximation, respectively. Putting

$$w = \int \frac{c(N)}{N} dN \tag{100}$$

and referring to

$$\frac{\partial w}{\partial x} = \frac{c}{N}\frac{\partial N}{\partial x} , \qquad \frac{\partial w}{\partial t} = \frac{c}{N}\frac{\partial N}{\partial t} , \tag{101}$$

Eqs.(97) and (98) are rewritten respectively as

$$\frac{\partial w}{\partial t} + v\frac{\partial w}{\partial x} + c\frac{\partial v}{\partial x} = 0 , \tag{102}$$

$$\frac{\partial v}{\partial t} + v\frac{\partial v}{\partial x} + c\frac{\partial w}{\partial x} = 0 . \tag{103}$$

Further, adding or subtracting both sides of Eqs.(102) and (103) respectively, we obtain a compact form

$$\left[\frac{\partial}{\partial t} + (v \pm c)\frac{\partial}{\partial x}\right](v \pm w) = 0 . \tag{104}$$

Setting $w = \pm v$, both Eqs.(102) and (103) reduce to an identical form and then Eq.(104) can also be written in the form

$$\left[\frac{\partial}{\partial t} + (v \pm c)\frac{\partial}{\partial x}\right] v = 0 \tag{105}$$

and its transformation into a system of coordinates moving with the velocity $\pm$ c recovers Eq.(7) or (10) for simple waves in Section 3.

9. NONLINEAR PARABOLIC EQUATIONS AND SELF-MODULATION OF WAVEPACKETS

The nonlinear Schrödinger equation (78) is reduced, on transforming its space-coordinate, to the fundamental form of nonlinear parabolic equations

$$i\frac{\partial \psi}{\partial t} + \frac{\partial^2 \psi}{\partial x^2} + \beta|\psi|^2\psi = 0 . \tag{106}$$

Then the nonlinear dispersion relation (87), trajectory equations (89) and (90) for quasi-particles and the kinetic equation (88)and (92) for their phase space density are all simplified, setting $v'_{g0}/2 \to 1$ and ignoring the second-order differential with respect to the amplitude a, in the forms

$$\hat{\omega} = k^2 - \beta a^2 , \tag{107}$$

$$\dot{x} = \frac{\partial \hat{\omega}}{\partial k} = 2k \quad , \tag{108}$$

$$\dot{k} = -\frac{\partial \hat{\omega}}{\partial x} = \beta \frac{\partial a^2}{\partial x} \quad , \tag{109}$$

$$\frac{da^2}{dt} = \frac{\partial a^2}{\partial t} + v_g \frac{\partial a^2}{\partial x} = -a^2 \frac{\partial v_g}{\partial x} \quad , \tag{110}$$

$$\frac{\partial F}{\partial t} + 2k \frac{\partial F}{\partial x} + \beta \frac{\partial a^2}{\partial x} \cdot \frac{\partial F}{\partial k} = 0 \quad . \tag{111}$$

We now go back to Eq.(82) and apply to a monochromatic wave with amplitude a_0 and wavenumber k_0 a small perturbation

$$a = a_0 + a_1 e^{i(\tilde{k}x - \tilde{\omega}t)}, \quad k = k_0 + k_1 e^{i(\tilde{k}x - \tilde{\omega}t)} \quad (a_1 << a_0 \,, \; k_1 << k_0) \tag{112}$$

where $\tilde{\omega}$ ($<< \omega_0$) and $\tilde{k}$ ($<< k_0$) are the angular frequency and wavenumber of a modulation. Then in a linear approximation we obtain from Eqs.(109) and (110) the dispersion relation for modulated waves

$$\tilde{\omega} = \tilde{k}[v_{g0} \pm i\sqrt{\beta v'_{g0} a_0^2}\,] \ . \tag{113}$$

It is seen that when $\beta v'_{g0} > 0$ the wave becomes modulationally unstable and tends to break into wave-packets, causing their self-contraction, as discussed hereafter in detail.

Figures 14 to 19 represent some examples of nonstationary numerical solutions of the nonlinear Schrödinger equation partly with dissipation in a general form

$$i\frac{\partial \psi}{\partial t} + \frac{\partial^2 \psi}{\partial x^2} + \beta|\psi|^2\psi + i\gamma\psi = 0 \tag{114}$$

subject to the initial condition of a modulated wave

$$\Psi(x, 0) = (1 + \rho(o) \cos kx) \cos k_0 x \ , \tag{115}$$

namely the temporal evolution of the waveform and its Fourier components, carrier and side band. Fig. 14 shows an example for the strong nonlinear regime ($\beta = 0.045 > \beta_c \simeq k^2/2 = 0.03$; β_c: critical nonlinear coefficient) without dissipation. It is seen that the wave becomes modulationally unstable and tends to break into wave-packets, causing their self-contraction and forming a pair of envelope solitons. It is also seen in Fourier spectra that with temporal gradual decrease in carrier wave amplitude, the first upper and lower side-band wave amplitudes increase gradually and higher side band waves newly appear successively, causing energy transfer from the carrier to side band waves with growing self-modulation until a half of the recurrence time where the envelope soliton amplitude reaches a maximum. After that, the carrier wave amplitude begins to increase gradually and the side band wave amplitudes decrease in such a way that all the phenomena trace back along the completely inverse processes, eventually almost reconstructing the initial state at the recurrence time $t_R = 125$.

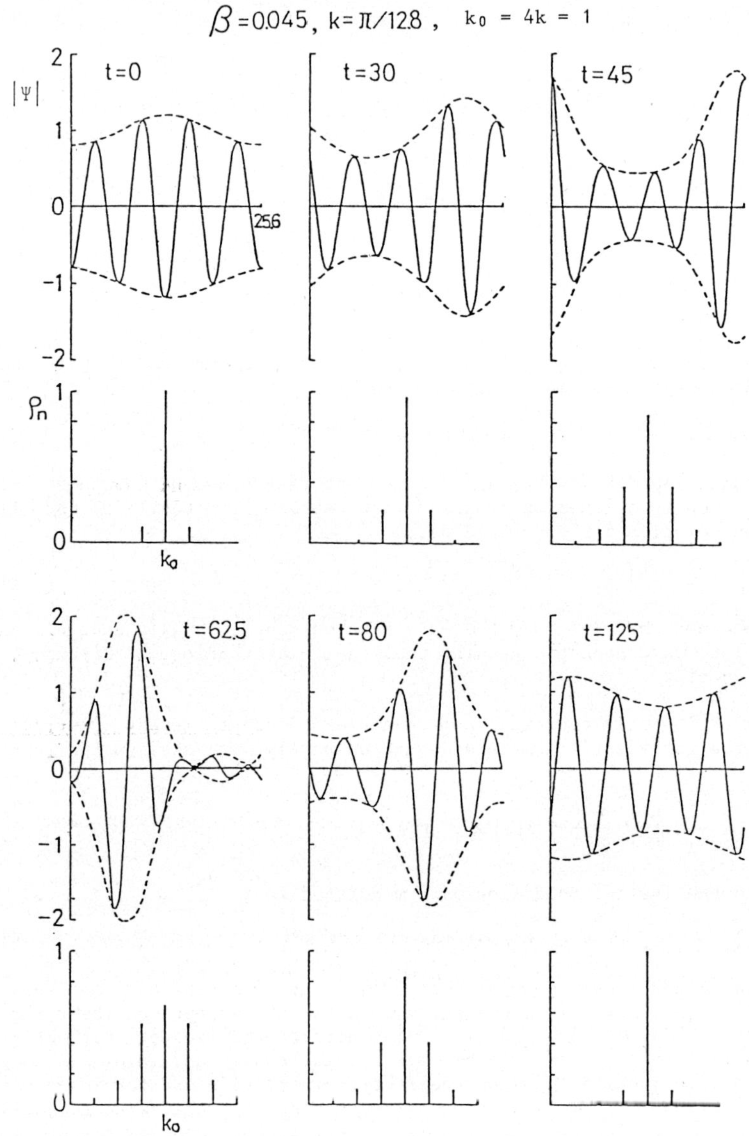

Fig. 14. The temporal evolution of the waveform described by the nonlinear Schrödinger equation and its Fourier components.

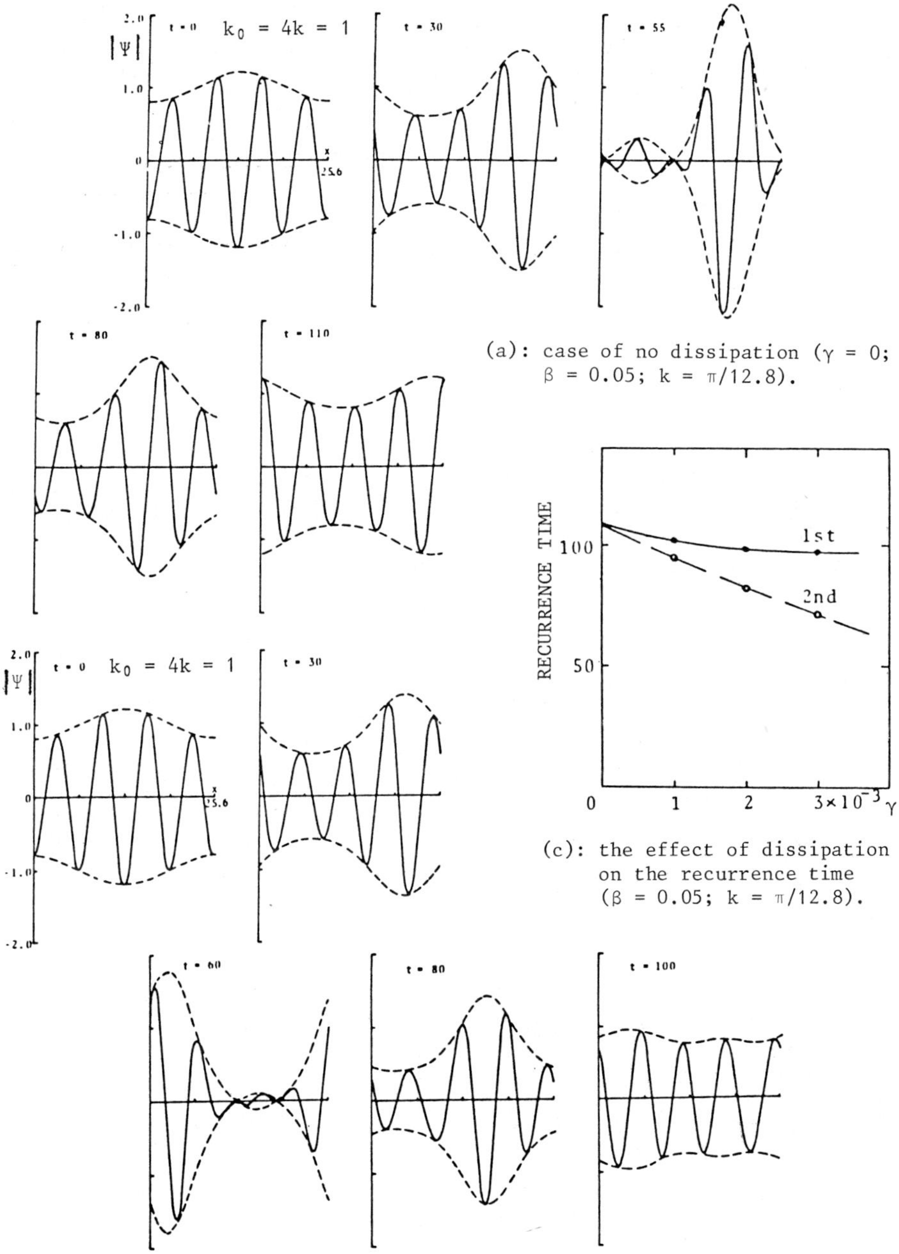

(a): case of no dissipation ($\gamma = 0$; $\beta = 0.05$; $k = \pi/12.8$).

(c): the effect of dissipation on the recurrence time ($\beta = 0.05$; $k = \pi/12.8$).

(b): case of dissipation ($\gamma = 2 \times 10^{-3}$; $\beta = 0.05$; $k = \pi/12.8$).

Fig. 15. The temporal evolution of the waveform described by the nonlinear Schrödinger equation with dissipation and its recurrence time.

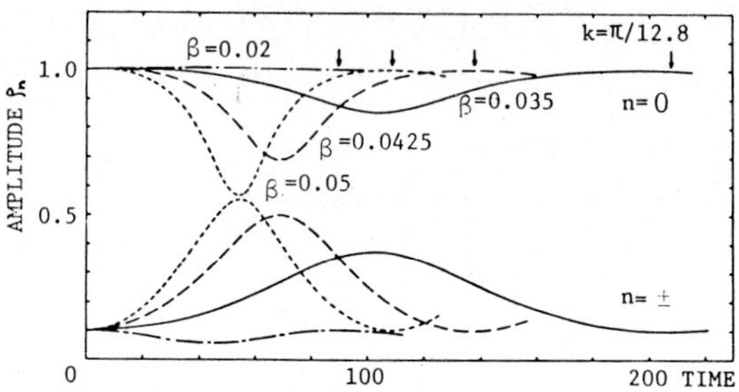

(a): amplitudes for no dissipation (γ = 0, k = π/12.8).

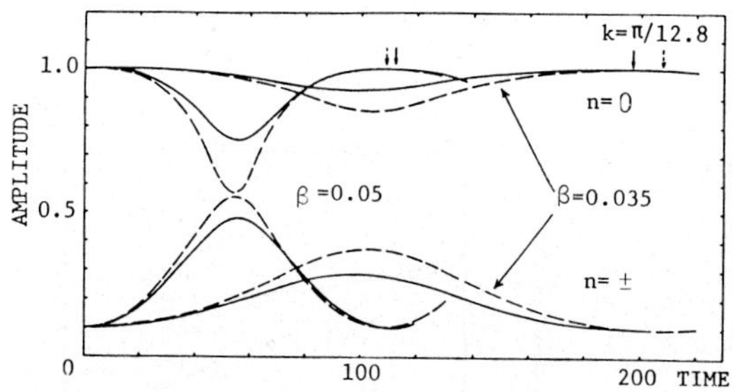

(b): analytical solutions for amplitudes (γ = 0, k = π/12.8); solid and dashed curves refer to analytical and numerical solutions, respectively.

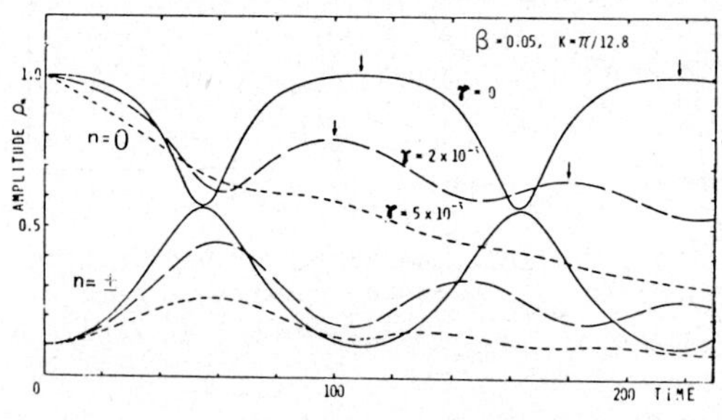

(c): amplitudes for dissipation (β = 0.05, k = π/12.8); solid, dashed and dotted curves refer to $\gamma = 0$, $\gamma = 2 \times 10^{-3}$, and $\gamma = 5 \times 10^{-3}$, respectively.

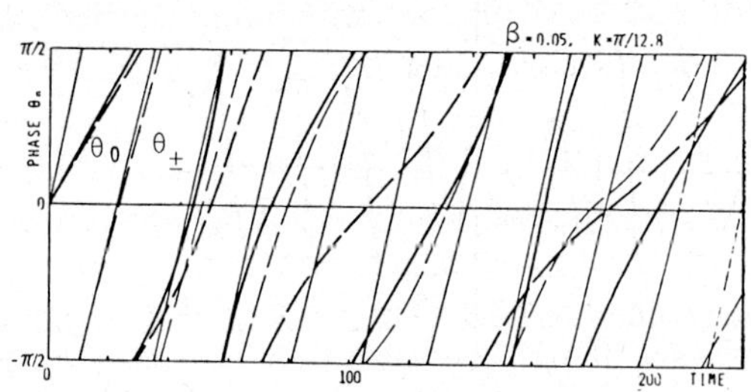

(d): phase angles for no and weak dissipation (β = 0.05, k = π/12.8); solid and dashed curves refer to γ = 0, γ = 2×10^{-3}, respectively.

Fig. 16. The temporal evolution of Fourier components, carrier and side band. Arrows indicate recurrence times. $k_0 = 4k = 1$.

Fig. 15 shows another example, particularly to see the effect of dissipation on the waveform (a, b) and recurrence time (c) by comparing both cases for no and weak dissipation for somewhat larger nonlinear coefficient ($\beta = 0.05$). The temporal evolution of the waveform for no dissipation (a) is almost the same as for the previous case (Fig. 14) except that the recurrence time is somewhat shorter ($t_R = 110$) in the present case. By comparing (b) for weak dissipation ($\gamma = 0.002$: dissipation coefficient) with (a) ($\gamma = 0$), it is seen that the waveform itself is very much similar except at the recurrence time which becomes shorter ($t_R = 100$) and where the waveform in (b) differs somewhat from that for the initial state. This is thought to be because dissipation affects not only the wave amplitude but also the phase, as will be seen in Fig. 16(d). Fig.(c) shows the effect of dissipation on the recurrence time that may be defined in general as the time when the carrier wave amplitude takes a maximum. With increasing dissipation, the recurrence time decreases and the second recurrence time becomes further shorter than the first one. This is in contrast to a general increase in recurrence time with increasing dissipation for the case of K-dV-Burgers equation (Figs. 9(b) and 10(d)).

Fig. 16 represents the temporal evolution of Fourier components, carrier and side band in amplitude (a, b, c) and phase (d) for various nonlinear and dissipative coefficients. For the case of no dissipation in (a) (numerical solutions) and (b) (analitical solutions to be discussed later), it is seen that both carrier and side band waves recover their initial states completely at the same recurrence time indicated by arrows which decreases with increasing nonlinearities. This is in contrast to the increase in recurrence time with increasing nonlinearities for the case of KdV equation (Fig. 12). While the amplitudes of the first upper and lower side band waves are always identical, their growth rate and a maximum amplitude also increase by the modulational instability with increasing nonlinearities. In the weakly nonlinear regime ($\beta < \beta_c \simeq 0.03$), the amplitude of the first side band waves rather decreases as seen for $\beta = 0.02$ in (a) and then the degree of modulation becomes minimum at a half of the recurrence time ($t_R/2 = 44.5$). For the case of dissipation in (c) ($\beta = 0.05$; $\gamma = 2 \times 10^{-3}$, 5×10^{-3}), the recurrence time is seen to decrease as well as the amplitudes of carrier and side band waves with increasing dissipation, the second recurrence time tending to become shorter than the first one. With further increasing dissipation as seen for the case, $\gamma = 5 \times 10^{-3}$, the carrier wave amplitude tends not to possess a maximum, so that the recurrence phenomena become unclear. Fig.(d) shows the effect of dissipation on the phases of carrier and side band waves: solid and dashed lines refer to no ($\gamma = 0$) and weak dissipation ($\gamma = 2 \times 10^{-3}$), respectively; thick and thin lines refer to carrier and side band waves, respectively. For both cases, the phase angles are seen to advance and consequently the nonlinear phase velocity becomes smaller than the linear velocity. For the case of dissipation, the degree of advancement of the phase angles becomes slower than for the case of no dissipation and also decreases with time, resulting in a gradual temporal increase in the phase velocity.

Figures 17 and 19 represent another examples of the temporal evolution of the waveform and its Fourier components for a special case, $k_0 = 0$, $\rho(0) = 0.2$ in the initial condition (115). Fig. 17 presents a case for $\beta = 0.04$ in the strongly nonlinear regime where energy transfer from the carrier to side band waves takes place, not only amplifying the fundamental component but also generating new higher harmonic waves propagating forwards and backwards successively. Consequently, the waveform is stationary but steepened symmetrically until a half of the recurrence time ($t_R/2 = 77.8$) where the amplitude of a soliton-like but standing waveform reaches a maximum. Beyond that, the waveform traces back along the inverse processes, completely reconstructing the initial state as in

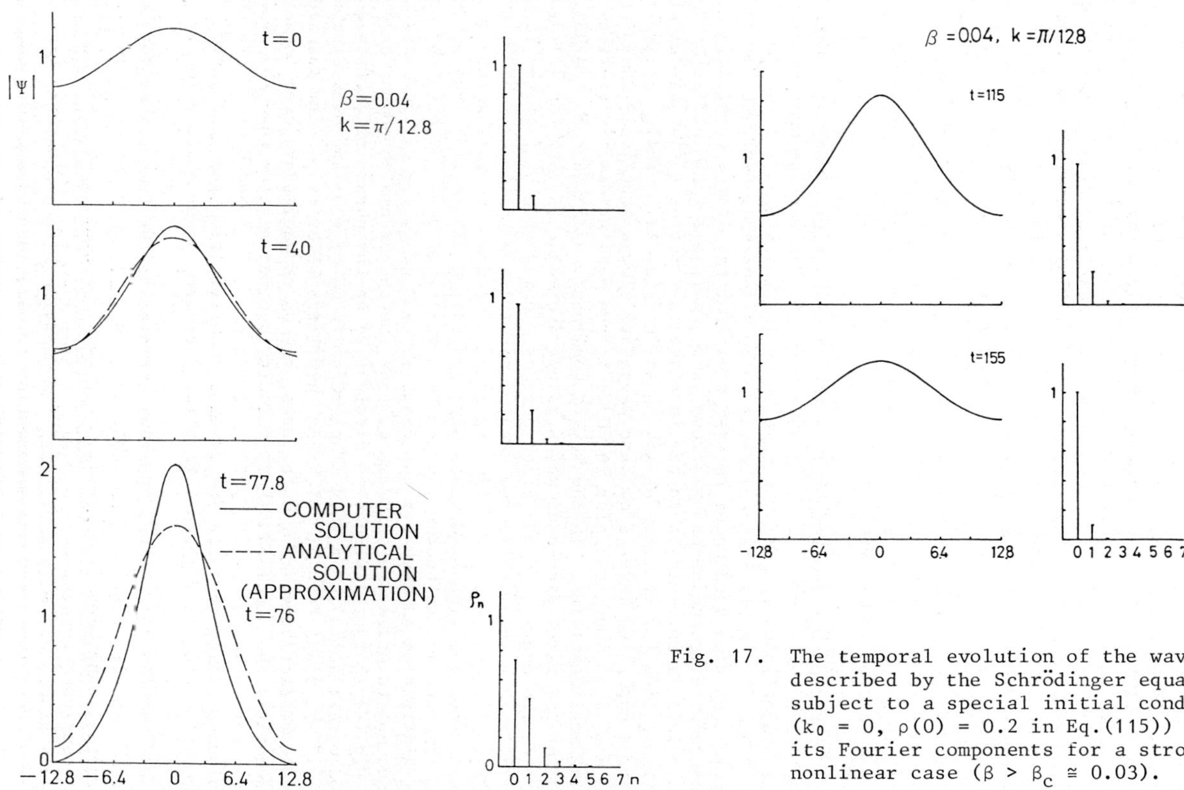

Fig. 17. The temporal evolution of the waveform described by the Schrödinger equation subject to a special initial condition (k_0 = 0, $\rho(0)$ = 0.2 in Eq.(115)) and its Fourier components for a strongly nonlinear case ($\beta > \beta_c \cong 0.03$).

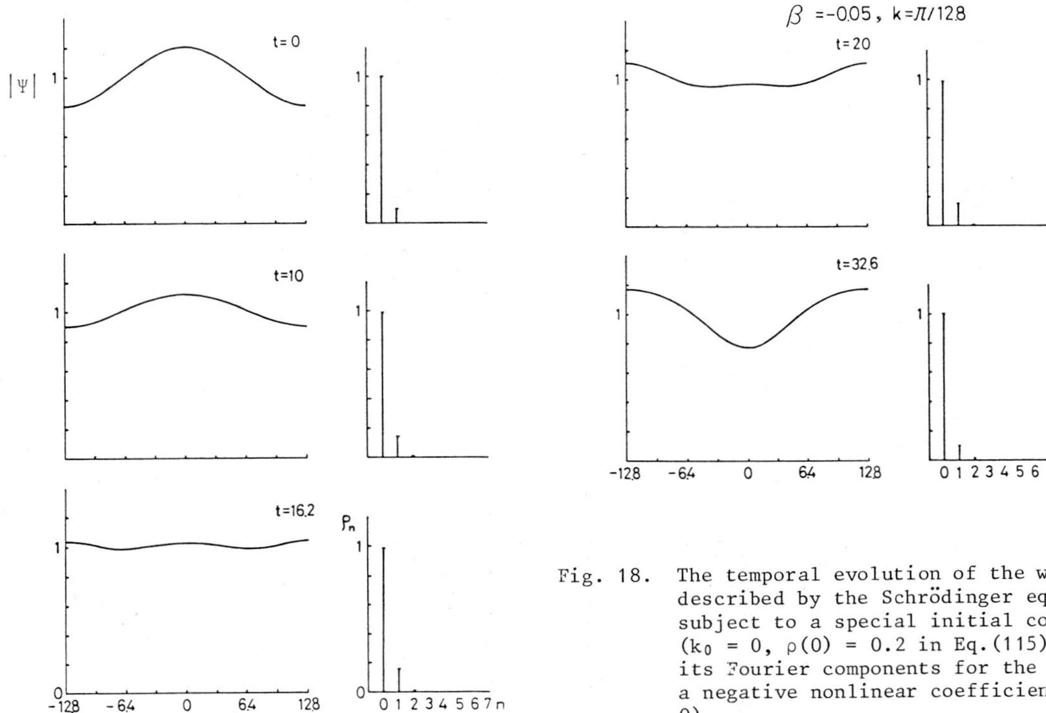

Fig. 18. The temporal evolution of the waveform described by the Schrödinger equation subject to a special initial condition ($k_0 = 0$, $\rho(0) = 0.2$ in Eq.(115)) and its Fourier components for the case of a negative nonlinear coefficient ($\beta < 0$).

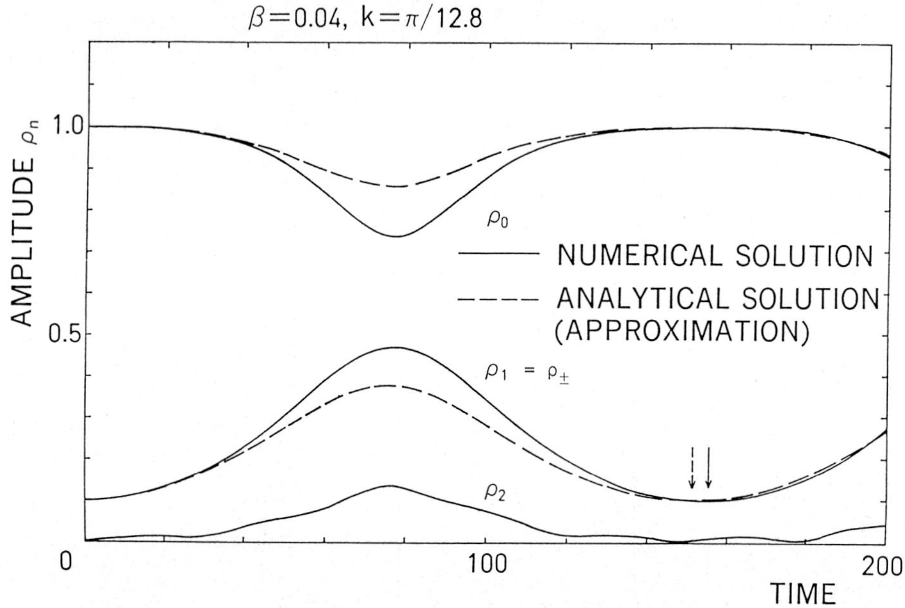

(a): amplitudes, particularly including the second side band or harmonics.

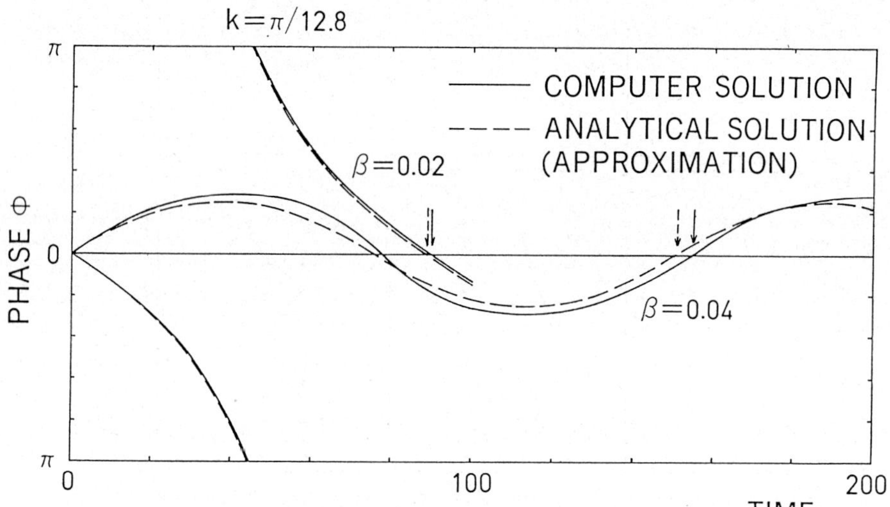

(b): characteristic phase angle .

Fig. 19. The temporal evolution of Fourier components for the case of Fig. 18. Solid and dashed curves refer to numerical and analytical solutions, respectively. Arrows indicate recurrence times.

the previous examples (Figs. 14 and 15(a),(b)). Fig. 18 presents another case for negative nonlinear coefficient (β = - 0.05) where energy transfer from the carrier to side band waves is not significant because of weak nonlinearity but the waveform itself is seen to be flattened and to produce two tiny dark solitons with a slight dent at a half of the recurrence time ($t_R/2$ = 16.2). After that, the waveform traces back through the inverse processes, almost reconstructing the initial state just as in Fig. 17. With increasing nonlinearities ($|\beta|$) in the region of negative nonlinear coefficient, energy transfer from the carrier to side band waves increases as well as the amplitudes of dark solitons and the phenomena seem to be similar in some respects to the creation and annihilation of solitons described by the KdV equation subject to the initial condition of a sinusoidal wave. Fig. 19 presents a comparison of numerical and analytical solutions for the temporal evolution of the amplitudes of Fourier components (a) and a characteristic phase angle Φ (b). This angle is a phase shift due to nonlinear and dispersive effects to be defined explicitly in the following analysis. Since the amplitude and phase of each Fourier component are thought to be independent on the carrier wavenumber k_0, Fig.(a) is essentially the same as Fig. 16(a, b) except that the second side band or harmonics are newly included here but for a different nonlinear coefficient (β = 0.04). It is seen that the amplitude of the second side band or harmonics also takes a maximum as well as the first side band or fundamentals at a half of the recurrence time ($t_R/2$ = 76). Fig.(b) is a presentation of the characteristic phase Φ different from the phase angles of Fourier components shown in Fig. 16(d). Since Φ = 0 at t = 0, a necessary condition for the recurrence time is written as $\Phi(t_R)$ = 0 as indicated by arrows but this condition is not always sufficient. For instance, we have $\Phi(t_R/2)$ = 0 at a half of the recurrence time for β = 0.04 (> $\beta_c \cong$ 0.03), since Φ oscillates so as to become positive or negative. For β = 0.02 (< $\beta_c \cong$ 0.03) in the weak nonlinear regime, we have $\Phi(t_R/2)$ = 0 at a half of the recurrence time ($t_R/2$ = 44.5) where a conversion takes place from amplitude to phase modulation.

Numerical solutions presented above have been obtained by means of Fourier analysis through the following procedure. In Eq.(114), we expand the function Ψ in a Fourier series

$$\psi(x, t) \cong \sum_{n=0,\pm 1}^{N} A_n(t) e^{i\phi_n(x, t)}, \tag{116}$$

where n = 0, + and - refer to the carrier, upper side band and lower side band waves, respectively, N the number of terms in truncation and

$$A_n(t) = a_n(t) + ib_n(t) = \rho_n e^{i\theta_n}, \quad \phi_n(x, t) = k_n x - \omega_n t. \tag{117}$$

Substituting Eq.(116) in Eq.(114) and using the linear dispersion relation $\omega = k^2$, we find a system of linear ordinary differential equations with respect to a_n and b_n, expressing couplings between carrier and side band waves. Performing numerical integration with respect to time t, we finally obtain Fourier components $A_n = a_n + ib_n$.

In the initial condition (115), we have chosen $\rho(0)$ = 0.2, k = $\pi/12.8$ = 0.25 for all calculations, k_0 = 4k for Figs. 14 to 16 and k_0 = 0 for Figs. 17 to 19.

We finally proceed to analytical solutions for Eq.(114) subject to the initial condition (115). To effect this, the nonlinearity is assumed to be not so strong that the second and higher side band waves can be neglected. We have only to consider carrier and first side band waves. Then the complex amplitudes of the upper and lower side band waves are also assumed to be equal to each other from their initial conditions and conservation of energy. Let

$$A_0 = \rho_0 e^{i\theta_0}, \quad A_\pm = \rho_\pm e^{i\theta_\pm}, \quad (\rho_\pm = \rho_+ = \rho_-, \quad \theta_\pm = \theta_+ = \theta_-) \tag{118}$$

where ρ_0, θ_0 and $\rho_\pm$, $\theta_\pm$ are the amplitudes and phases of the carrier and first side band waves. Substitution of Eq.(116) thus simplified on referring to Eqs. (118) leads to a system of coupled equations with respect to ρ_0, θ_0, $\rho_\pm$ and $\theta_\pm$

$$\frac{d\rho_0}{dt} = -2\beta\rho_0\rho_\pm^2 \sin\Phi, \tag{119}$$

$$\frac{d\theta_0}{dt} = \beta\{\rho_0^2 + 2\rho_\pm^2 (2 + \cos\Phi)\}, \tag{120}$$

$$\frac{d\rho_\pm}{dt} = \beta\rho_0^2\rho_\pm \sin\Phi, \tag{121}$$

$$\frac{d\theta_\pm}{dt} = \beta\{\rho_0^2 (2 + \cos\Phi) + 3\rho_\pm^2\}, \tag{122}$$

where Φ is the characteristic phase angle defined as

$$\Phi = 2(\theta_\pm - \theta_0) - \Omega t \tag{123}$$

indicating a phase shift due to nonlinear and dispersive effects, while Ω is defined as

$$\Omega = 2k^2 = \omega_+ + \omega_- - 2\omega_0, \tag{124}$$

indicating a frequency shift due to dispersion. In addition, conservation of energy can be written as

$$\rho_0^2 + 2\rho_\pm^2 = a = \text{const.}, \tag{125}$$

where a is the initial value of energy. Eliminating ρ_0, θ_0, $\theta_\pm$ and Φ from Eqs. (119) to (122) and putting $\rho = \rho_\pm^2$, we obtain the differential equation with respect to ρ

$$\frac{d\rho}{dt} = \beta[f(\rho)]^{1/2}, \tag{126}$$

where $f(\rho)$ is a polynomial of the fourth order expressed as

$$f(\rho) = 7\rho^4 - 2(2a + \frac{3\Omega}{\beta})\rho^3 - \{\frac{\Omega}{\beta}(\frac{\Omega}{\beta} - 4a) + 12\Gamma\}\rho^2 - 8\Gamma(\frac{\Omega}{2\beta} - a)\rho - 4\Gamma^2$$

$$= 7(\rho - \alpha_1)(\rho - \alpha_2)(\rho - \alpha_3)(\rho - \alpha_4), \quad (\alpha_1 > \alpha_2 > \alpha_3 > \alpha_4), \tag{127}$$

α_1, α_2, α_3 and α_4 are four real roots of $f(\rho) = 0$ and Γ is an integral constant, being determined by the initial condition as

$$\Gamma = \rho_\pm^{\,2}(o)(2 + \frac{1}{2}\rho_\pm^{\,2}(o) - \frac{\Omega}{2\beta}) \ . \tag{128}$$

Then the solution of Eq.(126) ρ, namely the side-band wave amplitude is obtained in terms of Jacobi's elliptic function sn (e.g. Dwight, 1961) as

$$\rho_\pm = [\frac{\alpha_2(\alpha_1-\alpha_3) - \alpha_1(\alpha_2-\alpha_3)\ \mathrm{sn}^2(pt)}{(\alpha_1-\alpha_3) - (\alpha_2-\alpha_3)\ \mathrm{sn}^2(pt)}]^{1/2} \quad \text{for } 0 < \beta < \beta_c \tag{129}$$

$$= [\frac{\alpha_3(\alpha_2 - \alpha_4) - \alpha_4(\alpha_2 - \alpha_3)\ \mathrm{sn}^2(pt)}{(\alpha_2 - \alpha_4) - (\alpha_2 - \alpha_3)\ \mathrm{sn}^2\ (pt)}]^{1/2} \quad \text{for } \beta < 0,\ \beta > \beta_c\ , \tag{130}$$

where

$$p = \frac{\beta}{2}\ [7(\alpha_1-\alpha_3)(\alpha_2-\alpha_4)]^{1/2} \tag{131}$$

and the characteristic phase is written as

$$\Phi = \cos^{-1}[\{\rho(\frac{3}{2}\rho + \frac{\Omega}{2\beta} - a) + \Gamma\}/\rho(a - 2\rho)],\ \rho = \rho_\pm^{\,2}\ . \tag{132}$$

The carrier wave amplitude can also be obtained immediately from Eq.(125). In addition, the recurrence time t_R is expressed in terms of the complete elliptic integral of the first kind K (e.g. Dwight, 1961) as

$$t_R = \frac{4K(k_e)}{\beta p}\ , \qquad k_e^{\,2} = \frac{(\alpha_1-\alpha_4)(\alpha_2-\alpha_3)}{(\alpha_1-\alpha_3)(\alpha_2-\alpha_4)}\ , \tag{133}$$

where k_e is the modulus of complete elliptic integrals of the first kind.

From Eqs.(129), (130) and (133), it is evident that nonlinear characteristics of phenomena can be divided into three regions with respect to the nonlinear coefficient β, namely the regions of negative and positive β and for positive β, the strongly and weakly nonlinear regions whose critical β_c in the marginal state is given by

$$\beta_c = \frac{\Omega}{4 - 6\rho_\pm^{\,2}(0)} \cong \frac{\Omega}{4} = \frac{k^2}{2}\ . \tag{134}$$

This has already been recognized by numerical solutions presented so far and some of them has been compared with the present theory, analytical but approximate, as was seen in Fig. 17 on the temporal evolution of the waveform, in Figs. 16(b) and 19(a) on the amplitudes of Fourier components, and in Fig. 19(b) on the characteristic phase. From these, we have seen a good agreement between analytical and numerical solutions and finally present additional results on the dependence of the recurrence time on nonlinearity in Figure 20 that may provide further confidence in analytical solutions. Solid curves refer to analitical solutions and solid circles or dashed curves connecting them refer to numerical solutions, both being in fairly good agreement. Fig.(a) represents the recurrence time in the region of positive nonlinear coefficients that is also devided into the weakly and strongly nonlinear regions. In the strongly nonlinear regime ($\beta > \beta_c \cong 0.03$), the recurrence time decreases with increasing nonlinearity, while in the weakly nonlinear regime ($0 < \beta < \beta_c$), it decreases with decreasing nonlinearity. At the critical β_c, the recurrence time goes to infinity

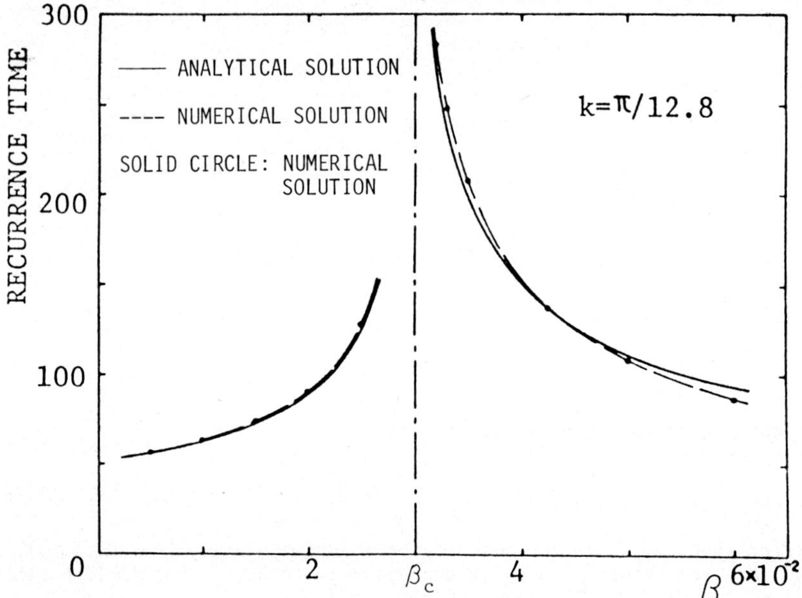

(a): case β > 0.

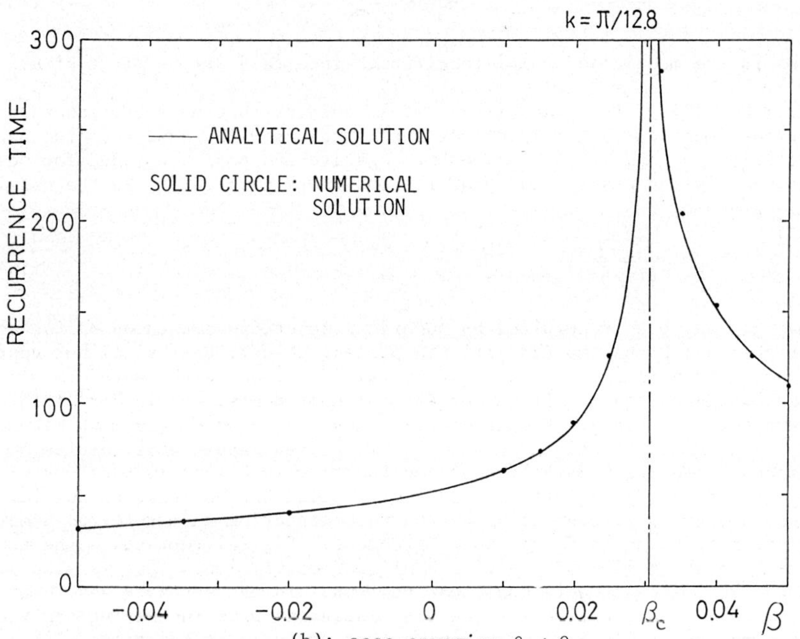

(b): case covering β < 0.

Fig. 20. Dependence of the recurrence time on nonlinearity.

and Fig.(b) shows its extension to the region of negative nonlinear coefficients. The recurrence time is continuous at $\beta = 0$ and decreases with increasing $|\beta|$ in the region $\beta < 0$.

The results indicate that the first side band waves play a crucial role in determining the mechanisms of recurrence phenomena. It is thus generally sufficient, except for very strong nonlinearities, to consider the carrier and first side band waves in the Fourier spectrum, making an analytical treatment rather simple and feasible.

10. CONCLUSION

We have thus attemped to treat a variety of problems in nonlinear electromagnetics in terms of quasi-particles, solitons and/or shocks and to apply such a concept to physical interpretations for several examples of numerical solutions or temporal evolution of the waveform described by the Korteweg-de Vries-Burgers and nonlinear Schrödinger equations subject to an initial condition. Although the quasi-particle approach will prove its real value particularly for a system of many quasi-particles from the mathematical point of view, practically by solving their trajectory equations and kinetic equation, its concept is always valuable and useful for obtaining a unified view and physical interpretations of nonlinear wave phenomena, simply based upon dispersion relations. For instance, the generations of new side band waves by self-modulation of a wave-packet described by the nonlinear Schrödinger equation can be regarded as the generation of quasi-particles that die out completely at the recurrence time, as discussed in Section 9.

Another thing to be emphasized is the existence of recurrence phenomena for both cases of the KdV and nonlinear Schrödinger equations that has been proved by numerical solutions as well as analytical but approximate solutions. Although it has only been indicated that the so-called "dark solitons" may be produced in the region of negative nonlinear coefficients in the Schrödinger equation as seen in Fig. 18, further studies for negative but large nonlinear coefficients have confirmed the generation of definite dark solitons which will be discussed elsewhere.

From the point of view of applications, there seems to be a growing importance for nonlinear electromagnetic pulses (NEMP) in a broad sense in general and in the field of environmental electromagnetics in particular, since propagation or transmission media often exhibit nonlinear characters for large amplitude waves or pulses with short duration such as "Optical EMP", "Atmospheric or lightning EMP" and "Nuclear EMP". Applications of the present approach to specific problems of environmental electromagnetics or EMP will be a largely open task of interest in the future.

ACKNOWLEDGEMENTS

One of the authors (Kikuchi) expresses his sincere thanks to Professor N. Marcuvitz who directed his interest to the quasi-particle approach to nonlinear wave problems while he was associated with New York University, Bronx, New York. Consequently, the descriptions of this concept come largely from him (Marcuvitz, 1970, 1980, this issue).

REFERENCES

Dwight, H. B., Tables of Integrals and Other Mathematical Data (Macmillan, Toronto, 1961), p. 177.

Gardner, C. S., Greene, J. M., Kruskal, M. D., and Miura, R. M., Phys. Rev. Letters 19, 1097 (1967).

Kadomtsev, B. B. and Karpman, V. I., Usp. Fiz. Nauk 103, 193 (1971): Soviet Phys. USPEKHI 14, 40 (1971).

Karpman, V. I., Nonlinear Waves in Dispersive Media - I (Translation: Pergamon, Oxford, 1975).

Kikuchi, H., J. Atmos. Terr. Phys. 38, 1055 (1976).

Marcuvitz, N., IEEE Trans. Electron Devices ED-17, 252 (1970).

Marcuvitz, N., Proc. IEEE 68, 1380 (1980).

Whitham, G. B., Linear and Nonlinear Waves (Wiley, New York, 1974).

Zabusky, N. J. and Kruskal, M. D., Phys. Rev. Letters 15, 240 (1965).

Nonlinear and Environmental Electromagnetics, edited by H. Kikuchi
Elsevier Science Publishers B.V., Amsterdam, 1985 – Printed in The Netherlands

SOLITONS IN ONE AND TWO DIMENSIONAL TRANSMISSION SYSTEMS

Akira NOGUCHI

Department of Electrical Engineering,
Faculty of Science and Technology, Keio University
3-14-1 Hiyoshi Kohokuku Yokohama 223 Japan

ABSTRACT

The basic properties of solitons are observed in nonlinear and dispersive electrical transmission systems. When the frequency of the input signal is low, the line equation is reduced to the Korteweg-de Vries equation or the modified KdV equation and compressive or rarefactive solitons propagate along the line. At sufficiently high frequency the nonlinear Schroedinger equation is derived. The modulational instability and the envelope solitons are observed. Numerical experiments on the collisions of two dimensional solitons in the transmission system are shown and the exact-solution for an irregular collision between two solitons is presented.

1. INTRODUCTION

In general, large amplitude pulse changes its waveform not only by dispersive effect but also by nonlinear effect, when it propagates in a medium. But Korteweg and de Vries [1] showed that a certain kind of balance between two effects yields a stable solitary pulse for a nonlinear partial differential equation which is now called the Korteweg-de Vries (KdV) equation. On the other hand Fermi, Pasta and Ulam [2] made a numerical study of thermalization of energy in a nonlinear lattice and observed a recurrence phenomenon contrary to their expectation.

In 1965 Zabusky and Kruskal [3] made the famous numerical experiments for the KdV equation. They found that very stable stationary pulses are generated and travel like particles from the original sinusoidal wave and that these facts contribute to the recurrence phenomenon. They named the solitary waves solitons. Since this discovery of solitons, a large number of studies have been made in various branches of physics, mathematics and engineering [4]-[9]. Many nonlinear phenomena can be described by the concept of solitons. Numerous nonlinear evolution equations which have soliton solutions have been proposed and investigated besides the KdV equation.

With transmission line, we can have simple models for physical phenomena and moreover have some applications of them. Some studies on solitons have been made by using nonlinear transmission lines [10]-[20]. In this article, we consider the relation between nonlinear transmission system equations and some soliton equations, and show several experimental results to observe properties of solitons.

2. TRANSMISSION LINE AND THE KORTEWEG-DE VRIES EQUATION

2.1 Theory

In this section, we derive the KdV equation from a nonlinear dispersive transmission line in a long wave approximation [11]. The transmission line consists of a number of identical sections that contain nonlinear elements. A typical equivalent circuit of them is shown in Fig.1. The transmission line equation is written as follows:

$$\frac{\partial^2 Q(V)}{\partial t^2} = \frac{1}{L}\frac{\partial^2 V}{\partial x^2} + C'\frac{\partial^4 V}{\partial x^2 \partial t^2} . \tag{1}$$

In this line the capacitor $C_n(V)$ is nonlinear and its stored charge $Q(V)$ is a nonlinear function of the applied voltage V. We assume the nonlinear relation

$$Q(V) = C(V - \alpha V^2) . \tag{2}$$

The capacitor C' makes the line dispersive. The dispersion relation can be obtained by rewriting partial differential operators

$$\frac{\partial}{\partial t} \to i\omega , \quad \frac{\partial}{\partial x} \to - i k , \tag{3}$$

as in the linear case. It yields

$$\omega (1 - 2\alpha V)\omega + \frac{C'}{C}\omega^2 k^2 = \frac{k^2}{L C} . \tag{4}$$

In a long wave and a small but finite amplitude approximation, eq.(4) is

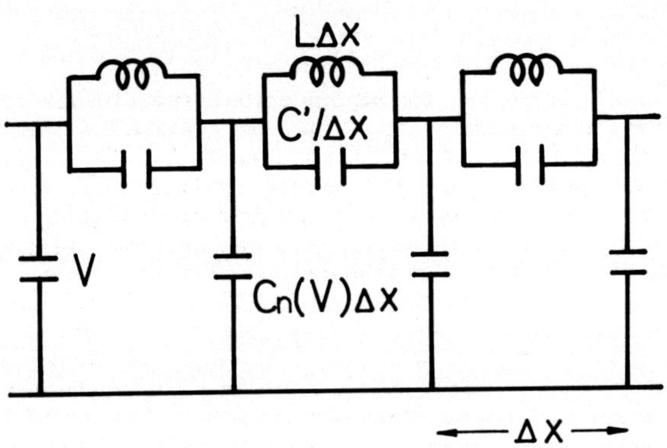

Fig.1. Equivalent circuit of the nonlinear dispersive transmission line.

written as

$$\omega = \pm \left(1 + \alpha V - \frac{C'}{2C} k^2 \right) \frac{k}{\sqrt{LC}} . \tag{5}$$

The positive and negative signs of eq.(5) describe the waves propagating towards the positive and the negative directions respectively. We can take the positive sign in eq.(4) without loss of generality. We now take the inverse transformation of eq.(5) by rewriting

$$\omega \to -i \frac{\partial}{\partial t} , \quad k \to i \frac{\partial}{\partial x} . \tag{6}$$

and operate it on V. We obtain

$$\frac{\partial V}{\partial t} + \frac{1}{\sqrt{LC}} \left(\frac{\partial V}{\partial x} + \alpha V \frac{\partial V}{\partial x} + \frac{C'}{2C} \frac{\partial^3 V}{\partial x^3} \right) = 0 . \tag{7}$$

This is the approximate equation for the waves propagating to the positive direction in the transmission line. We introduce a coordinate system moving with the velocity $1 / \sqrt{LC}$

$$\xi = \sqrt{C' / 2C} \, (x - t / \sqrt{LC}) , \qquad \tau = t , \tag{8}$$

and transform V by the variable

$$u = \frac{\alpha}{6 \sqrt{LC}} V . \tag{9}$$

Then we obtain

$$\frac{\partial u}{\partial \tau} + 6 u \frac{\partial u}{\partial \xi} + \frac{\partial^3 u}{\partial \xi^3} = 0 . \tag{10}$$

Eq.(10) is the Korteweg-de Vries (KdV) equation which was investigated numerically by Zabusky and Kruskal [3]. The one-soliton solution of eq.(10) is given by

$$u = u_o \, \mathrm{sech}^2 \sqrt{\frac{u_o}{2}} \left(\xi - \frac{u_o}{3} \tau \right) . \tag{11}$$

From this expression, we can see that the velocity of soliton is proportional to the amplitude and that the width is inversely proportional to the square of the amplitude. It has been shown that we can obtain the exact solution that describes the interaction among multi solitons (N-soliton solution).

We consider the case when we have the nonlinear charge-voltage relation

$$Q(V) = C (V - \alpha V^3) , \qquad (12)$$

instead of eq.(2). With the same procedure which derives eq.(10), the transmission line equation (1) becomes the modified Korteweg-de Vries (mKdV) equation

$$\frac{\partial u}{\partial \tau} + 6 u^2 \frac{\partial u}{\partial \xi} + \frac{\partial^3 u}{\partial \xi^3} = 0 . \qquad (13)$$

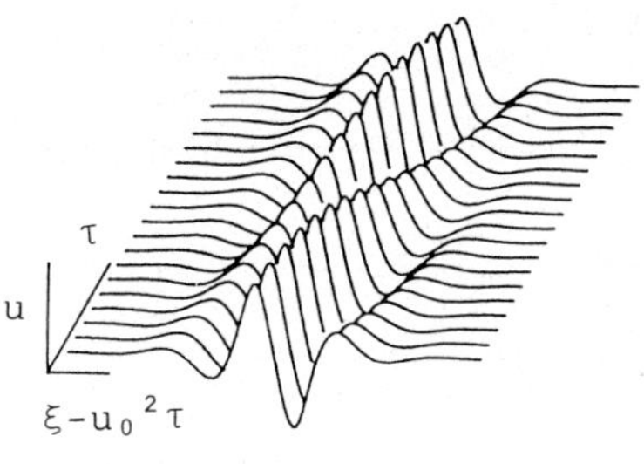

Fig.2. Breather soliton.

The one-soliton solution is

$$u = \pm u_o \operatorname{sech} u_o (\xi - u_o^2 \tau) . \qquad (14)$$

We have a rarefactive soliton as well as a compressive one for eq.(13). The mKdV equation also has a oscillating pulse solution

$$u = 2 \frac{\partial}{\partial \xi} \tan^{-1} \left[\frac{b}{a} \frac{\cos \{ a\xi + a (a^2 - 3 b^2) \tau \}}{\cosh \{ b\xi + b (b^2 - 3 a^2) \tau \}} \right] . \qquad (15)$$

This is a kind of soliton called breather soliton. It is shown in fig.2. We will see that at high frequency the breather soliton is degenerate into the envelope soliton in section 3.

2.2 Experiments

The fundamental properties of the KdV soliton can be observed experimentally in the transmission line [11],[20]. In the experiments we used reverse biased diodes (varicaps) to obtain the nonlinear relation (2). However to carry out the experiments of the mKdV solitons we use the line with saturating ferromagnetic inductors as the nonlinear elements. There are no essential difference in the derivation of the mKdV equation. It is known that solitons are very stable pulses. Any shape of the original input is decomposed into a number of solitons and a small amplitude oscillating tail. In fig.3 we show the wave evolutions from the initial box-car inputs. The number and the amplitudes of solitons are determined by the amplitude and the shape of the input signal. We can observe one soliton in fig.3-b and six in fig.3-a. There is no soliton but only oscillatary tail in fig.3-c. This is because we have no rarefactive soliton solutions for the KdV eq.(10). In fig.3-d rarefactive and compressive solitons can be found in the case of the mKdV line. Experiments about interactions of solitons are indicated in fig.4. Fig.4-a shows two different amplitude solitons propagating in the same direction. Since the larger amplitude soliton propagates faster, it overtakes the smaller one. We can find nonlinear behavior during interaction; the amplitude at n=14 is smaller than that of linear summation of two individual solitons and the position of the smaller soliton after n=14 is shifted to the right in the oscillograms. In the case that two solitons propagate in opposite directions

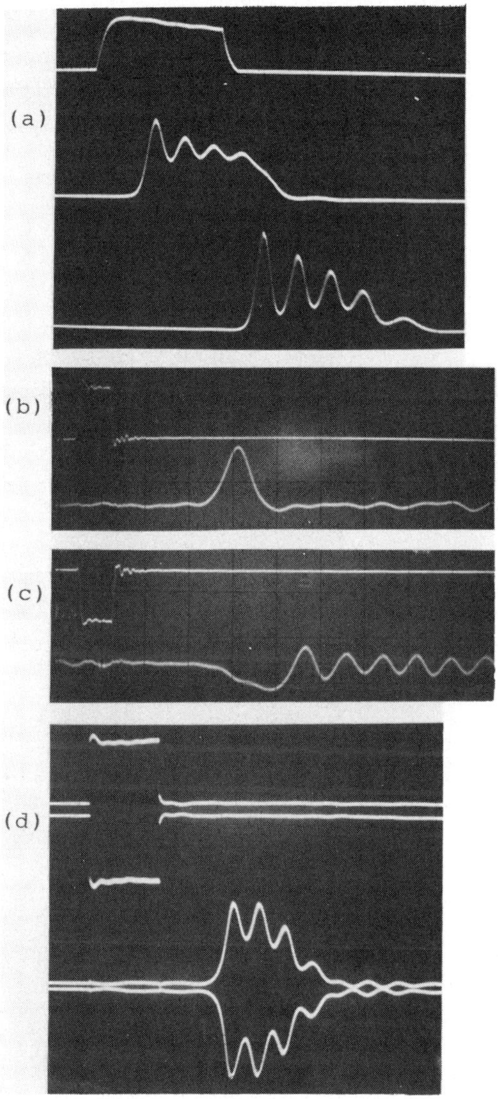

Fig.3. Observation of formation of solitons along
:(a)-(c) the KdV line
;(d) the mKdV line.

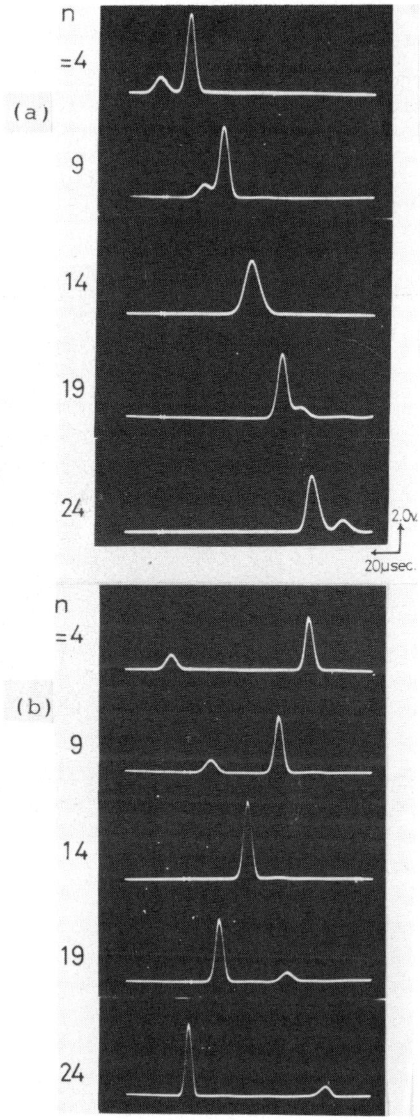

Fig.4. Interaction of two solitons along the transmission line.
(a) Overtaking interaction.
(b) Head-on collision.

each other, the experimental results in fig.4-b show little nonlinear behavior. The KdV equation does not describe this case.

The recurrence phenomena are shown in fig.5. Since the larger amplitude part of the wave propagates faster the small amplitude part, the input sinusoidal wave steepens and generates solitons as it propagates. At n=24 the initial sinusoidal wave is restored and the steepening resumes. The amplitude of the fundamental sinusoid and that of the higher harmonics along the line are plotted in fig.4-b. It shows clearly the generation of the higher harmonics and the recurrence into the original sinusoidal wave. On the mKdV transmission line we can observe the same recurrence phenomenon in figs.4-c and 4-d.

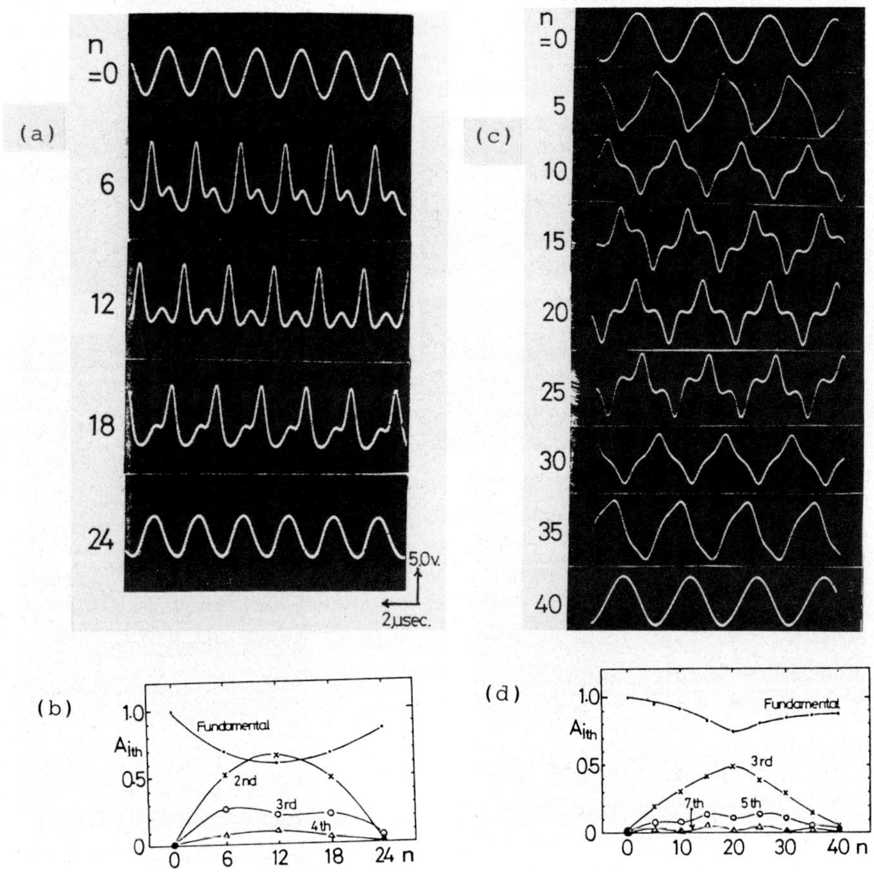

Fig.5. Observation of the recurrence phenomena along
:(a) the KdV line ;(c)the mKdV line.
(b) and (d) show spatial evolution of each harmonics.

3. ENVELOPE-SOLITONS AND MODULATIONAL INSTABILITY

3.1 Derivation of the Nonlinear Schroedinger Equation

In this section, we assume the dispersive property of the line is so strong that the formation of solitons mentioned at the previous section is suppressed. Therefore, we need not consider the generation of the higher order harmonic wave by the nonlinear elements. We put the applied voltage V in eq.(1) as a form

$$V(x,t) = \Phi(x,t) \exp\{ i(\omega_0 t - k_0 x) \} + \text{complex conjugate} . \tag{16}$$

Here ω_0 is the angular frequency and k_0 is wave number of the carrier wave. $\Phi(x,t)$ is the complex amplitude; $|\Phi|$ is the envelope of V(x,t) and arg(Φ) is the phase difference between V(x,t) and the sinusoidal carrier signal. Nonlinear element relation is to be expressed by eq.(12). In this case, the dispersion relation becomes

$$\omega^2(1 - 3\alpha|\Phi|^2) + \frac{C'}{C} k^2 = \frac{k^2}{LC} . \tag{17}$$

With the Taylor expansion of eq.(17) around $k = k_0$, $\omega = \omega_0$, $|\Phi|^2 = |\Phi_0|^2$, we have

$$\omega = \omega_0 + \left(\frac{\partial\omega}{\partial k} \right)_0 (k-k_0) + \frac{1}{2}\left(\frac{\partial^2\omega}{\partial k^2} \right)_0{}^2 (k-k_0)^2 + \left(\frac{\partial\omega}{\partial|\Phi|^2} \right)_0 (|\Phi|^2 - |\Phi_0|^2) . \tag{18}$$

Taking the transformation $\omega \to -i \frac{\partial}{\partial t}$, $k \to i\frac{\partial}{\partial x}$ in this expression and operating it on V(x,t) in eq.(18), we can derive the following equation [14],

$$i \frac{\partial\Phi}{\partial t} + iv_g \frac{\partial\Phi}{\partial x} - \frac{1}{2} v_g' \frac{\partial^2\Phi}{\partial x^2} + \gamma (|\Phi|^2 - |\Phi_0|^2) = 0 , \tag{19}$$

where

$$v_g = \left(\frac{\partial\omega}{\partial k} \right)_0 , \quad v_g' = \left(\frac{\partial^2\omega}{\partial k^2} \right)_0 \text{ and } \gamma = \left(\frac{\partial\omega}{\partial|\Phi|^2} \right)_0 .$$

The first and the second terms describe the envelope of the wave moving with the group veocity v_g. The third term makes the wave packet dispersive, and the last nonlinear term also deforms it. It is expected that on the balance between the group-dispersion and the nonlinear effect, there may be a steady state envelope wave. We introduce the coordinate system defined by

$$\xi = \sqrt{\frac{v_g'}{2}} (x - v_g t) \text{ and } \tau = t \,\text{sign}(v_g') , \tag{20}$$

and rewrite eq.(19) as

$$i\frac{\partial\Phi}{\partial\tau} - \frac{\partial^2\Phi}{\partial\xi^2} - \kappa(|\Phi|^2 - |\Phi_0|^2) = 0 , \tag{21}$$

where $\kappa = \gamma\,\mathrm{sign}(v_g')$. This is the nonlinear Schroedinger equation which has been derived in various branches of physics such as nonlinaer optics, deep water waves, plasma physics, superconductivity, vortex motion, and so on. Generally, eq.(21) describes the evolution of the wave envelope of a nonlinear dispersive system. When $\kappa > 0$ and $|\Phi_0| = 0$, the nonlinear Schroedinger equation has a stationary wave packet solution as follows:

$$\Phi = \sqrt{2\kappa}a \exp\{ \cdot 4i(b^2 - a^2)\tau - 2ib\xi \} \,\mathrm{sech}(2a\xi - 8ab\tau) . \tag{22}$$

Where a and b are arbitrary constants. It is a kind of soliton whose properties are similar to those of the KdV soliton. The N-soliton solution has been obtained analytically. When $\kappa < 0$, the nonlinear Schroedinger equation has a dark soliton or a envelope hole soliton. Another interesting property of the nonlinear Schroedinger equation is modulational instability. It is known that a wave train is unstable to infinitesimal modulational perturbation.

3.2 Experiments

Most experiments in this section have been performed using the same transmission line to obtain the mKdV solitons in section 2.2. The input signal is sinusoidal wave with amplitude modulation. The situation differs from that in the previous section, since the frequency or the wave-number of the signal is large. Because of the strong dispersion, the generation of the higher harmonics and the creation of solitons are suppressed. In fig.6, we show the

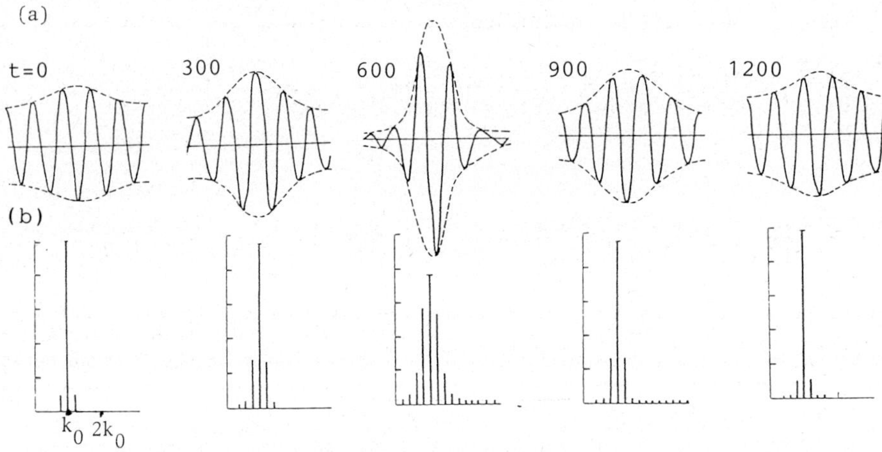

Fig.6. Numerical result of evolution for modulated wave.
(a) waveforms (b) Fourier spectra

results of the numerical experiment for the modulational instability [14]. From the waveforms and their Fourier spectra, we can see the increase in modulational amplitude as the wave propagates along the transmission lines. It is found that the decrease in modulational amplitude begins after the time t=600 and almost the same envelope-wave is recovered at t=1200. This fact can be interpreted as the recurrence phenomena in the nonlinear Schroedinger equation.

The modulational instability is observed by the experimental results shown in fig.7 [15]. Input signal is amplitude-modulated wave with modurational ratio 0.1. We can see that, though the carrier amplitude decreases due to loss, the index of amplitude modulation gradually increases. This modulational instability tends to form envelope solitons. Fig.8 shows the formation of envelope solitons at n=22nd stage. However the recurrence phenomenon is not evident because of loss. In fig.9 we can observe dark soliton or envelope-hole soliton in the transmission line whose group-dispersion v_g' is positive. Input modulated wave (modulational ratio 2.0) is gradually changing into dark soliton.

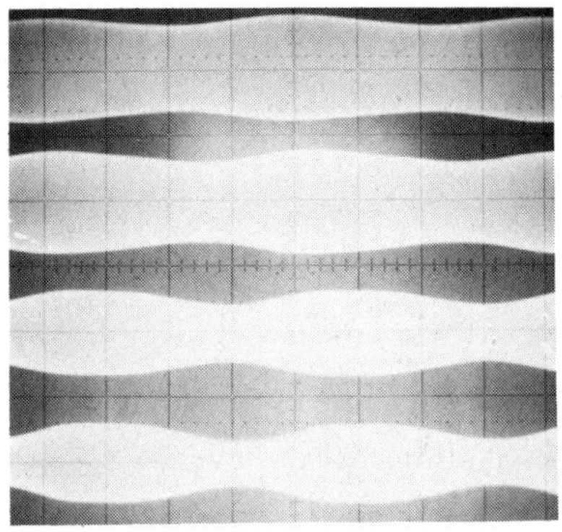

Fig.7. Wave response to the modulated wave showing the modulational instability.

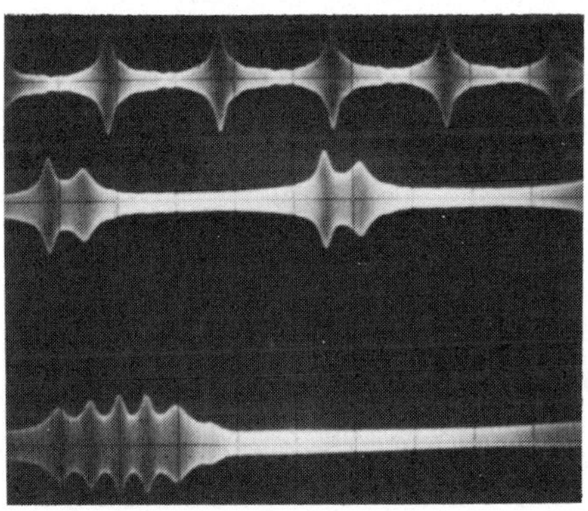

Fig.8. Observation of envelope soliton at n=22.

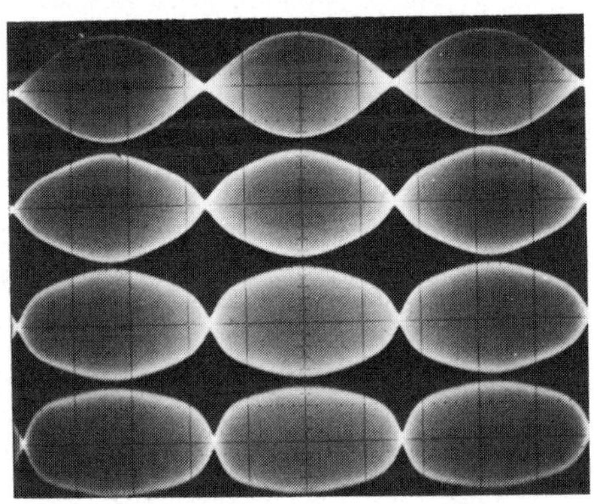

Fig.9. Observation of formation of dark soliton.

4. TWO DIMENSIONAL SOLITONS

4.1 The Kadomtsev-Petviashvili Equation

In order to investigate the propertie; of soliton in two dimensional space, we consider the nonlinear transmission network shown in fig.10. When the transmission system consists of a number of lumped circuit elements it becomes dispersive. As the continuum approximation for the network equation, we obtain the partial differential equation as

$$\frac{\partial^2 Q(V)}{\partial t^2} = \frac{1}{L} \nabla^2 V + \frac{h^2}{12} \nabla^4 V , \tag{23}$$

where

$$\nabla^2 = \frac{\partial^2}{\partial x^2} + \frac{\partial^2}{\partial y^2} .$$

In eq.(23) V_{ij} is considered as a function $V(x,y)$ of the two space coodinates ($x = ih$ and $y = jh$). Without the inductors between two diagonal nodes the wave equation of the system is anisotropic. When we assume that $Q(V)$ have quadratic nonlinearlity (12), the dispersion relation of eq.(23) is written by

$$\omega^2 = \frac{k_x^2 + k_y^2 + \frac{h^2}{12} (k_x^2 + k_y^2)^2}{\sqrt{LC\{1 - 2\alpha V\}}} \tag{24}$$

where k_x and k_y denote wave numbers in x and y directions respectively. We suppose that the wave propagates nearly to the x direction. In the long wave and the small signal approximations as in section 2.1, we obtain the next dispersion relation which describes waves propagating almost to the positive x

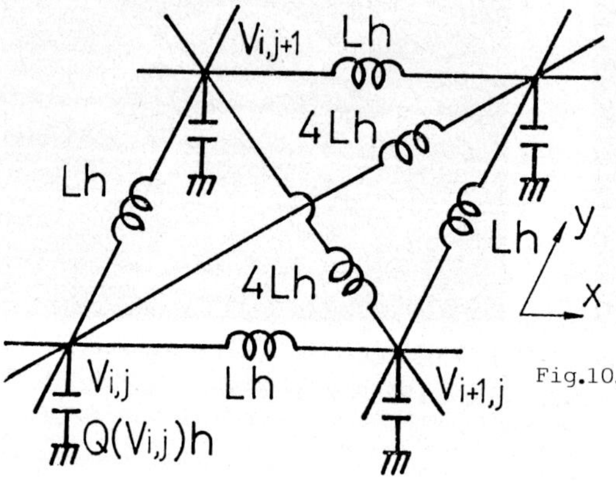

Fig.10. A section of two dimensional nonlinear transmission system. $\Delta x = \Delta y = h$

direction.

$$\omega k_x = \frac{1}{\sqrt{L\,C}} \{ k_x^2 + k_x \alpha V k_x + \frac{1}{2} k_y^2 + \frac{h^2}{24} k_x^4 \} . \tag{25}$$

Taking the inverse transform $\omega \to -i \frac{\partial}{\partial t}$, $k_x \to i \frac{\partial}{\partial x}$, $k_y \to i \frac{\partial}{\partial y}$ and operating it on V, we derive the expression

$$\frac{\partial}{\partial x} \left[\frac{\partial V}{\partial t} + \frac{1}{\sqrt{L\,C}} \left(\frac{\partial V}{\partial x} + \alpha V \frac{\partial V}{\partial x} + \frac{h^2}{24} \frac{\partial^3 V}{\partial x^3} \right) \right] = - \frac{1}{2} \frac{\partial^2 V}{\partial y^2} . \tag{26}$$

Introducing the coordinate system

$$\xi = \frac{h}{\sqrt{24}} \left(x - \frac{t}{\sqrt{L\,C}} \right) ,$$

$$\eta = 2y \quad \text{and} \quad \tau = t ,$$

and putting the variable

$$u = \frac{\alpha}{6\sqrt{L\,C}} V ,$$

we obtain the two dimensional KdV equation as

$$\frac{\partial}{\partial \xi} \left[\frac{\partial u}{\partial \tau} + 6 u \frac{\partial u}{\partial \xi} + \frac{\partial^3 u}{\partial \xi^3} \right] = - \frac{\partial^2 u}{\partial \eta^2} , \tag{27}$$

The equation above is also called the Kadomtsev-Petviashvili (KP) equation and has one soliton solution [21]

$$u = K^2 \operatorname{sech}^2 \{ K (\xi + l \eta) - \Omega \tau \} , \tag{28}$$

where

$$\Omega = K (4 K^2 + l^2) . \tag{29}$$

It should be mentioned that the solution eq.(28) is the oblique plane soliton and not fully localized one in two dimensional space. Some properties of soliton in two space dimensions are the same as those in one dimension. The exact solution which describes oblique interactions among multi-solitons has been solved analytically [22]. However it has been pointed out that the two-soliton solution becomes singular, if the angle of intersection of two soliton is smaller than a certain critical angle [23],[24]. As an actual physical situation it must not take place. In the next section, we investigate this problem numerically and present the analytical solution of this irregular interaction.

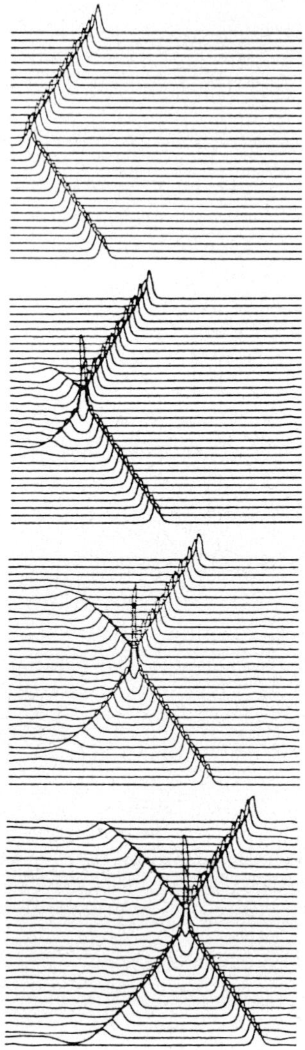

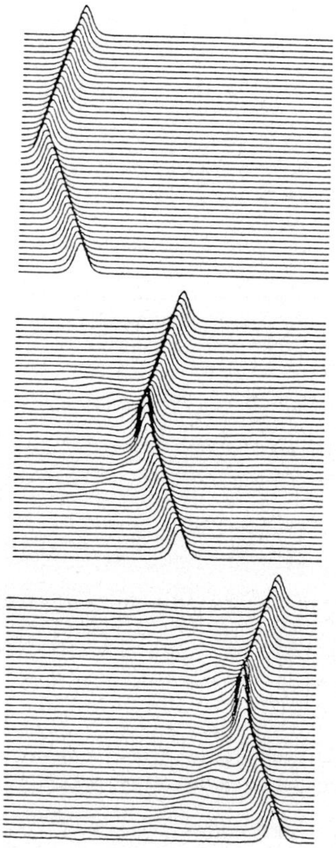

Fig.12. Irregular interaction of two solitons in two dimensional space.

Fig.11. Regular interaction of two solitons in two dimensional space

4.2 Numerical experiments

Two oblique plane soliton intersect each other somewhere because they are of infinite extent in two dimensional space. It makes difficult to distinguish the effect resulting after interaction from that due to the initial and boundary condition in numerical experiments [24]. Therefore we chose two planer solitons of semi-infinite lengthes as the initial condition. Fig.11 shows regular interaction between two solitons. Both solitons are propagating almost to the right direction, and we can observe the result of the interaction from this figure. The position shifts can be found as the results of nonlinear collision. In fig.12 we show the numerical experiment of an interaction in the case of small angle intersection . By analytic solution, two singular solitons with $-\mathrm{cosech}^2$ profile generate in the backward of the intersecting point. The numerical results are very interesting. At the point of the collision between two initial solitons, a new soliton seems to be generated. The third soliton which propagates faster than original solitons develops gradually. And additional two solitons are created from the interactions between the third soliton and the original ones. With careful consideration, we conclude that the irregular interaction as in fig.12 can be described by three-soliton solution with resonances not by two-soliton solution. This solution is written as

$$u = 2 \frac{\partial^2}{\partial \xi^2} \log f ,$$

$$f = 1 + \exp 2\theta_3 + \exp 2(\theta_1 + \theta_2) + \exp 2(\theta_3 + \theta_1) + A_{12} \exp 2(\theta_1 + \theta_2 + \theta_3) \tag{30}$$

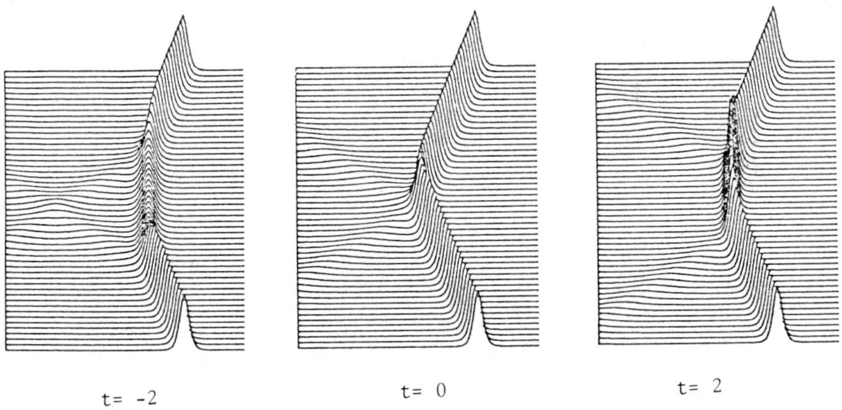

Fig.13 Three-soliton solution with resonances.

where

$$\theta_i = K_i\ (\xi + l_i \eta\ ) - \Omega_i \tau$$

$$D(K_i,\ K_i\ l_i,\ \Omega_i) = \Omega_i - K_i\ (\ 4\ K_i^{\ 2} + l_i^{\ 2}\ ) = 0$$

$$A_{ij} = \frac{D(K_i - K_j,\ K_i\ l_i - K_j\ l_j,\ \Omega_i - \Omega_j)}{D(K_i + K_j,\ K_i\ l_i + K_j\ l_j,\ \Omega_i + \Omega_j)}\ .$$

Fig.13 illustrated it. We see that the numerical results correspond with analytic ones after the interaction between two solitons.

5. CONCLUSION

In this paper we have attempted to describe how waves in the nonlinear transmission systems are related to the solitons. In sections 2 and 3 some fundamental properties of the KdV soliton and the envelope soliton are observed in the transmission systems. We have not refer to solitons in nonlinear lattice network [10] and Josephson junction transmission line [25]. Toda lattice equation [26] and the sine-Gordon equation, which describe those transmission systems respectively, also play an important role for the studies of solitons. Such one-dimensional soliton phenomena have been almost understood in this two decades of years. However it is not well investigated what is the higer dimensional soliton. Creation of new solitons by the interaction between original solitons, which has been shown numerically in section 4.2, seems to be one of the fundamental properties of higher-dimensional soliton.

ACKNOWLEDGMENTS

The auther is grateful to H. Watanabe, T. Makino and Y. Uehara for preparing the manuscript.

REFERENCES

[1] Korteweg,D.J. and de Vries,G.: 1895, On the change of form of long waves advancing in a rectangular canal, and on a new type of long stationary waves, Phil.Mag., 39, 422-443.

[2] Fermi,E., Pasta,J.R. and Ulam,S.M.: 1955, Collected Works of E.Fermi, vol.II, Univ. Chicago Press, 978-988.

[3] Zabusky,N.J. and Kruskal,M.D.: 1965, Interaction of solitons in a collisionless plasma and the recurrence of initial states. Phys.Rev.Lett., 15, 240-243.

[4] Scott,A.C., Chu,F.Y.F. and McLaughlin,D.W.: 1973, The soliton, a new concept in applied science. Proc.IEEE, 61, 1443-1483.

[5] Bullough,R.K. and Caudrey,P.J.(Eds.) 1980, "Solitons", Springer-Verlag.

[6] Lonngren,K. and Scott,A.(Eds.) 1978, "Solitons in Action", Academic Press.

[7] Bishop,A.R. and Schneider,T.(Eds.) 1978, "Solitons and Condensed Matter Physics", Springer-Verlag.

[8] Lamb,G.L.: 1980, "Elements of Soliton Theory", Wiley-Interscience.

[9] Dodd,R.K., Eilbeck,J.C., Gibbon,J.D. and Morris,H.C.: 1982, "Solitons and Nonlinear Wave Equations", Academic Press.

[10] Hirota,R. and Suzuki,K.: 1973, Theoretical and experimental studies of lattice solitons in nonlinear lamped networks, Proc.IEEE, 61, 1483-1491.

[11] Noguchi,A.: 1974, Solitons in a nonlinear transmission line, Elec. and Comm. in Japan, 57-A, 9-13.

[12] Ostrovskii,L.A., Papko,V.V. and Pelinovskii,E.N.: 1972, Solitary electromagnetic waves in nonlinear lines, Izvestiya Vysshikh Uchebnykh Zavedenii,Radiofisika, 15, 580-591. (English translation: Radiophysics and Quantum Electronics, 15, 438-446.)

[13] Kolosick,J.A., Landt,D.L., Hsuan,H.C.S. and Lonngrem,K.E.: 1974, Properties of solitary waves as observed on a nonlinear dispersive transmission line, Proc.IEEE, 62, 578-581.

[14] Mizumura,M. and Noguchi.A.: 1975, Modulational instability and envelop-soliton in a nonlinear transmission line, Trans.IECE Japan, 58-A, 704-711. (in Japanese)

[15] Yagi,T. and Noguchi,A.: 1976, Experimental studies on modulational instability by using nonlinear transmission lines, Elec. and Comm. in Japan, 59-A, 1-6.

[16] Gorshkov,K.A., Ostrovskii,L.A. and Papko,V.V.: 1976, Interactions and bound state of solitons as classical particles, Zh.Eksp.Teor.Fiz., 71, 585-593. (English translation: 1977, Sov.Phy.JETP, 44, 306-311.)

[17] Yagi,T., Hasegawa,M. and Noguchi,A.: 1977, Characteristics of AM-PM transformation of Modulated waves in nonlinear transmission lines, Elec. and Comm. in Japan, 60-A, 9-15.

[18] Yagi,T. and Noguchi,A.: 1977, Gyromagnetic nonlinear element and its application for a pulse shaping transmission line, Electronics Letters, 13, 683-685.

[19] Yagi,T., Watanabe,H. and Noguchi,A.: 1978, Growth and damping of solitons in a nonlinear transmission line with moving parameters, Proc.IEEE, 64, 702-703.

[20] Yagi,T. and Noguchi,A.: 1980, Harmonic frequency generation in nonlinear recurrence networks, Trans.IECE Japan, E63, 275-276.

[21] Kadomtsev,B.B. and Petviashvili,V.I.: 1970, The stability of solitary waves in weakly dispersive media, Dokl.Akad.Nauk SSR, 192, 753-756.

[22] Satsuma,J.: 1976, N-soliton solution of the two dimensional KdV equation, J.Phys.Soc.Japan, 40, 286-290.

[23] Miles,J.W.: 1977, Resonantly interacting solitary waves, J.Fluid.Mech, 79, 171-179.

[24] Kako,F. and Yajima,N.: 1980, Interaction ion acoustic soliton in two dimensional space, J.Phy.Soc.Japan, 49, 2063-2071.

[25] Scott, A.C., Chu,F.Y.F, and Reible,S.A.: 1976, Magneticflux propagation on a Josephson transmission line, J.Appl.Phys., 47 , 3272-3286.

[26] Toda,M.: 1981, "Theory of Nonlinear Lattice", Springer-Verlag.

Nonlinear and Environmental Electromagnetics, edited by H. Kikuchi
Elsevier Science Publishers B.V., Amsterdam, 1985 – Printed in The Netherlands

ION-ACOUSTIC SOLITONS IN A MULTI-COMPONENT PLASMA WITH NEGATIVE IONS

Yoshiharu Nakamura

Institute of Space and Astronautical Science
Komaba, Meguro-ku, Tokyo 153, Japan

ABSTRACT

The propagation of nonlinear ion-acoustic waves in a multi-component plasma with negative ions is investigated theoretically and experimentally. When the concentration of the negative ions is smaller than a critical value, compressive solitons are formed from a positive initial pulse and a negative pulse evolves into a wave train. At the critical concentration, both compressive and rarefactive solitons which are described by the modified Korteweg-de Vries equation are observed. When the concentration is larger than the critical, rarefactive solitons develop from a negative initial pulse and an initial positive pulse of small but finite amplitude evolves into a wave train. In this case, a positive pulse of large amplitude becomes a solitary wave whose characteristics agree with predictions using the pseudo-potential method.

1. INTRODUCTION

Solitons are now actively investigated in a variety of fields, for example, applied mathematics, fluid mechanics, nonlinear electric circuits, solid state physics and nonlinear optics. Solitons which were first named by Zabsuky and Kruskal(1965) are defined as localized waves which have a permanent form or shape and which stably interact with another soliton and retain their own shape after the collision(Drazin 1983). The Korteweg-de Vries(KdV) equation and the modified KdV equation possess soliton solutions.

In plasmas, the magnetosonic wave(Berezin and Karpman 1964) and the ion-acoustic wave(Washimi and Taniuchi 1966) of small but finite amplitudes are described by the KdV equation. Ion-acoustic KdV solitons were first observed by Ikezi, Taylor and Baker(1970) using a novel chamber called a double-plasma device (Taylor, Mackenzie and Ikezi 1972). Since that experiment, characteristics of ion-acoustic solitons and soliton phenomena such as velocities, cylindrical and spherical solitons, soliton resonances, solitons in a plasma with two-species of positive ions and recurrence have been studied and are reviewed in several papers (Ichikawa and Watanabe 1977;Ikezi 1978;Tran 1979;Nakamura 1982;Lonngren 1983).

The purpose of this paper is to describe experiments on ion-acoustic solitons in a plasma which includes contaminating negative ions. Naturally solitons in a usual plasma with a single species of positive ions and electrons are also included. To understand observed phenomena theoretically, numerical simulations are performed using the KdV equation, the modified KdV equation or

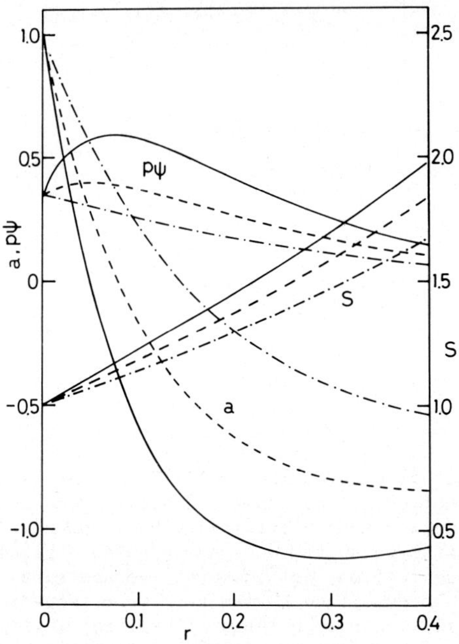

Figure 1. Examples of S, a and p when $\psi = 0.1$ as a function of r. The solid, dotted and dash-dotted curves are $\mu = 0.3$, 0.4 and 0.6, respectively.

original fluid equations.

2. THEORY AND NUMERICAL SIMULATIONS

2.1 Dispersion Relation

We consider a three-component plasma of positive and negative ions of a unit charge and Boltzmann electrons. We also assume a fluid model for the plasma. Then equations of motion, continuity equations and Poisson's equation are given as follows;

$$\frac{\partial v_+}{\partial \tau} + v_+ \frac{\partial v_+}{\partial \xi} = \frac{3T\partial n_+}{n_+ \partial \xi} - \frac{\partial \psi}{\partial \xi} , \tag{1}$$

$$\frac{\partial v_-}{\partial \tau} + v_- \frac{\partial v_-}{\partial \xi} = - \frac{3T\,\partial n_-}{n_- \partial \xi} + \frac{\partial \psi}{\mu \partial \xi} , \tag{2}$$

$$\frac{\partial n_+}{\partial \tau} + \frac{\partial n_+ v_+}{\partial \xi} = 0 , \tag{3}$$

$$\frac{\partial n_-}{\partial \tau} + \frac{\partial n_- v_-}{\partial \xi} = 0, \tag{4}$$

$$\frac{\partial^2 \psi}{\partial \xi^2} = \exp\psi + n_- - n_+ , \tag{5}$$

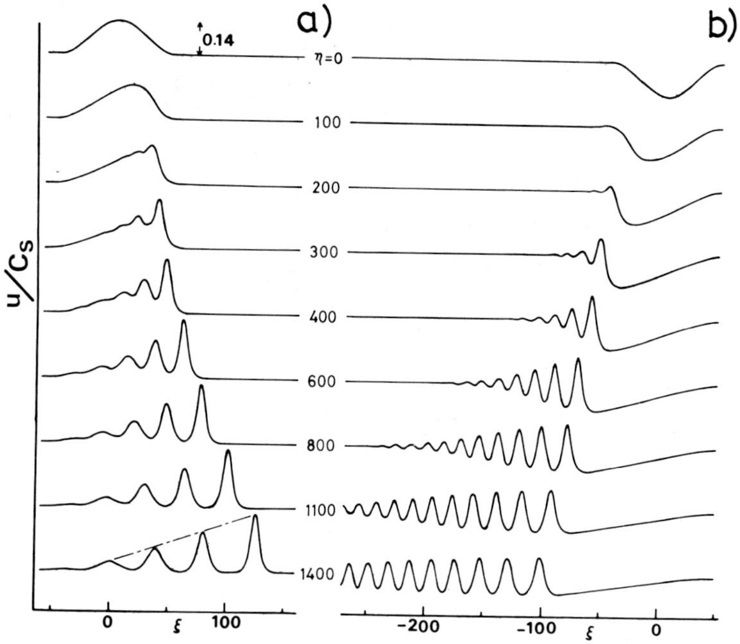

Figure 2. Numerical integrations of the KdV equation. $\eta=\tau$. The horizontal axis is the coordinate moving with C_s. (a) A positive initial pulse. (b) A negative initial pulse.

where the subscripts + and − denote the positive and negative ions, respectively, M is the mass, v is the fluid velocity normalized to $C_0=(\kappa T_e/M_+)^{1/2}$, n is the density normalized to the unperturbed electron density n, $\tau=\omega_{pi}t$, $\xi=x/\lambda_e$, $\omega_{pi}=(4\pi ne^2/M_+)^{1/2}$, $\lambda_e=(\kappa T_e/4\pi ne^2)^{1/2}$, $T=T_i/T_e$, T_e and T_i are the temperatures of electrons and ions, $\psi=e\phi/\kappa T_e$, ϕ is the electrostatic potential, and $\mu=M_-/M_+$.

Linealizing Eqs.(1)~(5), we obtain the following dispersion relation for perturbations of $\exp i(K\xi-\Omega\tau)$;

$$(1-r)(1+\frac{1}{K^2}) = \frac{1}{\Omega^2-3K^2T} + \frac{r}{\mu\Omega^2-3K^2T}, \tag{6}$$

where $r=n_{-0}/n_{+0}$, n_{-0} and n_{+0} are the unperturbed densities of the negative and positive ions. Here charge neutrality, that is, $n_{+0}=n_{-0}+n$, is used. The above dispersion relation is similar to the plasma with two positive ion species (Nakamura, Nakamura and Itoh 1976). The ion-acoustic velocity C_s which is the phase velocity for $K \ll 1$ obtained from Eq.(6) has two values. When $\mu > 3T$, which applies in the present experimental condition described later, one mode called a fast mode becomes the ion-acoustic wave of the positive ion when $r=0$ and the other is a virtual mode with $\Omega/K=(3T/\mu)^{1/2}$ at the limit $r=0$ (D'Angelo, Goeler and Ohe 1966). The ion-acoustic velocity C_s normalized to C_0 when $T=0$ is given by

$$S=\left(\frac{1+r/\mu}{1-r}\right)^{1/2} \tag{7}$$

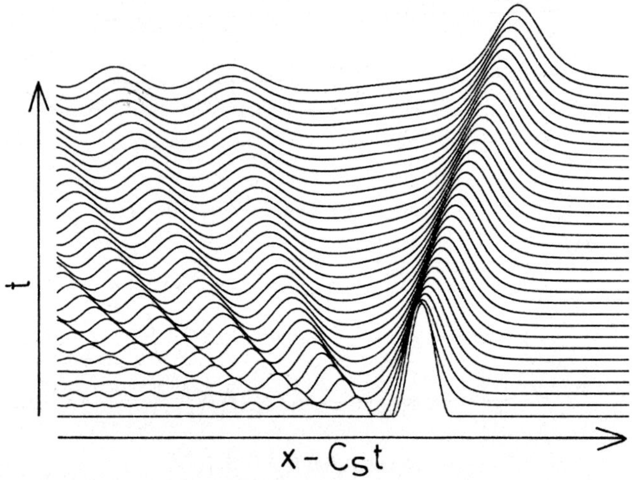

Figure 3. Propagation of a positive pulse whose width is narrower than the soliton of the same height.

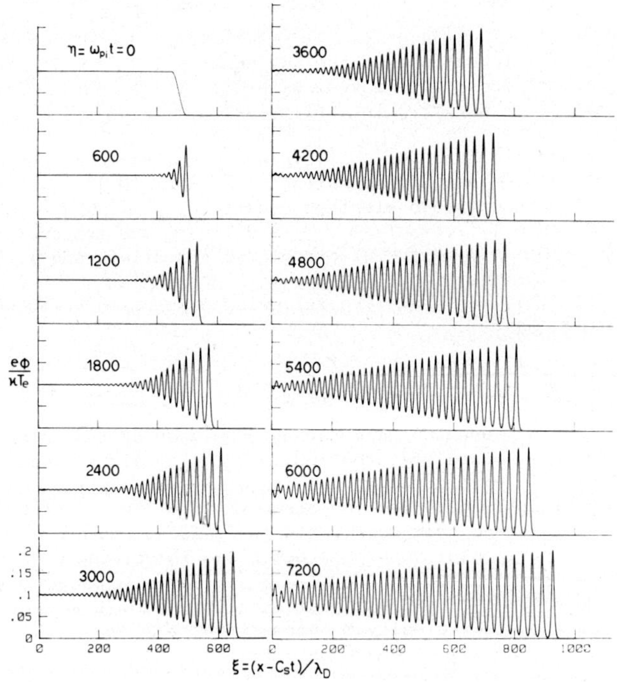

Figure 4. Propagation of a step-like perturbation.

and is shown in Fig. 1 for different μ. The fast mode has been observed experimentally, and its C_s was found to agree with the prediction of Eq.(6)(Wong, Mamas and Arnush 1975;Ludwig, Ferreira and Nakamura 1984).

2.2 KdV and Modified KdV Equations

From Eqs.(1)~(5), the following nonlinear equation is obtained for perturbations of small amplitude and long wavelength when T=0 by the perturbation method (Das 1979;Watanabe 1984).

$$\frac{\partial\psi}{S\partial\tau} + (1 + a\psi + p\psi^2)\frac{\partial\psi}{\partial\xi} + \frac{\partial^3\psi}{2\,\partial\xi^3} = 0 , \tag{8}$$

$$a = \frac{3}{2(1-r)S^4}\left(1 - \frac{r}{\mu^2}\right) - \frac{1}{2} , \tag{9}$$

$$p = \frac{15}{4(1-r)S^6}\left(1 + \frac{r}{\mu^3}\right) - \frac{1}{4} . \tag{10}$$

Values of a and $p\psi$ when $\psi = 0.1$ are also shown in Fig. 1 as a function of r for different μ. Although p is positive, a becomes zero at a certain critical value r_c which is smaller for a smaller μ. Equation(8) has soliton solutions(Wadati 1975). However, since $\psi \ll 1$, $p\psi^2$ is negligible compared with $a\psi$ when $r = 0$, which results that Eq.(8) becomes the KdV equation($a = 0$, $p = 0$). The heigher order term has been found to explain experimental results better than the KdV equation (Watanabe 1978). When $r = r_c$, $a = 0$, which results that Eq.(8) becomes the modified KdV equation($a = 0$, $p = 0$).

The single-soliton solution of the KdV equation is

$$\psi = \psi_0 \mathrm{sech}^2\left[\left(\frac{a\psi_0}{6}\right)^{1/2}\left(\xi - S\tau - \frac{aS\psi_0\tau}{3}\right)\right] , \tag{11}$$

where ψ_0 is the height or depth of solitons. The Mach number M of KdV solitons, i.e., the velocity divided C_s, is

$$M = 1 + a\psi_0/3 \tag{12}$$

which indicates that the velocity increment from C_s is proportional to the amplitude. The width of solitons normalized to λ_e satisfies the relation

$$(D/\lambda_e)^2\psi_0 = 6/a. \tag{13}$$

Equation(13) shows that the square of the width times the amplitude is constant.

The single-soliton solution of the modified KdV equation is

$$\psi = \psi_0 \mathrm{sech}\left[\left(\frac{p}{3}\right)^{1/2}\psi_0\left(\xi - S\tau - \frac{pS\psi_0^2\tau}{6}\right)\right] . \tag{14}$$

Hence the Mach number M and width D of modified KdV solitons are given by

$$M = 1 + p\psi_0^2/6 \tag{15}$$

and

$$(D/\lambda_e)|\psi_0| = (3/p)^{1/2}. \tag{16}$$

The Mach number M minus unity and D are proportional to the square of ψ_0 and inversely proportional to $|\psi_0|$, which are different characteristics from KdV solitons. It is noted that the modified KdV equation is independent of the sign of ψ, in other words, both compressive and rarefactive solitons can propagate. However, for the KdV equation, either compressive($a>0$) or rarefactive($a<0$) solitons can exist.

In the experiment described later, a sinusoidal pulse instead of a soliton is used as an initial signal. Therefore, spatial variation of the sinusoidal pulse has to be compared with the theory (Okutsu and Nakamura 1979). An example of the development of a waveform obtained by integrating Eq.(8)($a=1$, $p=0$) numerically is presented in Fig.2(a) for an initial compressive sinusoidal pulse. In this case, since the width of the pulse is wider than that of the soliton of the same amplitude, effect of dispersion is smaller than nonlinearity at the initial stage of propagation. Therefore, as the pulse propagates, the leading edge of the pulse steepens to make the role of dispersion larger and the first soliton is created when nonlinearity balances the dispersion. At the same time, the second soliton begins to emerge, and so on. Once a soliton is formed, it propagates without change of its form since the nonlinearity is balancing the dispersion. The solitons move to the positive ξ direction since their velocity is larger than C_s as given by Eq.(12). When all solitons separate, their peaks are on a straight line as shown when $\eta = 1400$. This is due to the fact that they were on about the same starting line and their speed is proportional to their amplitude. Numerical integrations for an initial pulse whose width is narrower than that of the soliton of the same amplitude are shown in Fig.3. In this case, the effect of dispersion in the early stages is larger than nonlinearity. As a result of this, oscillations are formed, which move with a speed slower than C_s. In the process of evolution, a soliton is formed from the leading part of the perturbation. A positive initial pulse, no matter how small its amplitude is, creates at least one soliton. Development of a step-like perturbation, i.e., non-localized initial condition is shown in Fig. 4. Irregular oscillations near the origin are due to the reflection at $\xi = 0$. It is evident from the figure that successive solitons whose amplitude is twice the height of the step appear from the perturbation (Khruslov 1975). This is also observed in a laboratory experiment(Tran et al. 1977).

No negative soliton($\psi<0$) can exist for a positive nonlinearity($a>0$) in the KdV equation for negative dispersion since both dispersion and nonlinearity broaden a localized pulse. Propagation of a negative initial pulse is shown in Fig. 2(b). In this case, not the leading or falling part but the rising part of the pulse steepens since the nonlinearity is positive. Then peaks appear due to dispersion, which is similar to Fig. 2(a). However, the number of peaks increases and their amplitude decreases as time goes on. They are not solitons but a wave train which will diverge out. It is predicted theoretically that solitons are not produced from a negative initial pulse(Lamb 1980).

2.3 Collisions of Solitons

The interaction of two solitons propagating in the same direction is observed in the numerical solution of the KdV equation for the initial condition of a continuous sinusoidal wave(Zabusky and Kruskal 1965). Numerical simulations of the KdV equation for the initial condition of two solitons are shown in Fig. 5. When the amplitude of an overtaking soliton is much larger than the soliton ahead

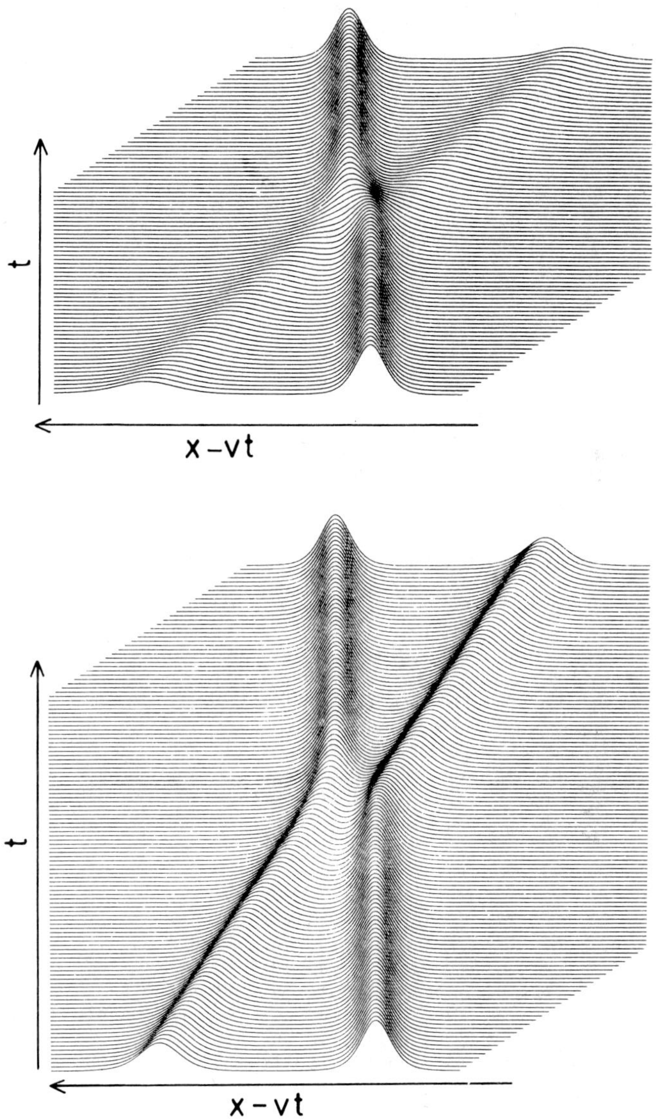

Figure 5. Overtaking collisions of two KdV solitons. The horizontal axis is the coordinate moving with the larger soliton. (1) The height of on soliton is much larger than the other. (2) The heights of both solitons are comparable.

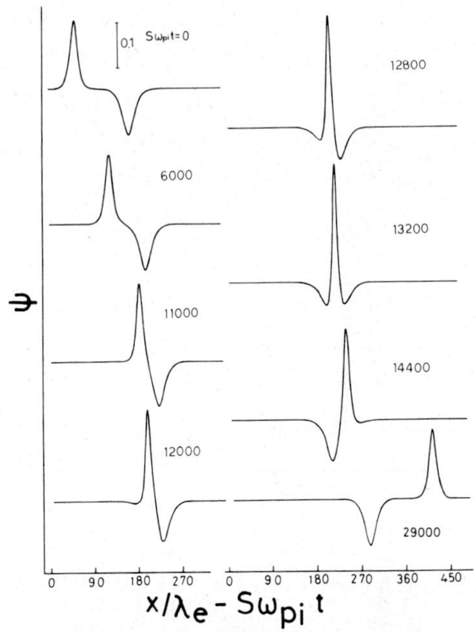

Figure 6. An overtaking collision of two modified KdV solitons.

of it, the larger soliton seems to pass through the smaller one(Fig. 5(a)). The amplitude of the perturbation when two solitons overlap completely is smaller than the linear sum of each amplitude. In the case that both amplitude are comparable(Fig. 5(b)), when the larger soliton catches up with the smaller one, the amplitude of the smaller one increases and exchanges amplitude. By this interaction, a phase advance and delay occur for the preceding and the following solitons, respectively. Therefore, the interaction is considered to be repulsive.

Overtaking collisions of two positive modified KdV solitons are found to be similar to those of KdV solitons. An overtaking collision of a negative and a positive soliton obtained by numerical integration of the modified KdV equation is shown Fig. 6. When the positive soliton approaches the foregoing negative soliton, a negative perturbation appears at the back of the positive soliton, which evolves into the negative soliton. It is noted that the amplitude of the positive soliton while it is interacting with the negative soliton is larger than its own amplitude. By the collision, a phase advance for the positive soliton and a phase delay for the negative soliton occur, which indicates that the interaction is also repulsive.

For head-on collisions of solitons, the interaction is of two solitons on different dispersion branches. However, the KdV and the modified KdV equations describe perturbations moving to one direction. Therefore, original fluid equations(1)~(5) are integrated numerically. As initial conditions, KdV solitons ($r = 0$) of an amplitude of ψ_0 given by Eq.(11) or modified KdV solitons($r = r_c$) of ψ_0 given by Eq.(14) are first assumed. Then initial profile of the electron density is given by $\exp\psi$. The initial densities and velocities of positive and

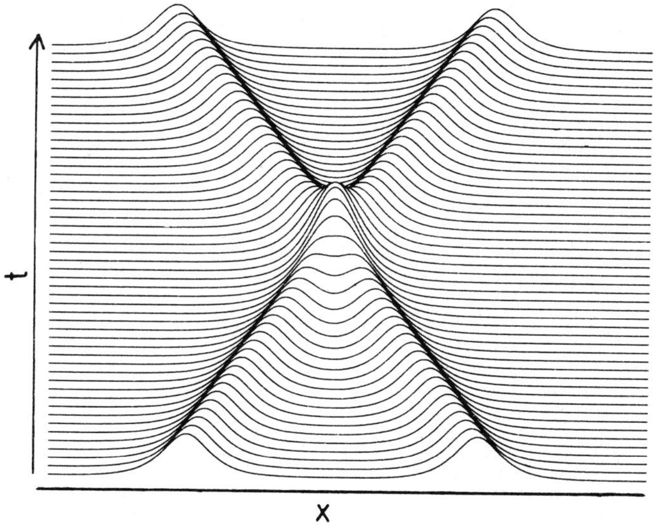

Figure 7. A head-on collision of two KdV solitons of equal height ($\psi_0 = 0.05$).

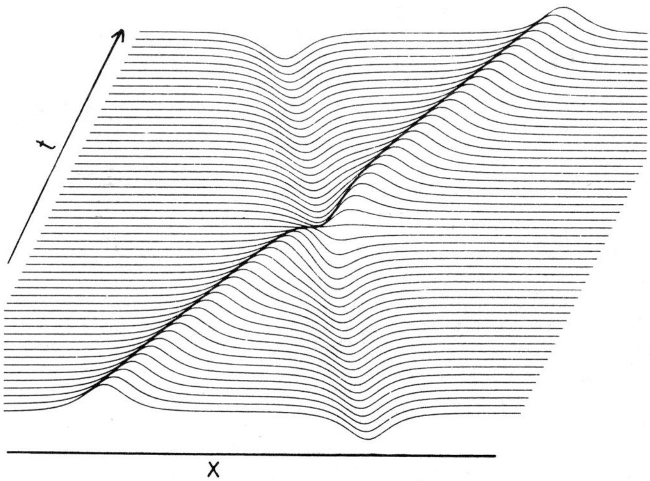

Figure 8. A head-on collision of two modified KdV solitons ($|\psi_0| = 0.05$).

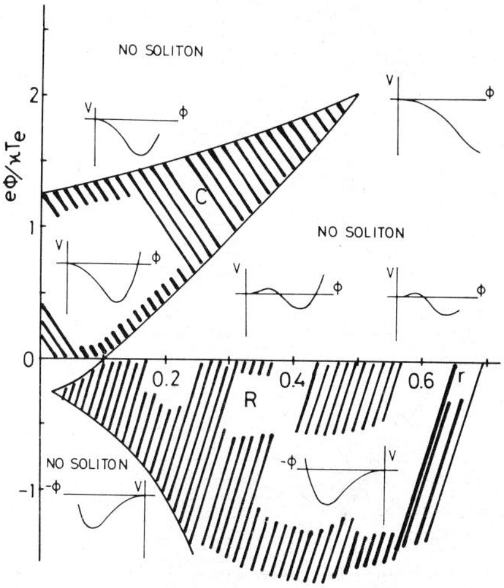

Figure 9. Regions of compressive or positive solitary waves (C) and rarefactive or negative solitary waves (R). $\mu = 0.476$.

negative ions are obtained from stationary, that is, the time derivatives in Eqs. (1)~(4) are zero, solutions and are given by the following relations for solitons moving to the positive direction,

$$n_+ = (1-r)^{-1}\left(1 - \frac{2\psi}{M^2S^2}\right)^{-1/2}, \tag{17}$$

$$n_- = r(1-r)^{-1}\left(1 + \frac{2\psi}{\mu M^2S^2}\right)^{-1/2}, \tag{18}$$

$$v_+ = MS - (M^2S^2 - 2\psi)^{1/2}, \tag{19}$$

$$v_- = MS - (M^2S^2 + 2\psi/\mu)^{1/2}, \tag{20}$$

where $\mu = 0.476$ and $S = 1.16$.

An example of a head-on collision of KdV solitons ($r = 0$) is drawn in Fig. 7. The result shows that the solitons pass through each other and keep their initial amplitudes except for a slight phase delay. Therefore, the interaction is considered to be attractive for the head-on collision. The amplitude of the peak when two solitons overlap is a little larger than the linear sum of each amplitude. An example of a head-on collision of a positive and a negative modified KdV soliton is shown in Fig. 8. It is noted that although $|\psi_0|$ for both solitons are equal, the corresponding amplitudes of $\delta n/n$ are not equal. When the interaction starts, their amplitudes become small but the perturbation does not become completely flat at the time of the collision. Then both solitons recover and propagate with the initial amplitudes. There appear a slight phase advance for the compressive soliton and a phase delay for the negative soliton. The collision properties described above have been observed using ion-acoustic solitons and KdV solitons in electric ciruits.

2.4 Pseudo-potential Method

Amplitudes of KdV and modified KdV solitons must be theoretically small, i.e., $|\psi| \ll 1$ since the perturbation method is used to derive those equations. Experimentally measured amplitudes were also small ($|\delta n/n| < 0.3$). If the amplitude were large, the perturbation method can not be applied and the original fluid equations (1)~(5) must be considered. However, when a wave shape is temporally stationary, the pseudo- or Sagdeev potential method given below predicts that there exist localized perturbations with a large amplitude (Sagdeev 1969;Tajiri and Toda 1985).

We neglect the ion temperature, i.e., $T = 0$ and assume the time derivatives in Eqs.(1)~(4) are zero. Then the following relation is obtained assuming $V = dV/d\psi = 0$ at $\psi = 0$.

$$\frac{d^2\psi}{d\xi^2} = -\frac{dV}{d\psi}, \tag{21}$$

$$V = 1 - \exp\psi + \frac{M^2S^2}{1-r}\left[1-(1-\frac{2\psi}{M^2S^2})^{1/2}\right] + \frac{\mu r M^2S^2}{1-r}\left[1-(1+\frac{2\psi}{\mu M^2S^2})^{1/2}\right]. \tag{22}$$

Equatios (21) is an equation of motion if ψ, ξ and V are thought to be displacement, time and a potential. The potential is called a pseudo- or Sagdeev potenial. The psedo-potentials for different negative ion concentrations and the Mach number M (the amplitude corresponds to the Mach number, i.e.,$\psi = e\phi/KT_e$ is obtained from M) are shown in Fig. 9. For example, ψ has a localized positive solution for the potential shown in the region C. The regions of ψ and r for the existence of solitary wave solutions are also shown in Fig. 9. When $r = 0$, which corresponds to no negative ions, only compressive solitons ($\psi > 0$) exist although their amplitude is limited to $\psi = 1.25$ whose Mach number ($=1.6$) is well known as the critical Mach number. When $0.017 < r \leq r_c = 0.10$, negative solitons can exist if ψ is within certain values. When $r > 0.10$ and $|\psi| \ll 1$, only rarefactive solitons can propagate. However, when $0.10 < r < 0.494$, compressive solitary waves of large amplitude can be excited.

3. EXPERIMENTAL PROCEDURE

Experiments were performed in a multi-dipole double-plasma machine shown in Fig. 10. The inner diameter (plasma diameter) of the device is 40cm and its total length is 90cm. The device is separated into a driver and a target section with a floating grid. The grid consists of a mesh with a transparency of 85%. The cathodes are 0.1mm diameter tungsten filaments placed 6cm from the surface of the anode wall. Each plasma section has fifteen filaments with a length of 5cm. The chamber was evacuated to 1×10^{-6} torr with a turbo-molecular pump. Argon and sulfer hexafluoride were introduced independently into the chamber under continuous pumping. The pressure of Ar was 1.5×10^{-4} torr and the partial pressure of SF_6 was changed from 0 to 4×10^{-5} torr. The dischrge voltage was 50V and the discharge current in each section was 70 - 150mA. Plasma parameters were measured by a plane Langmuir probe with a diameter of 5mm and a retarding ion energy analyzer. The back of the probe is covered with a ceramic insulator. The electron density $n = 10^8$-10^9 cm^{-3}, the electron temperature $T_e = 1.5$-2.3eV and the electron to positive ion temperature ratio $T_e/T_i = 10$-20.

Ion-acoustic perturbations were excited by either a positive or negative sinusoidal voltage pulse which was applied to the anode of the driver section.

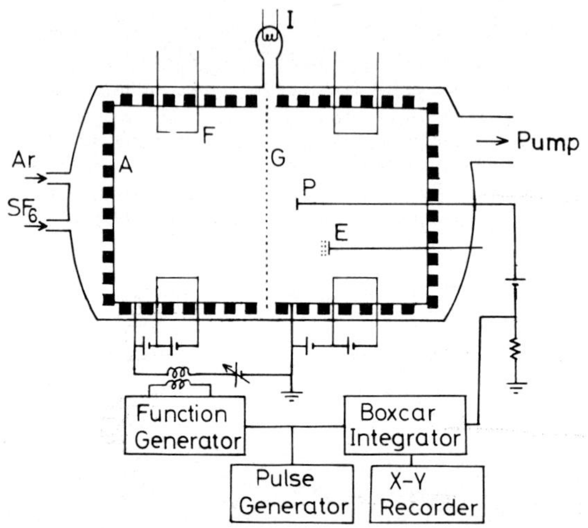

Figure 10. Schematic diagram of the experimental apparatus. A:magnets enclosed in thin aluminum pipes which are used as the anode. E:retarding energy analyzer. F:filaments. G:floating grid. I:ionization gauge. P:probe.

Signals were detected with the axially movable plane Langmuir probe. It was biased positively against the plasma potential to collect the electron saturation current and was sensitive to the perturbed electron density. The signals were fed into either an oscilloscope or a boxcar integrator whose output was recorded on an x-y recorder versus position.

4. EXPERIMENTAL RESULTS AND DISCUSSIONS

4.1 Dispersion Relation

Bombardment of SF_6 molecules by 50 eV primary electrons leads mainly to the formation of negative F^- ions due to dissociative attachment processes. Negative SF_6^- and SF_5^- ions are also formed by the capture of low-energy plasma electrons, and cross sections indicate that for an electron temperature of 2 eV the production rate of SF_5^- ions is much larger than SF_6^- ions. Furthermore, the ionization of SF_6 molecules by primary electrons leads to the formation of positive ions such as SF_5^+, SF_3^+, SF_4^-, etc. Therefore, the plasma consisted of several species of positive and negative ions. However, it is easily seen from the dispersion relation of ion-acoustic waves in a plasma which includes multi-component of ions that the phase velocity of the fast mode depends strongly on the light ion concentration, which has been confirmed by experiments (Tran and Coquerand 1976; Nakamura, Nakamura and Itoh 1976). The coefficient a of the nonlinear term in Eq.(8) also depends mainly on the light ion concentration (Ludwig et al. 1980). Hence we assume that the plasma was composed of Ar^+ ions, F^- ions and electrons and the contribution of other heavier positive and negative ions is neglected. The method of obtaining the concentration of F^- ions experimentally from a measured dispersion relation is described below.

An example of measured dispersion relations is shown in Fig. 11. The solid curve is obtained from Eq.(6) where r and T are chosen to give the best fit to the experimental result. The temperature and density of electrons used for the calculation were measured by the Langmuir probe. As a result of this, the F^-

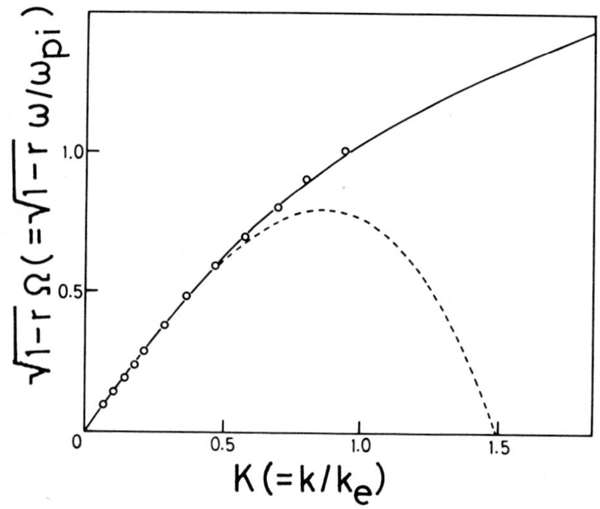

Figure 11. A measured dispersion relation. Argon pressure = 1.1×10^{-4} torr. SF_6 pressure = 1.5×10^{-5} torr. The solid curve is estimated fast mode from Eq.(6) for r = 0.34 and T = 0.07.

concentration r is obtained. The dotted curve is calculated from a linealized KdV equation which takes a finite ion temperature into consideration (Nakamura, Ferreira and Ludwig 1985) for the same parameters as the solid curve. The dispersion relation is negative like the one for a plasma without negative ions.

4.2 Korteweg-de Vries Solitons

Examples of received signals in a plasma without negative ions for a sinusoidal initial positive and negative pulse whose width is much wider than the soliton of the same height observed at several distances from the floating grid in the double-plasma device are shown in Fig. 11. Developments of the pulses are quite similar to the simulations shown in Fig. 2. In this case (r = 0), as the propagation property is normal, that is, the propagation velocity increases with the wave potential, steepening occurs at the leading part of the positive pulse (Fig. 11(a)) and at the rising part of the negative pulse. Then solitons appear from the positive pulse and a wave train is formed from the negative pulse due to the dispersion. The velocity and width of solitons as a function of the amplitude have been studied experimentally in detail and are found to agree with Eqs. (12) and (13), respectively. They are described in the review papers cited already. The shapes of the peaks in Fig. 12(b) are similar to the solitons in Fig. 12(a). Actually the measured relation between the height and the width of the first peak is also nearly given by Eq. (13). However, the velocity of the peak is slower than C_s. When the velocity is measured in the reference frame moving with the negative pulse, it is given by Eq. (12) as pointed out by Ikezi (1973) experimentally and by Yoshimura and Watanabe (1979) who also solved the KdV equation numerically. Therefore, the peak was mistaken to be a soliton. However, as noted in Fig. 2(a), its amplitude decays as it propagates and no stationary state forms.

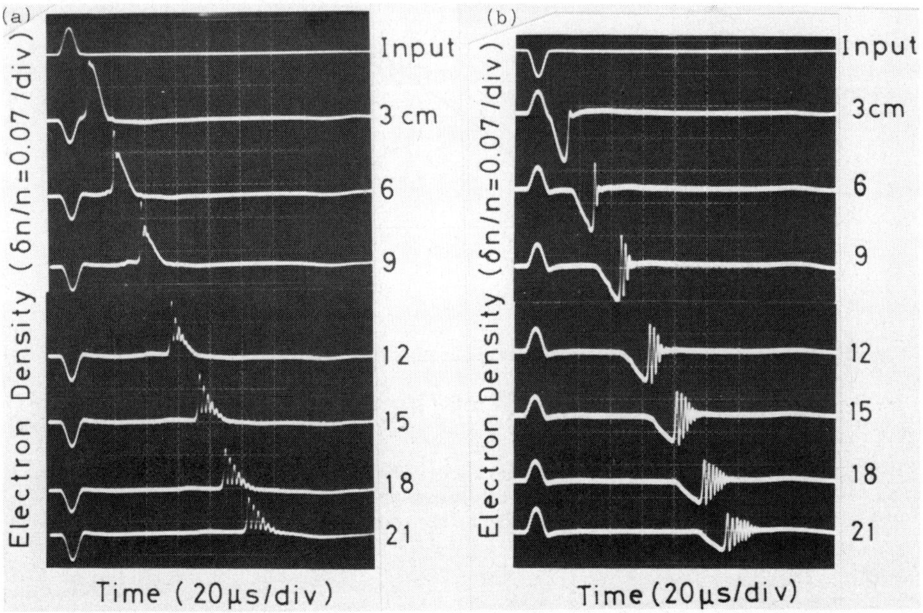

Figure 12. Observed signals at several distances from the grid. for sinusoidal initial pulses. r = 0. (a) a compressive initial pulse. (b) a rarefactive initial pulse.

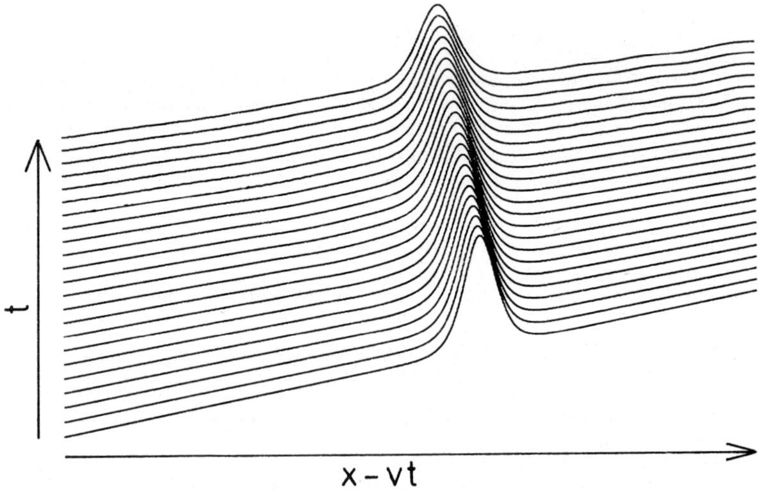

Figure 13. Interaction of a soliton with a negative wave of a long wavelength.

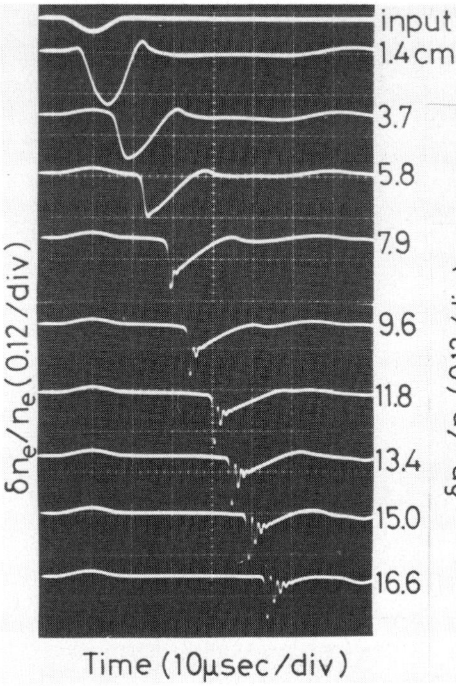

Figure 14. Electron density perturbations versus time with the distance from the grid as a parameter. r = 0.23. The top trace is the applied signal.

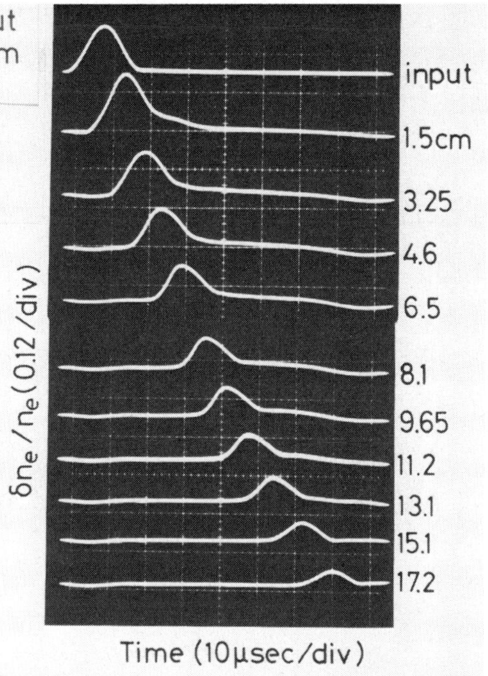

Figure 15. Wave response to a pulse excitation for several distances from the grid as a parameter. r = 0.14.

This is due to the interaction of the peak with a negative pulse or the density gradient on which the peak is moving. To simulate this situation, a soliton and a density gradient are used as the initial condition for the KdV equation and the result is shown in Fig. 13. The soliton moves to the left direction since the velocity of the density gradient in the coordinate moving with C_s is negative and its absolute value is larger than the soliton velocity. The gradient becomes gentler as it propagates due to the normal nonlinearity. The amplitude of the soliton becomes smaller, that is, it decays as it propagates and a residue is formed behind it.

An example of observed signals for a negative initial pulse whose width is much wider than λ_e when $r > r_c$ is shown in Fig. 14. When the picture is turned the upside down, development of the pulse is similar to the case of a compressive pulse in a plasma without negative ions, which is shown in Fig. 12(a). The leading part of the pulse steepens due to the anomalous propagation property, that is, the propagation speed is larger for a more negative perturbation. The steepening

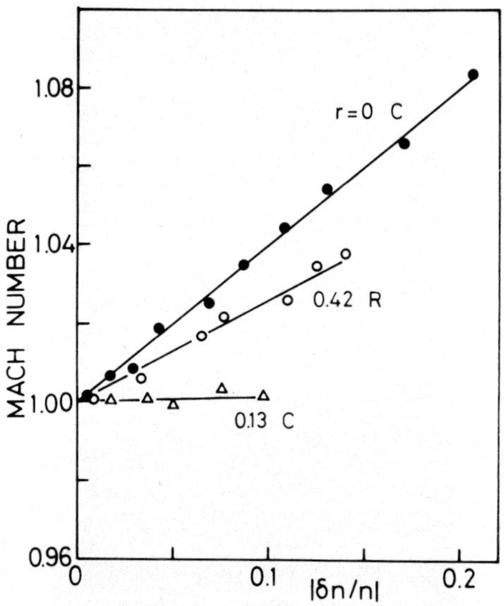

Figure 16. Measured Mach numbers of compressive solitons (r = 0), positive pulses (r = 0.13) and rarefactive solitons (r = 0.42) as a function of the normalized amplitude.

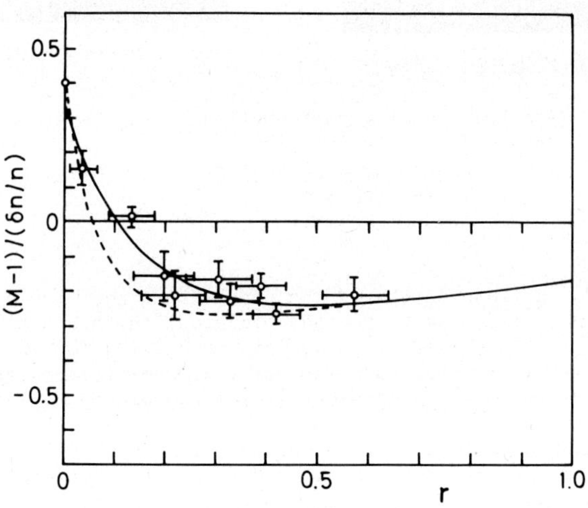

Figure 17. Measured (M - 1) divided by the normalized height or depth as a function of the negative ion concentration r. The solid and dotted curves are theoretical for T = 0 and 0.1, respectively.

caused by the nonlinearity makes the role of dispersion larger and the first soliton is created when the nonlinearity balances the dispersion. At the same time, the second soliton begins to emerge, and so on. The humps seen at the time of the applied pulse are directly coupled signals (Nakamura and Nomura 1980). Under the same plasma condition, a positive pulse was transmitted. As the pulse propagates, the leading part becomes gentle and steepening occurs not at the rising (leading) part but at the falling part of the pulse. As a result of this, oscillations instead of solitons appear eventually. This phenomenon is the counterpart of the propagation of a rarefactive pulse in a plasma with no negative ions, which is shown in Fig. 12(b). The result also suggests that the coefficient of the nonlinear term in the KdV equation is negative in the present plasma condition ($r = 0.23$ r_c).

Observation of the propagation of a positive pulse when $r = r_c$ is shown in Fig. 15. Although the amplitude ($\delta n/n \simeq 0.1$) of the pulse is not small enough to be considered linear, it does not show steepening at the leading edge but propagates like a linear wave without change of its shape except by damping since the width of the pulse is much larger than the Debye length and the effect of dispersion is small. This is because the nonlinear term of the KdV equation disappears. The measured critical concentration r_c ($= 0.14 \pm 0.04$) agrees with the prediction of Eq.(9), which confirms that the dominant species of ions are Ar^+ and F^-.

The measured velocities of compressive solitons when $r = 0$, rarefactive solitons when $r = 0.42$ and of the positive pulses when $r = 0.13$ are plotted in Fig. 16 as a function of the normalized amplitude. The velocity was measured only for the first soliton shown in Fig. 14 since the adjoining second and the third solitons are not separated completely yet (Okutsu and Nakamura 1979). The height or depth of solitons was varied with changing the excitation voltage applied to the anode of the source chamber. For both compressive and rarefactive solitons, the result indicates that $(M - 1)$ is linearly proportional to the amplitude, which is consistent with Eq. (12) and which has already been measured for compressive solitons in plasmas without negative ions (Ikezi 1973; Watanabe 1975; Okutsu and Nakamura 1979).

When $r = 0.13$ which is a little smaller than the experimental r_c ($= 0.14$), the measured wave is not a soliton but a positive pulse. However, the velocity increment is also proportional to its amplitude as explained below. Since the width of the pulse is much wider than the Debye length, the dispersion term in the KdV equation is much smaller than the nonlinear term within a distance of 15 cm, where the velocity was measured. As a result of this, the KdV equation predicts that the velocity of the peak of the pulse in the frame of reference moving with the ion-acoustic speed is proportional to its height.

The gradients of lines like those shown in Fig. 16, which are equal to $(M - 1)/(\delta n/n)$, are measured and summarized in Fig. 17. The open circle on the vertical axis is the case when $r = 0$, i.e., the plasma includes no negative ions. The solid and dotted curves are $a/3$ calculated from Eq. (9) when $T = 0$ and that of $T = 0.1$ calculated by Nakamura, Ferreira and Ludwig (1985). The non-zero ion temperature makes the ion-acoustic wave less dispersive, which results that the speed of solitons become faster for the same amplitude than when $T = 0$. The open circles in the upper plane correspond compressive solitons except the one for $r = 0.13$. The point for $r = 0.13$ corresponds to the positive pulse shown in Fig. 15. The data in the lower half plane are for rarefactive solitons, that is, their density perturbations $\delta n/n$ are negative. The experimental results are considered to agree with the theoretical prediction.

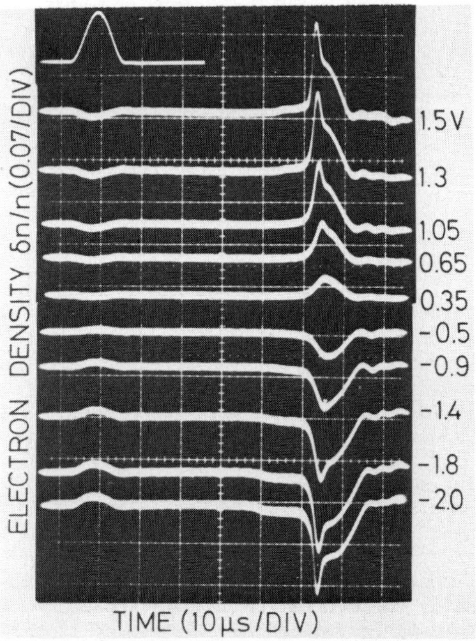

Figure 18.

Observed perturbations of the electron density at 14.8 cm from the grid for different excitation voltages. The positive and negative are respectively the height of the applied positive pulse (top trace) and the depth of the applied negative pulse which is the reverse of the top trace. $r = r_c = 0.1$.

Another characteristic which is often compared with the theory for the KdV solitons is the dependence of the width of solitons on the amplitude. Measurements indicate that the product of the square of the width and the amplitude is constant within experimental errors for a fixed r. The width is defined as the full width at the height of 0.786 times the amplitude of solitons (Eq.(11)). Measured products $(D/\lambda_e)^2(\delta n/n)$ versus the negative ion concentration agree with the theory in which a finite T is considered (Nakamura, Ferreira and Ludwig 1985).

4.3 Modified Korteweg-de Vries Solitons

When $r = r_c$, the nonlinear term of the KdV equation disappears, which results that waves behave like linear ones as seen in Fig. 15. However, for a wave of larger amplitude, the higher-order nonlinearity should be considered, which gives the modified KdV equation (Eq. (8) with $a = 0$ and $p \neq 0$). To see the effect of the higher-order nonlinearity, propagations of waves of larger amplitudes are studied under the condition that $r = r_c$. Signals detected at 14.8 cm from the grid for different excitation voltages are shown in Fig. 18. As the width of the pulse is much larger than λ_e, the dispersive effect is small in the initial state of propagation. When the amplitude is small (0.35 and -0.5V) but not too small to be considered linear, the pulse propagates without change of the shape except damping since the nonlinear term in the KdV equation vanishes and since the higher-order term is negligible as noted in Fig. 15. When the absolute value of the excitation voltage is increased (0.65 and -0.9 V), the leading parts of both positive and negative pulses steepen. This phenomenon is different from that given by the KdV equation as described in Sec. 4.2. The steepening makes the

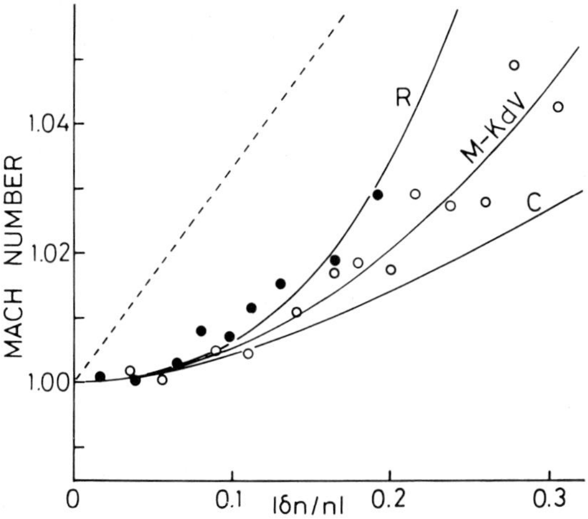

Figure 19.

The velocities normalized by C_s of compressive (open circles) and rarefactive (closed circles) solitons as a function of the amplitude. The curve M-KdV is Eq. (15) for p = 3.04. The curve R and C are calculated by the pseudo-potential method for rarefactive and compressive solitary waves, respectively.

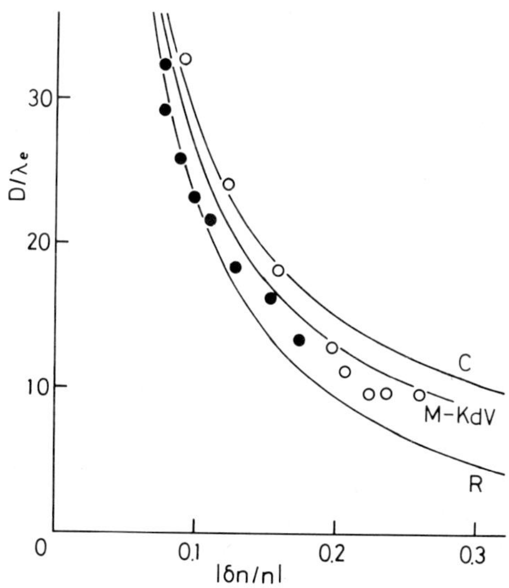

Figure 20.

The widths normalized by λ_e of compressive (open circles) and rarefactive (closed circles) solitons as a function of the amplitude. The curve M-KdV is Eq. (16) for p = 3.04. The curves R and C are calculated by the pseudo-potential method for rarefactive and compressive solitary waves, respectively.

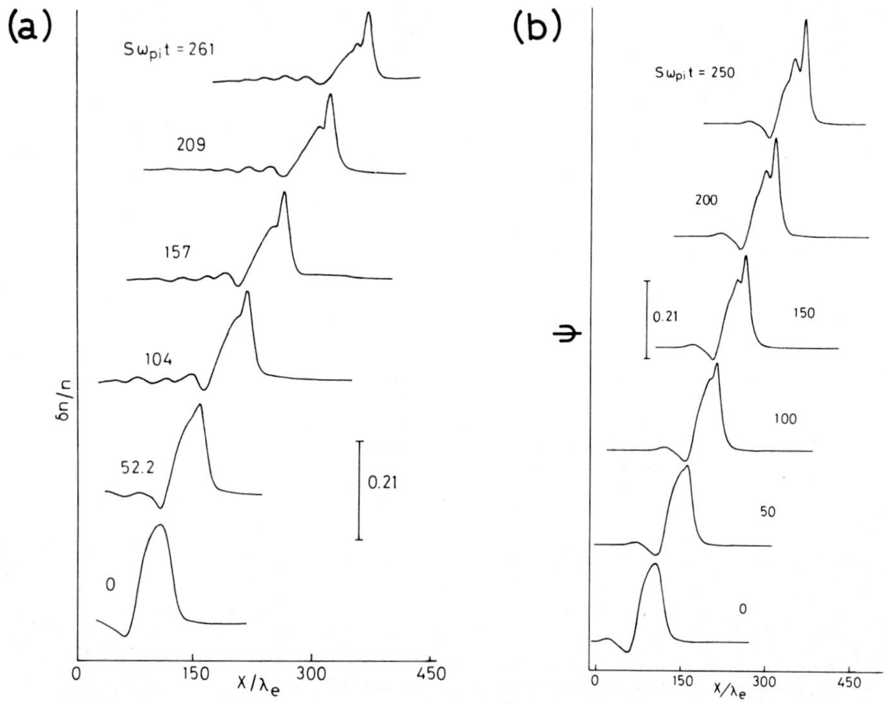

Figure 21. Development of a positive pulse when $r = r_c$. (a) Measured propagation of the positive pulse. (b) Calculated propagation using the modified KdV equation for $p = 2$ and the experimental initial wave form at $t = 0$.

effect of the dispersion larger and the crest and the sharp trough form at the leading part of the positive and the negative pulse, respectively (1.05 and -1.04 V). It is noted that the shape of the signals is same except for their sign for nearly equal amplitudes (1.5 and -2.0 V).

The velocities of the positive peaks and the negative troughs were measured at two positions between 10 and 15 cm. The result is shown in Fig. 19 as a function of the normalized height of the peak and the depth of the trough. The theoretical curve of Eq. (15) for $p = 3.04$ is also drawn in the figure. The Mach velocities estimated by the pseudo-potential method is also depicted for compressive and rarefactive solitary waves when $\mu = 0.476$ and $r_c = 0.102$. The Mach velocity obtained by the pseudo-potential method agrees with that given by the modified KdV equation for $\psi \ll 1$ since the modified KdV equation is derived under the assumption that perturbations are small. The experimental result that the velocity of negative solitons is larger than that of positive solitons agrees qualitatively with the theory. In a previous paper (Nakamura and Tsukabayashi 1984), the result has been only compared with the modified KdV equation.

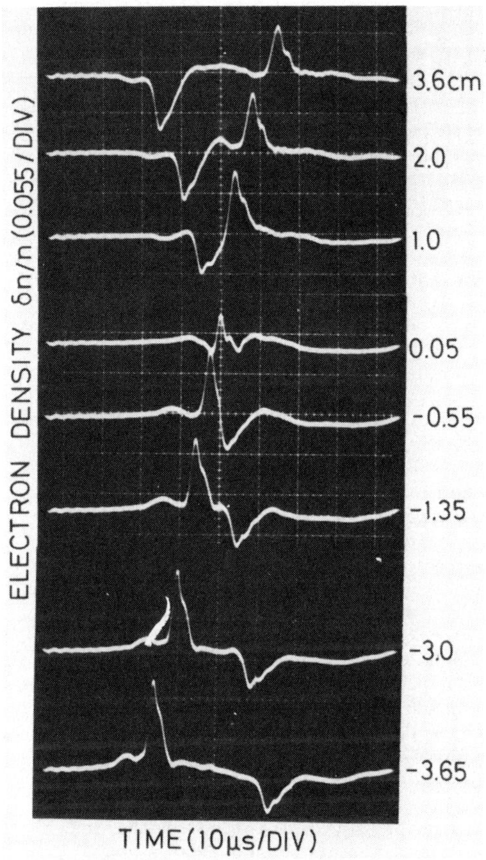

Figure 22.

Observed signals at several positions in the target chamber for the head-on collision of a compressive and a rarefactive soliton. The center between the grids is 0 cm.

Measured full widths at the half maximum of the positive and negative solitons normalized by λ_e are plotted in Fig. 20. The curve marked M-KdV is the theoretical value given by Eq. (16) for $\mu = 0.476$ and $r = 0.102$. The curves marked C and R are calculated by the pseudo-potential method for positive and negative solitary waves, respectively. Experimental data show that the product of the width and amplitude is constant, which is one of the characteristics of the modified KdV solitons. The experimental fact that the width of positive solitons is larger than negative solitons for the same amplitude is consistent with the prediction of the psedo-potential method.

Spatial development of a positive pulse when $r = r_c$ is shown in Fig. 21(a). Using a wave form near the grid ($t = 0$ in the figure) as an initial condition, the modified KdV equation (8)($a = 0$, $p \neq 0$) is numerically solved for different values of p. Similar wave shapes are obtained when $p = 2$, which are shown in Fig. 21(b). The value of p is smaller than the assumed value(3.04), which seems to indicate that the experimental mass ratio is larger than expected.

The KdV and modified KdV solitons are known to keep their identity after collisions so that the solitary waves are called solitons. An overtaking collision of the present solitary waves has been tried to excite two pulses at a certain time difference as was done for KdV solitons (Nakamura, Ferreira and Ludwig 1985). However, it could not be observed since their velocity difference was too small for the collision to occur in the present size of the device. Instead head-on collisions were studied by modifying the system into a triple-plasma device (Nakamura, Nomura and Itoh 1977). The system was constructed with addition of another source chamber and with replacement of the target chamber with a shorter one. The distance between the two grids was 25 cm. Two pulses were applied independently to the anodes of those source chambers from two function generators. An axially movable cylindrical probe was used to detect counter-propagating waves. Collisions of positive and positive, positive and negative, and negative and negative solitons were observed. The collision of compressive and rarefactive solitary waves is shown in Fig. 22. The left pulse in each trace is the wave before the collision and the right one is the wave which experienced the collision in the past. Although the solitary waves add to become small around the center, they recover and propagate keeping their shape. They also passed through without change of their shape for the collision of two solitons of the same sign.

4.4 Solitary Waves of Large Amplitude

Signals detected at 4.3 cm from the grid for different excitation voltages are shown in Fig. 23. When the negative pulse is applied (-1.5V), rarefactive solitons are formed since the wave is described by the KdV equation whose coefficient of the nonlinear term is negative as stated before. When a 1.5 V positive pulse is applied and when its amplitude is small, i.e., $\delta n/n = 0.25$, the leading part of the pulse become gentle, which implies that the velocity of the peak of the positive pulse is less than the ion-acoustic velocity C_s. The falling part steepens to form oscillations due to dispersion. This is similar to the propagation of a negative pulse in a plasma without negative ions. When the amplitude is increased (4 and 5 V) to reach a critical value ($\delta n/n = 0.76$), the leading part steepens to form a solitary peak due to dispersion. It is noted that the amplitude ($\delta n/n \simeq 1.2$) is very large and this is the largest solitary wave observed in a plasma.

Examples of the measured velocity of the peak as a function of the peak height are shown in Fig. 24. When the amplitude is increased from a small linear pulse, the velocity of the peak decreases due to the negative nonlinearity. After reaching a minimum, the velocity increases and becomes supersonic at a certain amplitude. This threshold amplitude is nearly equal to the one at which the steepening occurs at the leading part as noted in Fig. 23. Equation (21) predicts that the solitary waves are supersonic and the amplitude when M=1 gives the threshold amplitude shown in Fig. 9. The experimental results for M>1 are considered to agree with the theory.

Measured threshold amplitudes at which the wave becomes supersonic or the steepening occurs at the leading part of the pulse are plotted in Fig. 25 as a function of r. The vertical axis in Fig. 25 is not ψ but $\delta n/n$ whose theoretical value is obtained from the relation $\delta n/n = \exp\psi - 1$ using the critical ψ shown in Fig. 9. Measured critical depths for the formation of rarefactive solitary peaks when $r < r_c$ are also plotted in the same figure. Experimental results seem to agree with the theory and a small disagreement is thought to be due to neglecting the finite ion temperature and the other species of positive and negative ions.

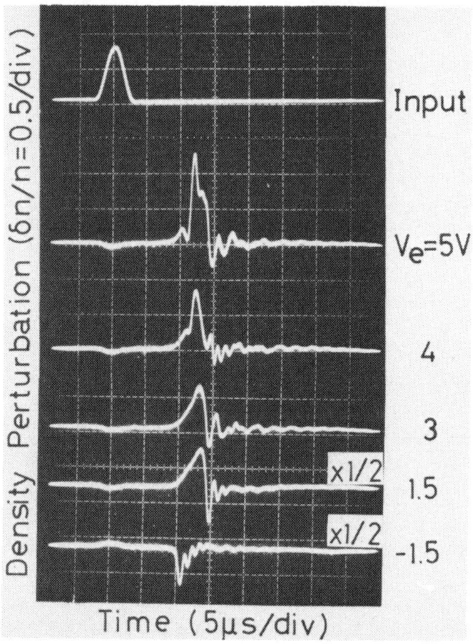

Figure 23.

Observed perturbations of the electron density at 4.3 cm from the grid for different peak amplitude voltages. $r = 0.25 > r_c$.

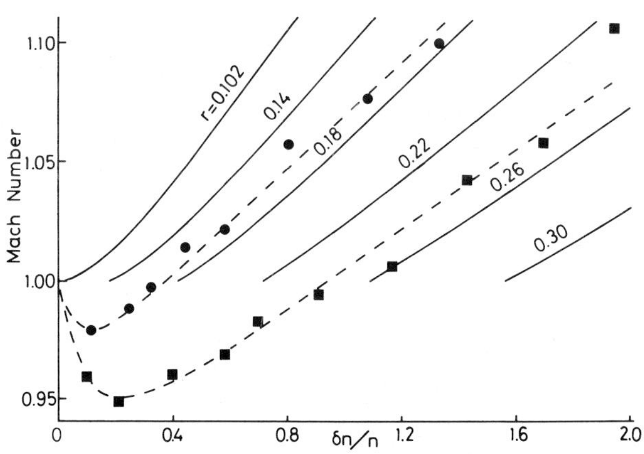

Figure 24. The dependence of the peak velocity on the peak height. Closed circles : $r = 0.17$. Closed squares : $r = 0.28$. The solid curves are calculated from Eq. (21) for $\mu = 0.476$.

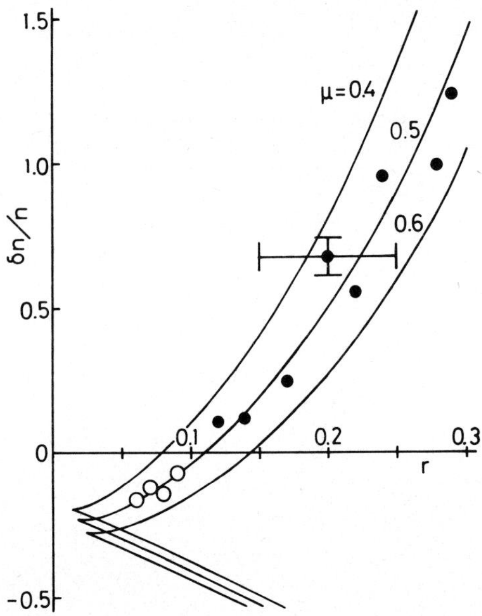

Figure 25.

The dependence of critical amplitudes on r. The open and closed circles are rarefactive and compressive waves, respectively. The curves are calculated from Eq. (21) for different r.

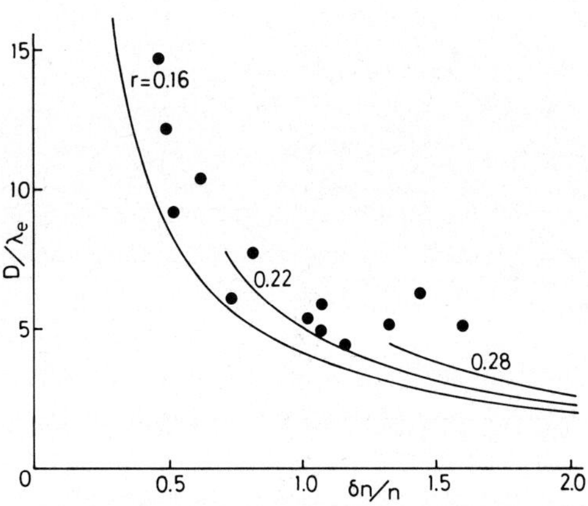

Figure 26.

Measured widths as a function of the amplitude when $r = 0.21$. The curves are estimated by the pseudo-potential method when $\mu = 0.476$.

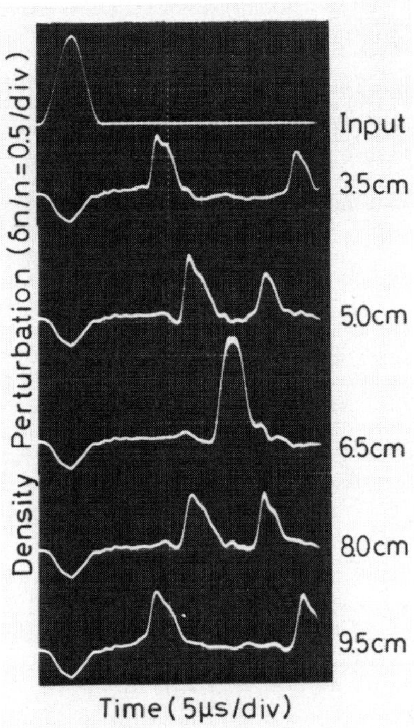

Figure 27.

A head-on collision of two large solitary waves.

Measured spatial widths at the half maximum of the solitary peak when r = 0.21 are plotted in Fig. 26. The three curves in the figure are obtained from the shape of solitary waves calculated by the pseudo-potential method. The minimum amplitude of each curve corresponds to the threshold value for the formation of the solitary waves, which is explained above. The experimental result is considered to agree approximately with the simple theory.

A head-on collision of two counter-propagating solitary waves of large amplitudes was observed in the triple-plasma device used for the head-on collision of the modified KdV solitons. An example is shown in Fig. 27. The solitary waves seem to pass through each other. However, in the case of head-on cllisions. the interaction time is short so that it may be difficult to observe phenomena which indicate the solitary waves are solitons or not. We could not succeed to observe overtaking collisions of these large solitary waves. Therefore, it is experimentally uncertain that these solitary waves are solitons or not. The characteristics of these waves are not studied theoretically. They remain to be a future problem both theoretically and experimentally.

Acknowledgements

The experiments began while the author was at Instituto de Pesquisas Espaciais, São José dos Campos, SP, Brazil and were performed with G. O. Ludwig, J. L. Ferreira and I. Tsukabayashi. The author would like to thank Profs. N. Yajima and S. Watanabe for discussions and encouragements.

REFERENCES

Berezin, Yu. A. and Karpman, V. I. (1964) Sov. Phys. - JETP 19, 1265.
D'Angelo, N., Goeler, S. V. and Ohe, T. (1966) Phys. Fluids 9, 1605.
Das, G. C. (1979) Plasma Phys. 21, 257.
Drazin, P. G. (1983) Solitons. Cambridge University Press.
Ichikawa, Y. H. and Watanabe, S. (1977) J. de Phys. C6-15, 38.
Ikezi, H., Taylor, R. J. and Baker, D. R. (1970) Phys. Rev. Lett. 25, 11.
Ikezi, H. (1973) Phys. Fluids 10, 1668.
Ikezi, H. (1978) Solitons in Action, K. E. Lonngren and A. Scott(eds), Academic Press, p. 153.
Khruslov, E. Ya. (1975) JETP Lett. 21, 217.
Lamb, G. L. Jr. (1980) Elements of Soliton Theory. John Wily and Sons, New York.
Lonngren, K. E. (1980) Plasma Phys. 25, 943.
Ludwig, G. O., Ferreira, J. L. and Nakamura, Y. (1984) Phys. Rev. Lett. 52, 275.
Nakamura, Y., Nakamura, M. and Itoh, T. (1976) Phys. Rev. Lett. 37, 209.
Nakamura, Y., Nomura, Y. and Itoh, T. (1977) Phys. Rev. Lett. 39, 1622.
Nakamura, Y. and Nomura, Y. (1980) Phys. Lett. 75A, 351.
Nakamura, Y. (1982) IEEE Trans. Plasma Sci. PS-10, 180.
Nakamura, Y. and Tsukabayashi, I. (1984) Phys. Rev. Lett. 52, 2356.
Nakamura, Y., Ferreira, J. L. and Ludwig, G. O. (1985) J. Plasma Phys.(to be published).
Okutsu, E. and Nakamura, Y. (1979) Plasma Phys. 21, 1053.
Sagdeev, R. Z. (1969) Reviews of Plasma Physics, vol. 4, Consultants Bureau, New York.
Tajiri, M. and Toda, M. (1985) J. Phys. Soc. Japan 54, 19.
Taylor, R. J., Mackenzie, K. R. and Ikezi, H. (1972) Rev. Sci. Instr. 43, 1675.
Tran, M. Q. and Coquerand, S. (1976) Phys. Rev. A 14, 2301.
Tran, M. Q., Appert, K., Hollenstein, Ch., Means, R. W. and Vaclawik (1977) J. Plasma Phys. 19, 381.
Tran, M. Q. (1979) Phys. Scripta 20, 317.
Wadati, M. (1975) J. Phys. Soc. Japan 38, 673, 681.
Washimi, H. and Taniuti, T. (1966) Phys. Rev. Lett. 17, 996.
Watanabe, S. (1975) J. Plasma Phys. 14, 353.
Watanabe, S. (1978) J. Phys. Soc. Japan 44, 611.
Watanabe, S. (1984) J. Phys. Soc. Japan 53, 952.
Wong, A. Y., Mamas, D. L. and Arnush, D. (1975) Phys. Fluids 18, 1489.
Yoshimura, K. and Watanabe, S. (1979) J. Phys. Soc. Japan 47, 998.
Zabusky, N. J. and Kruskal, M. D. (1965) Phys. Rev. Lett. 15, 240.

III. NONLINEAR AND CHAOTIC LINES

Nonlinear and Environmental Electromagnetics, edited by H. Kikuchi
Elsevier Science Publishers B.V., Amsterdam, 1985 – Printed in The Netherlands

THEORY OF UNDAMPED ELECTRIC OSCILLATIONS AND ELECTRICAL INSTABILITY IN A TRANSMISSION SYSTEM

Motinori Goto
Professor Emeritus, Meiji University

In 1927, anomalous electric oscillations occurred in the Inawashiro transmission system of the Tokyo Electric Light Company from the Inawashiro power stations to the Hatogaya substation. The present author analyzed these oscillations, as shown in his three reports[2),3),4)], where circuit resistances or magnetic saturation characteristics were simplified. In the present paper, an extended theory is presented so as to apply to any given values of circuit resistances and to any magnetic saturation characteristcs of main transformers expressed in graphs, power series or any other functions.

I. INTRODUCTION

In 1927, anomalous electric oscillations occurred in the Inawashiro 154 kV transmission system of the Tokyo Electric Light Company from the Inawashiro power station to Hatogaya substation. When the line voltage was raised near the nominal value in that line with a light load, there occurred strong mechanical vibrations of the generator and main transformers at the transmitting and receiving sites with a large continuous but intoned sound like "gu:n gu:n",---. While the rotating speed of the generator was kept constant, the wave forms of current and voltage were exceedingly unusual, as shown in Fig. 1.1 and these anomalous oscillations were termed "undamped electric oscillations". Although the generators had to be stopped under those circumstances, such oscillations disappeared for heavy load conditions.

The present author investigated these anomalous phenomena by means of a laboratory simulation model set at the Electrotechnical Laboratory. At that time one of his colleagues, Mr. Matsujiro Urushibata discovered another but related new phenomenon as shown in Fig. 1.2, where i_f is a field-exciting current. These anomalous oscillations in Fig. 1.1 and 1.2 belong to type A and B, respectively, as discussed hereafter and summarized in Table 3.1. Fig. 1.4 represents some applications of these undamped oscillations to a small size of simplified circuits such as frequency changers (down converters). (A) is the simplest one to generate f/2, f/3, f/5, and almost periodic oscillations. (B) is a circuit to easily generate f/2 and was used later for the so-called "parametron" circuit. (C) generates f/3 in three phases from f in single phase.

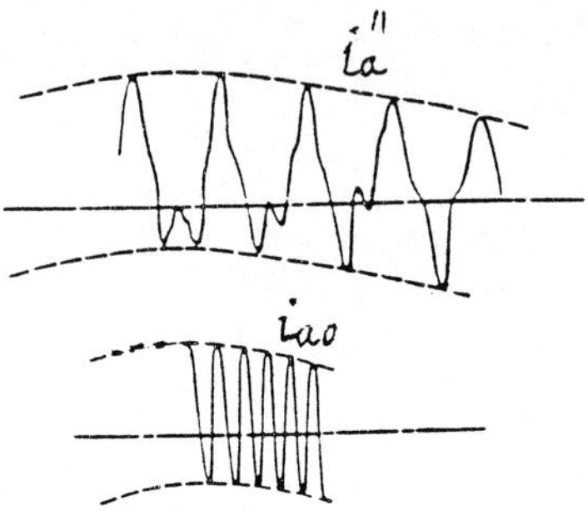

Fig. 1.1. Wave forms of the line current i''_a and delta component i_{ao} observed in the Inawashiro transmission system.

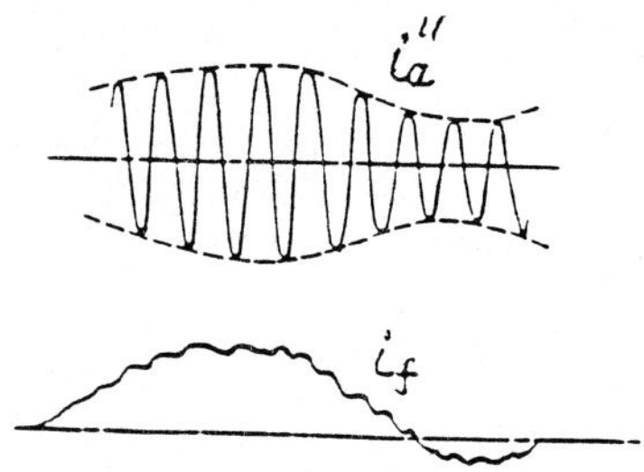

Fig. 1.2. Wave forms of currents in a laboratory simulation model.

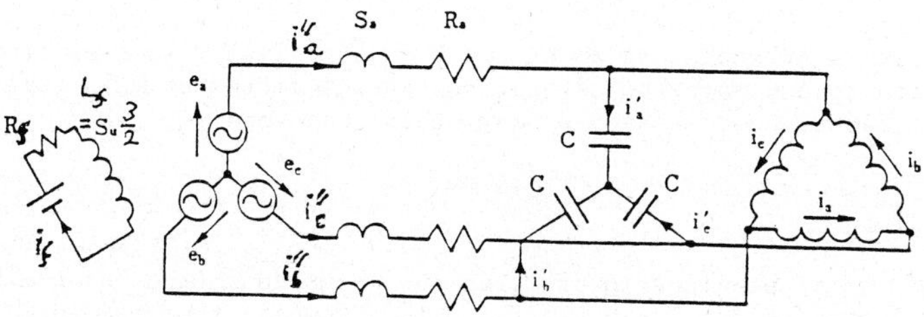

Fig. 1.3. A simplified equivalent circuit for a transmission system.

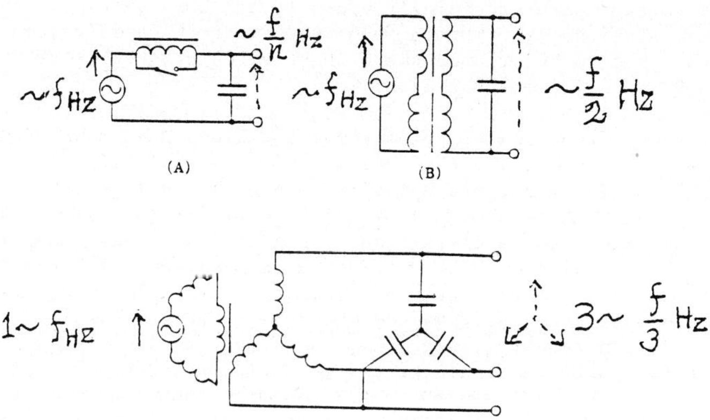

Fig. 1.4. Simplified frequency changers.

With regard to undamped oscillations mentioned above, the author published the case when magnetic saturation characteristics of transformers can be expressed in a cubic curve [3), 4)] or in a power series but for the resistance of a Δ-circuit $R' \fallingdotseq 0$ [2)]. In the present paper, an extended theory is presented so as to make an estimate of instability-state feasible for any magnetic saturation characteristics expressed in graphs, power series or any other functions and for any given values of circuit parameters. The symbols ! and !! express an "essential assumption" and an "essential relation", respectively.

II. CIRCUIT EQUATIONS FOR A TRANSMISSION SYSTEM

In Fig. 1.3, we have circuit equations

$$\left.\begin{aligned} e_f &= (R_f + pL_f)i_f + p(M_{af}i_a'' + M_{bf}i_b'' + M_{cf}i_c''),\\ v_a &= -\{(R_a + pL_a)i_a'' + pM_{ab}i_b'' + pM_{ca}i_c''\} - pM_{af}i_f,\\ v_b &= -\{(R_a + pL_b)i_b'' + pM_{bc}i_c'' + pM_{ab}i_a''\} - pM_{bf}i_f,\\ v_c &= -\{(R_a + pL_c)i_c'' + pM_{ca}i_a'' + pM_{bc}i_b''\} - pM_{cf}i_f; \end{aligned}\right\}, \qquad (2.1)!$$

where $L_a \equiv L_0 + 2L' \cos 2\theta$ ≡ self inductance of phase "a" of the armature circuit.

$$L_b \equiv L_0 + 2L' \cos\left(2\theta + \frac{2}{3}\pi\right),$$

$$L_c \equiv L_0 + 2L' \cos\left(2\theta - \frac{2}{3}\pi\right),$$

$M_{bc} \equiv M_0 + 2M' \cos 2\theta$ ≡ mutual inductance between phase "b" and "c" of the armature circuit.

$$M_{ca} \equiv M_0 + 2M' \cos\left(2\theta + \frac{2}{3}\pi\right),$$

$$M_{ab} \equiv M_0 + 2M' \cos\left(2\theta - \frac{2}{3}\pi\right).$$

$M_{af} \equiv M \cos\theta$ ≡ mutual inductance between the field circuit and phase "a" of the armature circuit.

$$M_{bf} \equiv M \cos\left(\theta - \frac{2}{3}\pi\right),$$

$$M_{cf} \equiv M \cos\left(\theta + \frac{2}{3}\pi\right),$$

R_f ≡ resistance of the field circuit.

R_a ≡ resistance of the armature per phase.

L_f ≡ self-inductance of the field circuit.

i_a'', i_b'', i_c'' ≡ instantaneous value of the armature current.

i_f ≡ instantaneous value of the field current.

v_a, v_b, v_c ≡ instantaneous value of armature potential.

e_f ≡ D.C. exciting voltage of the field circuit.

$$p \equiv \frac{d}{dt}, \quad \theta \equiv \omega t + \varphi,$$

where "e_f, R_a, ϕ, ω, R_f, L_f, L_o, L', M_o, M'" are constant with respect to "t".

In particular, for the case of the alternator with a sinusoidally distributed armature winding, we have

$$L' = M'. \tag{2.2}$$

Eqs. (2.1) may be rewritten in terms of symmetrical coordinates

$$\left.\begin{aligned} &i_{a0}'' \equiv \frac{1}{3}(i_a'' + i_b'' + i_c''), \quad i_{a1}'' \equiv \frac{1}{3}(i_a'' + a i_b'' + a^2 i_c'') \\ &i_{a2}'' = \overline{i_{a1}''}, \quad a \equiv \varepsilon^{j\frac{2}{3}\pi} \end{aligned}\right\} \tag*{(2.3)!}$$

as follows:

$$\left.\begin{aligned} &0 = (R_u + S_u p) i_u + Mp(\varepsilon^{-j\theta} i_{a1}'' + \varepsilon^{j\theta} i_{a2}''), \\ &v_{a0} = -(R_a + S_{a0} p) i_{a0}'', \\ &v_{a1} = -(R_a + S_a p) i_{a1}'' - S_r p \varepsilon^{2j\theta} i_{a2}'' - \frac{M}{2} p \varepsilon^{j\theta}(i_u + i_D), \\ &v_{a2} = \overline{v_{a1}} \end{aligned}\right\}, \tag*{(2.1')!!}$$

where

$$R_u \equiv \frac{2}{3} R_f, \quad S_u \equiv \frac{2}{3} L_f, \quad S_a \equiv L_0 - M_0, \quad S_{a0} \equiv L_0 + 2M_0, \quad S_u S_a - M^2 > 0, \quad i_D \equiv \frac{e_f}{R_f}, \quad i_f \equiv i_u + i_D, \quad \theta = \omega t + \varphi.$$

R_u, R_a, ϕ, ω, i_D, S_u, S_a, S_r, M, S_{a0} are constant with respect to time.

$S_u(S_a + S_r) - M^2 > 0$, $S_a - S_r > 0$.

In order to understand the whole aspect of phenomena, we first consider a simple case when magnetic saturation characteristics are expressed in a cubic curve. Then, putting

$$i_a \equiv c_1 f_a + c_3 f_a^3, \quad i_b \equiv c_1 f_b + c_3 f_b^3, \quad i_c \equiv c_1 f_c + c_3 f_c^3, \tag*{(2.4)!}$$

f_a, f_b, f_c ≡ number of flux interlinkages,

R' ≡ resistance of any one of the iron-core inductors,

we have from Eq. (2.1')

$$\left.\begin{aligned} &v_{a1} = \frac{i_{a1}}{pC} = \frac{p f_{a1} + R' i_{a1}}{a^2 - a} = -(R_a + S_a p) i_{a1}'' - S_r p \varepsilon^{2j\theta} i_{a2}'' - \frac{M}{2} p \varepsilon^{j\theta}(i_u + i_D), \\ &i_{a1}'' = \frac{pC}{a^2 - a}(p f_{a1} + R' i_{a1}) + (a - a^2) i_{a1}, \\ &v_{a0} = i_{a0}' = i_{a0}'' = 0, \\ &0 = (R_u + S_u p) i_u + Mp(\varepsilon^{-j\theta} i_{a1}'' + \varepsilon^{j\theta} i_{a2}''). \end{aligned}\right\} \tag*{(2.5)!!}$$

$$\left.\begin{array}{l} i_{a1}=c_1 f_{a1}+3c_3(f_{a1}^2 f_{a2}+f_{a2}^2 f_{a0}+f_{a0}^2 f_{a1}), \\ -pf_{a0}/R'=i_{a0}=c_3(f_{a0}^3+f_{a1}^3+f_{a2}^3)+f_{a0}(c_1+6c_3 f_{a1} f_{a2}). \end{array}\right\} \quad (2.6)!!$$

$$-j\sqrt{3}\,e_{a1}=\left[R_a+S_a p-\frac{M^2 p(p-j\omega)}{2\{R_u+S_u(p-j\omega)\}}\right]\{pC(pf_{a1}+R'i_{a1})+3i_{a1}\}$$
$$+pf_{a1}+R'i_{a1}-\varepsilon^{2j\theta}(p+2j\omega)\left[S_r-\frac{M^2(p+j\omega)}{2\{R_u+S_u(p+j\omega)\}}\right]\{pC(pf_{a2}+R'i_{a2})+3i_{a2}\}. \quad (2.7)!!$$

From Eqs. (2.5) to (2.7),

$$e_{a1}\equiv\frac{M}{2}\,j\omega i_D\varepsilon^{j\theta} \equiv \text{nominal induced E.M.F. due to exciting current } i_D \quad (2.8)!!$$

can be obtained.

In the ordinary steady state, there is a well-known term expressed by $\varepsilon^{j\omega t}$ in f_{a1}. Then i_{a1}, f_{f1} and f_{a0} in Eqs. (2.5), (2.6) may be expressed by

$$\left.\begin{array}{l} f_{a1}=\sum_{l=-\infty}^{\infty} Z^{(l)}\varepsilon^{j(6l+1)\omega t}, \\ i_{a1}=\sum_{l=-\infty}^{\infty} I^{(l)}\varepsilon^{j(6l+1)\omega t}, \\ f_{a0}\equiv\sum_{l=0}^{\infty}(f^{(l)}\varepsilon^{j(6l+3)\omega t}+\overline{f^{(l)}}\varepsilon^{-j(6l+3)\omega t}). \end{array}\right\} \quad (2.9)!!$$

where $Z^{(e)}$, $I^{(e)}$, $f^{(i)}$ are complex constants, and ω is a real constant.
In particular, for $\lim R' = 0$ (2.10) !
Eq.(2.9) is simplified, referring to Eq. (2.6), as

$$f_{a0}=0, \quad f_{a1}=Z^{(0)}\varepsilon^{j\omega t}\equiv z\varepsilon^{j\omega t}, \quad (2.11)$$

where Z is determined as $Z \geqq 0$ for simplicity and

$$i_{a1}=I^{(0)}\varepsilon^{jt}. \quad Z^{(r)}=I^{(r)}=0, \ (r\neq 0). \quad (2.12)$$
$$I^{(0)}=(c_1+3c_3z^2)z\equiv(c_1+x)z, \quad x\equiv 3c_3z^2. \quad (2.13)$$

Introducing (2.12), (2.13) into (2.7), we have

$$-j\sqrt{3}\,e_{a1}=[j\omega-\{R_a+j\omega(S_a-S_r\varepsilon^{2j\varphi})\}P+3(c_1+x)\{R_a+j\omega(S_a-S_r\varepsilon^{2j\varphi})\}]z, \quad (2.14)$$

where $P \equiv \omega^2 C$
In the ordinary steady state for $\lim R' = 0$, we have

$$I^{(0)}=(c_1+x)z, \quad f^{(1)}=0, \quad Z^{(r)}=0. \ (r\neq 0);$$
(i) $R_a+\omega S_r \sin 2\varphi \neq 0$, $3(c_1+x)-P \neq 0$,
$$\varphi=\mathrm{Tan}^{-1}\left[\frac{\omega}{R_a}\left\{\frac{1}{3(c_1+x)-P}+S_a-S_r\right\}\right]+n\pi,$$

$$-\frac{\sqrt{3}}{2}\omega M i_D \frac{(-1)^n}{\sqrt{1+\tan^2\varphi}} = \left\{R_a + \omega S_r \frac{2\tan\varphi}{1+\tan^2\varphi}\right\}\{3(c_1+x)-P\}z;$$

(ii) $3(c_1+x)-P=0,\ \frac{\sqrt{3}}{2}Mi_D=z,\ \varphi=-\frac{\pi}{2}+2n\pi;$ (2.15)!!

(iii) $i_D=z=0,\ \varphi=$arbitrary.

(iv) $i_D=0,\ \varphi=\frac{1}{2}(-1)^n\,\mathrm{Sin}^{-1}\left(-\frac{R_a}{\omega S_r}\right)+\frac{n\pi}{2},$

$$1+\left\{S_a - S_r(-1)^n\sqrt{1-\frac{R_a^2}{\omega^2 S_r^2}}\right\}\{3(c_1+x)-P\}=0;$$

where $z \equiv Z^{(0)} \geqq 0$, n = arbitrary integer, $x \equiv 3c_3z^2$.

For

$$\lim_{R_a\to 0}\frac{\omega}{R_a}\left\{\frac{1}{3(c_1+x)-P}+S_a-S_r\right\}=\infty,\ i_D\geqq 0$$

and $\lim R_a = 0$, we have from (i) or (ii) of (2.15)

(A_1) $\varphi=-\frac{\pi}{2}+2m\pi,\ i_D=\frac{2z}{\sqrt{3}M}[1+(S_a+S_r)\{3(c_1+x)-P\}]\ \ (1+(S_a+S_r)\{3(c_1+x)-P\}>0).$

(A_2) $\varphi=\frac{\pi}{2}+2m\pi,\ i_D=\frac{-2z}{\sqrt{3}M}[1+(S_a+S_r)\{3(c_1+x)-P\}]\ \ (1+(S_a+S_r)\{3(c_1+x)-P\}<0).$

From (i) or (iv) of (2.15), we have

(B) $\varphi=\pm\frac{\pi}{2}+2m\pi,\ i_D=0\ (1+(S_a+S_r)\{3(c_1+x)-P\}=0).$ (2.16)!!

From (iv) of (2.15)

(C) $\varphi=m\pi,\ i_D=0\ \ (1+(S_a-S_r)\{3(c_1+x)-P\}=0),$

where m = arbitrary integer,

$$I^{(0)}=(c_1+x)z,\ f^{(1)}=0,\ Z^{(r)}=I^{(r)}=0,\ r\neq 0.$$

III. SMALL TRANSIENT PHENOMENA (case of δi_{a1}, δf_{a0}, δf_{ai})

For small deviations δi_{a1}, δf_{a0}, δf_{a1} etc. for i_{a1}, i_{a2}, f_{a0}, f_{a1}, f_{a2} in the preceding section, we have from (2.6), (2.7)

$$\left[R_a+S_a p-\frac{M^2 p(p-j\omega)}{2\{R_u+S_u(p-j\omega)\}}\right]\{pC(p\delta f_{a1}+R'\delta i_{a1})+3\delta i_{a1}\}$$
$$+p\delta f_{a1}+R'\delta i_{a1}-\varepsilon^{2j\theta}(p+2j\omega)\left[S_r-\frac{M^2(p+j\omega)}{2\{R_u+S_u(p+j\omega)\}}\right]\{pC(p\delta f_{a2}$$
$$+R'\delta i_{a2})+3\delta i_{a2}\}=0\ , \qquad (3.1)!!$$

$$\delta i_{a1}=(c_1+6c_3 f_{a1} f_{a2}+3c_3 f_{a0}^2)\delta f_{a1}+3c_3(f_{a1}^2+2f_{a0}f_{a2})\delta f_{a2}+3c_3(f_{a2}^2+2f_{a0}f_{a1})\delta f_{a0}\ , \qquad (3.2)!!$$

$$-p\delta f_{a0}=R'[3c_3(f_{a1}^2+2f_{a0}f_{a2})\delta f_{a1}+3c_3(f_{a2}^2+2f_{a0}f_{a1})\delta f_{a2}+(c_1+6c_3 f_{a1}f_{a2}+3c_3 f_{a0}^2)\delta f_{a0}], \qquad (3.3)!!$$

where i_{a1}, f_{a0}, f_{a1} ≡ value in a steady state (just before the transient). If δf_{a1} includes the factors $\varepsilon^{j\omega t}$ and $\varepsilon^{(x+j\omega t)}$, we obtain in the same way as for Eq. (2.9)

$$\left.\begin{aligned}
\delta f_{a1}&=\sum_X\sum_{l=-\infty}^{\infty}[Z_1^{(l)}\varepsilon^{\{X+j(6l+1)\omega\}t}+Z_{-1}^{(l)}\varepsilon^{\{\bar{X}+j(6l+1)\omega\}t}],\\
\delta i_{a1}&=\sum_X\sum_{l=-\infty}^{\infty}[I_1^{(l)}\varepsilon^{\{X+j(6l+1)\omega\}t}+I_{-1}^{(l)}\varepsilon^{\{\bar{X}+j(6l+1)\omega\}t}],\\
\delta f_{a0}&=\sum_X\sum_{l=0}^{\infty}[f_1^{(l)}\varepsilon^{\{X+j(6l+3)\omega\}t}+\overline{f_1^{(l)}}\varepsilon^{\{\bar{X}-j(6l+3)\omega\}t}+f_{-1}^{(l)}\varepsilon^{\{\bar{X}+j(6l+3)\omega\}t}+\overline{f_{-1}^{(l)}}\varepsilon^{\{X-j(6l+3)\omega\}t}].
\end{aligned}\right\} \qquad (3.4)!!$$

For sufficiently small R_a, R_u and R', putting $R_a = dR_a$, $R_u = dR_u$, $R' \equiv dR'$, one can write

$$X=(X)_{\lim R_a=0,\ \lim R_u=0,\ \lim R'=0}+\left(\frac{\partial X}{\partial R_a}\right)_1 dR_a+\left(\frac{\partial X}{\partial R'}\right)_1 dR'+\left(\frac{\partial X}{\partial R_u}\right)_1 dR_u,$$
$$I_{\pm 1}^{(l)}=\cdots\cdots,\quad f_{\pm 1}^{(l)}=\cdots\cdots,\quad Z_{\pm 1}^{(l)}=\cdots\cdots\ . \qquad (3.5)$$

We find solutions for small transient phenomena for $\lim R_a = 0$, $\lim R_u = 0$ and $\lim R' = 0$. From Eq. (3.3), we have three cases, setting $p\delta f_{a0} = 0$,

$$\left\{\begin{aligned}
&\text{(i)}\ \ X'+j3\omega=0,\ f_{\pm 1}^{(l)}=0\ (\text{except } f_1^{(k)})\\
\text{or}\ &\text{(ii)}\ \ \bar{X}'+j3\omega=0,\ f_{\pm 1}^{(l)}=0\ (\text{except } f_{-1}^{(\bar{k})})\\
\text{where}\ &X'\equiv X+j6k\omega.\\
\text{or}\ &\text{(iii)}\ \ f_{\pm 1}^{(l)}=0.
\end{aligned}\right. \qquad (3.6)$$

For simplicity, sliding the value of "k" to "0", we have

$$\left.\begin{aligned}
&\text{(i')}\ \ X=-j3\omega,\ f_{\pm 1}^{(l)}=0\ (\text{except } f_1^{(0)})\\
\text{or}\ &\text{(ii')}\ \ X=j3\omega,\ f_{\pm 1}^{(l)}=0\ (\text{except } f_{-1}^{(0)})\\
\text{or}\ &\text{(iii')}\ \ f_{\pm 1}^{(l)}=0,
\end{aligned}\right\} \qquad (3.6')!!$$

where X for case (iii) is obtained from (3.1), (3.2) as

$$\delta i_{a1}=(c_1+6c_3 z^2)\delta f_{a1}+3c_3 z^2\delta f_{a2}. \qquad (3.7)$$

Substituting (3.4) and (iii') into (3.7), we have

$$\left.\begin{aligned}
I_1^{(l)}&=(c_1+2x)Z_1^{(l)}+z\overline{Z_{-1}^{(-l)}},\quad f_{\pm 1}^{(l)}=0.\\
\bar{I}_{-1}^{(-l)}&=(c_1+2x)\overline{Z_{-1}^{(-l)}}+xZ_1^{(l)}.
\end{aligned}\right\} \qquad (3.8)$$

Substituting

$$\delta i_{a1}=I_1{}^{(0)}\varepsilon^{(X+j\omega)t},\ \delta f_{a1}=Z_1{}^{(0)}\varepsilon^{(X+j\omega)t} \text{ and } \varepsilon^{2j\theta}\delta i_{a2}=\overline{I_{-1}{}^{(0)}}\varepsilon^{2j\varphi+(X+j\omega)t},\ \varepsilon^{2j\theta}\delta f_{a2}=\overline{Z_{-1}{}^{(0)}}\varepsilon^{2j\varphi+(X+j\omega)t}$$

into (3.1), we obtain

$$\begin{aligned}
&[2R_a(R_u+XS_u)+2R_uS_a(X+j\omega)+SX(X+j\omega)][(X+j\omega)C\{(X+j\omega)Z_1{}^{(0)}+R'I_1{}^{(0)}\}+3I_1{}^{(0)}]\\
&\quad+2(R_u+XS_u)\{(X+j\omega)Z_1{}^{(0)}+R'I_1{}^{(0)}\}-\varepsilon^{2j\varphi}[2R_uS_r(X+j\omega)+TX(X+j\omega)]\\
&\quad\cdot[(X-j\omega)C\{(X-j\omega)\bar{Z}_{-1}{}^{(0)}+R'\bar{I}_{-1}{}^{(0)}\}+3\bar{I}_{-1}{}^{(0)}]=0,\\
&[2R_a(R_u+XS_u)+2R_uS_a(X-j\omega)+SX(X-j\omega)]\,[(X-j\omega)C\{(X-j\omega)\bar{Z}_{-1}{}^{(0)}+R'\bar{I}_{-1}{}^{(0)}\}+3\overline{I_{-1}{}^{(0)}}]\\
&\quad+2(R_u+XS_u)\{(X-j\omega)\bar{Z}_{-1}{}^{(0)}+R'\bar{I}_{-1}{}^{(0)}\}-\varepsilon^{-2j\varphi}[2R_uS_r(X-j\omega)+TX(X-j\omega)]\\
&\quad\cdot[(X+j\omega)C\{(X+j\omega)Z_1{}^{(0)}+R'I_1{}^{(0)}\}+3I_1{}^{(0)}]=0\ ,
\end{aligned}\tag{3.9}$$

where $P\equiv\omega^2C,\ S\equiv 2S_aS_u-M^2,\ T\equiv 2S_rS_u-M^2.$

In (3.9), $\lim R'=\lim R_a=\lim R_u=0$ leads to

$$\left.\begin{aligned}
&\begin{cases}\{(X+j\omega)^2SC+2S_u+3S(c_1+2x)-3Tx\varepsilon^{2j\varphi}\}Z_1{}^{(0)}-\{(X-j\omega)^2TC\varepsilon^{2j\varphi}-3Sx+3T(c_1+2x)\varepsilon^{2j\varphi}\}\bar{Z}_{-1}{}^{(0)}=0\\ \text{or}\quad X+j\omega=0\quad\text{or}\quad X=0.\end{cases}\\
&\text{and}\\
&\begin{cases}\{(X-j\omega)^2SC+2S_u+3S(c_1+2x)-3Tx\varepsilon^{-2j\varphi}\}\bar{Z}_{-1}{}^{(0)}-\{(X+j\omega)^2TC\varepsilon^{-2j\varphi}-3Sx+3T(c_1+2x)\varepsilon^{-2j\varphi}\}Z_1{}^{(0)}=0\\ \text{or}\quad X-j\omega=0\quad\text{or}\quad X=0\ .\end{cases}
\end{aligned}\right\}\tag{3.10}$$

Eliminating $Z_1^{(0)}$ and $Z_{-1}^{(0)}$ in (3.10), we have

$$\begin{aligned}
&\Phi(X)\equiv(S^2-T^2)\{(X^2+\omega^2)^2C^2+6(c_1+2x)(X^2-\omega^2)C\\
&\quad+9(c_1+2x)^2\}+4SS_u\{(X^2-\omega^2)C+3(c_1+2x)\}+(2S_u+3xT)^2\\
&\quad-9x^2S^2=0\quad\text{or}\quad X^2+\omega^2=0\quad\text{or}\quad X^2=0\quad\text{or}\quad Z_{\pm1}{}^{(0)}=0.
\end{aligned}\tag{3.11)!!$$

Solving (3.11), we obtain five sets of X^2C

$$\left.\begin{aligned}
&X_1{}^2C=\frac{-[(S^2-T^2)\}P+3(c_1+2x)\}+2SS_u]+\sqrt{D}}{S^2-T^2},\\
&X_2{}^2C=\frac{-[(S^2-T^2)\{P+3(c_1+2x)\}+2SS_u]-\sqrt{D}}{S^2-T^2},\\
&X_3{}^2C=-P,\ X_4{}^2C=0,\ X_5{}^2C=-9P.
\end{aligned}\right\}\tag{3.11')!!$$

where

$$D\equiv 4P(S^2-T^2)\{3(S^2-T^2)(c_1+2x)+2SS_u\}+\{2S_uT-3x(S^2-T^2)\}^2,\ S>0,\ S>T,\ S^2>T^2.$$

For $X_n^2C > 0$, it follows that

$$(X_1{}^2C\geqq 0)\leftrightarrow(P_{41}\leqq P\leqq P_1)\vee(P_1\leqq P\leqq P_{41}).\tag{3.12)!!$$

$$(X_1{}^2C\geqq -P)\leftrightarrow(P\geqq P_{31}).\tag{3.13)!!$$

$$(X_1{}^2C\geqq -9P)\leftrightarrow(P\geqq P_{51}).\tag{3.14)!!$$

$$\lim_{P\to\infty}\frac{X_1{}^2C}{P}=-1+0.\tag{3.15)!!$$

$$(X_2{}^2C<0)\leftrightarrow(P\ \text{arbitray}).\tag{3.16)!!$$

$$(X_2{}^2C\geqq -9P)\leftrightarrow(P\geqq P_{52}).\tag{3.17)!!$$

$$\lim_{P\to\infty}\frac{X_2{}^2C}{P}=-1-0.\tag{3.18)!!$$

where ↔ indicates "necessary and sufficient", V indicates "or" and

$$P_1 \equiv \frac{2S_u}{S-T}+3(c_1+x)\equiv\frac{1}{S_a-S_r}+3(c_1+x),$$

$$P_{31} \equiv \frac{\{3(c_1+x)(S-T)+2S_u\}\{3(c_1+3x)(S+T)+2S_u\}}{4\{3(c_1+2x)(S^2-T^2)+2SS_u\}}$$

$$P_{41} \equiv \frac{2S_u}{S+T}+3(c_1+3x)\equiv\frac{S_u}{(S_a+S_r)S_u-M^2}+3(c_1+3x),$$

$$P_{51} \equiv \frac{15(c_1+2x)(S^2-T^2)+10SS_u-\sqrt{\Delta}}{32(S^2-T^2)},$$

$$P_{52} \equiv \frac{15(c_1+2x)(S^2-T^2)+10SS_u+\sqrt{\Delta}}{32(S^2-T^2)}.$$

$$\Delta \equiv 9\{3(c_1+2x)(S^2-T^2)+2SS_u\}^2+16\{2S_uT-3x(S^2-T^2)\}^2.$$

The relations of $I_{\pm1}^{(\ell)}$ and $Z_{\pm1}^{(\ell)}$ for various X_n are as follows:

(i) X=(one of X_1)

$$\left.\begin{aligned}&[(X_1+j\omega)^2SC+2S_u+3S(c_1+2x)+3Tx]Z_1^{(0)}\\&\quad+[(X_1-j\omega)^2TC+3Sx+3T(c_1+2x)]\bar{Z}_{-1}^{(0)}=0,\\&\text{or}\quad[(X_1+j\omega)^2TC+3Sx+3T(c_1+2x)]Z_1^{(0)}\\&\quad+[(X_1-j\omega)^2SC+2S_u+3S(c_1+2x)+3Tx]\bar{Z}_{-1}^{(0)}=0.\end{aligned}\right\} \tag{3.19}$$

$$Z_{\pm1}^{(l)}=0\ (l\neq0). \tag{3.20}$$

$$I_1^{(l')}=(c_1+2x)Z_1^{(l')}+x\bar{Z}_{-1}^{(-l')},$$

$$\bar{I}_{-1}^{(l')}=(c_1+2x)\bar{Z}_{-1}^{(-l')}+xZ_1^{(l')},\quad f_{\pm1}^{(l')}=0\ (l'=\text{arbitray integer}). \tag{3.21}$$

(ii) X=(one of X_2)

Replacing X_1 in (3.19) by X_2, we have

(iii) X=(one of X_3) $=-j\omega$

$$3\{Sx+T(c_1+2x)\}Z_1^{(0)}+\{-4PS+2S_u+3S(c_1+2x)+3Tx\}\bar{Z}_{-1}^{(0)}=0. \tag{3.22}$$

$$Z_{\pm1}^{(l)}=0\ (l\neq0). \tag{3.23}$$

$$I_1^{(l')}=(c_1+2x)Z_1^{(l')}+x\bar{Z}_{-1}^{(-l')},\quad \bar{I}_{-1}^{(-l')}=(c_1+2x)\bar{Z}_{-1}^{(-l')}+xZ_1^{(l')},$$

$$f_{\pm1}^{(l')}=0\ (l'=\text{arbitrary integer}). \tag{3.24}$$

(iv) X=(one of X_4)=0

is the same as (3.21).

(v) X=(one of X_5)=$-3j\omega$

$$\left.\begin{aligned}\bar{Z}_{-1}^{(-1)}=Z_1^{(0)}&=\frac{[\{16P-3(c_1+2x)\}(S^2-T^2)-2SS_u]3xf_1^{(0)}}{64(S^2-T^2)P^2-20P\{3(c_1+2x)(S^2-T^2)+2SS_u\}+\{3(c_1+3x)(S+T)+2S_u\}\{3(c_1+x)(S-T)+2S_u\}},\\Z_1^{(1)}=\bar{Z}_1^{(0)}&=\frac{\{3x(S^2-T^2)-2S_uT\}3xf_1^{(0)}}{64(S^2-T^2)P^2-20P\{3(c_1+2x)(S^2-T^2)+2SS_u\}+\{3(c_1+3x)(S+T)+2S_u\}\{3(c_1+x)(S-T)+2S_u\}}.\end{aligned}\right\} \tag{3.25}$$

$$Z_{\mp1}^{(\pm1)}=Z_{\pm1}^{(r)}=0\ (r\neq0,\pm1). \tag{3.26}$$

$$\left.\begin{aligned} I_1^{(0)} &= (c_1+2x)Z_1^{(0)} + x\bar{Z}_{-1}^{(0)} + xf_1^{(0)}, & \bar{I}_{-1}^{(0)} &= (c_1+2x)\bar{Z}_{-1}^{(0)} + xZ_1^{(0)}, \\ I_1^{(1)} &= (c_1+2x)Z_1^{(1)} + x\bar{Z}_{-1}^{(-1)}, & \bar{I}_{-1}^{(-1)} &= (c_1+2x)\bar{Z}_{-1}^{(-1)} + xZ_1^{(1)} + xf_1^{(0)}, \\ I_1^{(r)} &= (c_1+2x)Z_1^{(r)} + x\bar{Z}_{-1}^{(-r)}, & \bar{I}_{-1}^{(-r)} &= (c_1+2x)\overline{Z_{-1}^{(-r)}} + xZ_1^{(r)},\ (r \neq 0, \pm 1) \end{aligned}\right\} \tag{3.27}$$

$$f_{\pm 1}^{(1)} = 0 \text{ (except } f_1^{(0)}\text{)}. \tag{3.28}$$

Ranges of Electrical Instabilities for small R_a, R_u, R'.

The existence of the positive real part of X in ε^{Xt} indicates an electrical instability. To find this range, we seek a region where the following partial derivatives are positive:

$$\frac{\partial X_n}{\partial R_a},\ \frac{\partial X_n}{\partial R_u},\ \frac{\partial X_n}{\partial R'}$$

Therefore, we have only to calculate the real part of

$$X = (X)_{\lim R_a = \lim R_u = \lim R' = 0} + \left(\frac{\partial X}{\partial R_a}\right)_1 dR_a + \left(\frac{\partial X}{\partial R_u}\right)_1 dR_u + \left(\frac{\partial X}{\partial R'}\right)_1 dR'. \tag{3.29}$$

The first term of the right-hand side of (3.29) is already known in (3.12) - (3.18). We find other terms from (3.1) - (3.3) as

$$\left(\frac{\partial X}{\partial R_a}\right)_{X=X_1} = \frac{2S_u^2[4(S^2-T^2)S^2P + 2ST\{2S_uT - 3x(S^2-T^2)\} - (S^2+T^2)\sqrt{D}]}{(S^2-T^2)[3(c_1+2x)(S^2-T^2) + 2S_uS - \sqrt{D}]\sqrt{D}} \geqq 0,$$

$$X_1^2 \neq 0,\quad X_1^2C \neq -P;\quad \leftrightarrow (P_{31} < P \leqq P_{1a},\ P \neq P_1,\quad (c_1+2x)T + xS \neq 0). \tag{3.30)!!$$

$$\left(\frac{\partial X}{\partial R_u}\right)_{X=X_1} = \frac{-M^2\{2(S^2-T^2)P - 2S_uT + 3x(S^2-T^2) - \sqrt{D}\}}{4\{S_u(S_a+S_r) - M^2\}^2 X_1^2 C\sqrt{D}} > 0,$$

$$X_1^2 \neq 0;\quad \leftrightarrow (P_{41} < P,\quad P \neq P_1). \tag{3.31)!!$$

$$\left(\frac{\partial X}{\partial R'}\right)_{X=X_1} = \left[\frac{3f_1(Y,P)}{2Y(Y+8P)\sqrt{D}}\right]_{Y=X_1^2C+P} \geqq 0,\ X_1^2C \neq 0,\ -P,\ -9P;$$

$$\leftrightarrow [P_{51} < P \leqq P_{1R}) \vee (P_{31} < P \leqq P'_{1R})],\ (P \neq P_1),\ (P \neq P_{41}). \tag{3.32)!!$$

$$\left(\frac{\partial X}{\partial R_a}\right)_{X=X_2} = \frac{2S_u^2[4(S^2-T^2)S^2P + 2ST\{2S_uT - 3x(S^2-T^2)\} + (S^2+T^2)\sqrt{D}}{-(S^2-T^2)[3(c_1+2x)(S^2-T^2) + 2S_uS + \sqrt{D}]\sqrt{D}} < 0,$$

$$P \neq P_1;\quad \leftrightarrow (P \neq P_1). \tag{3.33)!!$$

$$\left(\frac{\partial X}{\partial R_u}\right)_{X=X_2} = \frac{M^2\{2(S^2-T^2)P - 2S_uT + 3x(S^2-T^2) + \sqrt{D}\}}{4\{S_u(S_a+S_r) - M^2\}^2 X_1^2 C\sqrt{D}} < 0,\quad P \neq P_1;\quad \leftrightarrow (P \neq P_1). \tag{3.34)!!$$

$$\left(\frac{\partial X}{\partial R'}\right)_{X=X_2} = \left[\frac{-3f_1(Y,P)}{2Y(Y+8P)\sqrt{D}}\right]_{Y=X_2^2C+P} \geqq 0,\quad X_2^2C \neq -9P;\quad \leftrightarrow (P_{52} < P \leqq P_{2R}). \tag{3.35)!!$$

$$\left(\frac{\partial X}{\partial R_a}\right)_{X=X_2} = \frac{6R_u\{3S(c_1+x)(c_1+3x) + 2S_u(c_1+2x) - 4(c_1+2x)SP\}}{4P\{3(c_1+2x)(S^2-T^2) + 2SS_u\} - \{2S_u + 3(c_1+x)(S-T)\}\{2S_u + 3(c_1+3x)(S+T)\}}$$

$$\geqq 0,\quad X_1^2C \neq -P,\quad c_1+2x \neq 0;$$

$$\leftrightarrow (P_3 \leqq P < P_{31},\quad c_1+2x \neq 0). \tag{3.36)!!$$

$$\left(\frac{\partial X}{\partial R_a}\right)_{X=X_2} = 0,\quad X_1^2C \neq -P,\quad c_1 = x = 0,\ P \neq P_1;\ \leftrightarrow \left(P \neq \frac{S_u}{2S},\ P \neq \frac{2S_u}{S-T}\right) \tag{3.37)!!$$

$$\left(\frac{\partial X}{\partial R_u}\right)_{X=X_2} = 0;\quad \leftrightarrow P = \text{arbitrary}. \tag{3.38)!!$$

$$\left(\frac{\partial X}{\partial R'}\right)_{X=X_1}=\frac{1}{3}\left[\left(\frac{\partial X}{\partial R_a}\right)_{X=X_1}\qquad\right]\geqq 0,\quad X_1{}^2C\doteqdot -P,\quad c_1+2x\doteqdot 0;$$

$$\leftrightarrow(P_3\leqq P<P_{31},\quad c_1+2x\doteqdot 0). \tag{3.39}$$!!

$$\left(\frac{\partial X}{\partial R'}\right)_{X=X_1}=0,\quad X_1{}^2C\doteqdot -P,\quad c_1=x=0,\quad P\doteqdot P_1;\quad \leftrightarrow\left(P\doteqdot\frac{S_u}{2S},\quad P\doteqdot\frac{2S_u}{S-T}\right). \tag{3.40}$$!!

$$\left(\frac{\partial X}{\partial R_a}\right)_{X=X_4}=0;\quad \leftrightarrow P=\text{arbitrary}. \tag{3.41}$$!!

$$\left(\frac{\partial X}{\partial R_u}\right)_{X=\text{one of }X_4}=-\frac{2[(S_a+S_r)\{3(c_1+3x)-P\}+1]}{(S+T)\{3(c_1+3x)-P\}+2S_u}\geqq 0,\quad P\doteqdot P_{41};\ \leftrightarrow P_4\leqq P<P_{41}. \tag{3.42}$$!!

$$\left(\frac{\partial X}{\partial R_u}\right)_{X=\text{the other }X_4}=-\frac{1}{S_u}<0;\quad \leftrightarrow P=\text{arbitrary}. \tag{3.43}$$!!

$$\left(\frac{\partial X}{\partial R'}\right)_{X=X_4}=0;\quad \leftrightarrow P=\text{arbitrary}. \tag{3.44}$$!!

$$\left(\frac{\partial X}{\partial R_u}\right)_{X=X_5}=0;\quad \leftrightarrow P=\text{arbitrary}. \tag{3.45}$$!!

$$\left(\frac{\partial X}{\partial R_u}\right)_{X=X_5}=0;\quad \leftrightarrow P=\text{arbitrary}. \tag{3.46}$$!!

$$\left(\frac{\partial X}{\partial R'}\right)_{X=X_5}=-\frac{f(P,x)}{\varphi(P,x)}\geqq 0,\quad X_5\doteqdot X_1,\quad X_5\doteqdot X_2;\quad \leftrightarrow(P_5\leqq P<P_{51})\vee(P_5'\leqq P<P_{52}), \tag{3.47}$$!!

where

$$D\equiv 4P(S^2-T^2)\{3(S^2-T^2)(c_1+2x)+2SS_u\}+\{2S_uT-3x(S^2-T^2)\}^2,$$

$$P_{31}\equiv\frac{[3(c_1+x)(S-T)+2S_u]\,[3(c_1+3x)(S+T)+(2S_u]}{4\{3(c_1+2x)(S^2-T^2)+2SSu\}},$$

$$P_{1a}\equiv\frac{3(c_1+2x)(S^2+T^2)^2+12xS^3T+2SS_u(S^2-T^2)+\sqrt{D'}}{8S^4},$$

$$D'\equiv\{3(c_1+2x)(S^2+T^2)^2+12xS^3T+2SS_u(S^2-T^2)\}^2+4S^4\{2S_uT-3x(S^2-T^2)\}^2,$$

$$P_1\equiv\frac{2S_u}{S-T}+3(c_1+x)\equiv\frac{1}{S_a-S_r}+3(c_1+x),$$

$$P_{41}\equiv\frac{2S_u}{S+T}+3(c_1+3x)\equiv\frac{S_u}{(S_a+S_r)S_u-M^2}+3(c_1+3x),$$

$$f_1(Y,P)\equiv Y^2A+2PYB-16BP^2+2EY+2FP,$$

$$A\equiv(S^2-T^2)\{(c_1+2x)^2+2x^2\},\quad B\equiv 2(S^2-T^2)(c_1+2x)^2,$$

$$E\equiv S_u\{S(c_1+2x)^2+2Sx^2+2Tx(c_1+2x)\}+\frac{3}{2}(c_1+2x)^3(S^2-T^2),$$

$$F\equiv 8[S_u\{S(c_1+2x)^2+Sx^2+2Tx(c_1+2x)\}+\frac{3}{2}(c_1+2x)(S^2-T^2)\{(c_1+2x)^2-x^2\}],$$

$$P_{51}\equiv\frac{15(S^2-T^2)(c_1+2x)+10SS_u-\sqrt{\Delta}}{32(S^2-T^2)},$$

$$\Delta\equiv 9\{3(c_1+2x)(S^2-T^2)+2SS_u\}^2+16\{2S_uT-3x(S^2-T^2)\}^2,$$

$$\begin{aligned}f(P,x)\equiv{}&4(S^2-T^2)(c_1+2x)(4P)^2-(4P)[10SS_u(c_1+2x)\\&+(S^2-T^2)\{15(c_1+2x)^2-24x^2\}]+4(c_1+2x)S_u^2\\&+12S_u\{(c_1+2x)^2S-x^2S+x(c_1+2x)T\}+9(c_1+2x)(S^2-T^2)\{(c_1+2x)^2-3x^2\},\end{aligned}$$

$$\begin{aligned}\varphi(P,x)\equiv{}&4(S^2-T^2)(4P)^2-(4P)\{15(c_1+2x)(S^2-T^2)+10SS_u\}\\&+4S_u^2+12S_u\{(c_1+2x)S+xT\}+9(S^2-T^2)\{(c_1+2x)^2-x^2\},\end{aligned}$$

$$\varphi_1(Y,P)\equiv\left\{Y+3(c_1+2x)+\frac{2S_uS}{S^2-T^2}\right\}^2-4P\left\{3(c_1+2x)+\frac{2SS_u}{S^2-T^2}\right\}-\left\{3x-\frac{2S_uT}{S^2-T^2}\right\}^2,$$

P_{1R} ≡ the smallest value of P's given by the two curves $[\varphi_1(Y,P)]_{Y=X_1^2C+P}=0$ and $[f_1(Y,P)]_{Y=X_1^2C+P}=0$, where 3x should be common to ϕ_1 and f_1

P'_{1R} ≡ the largest value of the above-mentioned P's

$$P_{52}\equiv\frac{15(S^2-T^2)(c_1+2x)+10SS_u+\sqrt{\Delta}}{32(S^2-T^2)},$$

P_{2R} ≡ positive value of P's give by the two curves $[\phi_1(Y,P)]_{Y=X_2^2+P}$ and $[f_1(Y,P)]_{Y=X_2^2+P}$, where 3x should be common.

$$P_3\equiv\frac{3(c_1+x)(c_1+3x)}{4(c_1+2x)}+\frac{S_u}{2S},$$

$$P_4\equiv3(c_1+3x)+\frac{1}{S_a+S_r},$$

$$P_5\equiv\frac{(S^2-T^2)\{15(c_1+2x)^2-24x^2\}+10SS_u(c_1+2x)-\sqrt{D''}}{32(S^2-T^2)(c_1+2x)},$$

$$D''\equiv9[(S^2-T^2)\{3(c_1+2x)^2-8x^2\}+2SS_u(c_1+2x)]^2+16(c_1+2x)^2\{2S_uT-3x(S^2-T^2)\}^2,$$

$$P_5'\equiv\frac{(S^2-T^2)\{15(c_1+2x)^2-24x^2\}+10SS_u(c_1+2x)+\sqrt{D''}}{32(S^2-T^2)(c_1+2x)}.$$

Since $X_1^2C \geqq 0$ for $P_{41} \leqq P \leqq P_1$ and $P_1 \leqq P \leqq P_{41}$ from (3.12), one of X_1 holds $X_1 \geqq 0$ and the other X_1 becomes $X_1 \leqq 0$. For other P, we always have $X_1^2C < 0$. Since X_2^2, X_3^2, X_5^2 are always negative, we have no real parts. Then, if

$$\text{Re}\left[\frac{\partial X}{\partial R_a}\delta R_a+\frac{\partial X}{\partial R_u}\delta R_u+\frac{\partial X}{\partial R'}\delta R'\right]>0$$

for δR_a, δR_u, $\delta R'$, the state becomes unstable or self-oscillatory. Table 3.1 represents the domain of the instability and its generating frequency. The coordinate 3x (≡ $9Z^2c_3$) in Fig. 3.1 is proportional to the square of the terminal voltage of the transformer in the ordinary steady state. P is proportional to the line eapacity. The type A indicates the undamped oscillations of unsymmetrical wave forms with respect to the zero line, while the type B refers to the undamped oscillations of symmetrical wave forms with respect to the zero line. The hatched region, diagonally striped represents the domain of undampted oscillations for either type A or B, while the region, vertically striped indicates the ordinarily unstable domain: f is the source frequency, δf is its small deviation, and Δf indicates its very small deviation or zero.

Table 3.1

	Domain of P	Kind of X	Main frequency of δi_{a1}	Main frequency of δf_{a0}	Type
$\frac{\delta X}{\delta R'} \geqq 0$	$P_5 - P_{51}$	X_5, X_5	$2f \mp \Delta f,\ 4f \pm \Delta f$	$6f \mp \Delta f,\ \pm \Delta f$	Undamped electric oscillation A unsymmetrical with respect to zero line
	$P_{51} - P_{1R}$	X_1, X_1	$2f \mp \delta f,\ 4f \pm \delta f$	$6f \mp \delta f,\ \pm \delta f$	
	$P_5' - P_{52}$	X_5, X_5	$2f \mp \Delta f,\ 4f \pm \Delta f$	$6f \mp \Delta f,\ \pm \Delta f$	
	$P_{52} - P_{2R}$	X_2, X_2	$2f \mp \delta f,\ 4f \pm \delta f$	$\delta f \mp \delta f,\ \pm \delta f$	
	$P_3 - P_{31}$	X_3, X_3	$2f - \Delta f,\ \Delta f$	$4f - \Delta f,\ 2f + \Delta f$	
	$P_{31} - P_{1R}'$	X_1, X_1	$2f - \delta f,\ \delta f$	$4f - \delta f,\ 2f + \delta f$	
$\frac{\partial X}{\partial R_a} \geqq 0$	$P_3 - P_{31}$	X_3, X_3	$2f - \Delta f,\ \Delta f$		
	$P_{31} - P_{1a}$	X_1, X_1	$2f - \delta f,\ \delta f$		
$X \geqq 0$	$P_1 - P_{41}$	X_1	f		Ordinarily unstable
$\frac{\partial X}{\partial R_u} \geqq 0$	$P_4 - P_{41}$	X_4	f		B Symmetrical
	$P_{41} - \infty$	X_1, X_1	$f - \delta f,\ f + \delta f$		

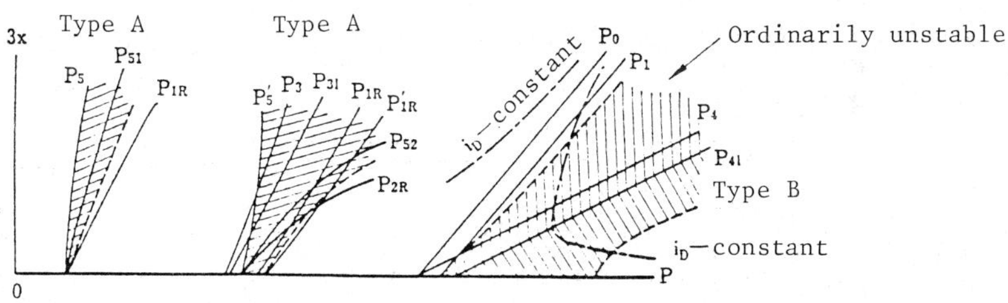

Fig. 3.1 Domain of undamped oscillations.

IV. SOLUTIONS FOR ORDINARY STEADY STATES CONSIDERING MAGNETIC SATURATION CURVES ($\delta f_{a0} \fallingdotseq 0$)

As shown in (4.1), i_a, i_b, i_c are expressed as the power series and their symmetrical components i_{a1}, i_{a0} are written as

$$
\begin{aligned}
i_{a1} &= \sum_{m=0}^{\infty} [f_{a1}^{6m+1}(\sum_{n=0}^{\infty} {}_{2n+6m+1}C_n \cdot c_{2n+6m+1} f_{a1}^{n} f_{a2}^{n}) + f_{a2}^{6m+5}(\sum_{n=0}^{\infty} {}_{2n+6m+5}C_n \cdot c_{2n+6m+5} f_{a1}^{n} f_{a2}^{n})], \\
i_{a0} &= \sum_{m=0}^{\infty} [(f_{a1}^{6m+3} + f_{a2}^{6m+3}) \sum_{0=u}^{\infty} {}_{2n+6m+3}C_n \cdot c_{2n+6m+3} f_{a1}^{n} f_{a2}^{n}] \\
i_a &= \sum_{n=0}^{\infty} c_{2n+1} f_a^{2n+1}, \quad i_b = \sum_{n=0}^{\infty} c_{2n+1} f_b^{2n+1}, \quad i_c = \sum_{n=0}^{\infty} c_{2n+1} f_c^{2n+1}.
\end{aligned}
\tag{4.1}
$$

For the ordinary steady state, if f_{a1} includes $\varepsilon^{j\omega t}$, then we write

$$
f_{a1} = \sum_{l=-\infty}^{\infty} Z^{(l)} \varepsilon^{j(6l+1)\omega t}, \quad i_{a1} = \sum_{l=-\infty}^{\infty} I^{(l)} \varepsilon^{j(6l+1)\omega t} \tag{4.2}
$$

if $\varepsilon^{j\omega t}$ is a main term and other terms are small, then we put

$$
I^{(0)} \doteqdot z(\sum_{n=0}^{\infty} {}_{2n+1}C_n \cdot c_{2n+1} z^{2n}) \equiv F(z), \quad z \equiv Z^{(0)}, \tag{4.3}
$$

where F(z) is determined from the magnetization curve of the transformer. For simplicity, let

$$
S_r = 0 \tag{4.4}
$$

Then, substituting $I^{(0)}$ and z into (2.7), we have

$$
\begin{aligned}
-j\sqrt{3}V\varepsilon^{j\varphi} &= (R_a + j\omega S_a)\{-Pz + (j\omega CR' + 3)I^{(0)}\} + j\omega z + R' I^{(0)}, \\
V\varepsilon^{j\varphi} &\equiv e_{a1} = -\frac{M}{2} j\omega i_D \varepsilon^{j\varphi}.
\end{aligned}
\tag{4.5}
$$

V. SOLUTIONS FOR SMALL TRANSIENT PHENOMENA CONSIDERING MAGNETIC SATURATION CURVES ($\delta f_{a0} \fallingdotseq 0$)

Taking into account magnetic saturation curves, the relations corresponding to (3.1) - (3.3) can be written, referring to $f_{a0} \fallingdotseq 0$ (assumption for i_{a1}, f_{a1}, f_{a0} in Section IV) and (4.1) for

$$
\underline{\delta f_{a0} \doteqdot 0} \tag{5.1}
$$

as

$$
\begin{aligned}
\delta i_{a1} = \sum_{m=0}^{\infty} [\{ & f_{a1}^{6m} (\sum_{n=0}^{\infty} (n+6m+1) \cdot {}_{2n+6m+1}C_n \cdot c_{2n+6m+1} f_{a1}^{n} f_{a2}^{n}) \\
& + f_{a2}^{6m+5}(\sum_{n=1}^{\infty} n \cdot {}_{2n+6m+5}C_n \cdot c_{2n+6m+5} f_{a1}^{n-1} f_{a2}^{n})\} \delta f_{a1} \\
& + \{f_{a1}^{6m+1}(\sum_{n=1}^{\infty} n \cdot {}_{2n+6m+1}C_n \cdot c_{2n+6m+1} f_{a1}^{n} f_{a2}^{n-1}) \\
& + f_{a2}^{6m+4}(\sum_{n=1}^{\infty} (n+6m+5) \cdot {}_{2n+6m+5}C_n \cdot c_{2n+6m+5} f_{a1}^{n} f_{a2}^{n})\} \delta f_{a2}]
\end{aligned}
\tag{5.2}
$$

If f_{a1} consists of the term $Z\varepsilon^{j\omega t}$ and δf_{a1} contains the terms, $\varepsilon^{j\omega t}$ and $\varepsilon^{(X+j\omega)t}$, then δf_{a1}, δi_{a1} are expressed by (3.4). Retaining principal terms, we write as

$$f_{a1}=z\varepsilon^{j\omega t},\quad \left.\begin{aligned}\delta f_{a1}&=Z_1\varepsilon^{(X+j\omega)t}+Z_{-1}\varepsilon^{(\bar{X}+j\omega)t},\\ \delta i_{a1}&=I_1\varepsilon^{(X+j\omega)t}+I_{-1}\varepsilon^{(\bar{X}+j\omega)t},\end{aligned}\right\} \tag{5.3}$$

where Z is assumed real. The main terms I_1 and I_{-1} are obtained from (5.2) for m = 0 as

$$\begin{aligned}I_1&=\sum_{n=0}^{\infty}(n+1)\cdot{}_{2n+1}C_n\cdot c_{2n+1}z^{2n}Z_1+\sum_{n=0}^{\infty}n\cdot{}_{2n+1}C_n\cdot c_{2n+1}z^{2n}\bar{Z}_{-1}\\ &=\frac{1}{2}\left[F'(z)+\frac{F(z)}{z}\right]Z_1+\frac{1}{2}\left[F'(z)-\frac{F(z)}{z}\right]\bar{Z}_{-1}\end{aligned} \tag{5.4}$$

$$I_{-1}=\frac{1}{2}\left[F'(z)+\frac{F(z)}{z}\right]Z_{-1}+\frac{1}{2}\left[F'(z)-\frac{F(z)}{z}\right]\bar{Z}_1\,. \tag{5.5}$$

Introducing (5.3) into (3.1), we have

$$\begin{aligned}&\left[R_a+S_a(X+j\omega)-\frac{M^2(X+j\omega)X}{2(R_u+S_uX)}\right][(X+j\omega)C\{(X+j\omega)Z_1+R'I_1\}+3I_1]+(X+j\omega)Z_1+R'I_1\\ &\quad-\varepsilon^{2j\varphi}(X+j\omega)\left[S_r-\frac{M^2X}{2(R_u+S_uX)}\right][(X-j\omega)C\{(X-j\omega)\bar{Z}_{-1}+R'\bar{I}_{-1}\}+3\bar{I}_{-1}]=0.\end{aligned} \tag{5.6}$$

Introducing (5.4), (5.5) into (5.6), we have

$$\begin{aligned}&\Bigg[\left[R_a+S_a(X+j\omega)-\frac{M^2(X+j\omega)X}{2(R_u+S_uX)}\right]\\ &\cdot\left[(X+j\omega)C\left\{(X+j\omega)+\frac{R'}{2}\left(F'(z)+\frac{F(z)}{z}\right)\right\}+\frac{3}{2}\left\{F'(z)+\frac{F(z)}{z}\right\}\right]\\ &+(X+j\omega)+\frac{R'}{2}\left\{F'(z)+\frac{F(z)}{z}\right\}\\ &-\varepsilon^{2j\varphi}(X+j\omega)\left[S_r-\frac{M^2X}{2(R_u+S_uX)}\right][(X-j\omega)CR'+3]\cdot\frac{1}{2}\left\{F'(z)-\frac{F(z)}{z}\right\}\Bigg]Z_1\\ &+\Bigg[\left[R_a+S_a(X+j\omega)-\frac{M^2(X+j\omega)X}{2(R_u+S_uX)}\right][(X+j\omega)CR'+3]\cdot\frac{1}{2}\left\{F'(z)-\frac{F(z)}{z}\right\}\\ &+\frac{R'}{2}\left(F'(z)-\frac{F(z)}{z}\right)\\ &-\varepsilon^{2j\varphi}(X+j\omega)\left[S_r-\frac{M^2X}{2(R_u+S_uX)}\right]\\ &\cdot\left[(X-j\omega)C\left\{X-j\omega+\frac{R'}{2}\left(F'(z)+\frac{F(z)}{z}\right)\right\}+\frac{3}{2}\left\{F'(z)+\frac{F(z)}{z}\right\}\right]\Bigg]\bar{Z}_{-1}=0.\end{aligned} \tag{5.7}$$

Similarly, substituting the terms containing $\varepsilon^{(X+j\omega t)}$ only into (3.1), we obtain formulas where X, Z_1, I_1, $\bar{Z}_{-1}$, $\bar{I}_{-1}$ are replaced by $\bar{X}$, Z_{-1}, I_{-1}, $\bar{Z}_1$, $\bar{I}_1$, respec-ively. Taking their conjugates, we obtain Z_{-1}, I_{-1}, Z_1, I_1. If we write (5.7) as

$$\Phi(X,j\omega)Z_1+\Psi(X,j\omega)\bar{Z}_{-1}=0 \tag{5.7'}$$

then we have the conjugate equation

$$\Phi(X, -j\omega)\bar{Z}_{-1}+\Psi(X, -j\omega)Z_1=0 \tag{5.8}$$

For

$$Z_1\bar{Z}_{-1}\neq 0 \tag{5.9}$$

(5.7'), (5.8) lead to

$$\Phi(X, j\omega)\Phi(X, -j\omega)-\Psi(X, j\omega)\Psi(X, -j\omega)=0. \tag{5.10}$$

This has real coefficient of X and is unchanged by the replacement of "j" to "-j". If there exists X whose real part is positive, then the state becomes unstable, leading to undamped oscillations.

VI. SOLUTIONS FOR SMALL TRANSIENT PHENOMENA CONSIDERING MAGNETIC SATURATION CURVES ($\delta f_{a0} \neq 0$)

The circuit is the same as the previous case but now $\delta f_{a0} \neq 0$. However, we assume $f_{a0} \fallingdotseq 0$ as in the ordinary steady state. Then, we have from the third equtation in Eq. (4.1)

$$\left.\begin{aligned}
\delta i_a &= \sum_{n=0}^{\infty}(2n+1)c_{2n+1}f_a^{2n}\delta f_a,\\
\delta i_b &= \sum_{n=0}^{\infty}(2n+1)c_{2n+1}f_b^{2n}\delta f_b,\\
\delta i_c &= \sum_{n=0}^{\infty}(2n+1)c_{2n+1}f_c^{2n}\delta f_c.
\end{aligned}\right\} \tag{6.1}$$

Putting $a\equiv\varepsilon^{j(2x/3)}$, the symmetrical componeuts of $\delta i_a, \delta i_b, \delta i_c$

$$\left.\begin{aligned}
&\delta i_{a1}=\frac{1}{3}(\delta i_a+a\delta i_b+a^2\delta i_c), \quad \delta i_{a2}=\overline{\delta i_{a1}},\\
&\delta i_{a0}=\frac{1}{3}(\delta i_a+\delta i_b+\delta i_c)
\end{aligned}\right\} \tag{6.2}$$

can be expressed, referring to $f_{a0}=0$, as

$$\begin{aligned}
\delta i_{a1}=&\Big[\sum_{n=0}^{\infty}(2n+1)\cdot {}_{2n}C_n\cdot c_{2n+1}f_{a1}^{\,n}f_{a2}^{\,n}\\
&+\sum_{m=1}^{\infty}(f_{a1}^{\,6m}+f_{a2}^{\,6m})\sum_{n=0}^{\infty}(2n+6m+1)\cdot {}_{2n+6m}C_n\cdot c_{2n+6m+1}f_{a1}^{\,n}f_{a2}^{\,n}\Big]\delta f_{a1}\\
&+\sum_{m=0}^{\infty}\Big[f_{a1}^{\,6m+2}\sum_{n=0}^{\infty}(2n+6m+3)\cdot {}_{2n+6m+2}C_n\cdot c_{2n+6m+3}f_{a1}^{\,n}f_{a2}^{\,n}\\
&+f_{a2}^{\,6m+4}\sum_{n=0}^{\infty}(2n+6m+5)\cdot {}_{2n+6m+4}C_n\cdot c_{2n+6m+5}f_{a1}^{\,n}f_{a2}^{\,n}\Big]\delta f_{a2}\\
&+\sum_{m=0}^{\infty}\Big[f_{a1}^{\,6m+4}\sum_{n=0}^{\infty}(2n+6m+5)\cdot {}_{2n+6m+4}C_n\cdot c_{2n+6m+5}f_{a1}^{\,n}f_{a2}^{\,n}\\
&+f_{a2}^{\,6m+2}\sum_{n=0}^{\infty}(2n+6m+3)\cdot {}_{2n+6m+2}C_n\cdot c_{2n+6m+3}f_{a1}^{\,n}f_{a2}^{\,n}\Big]\delta f_{a0}
\end{aligned} \tag{6.3}$$!!

$$\delta i_{a0}=\Big[\sum_{n=0}^{\infty}(2n+1)\cdot{}_{2n}C_n\cdot c_{2n+1}f_{a1}{}^{n}f_{a2}{}^{n}$$
$$+\sum_{m=1}^{\infty}(f_{a1}{}^{6m}+f_{a2}{}^{6m})\sum_{n=0}^{\infty}(2n+6m+1)\cdot{}_{2n+6m}C_n\cdot c_{2n+6m+1}f_{a1}{}^{n}f_{a2}{}^{n}\Big]\delta f_{a0}$$
$$+\sum_{m=0}^{\infty}\Big[f_{a1}{}^{6m+2}\sum_{n=0}^{\infty}(2n+6m+3)\cdot{}_{2n+6m+2}C_n\cdot c_{2n+6m+3}f_{a1}{}^{n}f_{a2}{}^{n}$$
$$+f_{a2}{}^{6m+4}\sum_{n=0}^{\infty}(2n+6m+5)\cdot{}_{2n+6m+4}C_n\cdot c_{2n+6m+5}f_{a1}{}^{n}f_{a2}{}^{n}\Big]\delta f_{a1}$$
$$+\sum_{m=0}^{\infty}\Big[f_{a1}{}^{6m+4}\sum_{n=0}^{\infty}(2n+6m+5)\cdot{}_{2n+6m+4}C_n\cdot c_{2n+6m+5}f_{a1}{}^{n}f_{a2}{}^{n}$$
$$+f_{a2}{}^{6m+2}\sum_{n=0}^{\infty}(2n+6m+3)\cdot{}_{2n+6m+2}C_n\cdot c_{2n+6m+3}f_{a1}{}^{n}f_{a2}{}^{n}\Big]\delta f_{a2}$$
(6.4)!!

We assume the main terms for f_{a1}, i_{a1}, f_{a0}, i_{a0} in ordinary steady states, neglecting higher harmonics, to be

$$f_{a1}\fallingdotseq z\varepsilon^{j\omega t},\ f_{a0}\fallingdotseq 0,\ i_{a1}\fallingdotseq I^{(0)}\varepsilon^{j\omega t},\ i_{a0}\fallingdotseq 0.$$
(6.5)!

$$I^{(0)}=z\Big(\sum_{n=0}^{\infty}{}_{2n+1}C_n\cdot c_{n+1}z^{2n}\Big)\equiv F(z),$$
(6.6)!

while the main terms for δf_{a1}, δi_{a1}, δf_{a0}, δi_{a0} are expressed as

$$\left.\begin{aligned}
&\delta f_{a1}\fallingdotseq Z_1\varepsilon^{(X+j\omega)t}+Z_{-1}\varepsilon^{(\bar{X}+j\omega)t}\\
&\delta f_{a0}\fallingdotseq f_1\varepsilon^{(X+j3\omega)t}+f_{-1}\varepsilon^{(\bar{X}+j3\omega)t}+\bar{f}_1\varepsilon^{(\bar{X}-j3\omega)t}+\bar{f}_{-1}\varepsilon^{(X-j3\omega)t},\\
&\delta i_{a1}\fallingdotseq I_1\varepsilon^{(X+j\omega)t}+I_{-1}\varepsilon^{(\bar{X}+j\omega)t},\\
&\delta i_{a0}\fallingdotseq J_1\varepsilon^{(X+j3\omega)t}+J_{-1}\varepsilon^{(\bar{X}+j3\omega)t}+\bar{J}_1\varepsilon^{(\bar{X}-j3\omega)t}+\bar{J}_{-1}\varepsilon^{(X-j3\omega)t}.
\end{aligned}\right\}$$
(6.7)!

Using F(x) in (6.6), $\Phi_1(z)$, $\Phi_2(z)$, $\Phi_3(z)$ are defined as

$$\Phi_1(z)\equiv\sum_{n=0}^{\infty}(2n+1)\cdot{}_{2n}C_n\cdot c_{2n+1}z^{2n}=\sum_{n=0}^{\infty}(n+1)\cdot{}_{2n+1}C_n\cdot c_{2+1}z^{2n}=\frac{1}{2}\Big[F'(z)+\frac{F(z)}{z}\Big],$$
(6.8)!

$$\Phi_3(z)\equiv\sum_{n=0}^{\infty}(2n+3)\cdot{}_{2n+2}C_n\cdot c_{2n+3}z^{2n+2}=\sum_{n=0}^{\infty}(n+3)\cdot{}_{2n+3}C_n\cdot c_{2n+3}z^{2n+2}$$
$$=\sum_{n=0}^{\infty}(n+1)\cdot{}_{2n+3}C_{n+1}\cdot c_{2n+3}z^{2n+2}=\sum_{n=1}^{\infty}n\cdot{}_{2n+1}C_n\cdot c_{2n+1}z^{2n}=\frac{1}{2}\Big[F'(z)-\frac{F(z)}{z}\Big]$$
(6.9)!

$$=\Phi_1-\frac{2}{z^2}\int_0^z\Phi_1 z\,dz.$$
(6.9′)

$$\Phi_5(z)\equiv\sum_{n=0}^{\infty}(2n+5)\cdot{}_{2n+4}C_n\cdot c_{2n+5}z^{2n+4}=\sum_{n=0}^{\infty}(n+5)\cdot{}_{2n+5}C_n\cdot c_{2n+5}z^{2n+4}$$
$$=\Phi_3-\frac{6}{z^4}\int_0^z\Phi_3 z^3\,dz.$$
(6.10)!

Then, using these functions and referring to Eq. (6.7), we obtain the following relations among principal terms in Eq. (6.3):

$$\left.\begin{aligned}
&I_1\fallingdotseq\Phi_1 Z_1+\Phi_3\bar{Z}_{-1}+\Phi_3 f_1,\\
&I_{-1}\fallingdotseq\Phi_1 Z_{-1}+\Phi_3\bar{Z}_1+\Phi_3 f_{-1},\\
&J_1\fallingdotseq\Phi_3 Z_1+\Phi_5\bar{Z}_{-1}+\Phi_1 f_1,\\
&J_{-1}\fallingdotseq\Phi_3 Z_{-1}+\Phi_5\bar{Z}_1+\Phi_1 f_{-1}.
\end{aligned}\right\}$$
(6.11)!!

Then, circuit equations are the same as (3.1), and are written as

$$\left[R_a+S_a p-\frac{M^2 p(p-j\omega)}{2\{R_u+S_u(p-j\omega)\}}\right]\{pC(p\delta f_{a1}+R'\delta i_{a1})+3\,\delta i_{a1}\}+p\delta f_{a1}+R'\delta i_{a1}-\varepsilon^{2j\theta}(p+2j\omega)$$

$$\cdot\left[S_r-\frac{M^2(p+j\omega)}{2\{R_u+S_u(p-j\omega)\}}\right]\{pC(p\delta f_{a2}+R'\delta i_{a2})+3\,\delta i_{a2}\}=0. \quad (6.12)!!$$

$$p\delta f_{a0}+R'\delta i_{a0}=0. \quad (6.13)!!$$

Introducing (6.7) into (6.12), we have

$$\left[R_a+S_a(X+j\omega)-\frac{M^2(X+j\omega)X}{2\{R_u+S_u X\}}\right][(X+j\omega)C\{(X+j\omega)Z_1+R'I_1\}+3I_1]+(X+j\omega)Z_1+R'I_1$$

$$-\varepsilon^{2j\varphi}(X+j\omega)\left[S_r-\frac{M^2X}{2(R_u+S_u X)}\right][(X-j\omega)C\{(X-j\omega)\bar{Z}_{-1}+R'\bar{I}_{-1}\}+3\bar{I}_{-1}]=0\,. \quad (6.14)$$

This can be rewritten as

$$Z_1\left[\left\{R_a+S_a(X+j\omega)-\frac{M^2(X+j\omega)X}{2(R_u+S_u X)}\right\}(X+j\omega)^2C+(X+j\omega)\right]$$

$$+\bar{Z}_{-1}\left[-\varepsilon^{2j\varphi}(X+j\omega)\left\{S_r-\frac{M^2X}{2(R_u+S_u X)}\right\}(X-j\omega)^2C\right]$$

$$+I_1\left[\left\{R_a+S_a(X+j\omega)-\frac{M^2(X+j\omega)X}{2(R_u+S_u X)}\right\}\{(X+j\omega)CR'+3\}+R'\right]$$

$$+\bar{I}_{-1}\left[-\varepsilon^{2j\varphi}(X+j\omega)\left\{S_r-\frac{M^2X}{2(R_u+S_u X)}\right\}\{(X-j\omega)CR'+3\}\right]=0. \quad (6.14')$$

If one uses Z_{-1}, $\bar{Z}_1$, I_{-1}, $\bar{I}_1$ instead of Z_1, $\bar{Z}_1$, I_1, $\bar{I}_{-1}$, then X, j should be changed to $\bar{X}$, $-j$. Namely, if one writes (6.14') as

$$Z_1G_1(X,j)+\bar{Z}_{-1}G_{-1}(X,j)+I_1H_1(X,j)+\bar{I}_{-1}H_{-1}(X,j)=0, \quad (6.14'')$$

then we have

$$\bar{Z}_{-1}G_1(X,-j)+Z_1G_{-1}(X,-j)+\bar{I}_{-1}H_1(X,-j)+I_1H_{-1}(X,-j)=0, \quad (6.15)$$

On the other hand, applying (6.11), (6.7) to (6.13), we have

$$\left.\begin{aligned}&(X+j3\omega)f_1+R'J_1=0,\\&(X-j3\omega)\bar{f}_{-1}+R'\bar{J}_{-1}=0.\end{aligned}\right\} \quad (6.16)$$

$$\therefore\ \left.\begin{aligned}&[(X+j3\omega)+R'\Phi_1]f_1+R'(\Phi_3Z_1+\Phi_5\bar{Z}_{-1})\doteqdot 0,\\&[(X-j3\omega)+R'\Phi_1]\bar{f}_{-1}+R'(\Phi_3\bar{Z}_{-1}+\Phi_5Z_1)\doteqdot 0.\end{aligned}\right\} \quad (6.17)$$

$$\therefore\ \left.\begin{aligned}&I_1\doteqdot\Phi_1Z_1+\Phi_3\bar{Z}_{-1}-\Phi_3\frac{R'(\Phi_3Z_1+\Phi_5\bar{Z}_{-1})}{(X+j3\omega)+R'\Phi_1},\\&\bar{I}_{-1}\doteqdot\Phi_1\bar{Z}_{-1}+\Phi_3Z_1-\Phi_3\frac{R'(\Phi_3\bar{Z}_{-1}+\Phi_5Z_1)}{(X-j3\omega)+R'\Phi_1}.\end{aligned}\right\} \quad (6.18)$$

Introducing (6.18) into (6.14"), (6.15), we obtain the relation between Z_1 and $\bar{Z}_{-1}$

$$Z_1\left[G_1(X,j)+\left\{\Phi_1-\frac{\Phi_3^2R'}{X+j3\omega+R'\Phi_1}\right\}H_1(X,j)+\left\{\Phi_3-\frac{\Phi_3\Phi_5R'}{X-j3\omega+R'\Phi_1}\right\}H_{-1}(X,j)\right]$$

$$+\bar{Z}_{-1}\left[G_{-1}(X,j)+\left\{\Phi_3-\frac{\Phi_3\Phi_5R'}{X+j3\omega+R'\Phi_1}\right\}H_1(X,j)\right.$$

$$\left.+\left\{\Phi_1-\frac{\Phi_3^2R'}{X-j3\omega+R'\Phi_1}\right\}H_{-1}(X,j)\right]\doteqdot 0. \qquad (6.19)$$

This equation can be simplified as

$$Z_1\Psi_1(X,j)+\bar{Z}_{-1}\Psi_{-1}(X,j)\doteqdot 0. \qquad (6.19')$$

Then, using (6.15), we also have

$$\bar{Z}_{-1}\Psi_1(X,-j)+Z_1\Psi_{-1}(X,-j)\doteqdot 0 \qquad (6.20)$$

Let

$$Z_1\bar{Z}_{-1}\neq 0 \qquad (6.21)$$

Then, eliminating Z_1, $\bar{Z}_{-1}$ in (6.19'), (6.20), we obtain the equation with real coefficients with respect to X

$$\Psi_1(X,j)\Psi_1(X,-j)-\Psi_{-1}(X,j)\Psi_{-1}(X,-j)\doteqdot 0. \qquad (6.22)$$

By solving Eq. (6.22) for given z (proportional to the line voltage) with respect to X, it can be seen whether it has a root whose real part is positive or not. We thus obtain the domain of instability for given circuit parameters. Many problems of nonlinear oscillations will be solved on the basis of Eq. (6.22).

VII. CONCLUSION

A large part of problems of nonlinear oscillations still remains to be solved partly because governing nonlinear differential equations are difficult to solve analytically. The problem of electrical instability and undamped oscillations encountered actually in the transmission system is rather old and the present author obtained a general solution analytically to find the domain of instability or self-oscillations for the cases when magnetic saturation characteristics of transformers can be expressed in a cubic curve[3),4)] or in a power series but for the resistance of a Δ-circuit $R' = 0$[2)] or $i_{a0} = 0$. In the present paper, the previous thory has been extended in order to make it available to obtain the domain of instability or generation of self-oscillations for any given values of circuit resistances and for cases when magnetic saturation characteristics of transformers are expressed in graphs, a power series or any other functions. In this connection, particular emphasis is placed on the use of the function F(z) in terms of $[F'(z) \pm \{F(z)/z\}]$ rather than an infinite series for expressions for nonlinear characteristics. This simplifies its solution significantly. We can thus treat the problam far easily.

REFERENCES

1. M. Kameda: "Phenomena of Special Electric Oscillations in a Power Transmission System", J. Inst. Electr. Eng. Japan, 50, 880 (1930).

2. M. Goto: "Undamped Electric Oscillation of an Alternator Connected to a Transmission Line", Research No. 281 of Electrotechnical Laboratory (ETL), Ministry of Communications, Japan, April (1930).

3. M. Goto: "Undamped Oscillations and Electrical Instability State in a Power Transmission System", J. Inst. Electr. Eng. Japan, 51, 759 (1931): Errata, 52, 560 (1932).

4. M. Goto, S. Morikawa, and G. Takeuchi: "Undamped Electric Oscillation and Electrical Instability of a Transmission System, Research No. 366 of ETL, June (1934).

5. G. Takeuchi: "Sub-Harmonic Generation by Iron-Core Reactor", Research No. 480 of ETL (1945).

6. M. Goto et al. : "Undamped Oscillations during Charges at the Nezame Power Station of the Tokyo Transmission Line", Bull. ETL, 15, 7 (1951).

7. T. Norimatsu: "Electrical Instability Phenomena due to Magnetic Saturation of the Transformer's Iron-Core in a Transmission System", Research No. 613 of ETL (1961).

Nonlinear and Environmental Electromagnetics, edited by H. Kikuchi
Elsevier Science Publishers B.V., Amsterdam, 1985 – Printed in The Netherlands

CHAOTIC OSCILLATIONS IN JOSEPHSON JUNCTION LINES

Kenjiro Maginu
Department of Mathematical Engineering and Instrumentation
Physics, Faculty of Engineering, University of Tokyo

ABSTRACT. We consider a homogeneous Josephson junction line which forms a ring of finite length L. When a suitable bias current I is supplied across the junction layer, and when the length L of the ring is not so large, the voltage across the junction layer oscillates periodically maintaining the spatial homogeneity. We show that this homogeneous oscillation becomes unstable when L exceeds a certain threshold length L_0. A spatially inhomogeneous periodic oscillation bifurcates from the homogeneous oscillation at this threshold length. (This is a period-doubling bifurcation.) The Josephson junction line can be simulated qualitatively by using coupled penduli model. The period-doubling bifurcation (from a spatially homogeneous rotating state of the penduli to a spatially inhomogeneous rotating state) is also observed in this model when a parameter (which corresponds to the length of the junction line) exceeds a certain threshold. This coupled-penduli model exhibits a chaotic motion when the parameter is taken in a suitable range.

1. Introduction

The Josephson junction line [1] consists of two superconductor strips separated by a thin insulator layer (the Josephson junction layer). When the layer is thin, it permits coupling of superconducting states of the two strips and causes a tunnel effect (the Josephson effect) [2]. As a result of this effect, the junction line supports steady propagation of active impulse waves [1] when a suitable bias current I is supplied across the junction layer. (The voltage impulse waves propagate along the line with a constant speed maintaining a fixed waveform.) We can observe another qualitatively different motion on the junction layer when I is taken in a suitable range. We can observe periodic oscillations of the voltage across the junction layer.

Dynamical behavior of the Josephson junction line is described by the following hyperbolic partial differential equation called the damped sine-Gordon equation [1]:

$$\phi_{tt} + \varepsilon\phi_t + \sin\phi = \phi_{xx} + I, \qquad x \in [0, L]. \tag{1.a}$$

Here ϕ represents the difference of (quantum methanical) phases of the superconducting states of the two superconductor strips. The parameters I and $\varepsilon > 0$ account for the constant bias current (which is supplied across the junction layer) and the energy loss due to non-superconducting current, respectively. The voltage V across the junction is associated with the phase difference ϕ by the relation $V(x,t) = \kappa(\partial/\partial t)\phi(x,t)$, where κ is a certain physical constant. For simplicity we assume that the junction line forms a ring of length L; namely we assume the following periodicity boundary conditions:

$$\phi(0,t) = \phi(L,t), \qquad \phi_t(0,t) = \phi_t(L,t). \tag{1.b}$$

Spatially homogeneous solutions of (1) obey the following ordinary differential equation (the space-clumped system):

$$\phi_{tt}(t) + \varepsilon\phi_t(t) + f(\phi) = I, \qquad f(\phi) = \sin\phi. \tag{2}$$

It is easy to see that this equation coincides with the equation of a pendulum under the supply of constant torque I; the variable ϕ represents the angle of its arm from the vertical line and $\varepsilon > 0$ accounts for the coefficient of friction. If the torque $I > 0$ is larger than a certain threshold (which depend on ε), the pendulum can rotate continuously and periodically. Namely the equation (2) has a periodic solution of the second kind $\phi(t)$ [3] which satisfies the periodicity condition $\phi(t+T) = \phi(t) + 2\pi$, where $T > 0$ is the period of the rotation. The oscillation of the voltage on the Josephson junction corresponds to this periodic rotation of the pendulum. (The rotation of the angle ϕ is observed as a oscillation of the voltage V because the voltage is associated with ϕ by the relation $V = \kappa(d/dt)\phi(t)$.)

When many penduli are arranged linearly and the neighboring ones are coupled by springs with Hooke's coefficient $k > 0$, the motion of the penduli is described by the system of differential equations

$$(d^2/dt^2)\phi_i + \varepsilon(d/dt)\phi_i + \sin\phi_i = I + k(\phi_{i+1} + \phi_{i-1} - 2\phi_i), \qquad 1 \leq i \leq N. \tag{3}$$

This system reduces to the partial differential equation (1.a) when $k = (N/L)^2$ and $N \to +\infty$. Namely the Josephson junction line can be simulated by the system of coupled penduli.

The equation (1) has a travelling wave solution, i.e. a solution of the form $\phi = \phi(x - ct)$, when $\varepsilon > 0$ and $I \neq 0$. This solution corresponds to steady spatial propagation of rotation of the penduli. Figure 1 shows waveform of the

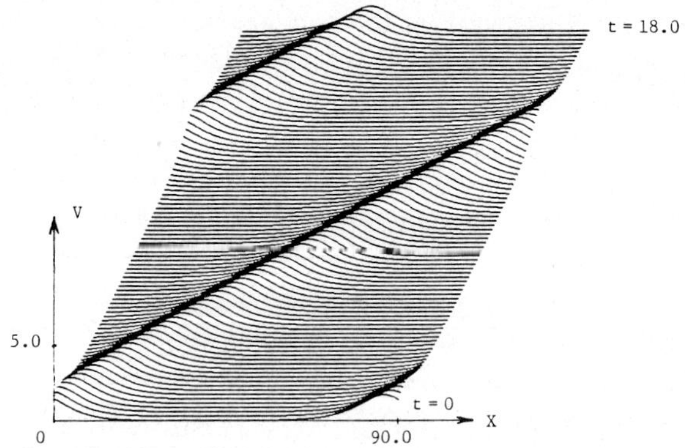

Figure 1.
The voltage impulse wave propagating on a ring of the Josephson junction line.

voltage impulse wave $V(x,t) = (\partial/\partial t)\phi(x-ct)$ when $\varepsilon = 0.2$, $I = 0.3$ and $L \doteq 90$. (The periodicity boundary conditions are assumed.) This voltage impulse wave is similar to the impulse wave of nerve excitation observed on the nerve axon [4]. The propagation speed c and the waveform is determined by the bias current I (which account for the energy supply to the junction layer) and the coefficeint $\varepsilon > 0$ (which account for the energy loss due to the non-superconducting current across the layer). The existence and stability of the impulse wave (and periodic trains of the impulse waves) were proved in [5].

When the equation (2) of the pendulum has a periodic solution of the second kind, $\phi = \phi(t)$ say, the equation (3) of the coupled penduli also has a periodic solution of the second kind $\phi_i = \phi(t)$, $1 \leqq i \leqq N$, which corresponds to the "on-phase" (or spatially synchronized) rotation of the penduli. The equation (1) of the Josephson junction line, in this case, also has a spatially homogeneous periodic solution of the second kind $\phi = \phi(t)$. Figure 2 illustrates the voltage $V = (\partial/\partial t)\phi$ across the junction layer when the junction line exhibits the spatially homogeneous rotation. The periodic solution $\phi(t)$ clearly does not depend on the choice of the length L of the junction line. However its stability crucially depends on L. We study the stability of $\phi(t)$ in the next section.

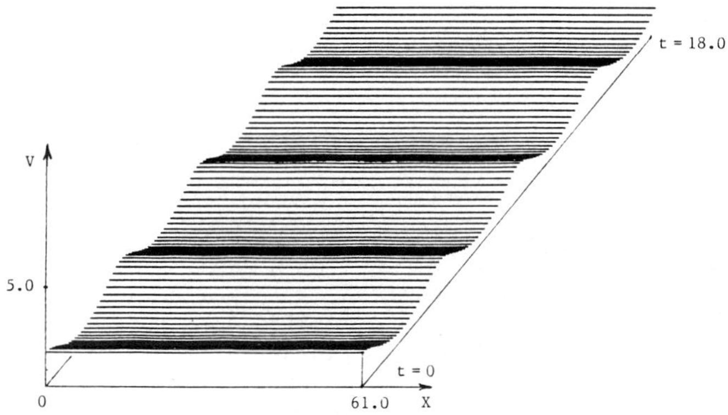

Fig.2. Stable spatially homogeneous oscillation of the voltage $V = (\partial/\partial t)\phi$ of the junction line when $L = 61.0$.

2. Spatially Homogeneous Oscillation of the Josephson Junction Line

Let $\phi(x,t)$ be a solution of (1) slightly deviating from the spatially homogeneous periodic solution (of the second kind) $\phi = \phi(t)$. The small deviation $z(x,t) = \phi(x,t) - \phi(x)$ obeys the following linearized perturbation equation of (1) with respect to $\phi(t)$:

(4.a) $$z_{tt} + \varepsilon z_t + g(t)z = z_{xx}, \qquad x \in [0, L],$$

(4.b) $$z(0,t) = z(L,t), \qquad z_t(0,t) = z_t(L,t),$$

where $g(t) = f'(\phi(t)) \equiv \cos(\phi(t))$. The function g(t) is a continuous periodic function with period T since $\phi(t)$ is a T-periodic solution of the second kind (i.e. since it satisfies the periodicity condition $\phi(t+T) = \phi(t) + 2\pi$).

Let us consider solutions of the perturbation equation of the form $z(x,t) = v(t)\cos(2k\pi x/L)$ and $v(t)\sin(2k\pi x/L)$ (which satisfies the periodicity boundary condition (4.b)). It is shown by substituting these solutions in (4) that the function $v(t)$ should satisfy the following linear ordinary differential equation with T-periodic coefficients:

$$(5) \qquad v_{tt} + \varepsilon v_t + \{\omega^2 + g(t)\}v = 0, \qquad \omega = 2k\pi/L.$$

This equation is written equivalently as

$$(5') \qquad (d/dt)\mathbf{p} = B(t;\omega)\mathbf{p},$$

$$\mathbf{p} = (v,\ v_t)', \qquad B(t;\omega) = \begin{bmatrix} 0 & 1 \\ -\omega^2 - g(t) & -\varepsilon \end{bmatrix}$$

Namely the perturbation equation (4) is unstable with respect to disturbances with spatial frequency $\omega = 2k\pi/L$ if (5') has an unstable solution when $\omega = 2k\pi/L$. In other words, the spatially homogeneous periodic solution $\phi(t)$ of (1) is unstable with respect to spatially inhomogeneous perturbations if the linear T-periodic system (5') is unstable for some $\omega = 2k\pi/L$, $k = 0,1, \cdots$. Figure 3 shows the process of growth of a spatially inhomogeneous disturbance added to the homogeneous periodic solution $\phi(t)$ of Fig.2 when $I = 0.3$, $\varepsilon = 0.2$ and $L = 70.0$. (We show in what follows that the periodic solution $\phi(t)$, in case $I = 0.3$ and $\varepsilon = 0.2$, becomes unstable when the length L of the junction line exceeds a critical length $L_0 = 62.0$.)

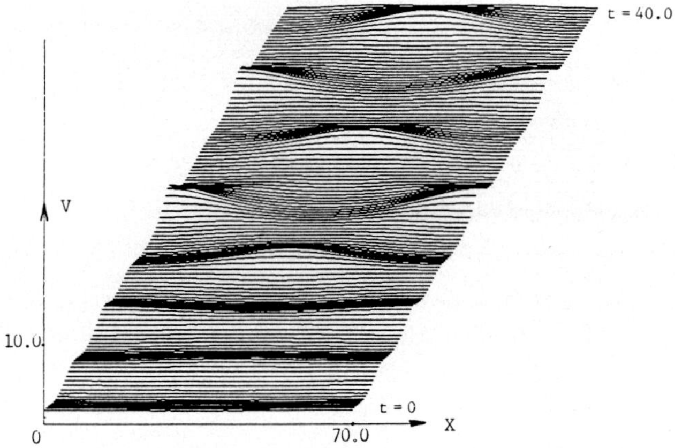

Fig.3. Growth of the inhomogeneous component of the perturbation added to the homogeneous periodic solution $\phi(t)$ when $L = 70.0$. The parameters I and ε are the same as in Fig.2.

Let $\Phi(t;\omega)$ denote a matrix solution of (5') subject to the initial condition $\Phi(0,\omega) = E$, where E is the 2×2 unit matrix. Further let $\mu_i(\omega)$, $i = 1,2$, denote eigenvalues of the matrix $\Phi(T;\omega)$. It is clear from the definition of $\Phi(t;\omega)$ that the T-periodic linear system (5') is unstable if and only if one of the eigenvalues of $\mu(T;\omega)$ satisfies the inequality $|\mu_i(\omega)| < 1$.

We know properties of the eigenvalues $\mu_i(\omega)$ extensively in case $\varepsilon = 0$ because equation (5) (which is equivalent to (5')) reduces in this case to the following equation to which we can apply the Sturm-Liouville theory [3]:

$$v_{tt} + \{\omega^2 + g(t)\}v = 0, \qquad g(t + T) = g(t). \tag{6}$$

Especially, when $g(t) = \sin(2\pi t/T)$, this equation reduces to well known Mathieu's equation for the parametric excitation. We obtain the next theorem as an extension of the Sturm-Liouville theory to the case $\varepsilon > 0$. Proof is given in [6].

Theorem. If $\varepsilon > 0$ is small, the eigenvalues $\mu_i(\omega)$ of $\Phi(T;\omega)$ satisfy

(a) $\mu_1(0) = 1$, (b) $|\mu_1(\omega)| < 1$, $0 < \omega < \lambda_1$,
(c) $\mu_1(\omega) < -1$, $\lambda_1 < \omega < \lambda_2$, (d) $|\mu_1(\omega)| < 1$, $\lambda_2 < \omega < \lambda_3$,
(e) $|\mu_2(\omega)| < 1$ for all ω,

where λ_1 and λ_2 $(0 < \lambda_1 < \lambda_2)$ are certain finite constants which are determined depending upon ε and I. If $\varepsilon > 0$ is not so small, λ_3 in (d) is equal to $+\infty$.

Remark: In case of Mathieu's equation (i.e. when $\varepsilon = 0$ and $g(t) = \cos(2\pi t/L)$), λ_1 and λ_2 in the above theorem are approximately equal to π/L because the parametric resonance in Mathieu's equation takes place when the frequency $2\pi/L$ of the time-varying term $g(t)$ is close to 2ω.

Let us consider the case where the coefficient $\varepsilon > 0$ of (1) is taken in a suitable range such that the λ_3 of the theorem is equal to $+\infty$. In this case, the eigenvalues $\mu_i(\omega)$ satisfy the inequality $|\mu_i(\omega)| < 1$, $i = 1,2$, for all ω in $(\lambda_2, +\infty)$. The spatial frequency ω of the inhomogeneous perturbation added to the homogeneous oscillation $\phi(t)$ of the junction line, on the other hand, can take the values

$$\omega_k = 2\pi k/L, \quad k = 0,1,2,\cdots. \tag{7}$$

They satisfy $\omega_k > \lambda_2$ for $k = 1,2,\cdots$ if $0 < L < L_0 = 2\pi/\lambda_2$. Hence, if the above assumption is satisfied and if the length L of the junction line is smaller than L_0, the Floquet multipliers $\mu_i(\omega_k)$ of the linear periodic system (4) satisfy $|\mu_i(\omega_k)| < 1$ for $k = 1,2,\cdots$ and $i = 1,2$. The 0-the spatial frequency $\omega_0 = 0$ satisfy, by the theorem, $\mu_1(\omega_0) = 1$ and $|\mu_2(\omega_0)| < 1$. These mean that the homogeneous oscillation $\phi(t)$ of the Josephson junction line is stable if the length L is smaller than L_0. The oscillation becomes unstable when L exceeds L_0 since the Floquet multiplier $\mu_1(\omega_1)$ becomes smaller than -1; namely, as shown in Fig.3, the inhomogeneous perturbation with the spatial frequency ω_1 is amplified by the rate $\mu_1(\omega_1)$ in each period T of the oscillation. The critical length L_0 is approximately equal to 62.0 when $I = 0.3$ and $\varepsilon = 0.2$. Hence the homogeneous oscillation $\phi(t)$ is stable when $L = 61.0$ and is unstable when $L = 70.0$, respectively, as shown in Fig.2 and Fig.3.

3. Spatially Inhomogeneous Periodic Oscillations on Josephson Junction Line

A spatially inhomogeneous periodic oscillation bifurcates from the homogeneous oscillation $\phi(t)$ when the length L of the junction line exceeds the critical length L_0. This bifurcation is a period-doubling bifurcation (i.e. a periodic solution of period 2T bifurcates from the T-periodic solution $\phi(t)$ at $L = L_0$) since the linear periodic system (4), which is the linearized perturbation equation of (1) with respect to $\phi(t)$, has the Froquet multiplier -1 when $L = L_0$. Rigorous proof for the existence and stability of the bifurcating periodic solution (which can be regarded as a sort of standing wave oscillation on the Josephson junction line) is given in [8].

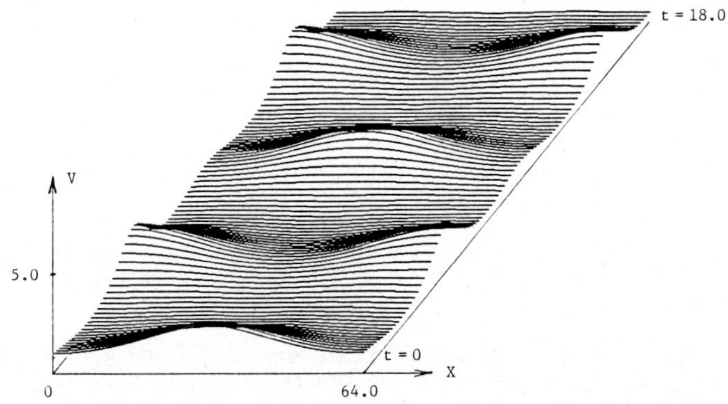

Fig.4. The standing wave oscillation on the Josephson junction line when L = 64.0. Parameters of the system is the same as in Fig.2.

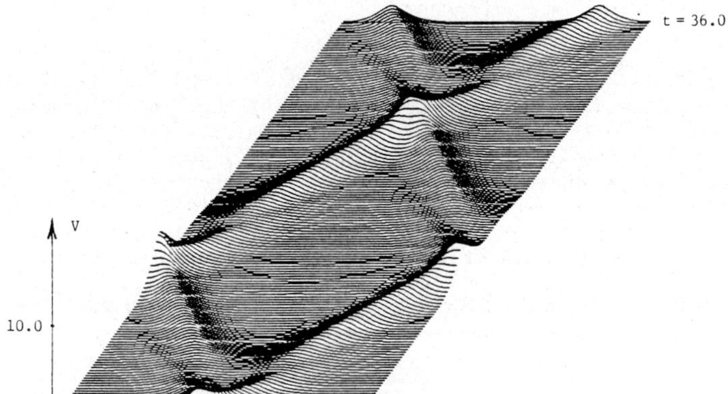

Fig.5. The standing wave oscillation when L = 160.0. Parameters of the system is the same as in Fig.2.

Figures 4 and 5 show examples of standing wave oscillations on the junction line. In the example of Fig.4, the length $L = 64.0$ of the junction line is close to the critical length $L_0 = 62.0$ (at which the bifurcation takes place), and therefore the deviation of the standing wave oscillation from the homogeneous oscillation $\phi(t)$ of Fig.2 is not so large. In the example of Fig.5, the length of the junction line is taken as $L = 160.0$. The pattern of this oscillation does not look like the standing wave oscillation of Fig.4. When L is much larger than L_0, the standing wave reduces to two impulse travelling waves (see Fig.1), one travelling to the right and the other travelling to the left, coming into collision and making nonlinear interaction with each other periodically in the course of time.

4. Motion of Coupled Penduli as a Model of the Josephson Junction Line

We noted in Section 1 that the system of coupled penduli, whose motion is described by the differential equations (3), can be regarded as a simulator of the Josephson junction line. The simplest case of the simulator consists of only two penduli, whose angles of the arms from the vertical line are denoted by ϕ_1 and ϕ_2. Its motion is described by

$$(8)\qquad \begin{aligned} (d^2/dt^2)\phi_1 + \varepsilon(d/dt)\phi_1 + \sin\phi_1 &= I + k(\phi_2 - \phi_1), \\ (d^2/dt^2)\phi_2 + \varepsilon(d/dt)\phi_2 + \sin\phi_2 &= I + k(\phi_1 - \phi_2). \end{aligned}$$

The homogeneous oscillation of the Josephson junction line corresponds to the "on-phase" oscillation (rotation) of the two penduli, i.e. the solution $\phi_1(t) = \phi_2(t) = \phi(t)$. It is shown in the same way as in Section 3 that this "on-phase" oscillation is stable when the Hooke's coefficient k of the coupling spring is in the range $k_0 < k < +\infty$, where $k_0 = (\lambda_2)^2/2$. The "on-phase" oscillation becomes unstable and an "off-phase" oscillation bifurcates from the "one-phase" oscillation when the coefficient k of the coupling spring becomes smaller than the critical value k_0. This bifurcation is, as in the case of the Josephson

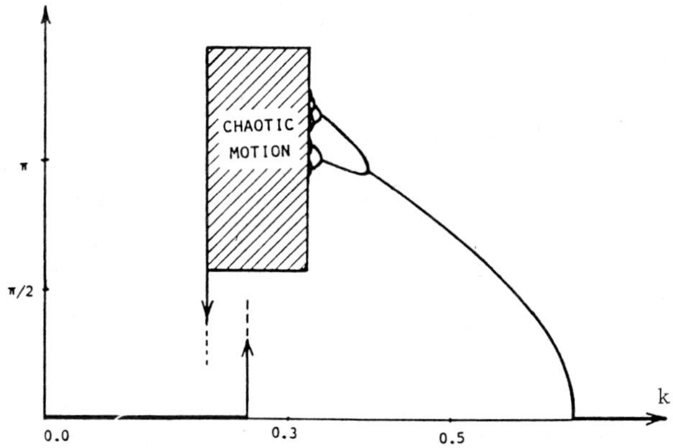

Fig.6. The amplitude of the phase difference of the two penduli $\phi_1(t) - \phi_2(t)$ for the "off-phase" oscillation of (8). The coefficient of the coupling spring k is taken as the parameter.

junction line, a period-doubling bifurcation. Figure 6 shows the amplitude of the "off-phase" component of the bifurcating oscillation of the coupled penduli. We can observe a chaotic motion in this coupled penduli system when the amplitude of the "off-phase" component becomes sufficiently large. Figure 7 shows a waveform of the chaotic oscillation observed in a numerical solution of (8). Similar chaotic motion can be observed in the system (3) in which many penduli are coupled to the neighboring ones by suitable springs.

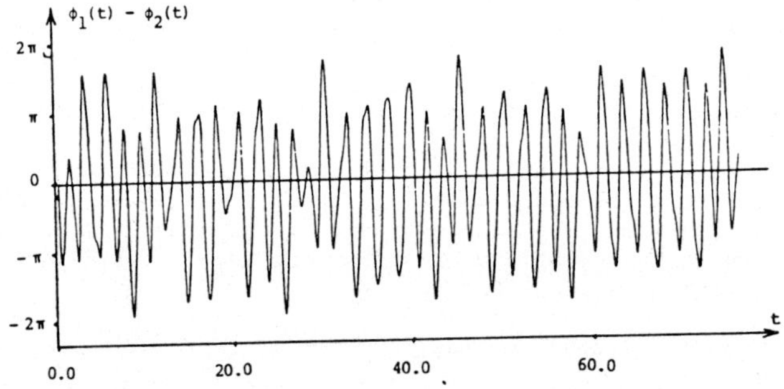

Fig.7. The chaotic motion of the coupled penduli system when $\varepsilon = 0.15$, $I = 0.25$ and $k = 0.3$. The phase difference of the two penduli is plotted as a function of time.

REFERENCES

[1] A. C. Scott, Propagation of maginetic flux on a long Josephson tunnel junction, Nuovo Cimento, B69 (1970), pp.241-260.

[2] B. D. Josephson, Supercurrents through barriers, Advances in Phys.,14 (1965), pp.419-451.

[3] E. A. Coddington and N. Levinson, Theory of Ordinary Differential Equations, McGraw-Hill, New York, 1955.

[4] J. Nagumo, S. Arimoto and S. Yoshizawa, An active pulse transmission line simulating nerve axon, Proc. I.R.E. 50 (1962), pp.2061-2070.

[5] K. Maginu, Stability of travelling wave solutions of the active Josephson junction line, J. Differential Equations, 37 (1980), pp.238-260.

[6] K. Maginu, Spatially homogeneous and inhomogeneous oscillations and chaotic motion in the active Josephson junction line, SIAM J. Appl. Math. 41 (1983), pp.225-243.

IV. ENVIRONMENTAL ELECTROMAGNETICS

Nonlinear and Environmental Electromagnetics, edited by H. Kikuchi
Elsevier Science Publishers B.V., Amsterdam, 1985 – Printed in The Netherlands

EFFECT OF THE LIGHTNING ELECTROMAGNETIC PULSE ON A TELECOMMUNICATION NETWORK : A REVIEW OF THE RESEARCH ACTIVITIES IN FRANCE

P. DEGAUQUE[*], J. HAMELIN[**], B. DEMOULIN[*]

* Université des Sciences et Techniques de Lille (USTL)
B.P. 36, F - 59650 VILLENEUVE D'ASCQ, FRANCE

** Centre National d'Etudes des Télécommunications (CNET)
B.P. 40 - 22301 LANNION CEDEX , FRANCE

1. INTRODUCTION

We have tried in the presentation which follows not to describe all relevant aspects of the influence of electromagnetic sources on a telecommunications network which are currently being studied in France, but, rather, to present work which seems to us to be representative or original.

Thus, we will only consider, in our discussion of sources, the characterization of the impulsive electromagnetic fields radiated by thunderstorm discharges. We will then develop our ideas dealing with coupling between an electromagnetic field and a cable or heterogeneous structures, such as buildings. Finally, we will present a new research tool recently developed in France ; a numerical recording system with which it is possible to determine, remotely, statistics on the overvoltages which appear at the inputs of automatic switching networks.

We hope, in this way, not only to indicate the present state of research being done in France in these subject areas but also to show the evolution of programs undertaken to better understand the perturbing electromagnetic sources and their consequences.

This work represents the efforts in France, not of one particular research entity, but, of several groups belonging to different universities or large organizations (see the list of the principal groups at the bottom of this page) which, while preserving some independence, have known how to participate collectively in joint experiments or detailed studies of delicate theoretical points.

Agencies and laboratories having an interest in research on lightning electromagnetic field characterization and coupling with telecommunications lines.
* Centre National d'Etudes des Télécommunications (CNET) * Commissariat à l'Energie Atomique (CEA) - Centre de Bruyère Le Chatel * Electricité de France (EdF) - Etudes et Recherches * Ecole Supérieure d'Electricité (Sup Elec) * Office National d'Etudes et de Recherches Aérospatiales (ONERA) * Université de Clermont II - Laboratoire d'électromagnétisme * Université des Sciences et Techniques de Lille - Laboratoire de radiopropagation * Université de Limoges - Laboratoire d'électromagnétisme

2. ELECTROMAGNETIC FIELD RADIATED BY LIGHTNING

2.1 Introduction

Work on characterizing the electromagnetic emissions produced by thunderstorm discharges is due, in France, to the occurrence of two events :

(a) The introduction, by the French Telecommunications service in the early 1970 s, of increasingly larger proportions of electronic systems. These systems were connected to inadequately protected networks and it was observed very quickly that they were very vulnerable to thunderstorm discharges.

(b) At the same time, Electricité de France (EdF) and the Commissariat à l'Energie Atomique (CEA), concerned with designing adequate protection against direct impacts of lightning, decided to undertake experimentation to characterize direct strikes, this, using a method suggested by P. Hubert for artificially triggering lightning. /1, 2, 3, 4/

These opportunities were such that, in 1975, a small research group was formed at the Centre National d'Etudes des Télécommunications (CNET) and, together, initially, with a group from the CEA laboratory familiar with electromagnetic detection, began studies with the goal to better characterize the electromagnetic radiation of thunderstorm discharges.

Following this point of view, our first concern was the study of the large amplitude electromagnetic impulses radiated by return strokes. Because of the energies involved, return strokes provoke the phenomena which cause the most damage to telecommunications equipment. It was important not only to determine the time varying properties but also the geometries which developed in space.

Subsequently, and in parallel with the development of models of the physics of the discharge processes and the associated electromagnetic emissions, we expanded the area of investigation to include leader, streamer and intracloud discharge processes, and broadened the frequency band studied to a few gigahertz.

In the same way, after having developed the necessary specialized measuring equipment and after having acquired an initial feel for triggered lightning electromagnetics, we studied the phenomenology of natural lightning discharges in both temperate and tropical environments.

2.2 Lightning events studied

The triggered lightning technique offers some specific advantages :
- The discharge is triggered within 2(-1, +3) seconds after the rocket is fired. This enables us to activate most of the experimental equipment during a specific period only.

- The ground impact point is generally known. Precise measurements of channel properties or elctromagnetic radiation analysis can therefore be achieved conveniently ; in particular, the interaction between the lightning flashes and any instrumented object can be studied with the proper diagnostics. In nearly 10 years of experiments about 200 lightning flashes have been triggered. They have enabled us to study various aspects of the physical nature of lightning such as:

a) lightning discharge channel properties : return-stroke current intensity, return-stroke propagation velocity, light emission and streamer phenomena associated with downward and upward leaders. /5, 6, 7/.

b) electrical field variations at ground and at altitude during lighting flashes. With several measurements performed at different ground locations, amplitude and localization of neutralized charges have been analyzed and correlated with the instantaneous current collected at the impact point. In some cases these parameters were analyzed as a function of local atmospheric properties such as precipitation densities and three dimensional wind fields in storm cells /8, 9, 10/.

c) electromagnetic signals radiated from the different phases of the lightning arcs. Evolutions of the electromagnetic waveforms have been observed at different distances from the ground impact and spatial evolution of the VHF electromagnetic sources associated with the different phases of a lightning discharge have been reconstructed and compared to the propagation path of the main channel. /11, 12, 13, 14, 15/.

We have also been able to study the effects of lightning on systems.

2.2.1. Lightning triggering techniques

Different systems for triggering lightning flashes have been used during successive campaigns between 1973 and 1983 ; they are presented in figure 1. For all of them, rockets of the anti-hail type (diameter 70mm, length 847mm, weight at launching 2.7 kg) have been used.

a) Classic triggering device

In this case, (fig. 1a) lightning strokes are triggered by the firing of a rocket fitted with a conducting wire, the end of which is connected directly to a coaxial shunt used for the current measurement. This type of device was employed successfully at St Privat d'Allier and at the Langmuir Laboratory, between 1973 and 1982 /1, 16/.

b) Device for triggering at altitude

In order to observe the lightning properties of a downward leader or to simulate a flying object struck by lightning, the triggering technique has to be modified (Fig. 1b). The principle of this device is to unwind a conducting wire, not connected electrically to the ground or to the lightning strike object. The flash travels freely from the lower end of the conducting wire down to the ground over a chosen distance between 50 and 200 m. This new technique (called Tipsy) was first successfully employed in 1980 at St Privat d'Allier and in 1981 at the Langmuir Laboratory.

c) New Triggering device (LRS patented)

In order to increase the reliability of the system and to reduce the cost of the equipment, a new technique has been developed and used successfully in Florida in 1983 and 1984. The wire is connected electrically to the ground but the spool is directly fitted to the rocket.

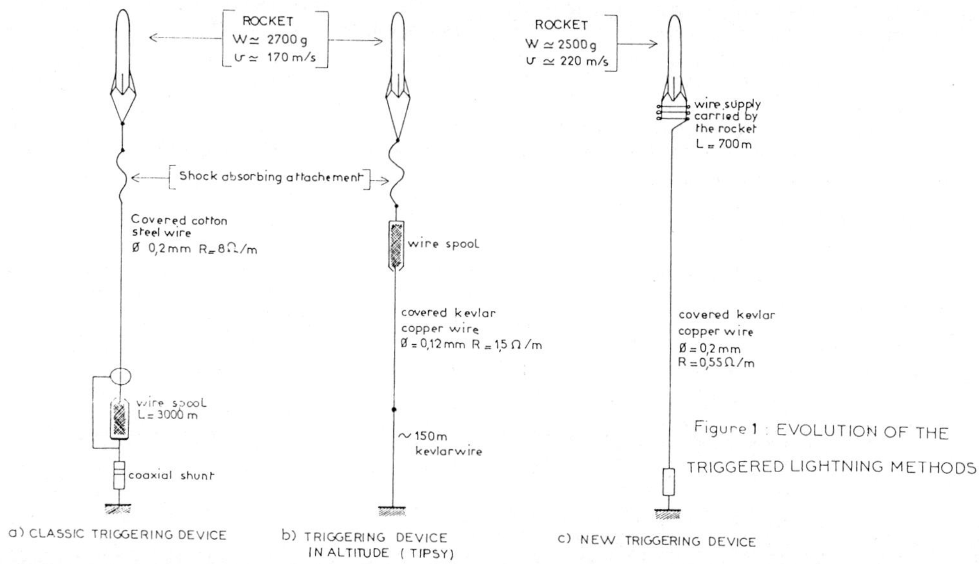

Figure 1 : EVOLUTION OF THE TRIGGERED LIGHTNING METHODS

2.2.2. Comparison of natural and triggered lightning

A comparison of natural and triggered lightning must take into consideration the great diversity of types in each category. For simplicity we will only examine the most frequent case in which negative charges are drained from the cloud to ground.

Triggered lightning is always of the ascendant type, as it is started by a positive upward leader which emerges from the upper extremity of the metallic wire where the intensification effect of the electric field is at a maximum. During its progression, this leader develops upwardly directed branches which, when viewed distantly, have the characteristic appearance of upward lightning strokes.

In a "classic" triggered event, the progression of the leader is accompanied by a continuous current which continues after the fusion of the wire with an intensity of several hundred amperes. This current transfers important quantities of charge ; sometimes more than a hundred coulombs.

Occasionally, the event ends with the extinction of the continuous current, in which case, we call this a "slow discharge". In more than 50 % of cases, however, the slow discharge is followed by the appearance of one or several discrete impulses which obey the same "dart leader-return stroke" mechanism found in the subsequent impulses of natural lightning. The risetimes of these impulses are often very rapid (durations less than a microsecond) and peak intensities can reach several tens of kiloamperes.

Figure 2 gives an exemple of the current variations for a classic triggered lightning stroke. This recording was made with an instrument whose bandwidth was inadequate to accurately reproduce the impulses. The latter have been sketched as vertical lines showing the time of occurence and the maximum amplitude.

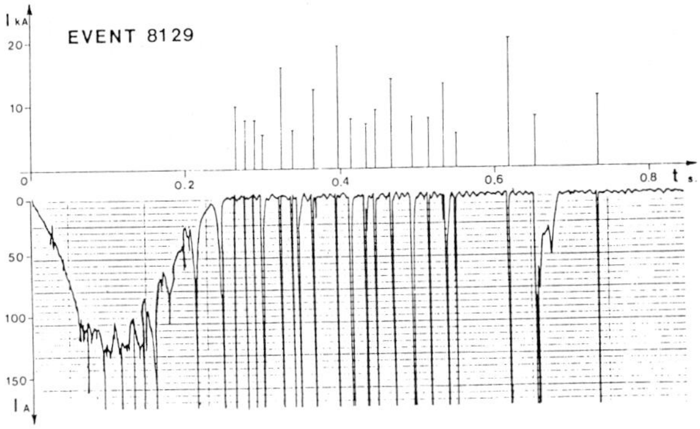

Figure 2: EXAMPLE OF ELECTRIC CURRENT VARIATION DURING A TRIGGERED EVENT

Figure 3: MAXIMUM INTENSITY, TOTAL CHARGE, AND OVERALL DURATION VALUES EXCEEDED WITH 90, 50, AND 10 PER CENT PROBABILITY RESPECTIVELY

		90%	50%	10%
I kA	A	2	12	29
	B	4	18	30
	C1	16	30	62
	C2	6	12	24
	D	0,04	0,25	1,5
TOTAL CHARGE Coul.	A	4	50	100
	B	6	30	175
	C	2	8	30
	D	2	10	70
OVERALL DURATION ms	A	70	350	850
	B	250	470	940
	C	40	180	800
	D	144	340	790

A - St PRIVAT
B - LANGMUIR LABORATORY
C - SAN SALVATORE DOWNWARD LIGHTNING
C1 - IS RELATED TO FIRST STROKE AND
C2 - TO SUBSEQUENT STROKES
D - SAN SALVATORE UPWARD LIGHTNING

On a quantitative level, figure 3 compares, in a simplified way, certain parameters which characterize natural and triggered lightning.

The properties of triggered lightning, similarities or differences with respect to natural discharges, and detailed descriptions of research techniques and results are given in numerous publications /4, 17/. We only intended, in the above paragraphs, to outline the principal points.

While triggered lightning served as a base for our initial measurements of the electromagnetic characteristics of thunderstorm discharges, we very quickly enlarged the field of study to include natural lightning in both a temperate environment (it was important to compare, using identical experimental means, natural and triggered lightning processes) and also tropical thunderstorm events /13/, where the intracloud activity becomes preponderant.

In 1981, on the Ivory Coast /18/, and in 1984, in southwest France, our measurements of natural thunderstorm discharge electromagnetic emissions were made in correlation, not only with the classical meteorological methods used to study thunderstorm cloud-mass dynamics (doppler radars, for example) but also together with systems permitting the localization of charges (a network of electric field mills and capacitive antennas (Ivory Coast) covering a 50 km x 50 km area) or streamer processes (2 300 Mhz interferometer stations located 10 km apart) within the cloud. Additionally, in 1984, measurements of cloud electrical and microphysical parameters were made using an instrumental aircraft (Transall).

2.3. Review of lightning return stroke properties

Electromagnetic measurements using triggered lightning have allowed us to characterize and correlate the principal properties of return strokes and their electromagnetic emissions. Examples include :

- current (shape and amplitude) measured at the ground strike point,
- the upward velocity of the return stroke current wave,
- the shape and amplitude of the radiated electromagnetic fields and their variation with distance.

The results of these measurements and the experimental methods used have been presented in a variety of publications /19,11,17/. Triggered lightning return stroke characteristics have been compared with measurements of natural lightning discharges /14/ made using essentially identical experimental equipment. Despite the inexistance in a triggered discharge, as in a naturally occurring, upwardly propagation return stroke, of a first return stroke with a rapid rise to a large amplitude peak, the differences between natural and triggered lightning electromagnetic emissions are minimal. To illustrate these similarities, we present in Figure 4, statistical distributions, based on 50 to 100 measurements, of triggered lightning return stroke electric and magnetic field risetimes and amplitudes (the measuring method used excluded values produced by events with currents less than (1KA).

Fig. 4a : Level of electric and magnetic fields at different distances $\gamma = 10^{-9}$ T

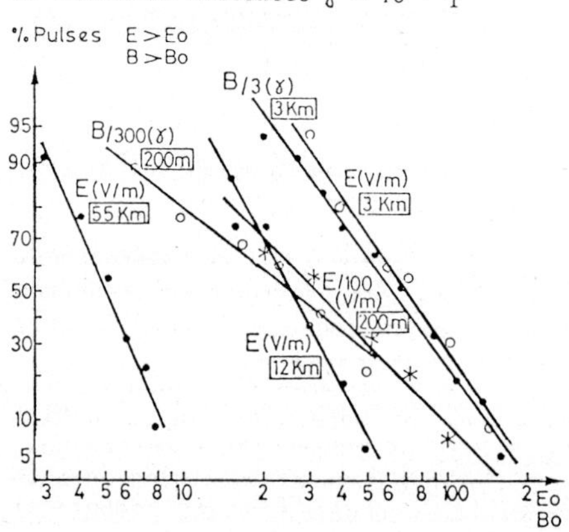

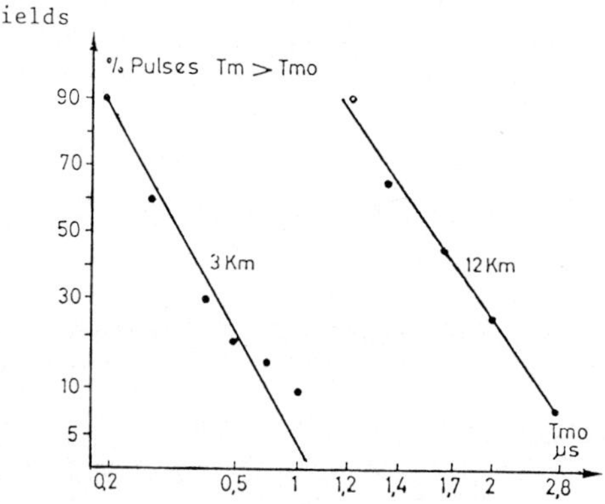

Fig. 4b : Rise time of electromagnetic impulses at 3 and 12 km

2.4. Recent research activity

Our goal will not be to analyse in great detail recent results which are described in detail in the following publications /22,23,24/. Rather, we will discuss three subjects which seem to us to be characteristic of our present research interests and activity.

- Investigations of the entire thunderstorm discharge phenomenon using measurements of the VHF and UHF radiation and measurements of submicrosecond field impulses. That is, the emissions associated with the formation of ionized channels.

- Time domain analysis ($f \leqslant 10$ MHz) of the most energetic intracloud discharge signals, their electromagnetic characteristics at the ground and classification of the principal types with, as goal, to be able eventually to formulate improved models.

- Finally, the analysis of rapid dE/dt variations of the electromagnetic impulses produced by return strokes. The object of these studies is to define more precisely the parameter dI/dt in current impulses striking objects. If the discrepancies, which exist at present between measured dI/dt values and dI/dt values inferred from measurements of dE/dt, are confirmed in future experiments, we will be forced to reexamine modelization of lightning return stroke currents and of the physics of the lightning striking process. With these motivations in mind, we plan to make simultaneous measurements of lightning dE/dt and dI/dt in Florida during the summer of 1985.

2.4.1. Phenomenology

To illustrate this aspect, we present one event which was recorded during a natural flash at St Privat d'Allier 23 August - 15 H 46'11" (see fig. 5). Beginning at the 50 ms/div time scale and moving upwards are presented one horizontal component of the magnetic field, RF signals at 6 different frequences (60, 100, 175, 300, 500, 900 MHz), a signal from a photoelectric detector oriented at an elevation angle of 5°, a 4.6 GHz signal, and the signal from a second optical detector pointed at 10° elevation angle.

It is convenient to divide this discharge into phases preceding and following the first return stroke. During the first phase the RF signals (especially between 60 and 175 MHz) are composed of bursts of closely spaced impulses ; each burst lasts for tens to hundreds of milliseconds. A second type of RF emission, single often intense impulses, occur near the ends of the bursts and especially during the second phase and at the end of the discharge. These latter signals are presumably the Q-noise mentioned by /25/, the solitary pulses observed by /26/ and the fast bursts defined by /27/. Of course, the appearance of these emissions is a function of the time scale on which they are reproduced and often contain effects introduced by the recording and reproduction systems.

Emissions at 60 through 500 MHz begin abruptly for the event in Figure 5 (point A). This is true for essentially all of the events we have studied. Except for a similar observation by /21/ we find no mention of this fact in the literature.

We were able, in essentially all cases, to identify lower frequency B

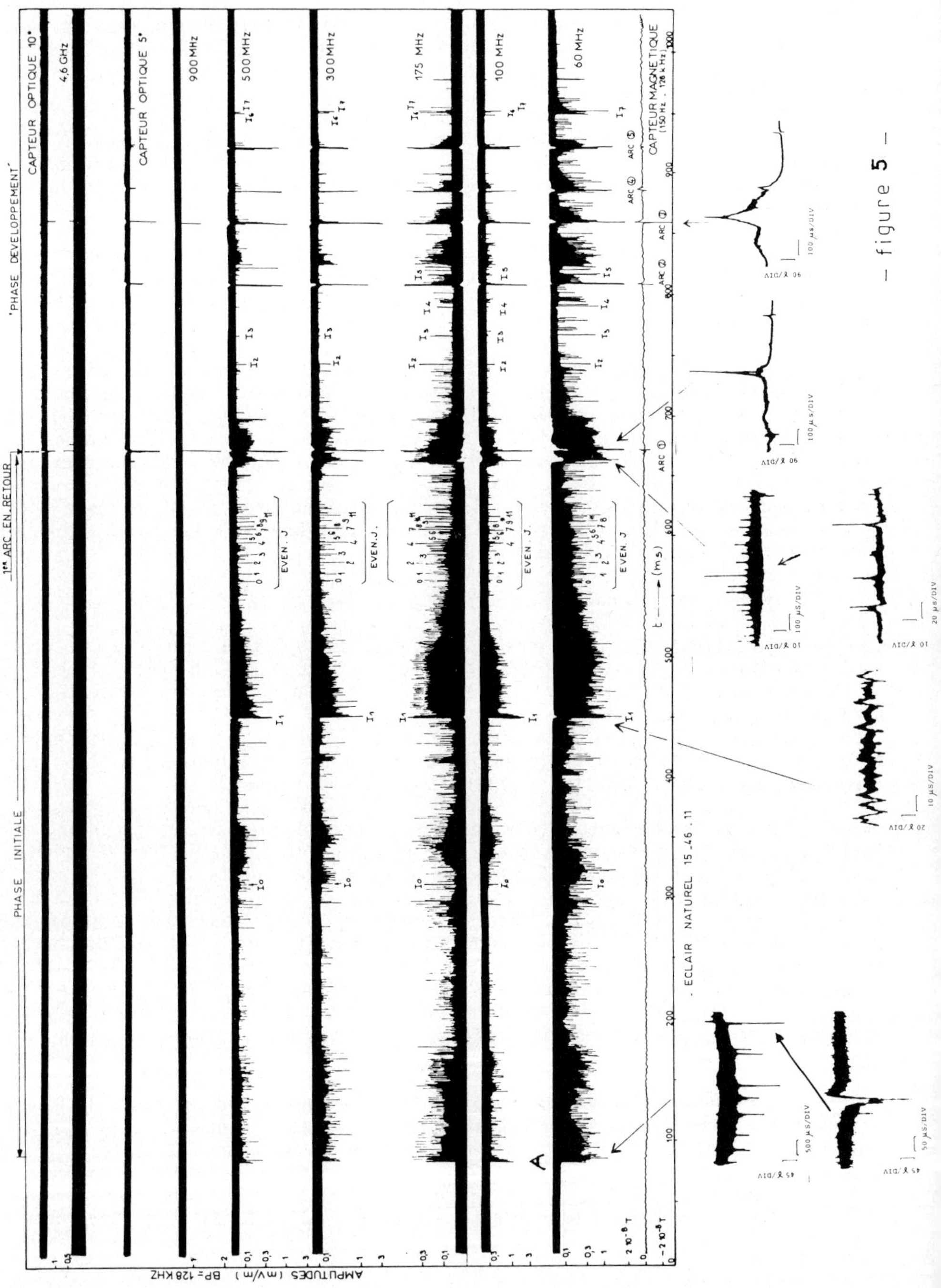
PHASE INITIALE
1er ARC EN RETOUR
'PHASE DEVELOPPEMENT'
CAPTEUR OPTIQUE 10°
4,6 GHz
CAPTEUR OPTIQUE 5°
900 MHz
500 MHz
300 MHz
175 MHz
100 MHz
60 MHz
CAPTEUR MAGNETIQUE
(150 Hz . 128 kHz)
AMPLITUDES (mV/m) BP = 128 KHZ
EVEN. J
ARC ①
ARC ②
ARC ③
ARC ④
ARC ⑤
t → (ms)
. ECLAIR NATUREL 15 .46 .11
A
100
200
300
400
500
600
700
800
900
1000
500 µS/DIV
50 µS/DIV
10 µS/DIV
100 µS/DIV
20 µS/DIV

– figure 5 –

field signals coinciding with the beginning of VHF/UHF activity. For the event in Figure 5 these signals consisted of a sequence of negative impulses, portions of which are shown on 500 μs/div and 50 μs/div time scales. (We have assumed that the return strokes which occured later in this discharge all lowered negative to ground. The resulting field change is defined to be positive). The total duration of this sequence was about 5 ms. The reverse, that is to consistently identify the beginning of VHF/UHF radiation using only the wideband B field records, was not possible.

RF emissions continue for the discharge during approximately 270 ms between points A and B. Weak magnetic field signals were also observed but association with details in the RF structure was not attempted.

Beginning at point B, there is a brutal reemergence of VHF/UHF activity which was accompanied by a 1 to 2 ms long sequence of rapid, "leader-like" magnetic field pulses. A portion of this sequence is shown on a 10 us/div time scale. The RF pulses, initially very tightly grouped, change character during the next 100-200 ms, until just before point CP individual pulses can be resolved.

The magnetic field signals which we found at point CP, that is, a periodic sequence of positive polarity impulsions (the same polarity as the return stroke), are very similar to the sequences of bipolar impulsions observed by /25/ tens of milliseconds before first return strokes and we have thus identified them as the "characteristic pulses" named by /29/.

Note however, that here the mean interval time (42 μs) and full pulse widths (20 μs) are appreciably less than the mean values (130 μs and 41 μs, respectively), given by /18/. It is interesting to note that the earlier sequence of impulses at point A are all of opposite polarity and the interval between impulses is longer and somewhat more irregular. /28/ found these same kinds of differences between the characteristic pulse sequences and the sequences of bipolar impulses which occured during isolated intracloud discharges.

Continuing our examination, strong radiation appears next at point RS at 60 through 500 MHz and also for the first time at 900 MHz and 4,6 GHz. The accompanying optical emissions at 5° and 10° and the observed magnetic field signal allow us to conclude that this was a return stroke.

There was a signal at 4,6 GHz although the discharge occured at about 110 degrees from the microwave antenna direction. The discharge was detected with one of its lobes. The calibration given of course is no longer valid.

During the remaining 300 ms of this event, we identified similarly five additional return strokes (SS). The third stroke was the only other event in this discharge to produce visible emissions at 900 MHz and 4,6 GHz. The magnetic field wave form for this return stroke is shown on a 100 μs/div time scale.

We have examined additional natural cloud-to-ground discharges. In each case, we found a similar sequence of characteristic pulses preceding the first return stroke. Measurements of different parameters which characterize this phenomenon such as the duration of the stepped leader (defined to be the time between point CP and the first return stroke) and the duration of the "preliminary variation", the time between the beginning of the VHF/UHF emissions and point CP have been reported in /21,30/.

2.4.2. Characteristics of intracloud discharge radiation

We will discuss several parameters of intracloud discharges.

The study of these components was carried out with the same means of measurement as for the return-stroke based on signals coming from magnetic sensors (BW 150 Hz - 20 MHz).

The majority of the measurements presented were recorded on the IVORY COAST (these discharges being predominant in the Tropics) during the COPT 81 experiments. (Convection Profonde Tropicale).

A network of field-mills operated by the ONERA permitted the simultaneous measurement of the electrostatic field variations by means of 10 measuring points. It therefore permitted to identify unipolar models - cloud to ground - and bipolar models - intracloud - which characterize each type of discharge.

Figure 6 gives an example of intra-cloud pulses. Pulses 1 has a duration of approximately 20-30 µs and is characterized by its double polarity.

We have frequently observed the presence of unipolar pulses -2- preceeding the bipolar pulse by 10-20 µs. This corresponds to different authors observations /18-29/.

Like pulses -1-, pulses -2- are seen in the intra-cloud process but also during the step leaders. These pulses of constant polarity appear in packets separated by a few microseconds and their number varies from a few microseconds and their number varies from a few to many tens of impulses.

Different parameters of wide bandwidth emissions of tropical discharges have been studied ; in particular the charateristics intracloud discharges.

The following parameters were studied :

1. Maximum amplitude
2. Slope from first transition
3. Peak duration at half amplitude
4. Pulses occurrences during the same discharge.

The distribution of amplitude corresponds to the whole of pulses collected in a given point.

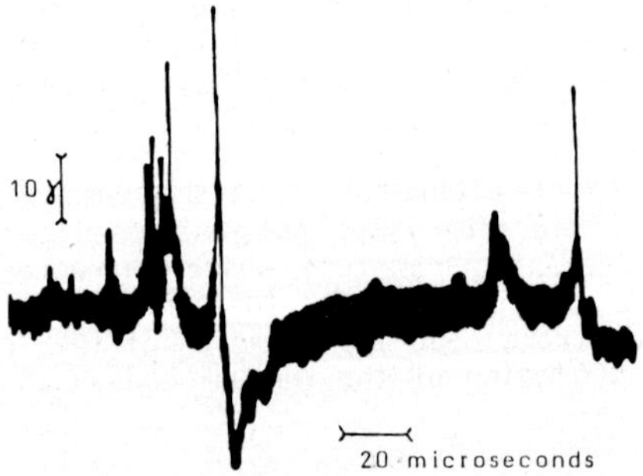

Fig. 6 : Fast unipolar pulse preceding for few micro-seconds) a bipolar pulse recorded during an intracloud stroke

1) Maximal amplitude

Table 1 gives the maximal amplitudes for all the discharges : 332 pulses filmed on oscilloscopes, which were recognized 40 I.C. and 40 C.G. by the proviously described process.

The most important conclusion is that the amplitude of intracloud discharges although on average inferior to that of cloud to ground discharges, is far from being negligible compared to this.

Table 1	90 %	50 %	10 %
I.C. (N=40)	50 γ	60 γ	150 γ
C.G. (N=40)	50 γ	140 γ	250 γ
Total events (N=332)	50 γ	55 γ	200 γ

2) Slope

Slope, during the rapid transition of the field, has been calculated for each curve by the least squared method applied between 10 % and 90 % of the maximum value. The values obtained to the I.C. and C.G. discharges are comparable (cf Table 2). The rapid transition of intracloud undertaken evolves like that of return stroke.

Table 2	90 %	50 %	10 %
I.C. (N=32)	45 γ/µs	75 γ/µs	120 γ/µs
C.G.(N=39)	50 γ/µs	90 γ/µs	175 γ/µs
Total events (N=317)	45 γ/µs	75 γ/µs	150 γ/µs

3) Duration at half amplitude

One of essential differences between cloud to ground and intracloud electromagnetic pulses is the duration considered at 50 % of the maximum amplitude. Table 3 gives the values obtained. For the return stroke we consider only the first peak (which is very restricting).

Table 3	90 %	50 %	10 %
I.C. (N=45)	0,8 µs	1,6 µs	5 µs
C.G. (N=40)	1,2 µs	3,5 µs	11 µs
Total events (N=331)	0,9 µs	2 µs	7 µs

4) Occurrence

The frequency of apparition of cloud to ground discharges is on average 33 ms /31/ between two discharges with an arc number of 4 to 15 /3/. During the intracloud discharges we were able to observe very close sequences of pulses separated by a few dozen, indeed even by a few microseconds, with a density which may reach 15 to 20 pulses by packet of a few hundreds of a microsecond.

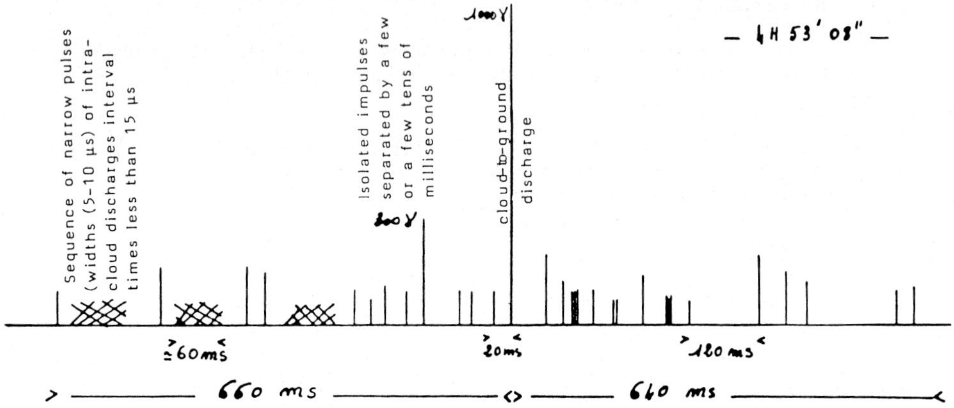

Figure 7 : Chronology for the discharge of 22 June 81 - 4h 53' 08''

To illustrate this we present a discharge recorded on the Ivory Coast 22 June 1981 at 4 H 53'08'' local time.

Figure 7 gives the chronology of the entire discharges. One can see a very strong pulse in the middle of the discharge : contrary to other impulses which correspond to intra-cloud discharges this is emitted by a cloud-to-ground discharge.

From these few indications of intracloud discharges, it is important to remember that these processes, largely present in tropical region, have certain characteristics comparable to those from cloud-to-ground discharges situated at a few kilometres distance, notably slope and amplitude, this being with a much more important periodicity.

Note : While paragraphs 2.4.1 and 2.4.2 summarize work performed by the CNET Electromagnetic Environment Group, the results which we will now discuss come from research performed by a member of the Electromagnetic Environment Group in 1984, under the direction of professeur KRIDER.
The current interest in and importance of these data justify, we believe, their inclusion in this report. The discussions in Sections 2.4.1 and 2.4.2, as well as that to follow, reproduce portions of /13/, /22/ and /24/, respectively.

2.4.3 The rapid variations in lightning fields

Recent measuremens have shown that the electric and magnetic fields that are radiated by leader steps and return strokes in cloud-to-ground lightning and by various intracloud discharge processes all contain large submicrosecond variations when there is minimal distortion in these fields due to ground wave propagation or the recording equipment /32,33/. Because submicrosecond fields have important implications for the design of lightning protection systems, a new experiment was conducted during the summer of 1984 to determine better the submicrosecond structure of E and dE/dt fields.

Lightning electric field, E, and dE/dt waveforms were recorded, together with the output of a wideband HF receiver, at an experiment site that was located on the eastern tip of Cape Canaveral, Florida.

The antennas and signal processing electronics were designed to provide a time-resolution of about 10 nsec on a dE/dt signal and 125 nsec on E. The HF receiver could be tuned to any frequency between 3 and 18 MHz, and the bandwidth was 1 MHz.

The experiment trailer was located as close as possible to the Atlantic Ocean so that the field from any lightning discharge that occurred over the ocean could be measured with minimum distorsion due to ground-wave propagation. The locations of the strike points of cloud-to-ground discharges were provided in real-time by a network of wideband magnetic direction finders.

Figure 8 shows an example of the dE/dt, E, and HF waveforms that were produced by a first return stroke in a cloud-to-ground discharge at a range of 36.2 km.

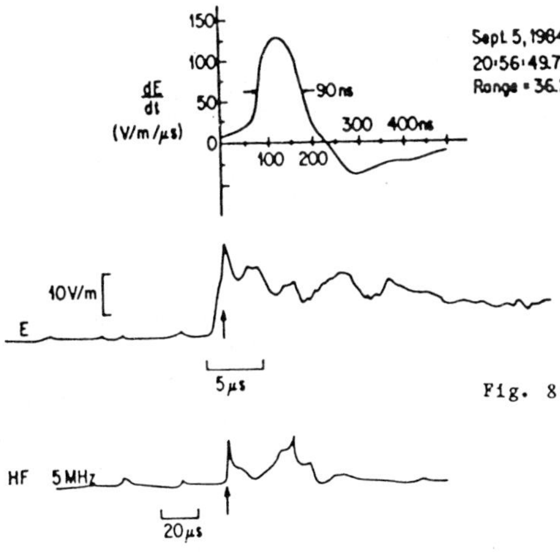

Fig. 8 Examples of the dE/dt signature (top), E waveform (middle), and HF emission produced by a first return stroke at 36.2 km.

The values of the maximum dE/dt during the fast field transition, the FWHM of dE/dt, and the electric field peak, E_p, are summarized in Table 4 for first return strokes, normal subsequent return strokes, and subsequent strokes that were preceded by a dart-stepped leader. In each entry, N is the total number of records that were analysed, and σ is the standard deviation. All values of dE/dt and E have been normalized to a range of 100 km, assuming that the measured amplitudes have an inverse distance dependence.

Type of Stroke	Max dE/dt (V/m/μsec)	FWHM (nsec)	Peak E* (V/m)
First	N = 62	N = 64	N = 48
Return	$\overline{dE/dt}$ = 45.4	$\overline{FWHM}$ = 97.2	$\overline{E_p}$ = 11.1
Strokes	σ = 13.4	σ = 18.4	σ = 3.7
Normal	N = 6	N = 7	N = 5
Subsequent	$\overline{dE/dt}$ = 40.6	$\overline{FWHM}$ = 81.4	$\overline{E_p}$ = 4.5
Strokes	σ = 18.0	σ = 10.7	σ = 1.9
Dart-Stepped	N = 7	N = 7	N = 6
Subsequent	$\overline{dE/dt}$ = 51.7	$\overline{FWHM}$ = 98.6	$\overline{E_p}$ = 8.7
Strokes	σ = 11.5	σ = 16.8	σ = 2.2

TABLE 4

Summary of the Maximum dE/dt, FWHM, and Peak E Values for Return Strokes with Range-Normalization of 100 km

* Some E fields exceeded the dynamic range of the recording system ; therefore, the values of Ep should be regarded as lower limits.

It should be noted, that the largest values of dE/dt are produced by subsequent return strokes that are preceded by dart-stepped leaders.

The mean maximum dE/dt that we have given for first return strokes is 45.4 V/m/μsec at 100 km. This value is about 33% larger than the 33/V/m/μsec previously reported by Weidman and Krider /34,33/ and clearly confirms the presence of large, submicrosecond variations in the fields produced by return strokes.

The implications that lightning fields have for lightning currents have been discussed previously by /24/ and others. If we assume that the initial, fast-field components that are produced by return strokes can all be described by a transmission-line model, then the maximum rate of change of the channel current is related to the maximum field derivative through the relation :

$$\frac{dI(t)}{dt} = \frac{2\pi D\varepsilon_0 c^2}{v} \frac{dE(t + D/c)}{dt},$$

where D is the distance of the discharge, v is the velocity of the current pulse, and where the channel has been assumed to be perpendicular to a flat, perfectly conducting ground /35/. This relation also assumes that the measured field is produced by a single current pulse that propagates without distortion up a single channel at a constant velocity.

If we assume that v for first return strokes near the ground is about 10^8 m/sec /36/, then a mean dE/dt of 45.4 $\pm$ 13.4 V/m/μsec at 100 km implies that the associated mean maximum dI/dt is 227 $\pm$ 67 kA/μsec, a value that is considerably larger than most measurements during strikes to instrumented towers /31,37/.

2.5. Conclusion

This presentation of some aspects of thunderstorm electromagnetic radiation shows that, if it is necessary, and we will discuss this further in the next chapter, to better understand coupling processes between fields and lines, cables and electronics systems, it is important to continue to define more precisely the nature of this environment.

Certain aspects of electronic components which in the past could be neglected, perhaps, for example, because the amplitudes concerned were small, now, with an evolution in technology, are capable, because of the rapid transients which they produce, of interfering with sensitive system elements.

3 - COUPLING TO CABLES

It is very important to get an idea of the voltages which could appear at the ends of a coaxial or multiwires cable, exposed to a disturbing field, in order to protect the electronic equipment against surges or to know the feasibility of a telecommunication. However, due to the complexity of the real problem, it is nearly impossible to determine mathematically the exact solution and thus simple types of coupling are studied successively.

The simplest case one can imagine is a wire parallel to a ground plane. The exterior problem which is the prediction of the current flowing on the outer shield is the first step. Then the interior problem can be solved if the transfer parameters of the cable are known. However presence of shield discontinuities may change dramatically the shielding effectiveness of the system. Furthermore disturbing electromagnetic waves penetrate also directly through the walls of telecommunication buildings. An estimation of the attenuation due to the building itself is also needed. We describe in this part of the paper the aspects which have been more specifically developed in various french laboratories.

3.1 Coupling to a wire situated near the ground surface

There are of course two approaches based either on the transmission line theory (T.L.) or on the antenna theory (A.T.). For the T.L. approach, the cable is first assumed to be infinite in extent and the boundary conditions at the ends of the cable are then introduced. If the ground is perfectly conducting, the waves propagating along the structure are T.E.M. waves and the determination of the parameters of the line are obvious. However, if the line is parallel to the earth surface for example, the ground conductivity must be taken into account since, for long cables, the attenuation is no more a negligible phenomena.

In this case both T.E. and T.M. waves are present and the generalization of the T.L. approach cannot be made. However if the conductivity of the ground is sufficently high and if the height of the wire is small compared to the wavelength, one can assume that the T.M. mode will be dominant. By introducing a transverse voltage between the wire and the earth surface, the T.L. approach will give results which are usually precise enough for practical cases.|38|

Various methods have been described to determine the series impedance and admittance of the line.

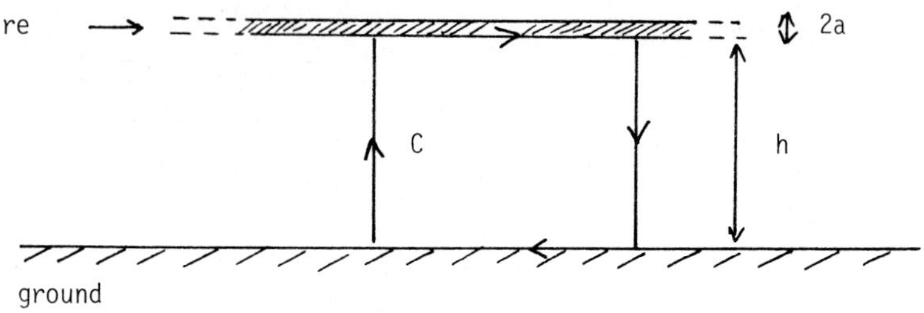

Figure 9 : Geometrical configuration

By applying Stokes' theorem and by integrating the first Maxwell's equation along the contour C in Figure 9 , Fontaine et al. |39| have shown that the series impedance is the sum of four terms : the internal impedance of the wire, the inductive reactance of the same wire but situated above a perfectly conducting ground, the internal impedance of the ground considered as a return conductor and lastly a corrective term taking into account the effect of the conductivity of the ground on the inductive terms. Comparison with an other approach made by Vance |38| is given in |39|. Such a comparison has also been made by Blumer and Ianovici |40|.

In fact, in the quasi T.E.M. approximation, thus when the transmission line theory is valid, all these approaches converge towards the same results since the corrective terms are almost due to the contribution of the T.E. mode which must be very small |41|. The advantage of the T.L. theory is of course its simplicity and it can be applied either for a cable in air or embedded in the ground. If the response in time domain is needed, F.F.T. subroutines are used.

As an example, let us consider a cable buried at a depth of 1 m, of radius a = 2 cm and with an insulating jacket with a relative permittivity $\varepsilon_r = 1.8$, the thickness of the jacket being 2 mm. The ground is characterized by a conductivity of 10^{-3} mho/m and a relative permittivity of 10. The disturbing source is assumed to be a plane wave propagating downwards in the ground and which has a wave form such that its mathematical expression is given by the difference of two exponential terms. In our example, the rising time of the incident pulse is 10 ns and the decay time is 100 ns.

The curve in Figure 10 shows the wave form of the current induced at the midpoint of the insulated cable, 300 m long, open circuited at both ends |42|. We see in this Figure that the response can be divided into two successively parts. During the first microseconds, the end effects do not produce any contribution and thus this part of the cable response will be the same as for an infinitely long cable. Then, after about 5 μs the reflections on the open ends give rise to opposite pulses of current.

As we have previously outlined, the T.L. theory is useful for simple configurations such as a wire parallel to the surface of the ground. However the mathematical analysis shows that the propagation constants associated with the current propagating on a wire above the air-ground interface are solutions of an implicit modal equation. Olsen and Chang |43|, |44| have pointed out that two guided modes exist along this structure thus propagating respectively with attenuation constants noted α_{TL} and α_{FW}. The first one α_{TL} has a value nearly equal to the one calculated with the quasi T.E.M. approximation and the associated mode is called transmission line mode. The other guided mode is called

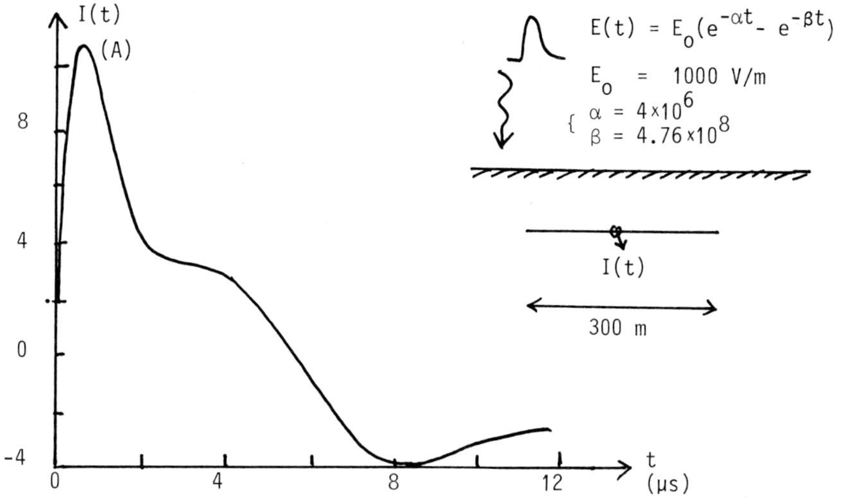

Figure 10 : Wave form of the current at the center of the cable.

wire-attached mode or fast-wave mode (F.W.) since its phase velocity is generally greater than the velocity of the T.L. Mode. The currents associated with these two modes decrease exponentially as a function of the distance between the disturbing source and the observation point.

Furthermore since the structure behaves as an open wave guide, another contribution to the total current I_t will be due to the radiated modes and the associated current is noted I_R. Therefore I_t is put in the following form : $I_t = I_{TL} + I_{FW} + I_R$.

It is thus important to know the limit of validity of the quasi T.E.M. approximation which assumes the presence of only one type of current ($I_{T.L.}$). If the disturbing source behaves as a delta-function voltage generator inserted in the line, it is possible to get an explicit form for the current distribution along the infinite line parallel to the ground surface |44| and results of extensive calculations are described in references |45| and |46|.

As an example, the curves in Figure 11 represent the current distribution along an overhead line situated at 9 m above the ground characterized by a conductivity of 10^{-2} mho/m and a relative permittivity of 5.

The frequency of the disturbing source is 1.5 MHz and thus the conditions of the quasi T.E.M. approximation, i.e. the height of the line much smaller than the wavelength and displacement currents in the ground negligible, are verified |47|.

We note on the curve of the total current that it decrases exponentially along the first 10 km since the transmission line mode is dominant. However, beyond this distance, the other modes cannot be neglected as it can be shown from Figure 11 where we have also represented the contribution of the transmission line mode only ($I_{T.L.}$) and the contribution of the two guided modes : $|I_{TL} + I_{FW}|$.

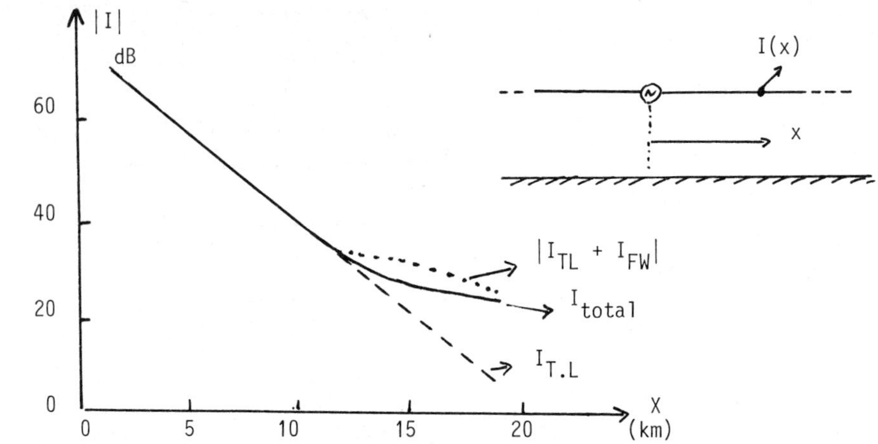

Figure 11 : Current distribution along an overhead line.

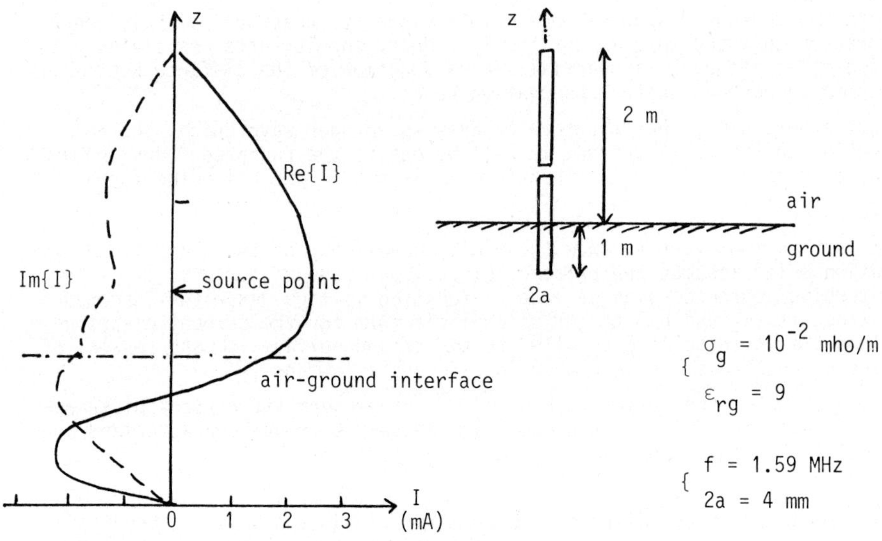

Figure 12 : Current distribution along the antenna.

From many other examples, one can estimate that the transmission line theory gives a good prediction of the current distribution along a distance such that the attenuation of the current does not exceed about 60 dB.

However this formalism is valid for horizontal wires and the coupling of electromagnetic waves to vertical parts of the line (between the wire and the ground plane for example) and the coupling between vertical and horizontal parts cannot be taken rigorously into account. This coupling can produce either resonances not predicted by the T.L. theory or a modification of the current distribution |41|. In this case a method based on the antenna theory is needed.

The current along the wire structure is solution of an integral equation which is solved using the method of moments either in frequency or in time domain |47|. Since this approach is quite general, one can simulate a good approximation of the structure. A simple application can be the evaluation of the input impedance of earth connections or the current distribution along a semi-embedded antenna. In Figure 12 we have assumed that this kind of antenna is driven by a delta-function voltage generator and the curves show the variation of the real and of the imaginary part of the current along the antenna |48|. Input impedances of earth connections are reported in reference |49|.

3.2 Voltages induced on telecommunication cables

When the current flowing along the outer shield of the cable is determined, the knowledge of the transfer parameters of the braid allows to calculate the induced voltages appearing at the ends of the line. We shall consider successively the case of a coaxial cable and of a two-wire shielded cable.

3.2.1 Voltages induced on a coaxial cable

The shielding effectiveness of the braid of a coaxial cable is usually characterized by the concept of transfer impedance Z_t. This parameter is defined, by unit length, as the ratio of change of open circuit voltage between the center conductor and the shield to the total shield current. If the shield is made with a homogeneous conducting material, Z_t represents the diffusion of the electric field component parallel to the shield axis. If the shield is made with braided wires a transfer reactance $jL_t\omega$ is added and characterizes the magnetic coupling through the shield. To take correctly into account the presence of the opening in the braid, a transfer capacitance corresponding to the electric coupling has to be introduced.

Usually time-harmonic measurements are made to get the amplitude of Z_t. However the various kinds of coupling can be clearly exhibited using measurements in time domain. Thus Demoulin et al |50| have developed this type of measurement and many experimentations have been made on various types of cables.

The cable under test is placed into a usual triaxial setup and a trapezoïdal voltage source is connected between the braid of the coaxial cable and the hollow metallic tube which acts as a disturbing line. The disturbing voltages are measured at both ends and digitalized, the variation of the transfer parameters versus frequency being obtained through F.F.T. subroutines. Demoulin et al |51| have shown that one of the main important parameter for the cable response is the weave angle Ψ between the axis of the cable and the strands.

Let us consider for example three cables with a high optical coverage ratio, the weave angle being equal respectively to $\Psi_1 = 50°$, $\Psi_2 = 59°$ and $\Psi_3 = 20°$.

Curves in Figure 13 represent the variation of the amplitude and phase angle of the transfer impedance versus frequency. We find the usual increase of $|Z_t|$ for frequencies above 1 MHz but $|Z_t|$ is proportional to f for cable 1 and 2 while, for cable 3, $|Z_t|$ is nearly proportional to the square root of frequency. Furthermore the phase angle tends to $\Pi/2$ for the first two cables, thus Z_t can be put in the following form : $Z_t \simeq jL_t\omega$; while for cable 3 the phase angle tends to about $-3\,\Pi/4$. Theoretical approaches have been given to try to explain this behavior and are described in references |51| and |52|.

A technique based again on measurements in time domain has also been tried to measure the transfer parameters above 100 MHz |53|. In this case, in order to avoid propagation effects on the cable 1m or 2m long, an additional shield made with a copper adhesive tape is put on the braid except on a length of few centimers in the middle of the cable. This cable under test behaves now as a well shielded cable except along the small discontinuity. With this method, transfer parameters have been determined until 700 MHz.

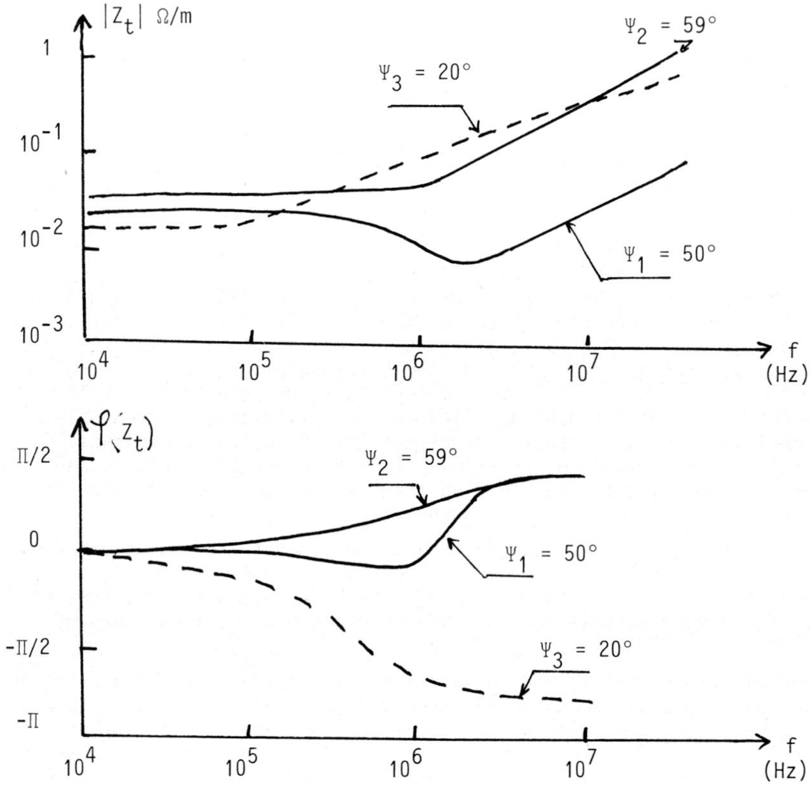

Figure 13 : Variation of the amplitude and phase angle of the transfer impedance.

3.2.2 Voltages induced on a multiconductors shielded cable

For multiconductors shielded cables, two kinds of disturbing voltages appear : a common mode voltage between the inner conductors and the shield and a differential mode voltage between two of these conductors. Both theoretical and experimental approaches have been undertaken to point out the main parameters which play a leading part for the differential voltages.

Demoulin and Degauque |54| have shown that the intrinsic differential transfer impedance characterizes the dissymmetry in the distances between conductors and the apertures of the braid. Furthermore a small dissymmetry either in the common mode input impedance of the electronic equipment connected to the lines or a dissymmetry in the position of the two-wire line inside the shield produce a differentiel voltage which could reach important values if resonances occur. As an example, curves in Figure 14 represent the variation of the common-mode Z_{tc} and differential mode Z_{td} transfer impedances versus frequency for a two-wire shielded cable.

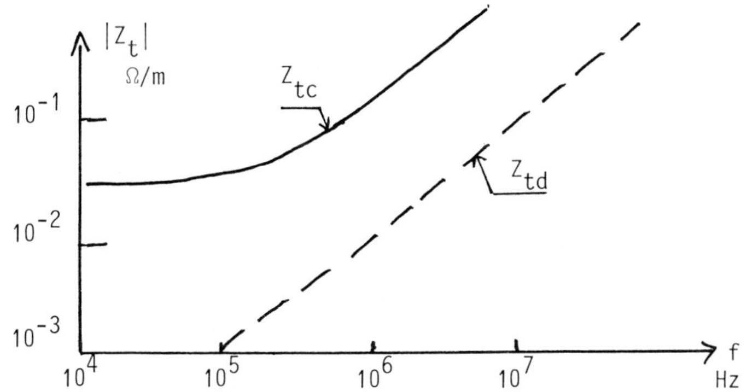

Figure 14 : Transfer impedances of a two-wire shielded cable.

We see that above 500 kHz, there is only a difference of 20 dB between the values of Z_{tc} and Z_{td}.

3.2.3 Influence of the shield discontinuities

To improve the shielding effectiveness double or triple shields can be used |55|. However the increase in the shielding of the cable itself must not be cancelled by other disturbing effects such as the influence of discontinuities at the ends of the cable. Indeed "pigtail" wires for example are used to connect the various shields of a multi-wire shielded cable |56| and it is thus important to know the influence of these wires on the shielding properties. It has been shown that if the connections are made on a distance of few centimeters, the direct coupling on the inner wires are negligible compared to the effect of the inductance of the pigtail wire. Other effects such that a connecting wire between two parts of a shield are described in |57| and |58|.

3.3 Penetration of electromagnetic waves into telecommunication buildings

A theoretical approach can be made by modelling the reinforced concrete building by interconnecting wires |59|. Another possibility is to make experiments by illuminating the building using transmitting antennas far enough from this building so that the incident wave is nearly a plane wave. The field amplitude is measured at many points inside and outside the building. Experiments have been made for frequencies ranging from 100 kHz to 1000 MHz |60|. At low frequencies, below 30 MHz, the attenuation of the magnetic field was measured while above 30 MHz, half wave antennas have been used.

As an example, the curves in Figure 15 represent the variations of the attenuation mean value for various rooms of an exchange building. We note a shielding efficiency anomaly at 60 MHz which could be explained by the fact that the walls are made with a juxtaposition of reinforced concrete panels with any electrical connections between each of them. Furthermore, we see that the mean value of the attenuation presented by the building is about 20 dB in a wide frequency range.

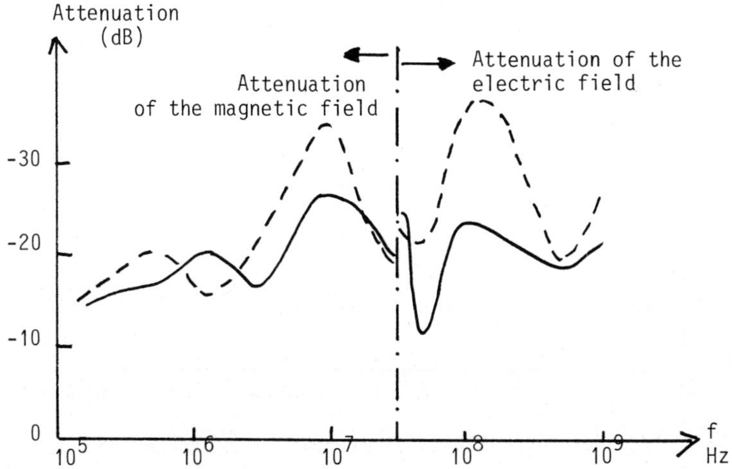

Figure 15 : Mean value of the attenuation presented by the exchange building and measured in two rooms.

3.4 Conclusion

In conclusion both theoretical and experimental approaches are needed to predict the disturbing voltages which can appear at various points of a telecommunication network. Numerical codes have been written while measurements of transfer parameters give the required input data.

To complete this work a statistical approach of surges appearing on the network has also been undertaken and will be briefly described in the next paragraph.

4 - COPELIA PROJECT (issue of /61/

The COPELIA project (in french : COnnaissance des PErturbations sur les LIgnes d'Abonnés) is the evaluation of electrical surges on the subscriber's lines.

It will allow the improvement of installations and the production of new equipments.

4.1 Principle

We register the electrical surges gathered on 80 lines spread across 10 geographical sites in France. These surges are recorded with 10 EPAL equipments (each have 8 channels) especially developped for random signals. In the 0-5 MHz frequency band the recordings are triggered : at low frequencies above a threshold value, at high frequency above thresold plus a specified slope value. The date at which the first signal exceeds the threshold is registered and is termed the "threshold date".

Each night the records are transfered to the control unit and stored in a memory (see figure 16).

4.2 Processing

The collected data are computer analysed to give statistical distributions. The maximum signal voltage amplitude and the frequency spectra shall be studied as a function of the soil conductivity, isokeraunic level, the length of lines and the proximity of electrical installations.

4.3 Characteristics of EPAL equipment

- 8 channels
- 100 µs LF sampling facility
- 100 ns HF sampling facility during 12.6 µs then 2 µs HF sampling facility during 256 µs
- recording of signals before threshold triggering 5.6 ms for LF signals (with a 100 µs step size) and 25.6 µs for HF signals (100 ns step size)
- possibility of information compression
- facilities for programming the following parameters : threshold, slope, compression or not, permissible amplitude level, and date (100 ns accuracy)
- memory capacity : 500 koctets either 950 HF signals without compression (295 µs for a signal), or 2700 HF signals with compression or 1900 LF signals (25.6 ms + signal duration).

4.4 First results

The figures 17 to 22 give a LF common mode voltage record and 5 HF common mode voltage records for the same thunder-storm. In the figure 18 the pulse is probably due to intra-cloud discharges and in the figures 19 to 22 the impulse is due to the lightning stroke (17).

At last we present (figure 24). The histogram of the peak amplitude values during two autumn months (Nov and Dec 82) for a transmission link at CAVAN, taken over a whole HF recording period. Since last summer's period the maximum amplitude has reached 2 kv.

A publication of the full results for three years will be presented soon.

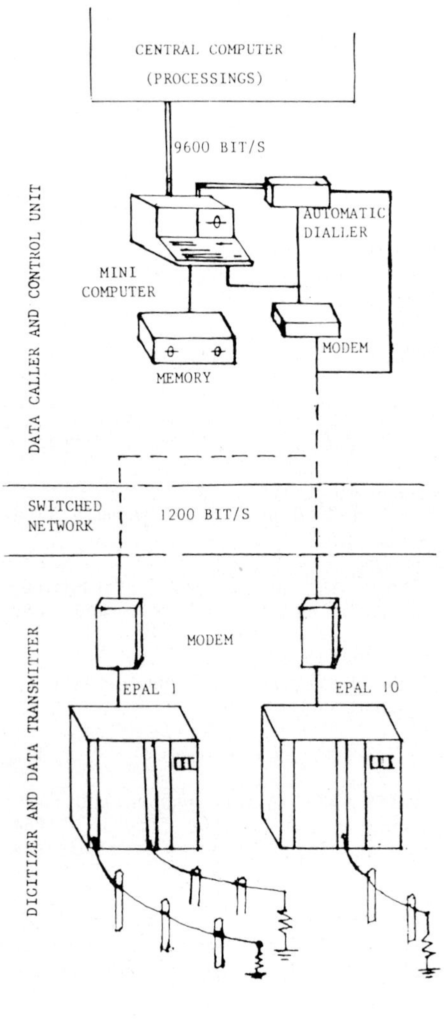

FIGURE 16 : COPELIA NETWORK

LF record : channel 6, amplitude 1 kV
threshold 80 V, duration 40 ms
threshold date 12H00 mn 38 s 28.9ms

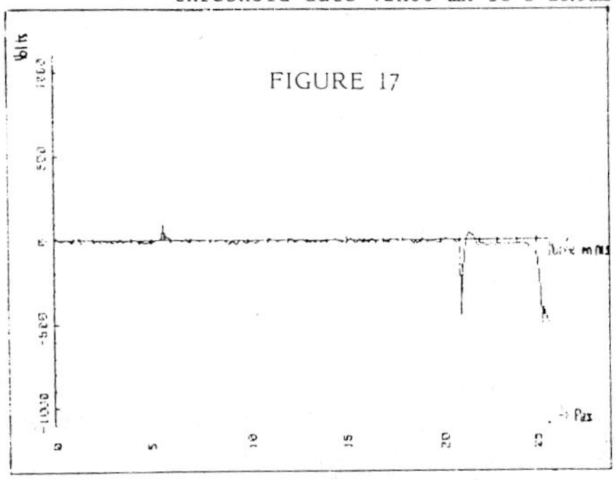

FIGURE 17

HF record : channel 6, amplitude 1 kV
threshold 80 V, slope 1 kV/ms
threshold date 12H00mn 38 s 29 ms 1 μs

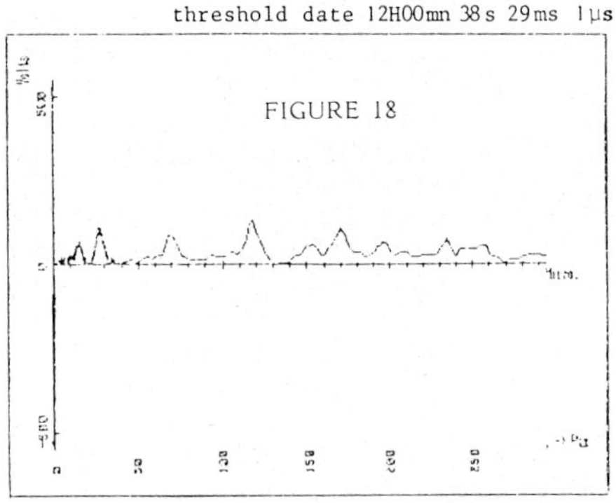

FIGURE 18

HF record : channel 6, amplitude 1 kV
threshold 80 V, slope 1 kV/ms
threshold date 12H00mn 38 s 44 ms 427 μs

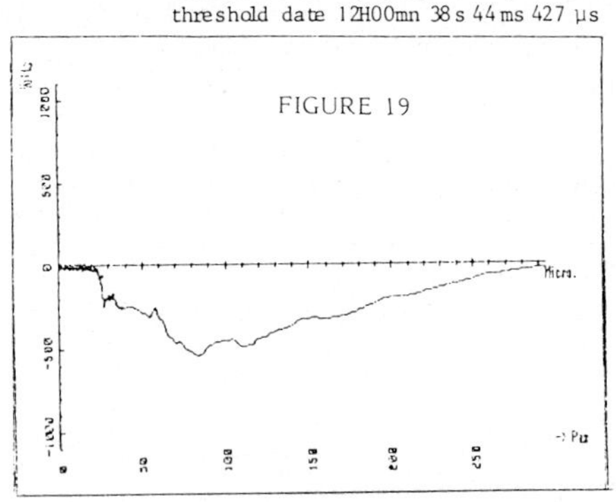

FIGURE 19

HF record : channel 6, amplitude 1 kV
threshold 80 V, slope 1 kV/ms
threshold date 12H00mn38s48ms390 µs

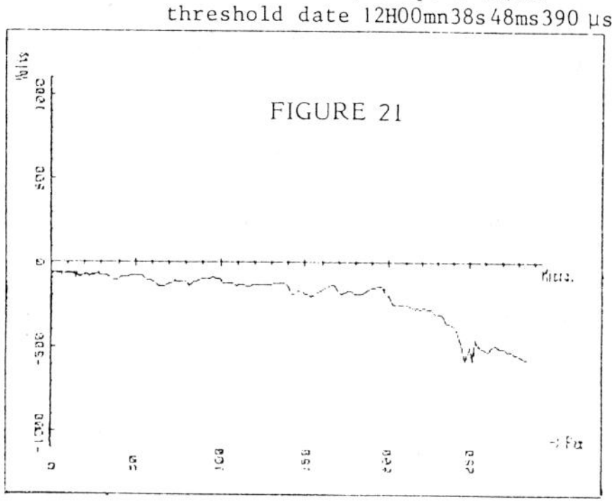

HF record : channel 6, amplitude 1 kV
threshold 80 V, slope 1 kV/ms
threshold date 12H00mn38s48ms106 µs

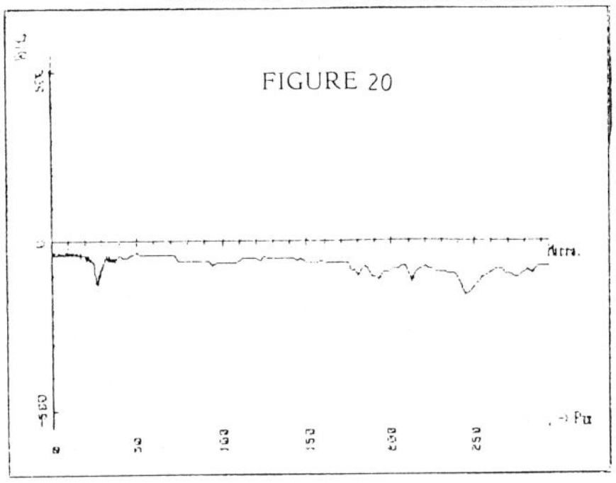

HF record : channel 6, amplitude 1 kV
threshold 80 V, slope 1 kV/ms
threshold date 12H00mn38s48ms658 µs

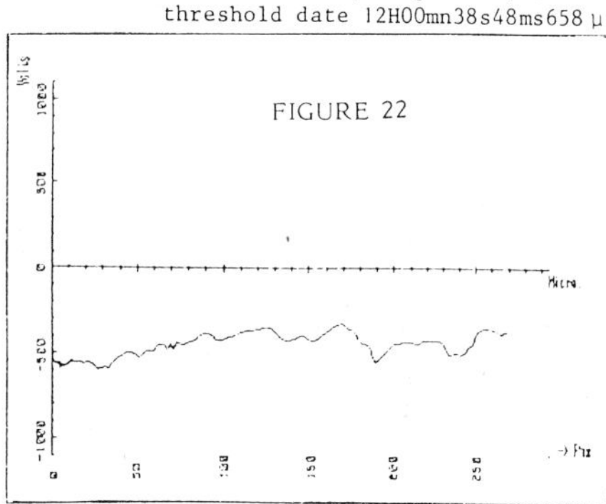

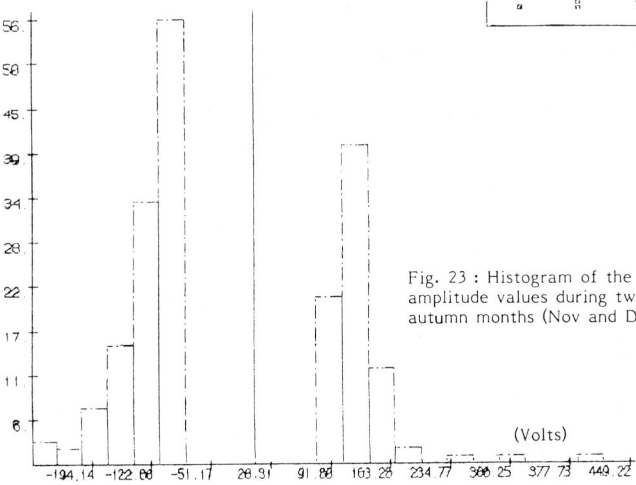

Fig. 23 : Histogram of the peak amplitude values during two autumn months (Nov and Dec 82)

5 - CONCLUSION

In france, work on the influence of the natural electromagnetic environment on telecommunications networks is characterized by research on the global problem ; study of atmospheric electromagnetic sources, study of the coupling mechanisms with above ground and buried cables, characterization of the transfer parameters of cables, characterization of the shielding of complex systems and finally statistical analyses of real case experiments. For each of these areas, experimental data have been compared with theoretical analyses. The increasing sophistication of telecommunications networks, which implies also that they are more sensitive (larger bandwidths, use of small signal amplitudes), forces us to follow this approach, in particular, for the analysis of rapid transient sources, even if of weak amplitude, and for the study of the coupling mechanisms between these submicrosecond impulses and telecommunications networks.

6 - BIBLIOGRAPHY

/1/ R. Fieux, C. Gary, P. Hubert, Artificially Triggered Lightning above Land Nature, 257, 212, 1975

/2/ Fieux R. et al, "Research on artificially triggered lightning in France, IEEE Transac on PAS, 97, 725-733, 1978

/3/ Saint-Privat d'Allier Research Group, "Eight years of lightning experiments at Saint-Privat d'Allier, RGE N° 9, 561-582, 1982

/4/ Saint-Privat d'Allier Research Group, "Application of triggered lightning in France : Possibilities and limitations", EMC ZURICH 10-C2, March 1985

/5/ P. Hubert, R. Fieux, Mesure des courants de foudre à la station de St-Privat d'Allier, Revue Gen. Elec., 5, 344-349, 1981

/6/ P. Hubert, G. Mouget, Return stroke velocity measurements in two triggered lightning flashes, J.G.R., 86, 5253-5261, 1981

/7/ V.P. Idone, R.E. Orville, P. Hubert, L. Barret, A. Eybert-Berard, Correlated observations of three triggered lightning flashes, J.G.R. ; vol. 89 D-1, 1385-1394, 1984

/8/ Walteufel, P., P. Metzger, J.L. Boulay, P. Laroche, and P. Hubert, Triggered lightning strokes originating in clear air, J.G.R., 85(5), 2861-2868, 1980

/9/ S. Chauzy, P. Raizonville - Space charge layers created by coronae at ground level below thnderclouds : measurement and modeling. J.G.R., 87, 3143-3148, 1982

/10/ P. Laroche, P. Metzger, J. Testud, S. Chauzy, Electric field variations produced by natural and triggered lightning and related radar measurements Proceeding in Atmospheric Electricity, Deepak Pub., 233-236, 1983

/11/ J. Hamelin, B. Djebari, J. Fontaine, R. Barreau, Electromagnetic field resulting from a lightning discharge, surges induced on over-head lines, mathematical model, EMC - ROTTERDAM, May 1979

/12/ C. Leteinturier, J. Hamelin, Analyse expérimentale des caractéristiques électromagnétiques de décharges orageuses dans la bande 200 Hz - 20 MHz, Annales des Télécommunications, mai/juin 1984

/13/ C. Leteinturier, C. Weidman, J. Hamelin, M. Le Boulch, The phenomenology of natural and triggered thunderstorm discharges, VIIth International Conference on Atmospheric Electricity - Albany, Juin 1984

/14/ M. Le Boulch, C. Weidman, J. Hamelin, C. Leteinturier, Characteristics and mechanisms of VHF/UHF radiation from lightning processes, VIIth International Conference on Atmospheric Electricity - Albany, juin 1984

/15/ Richard, P., G. Auffray, VHF-UHF interferometric imagery. Application to lightning, Radio-Science (to be published)

/16/ P. Hubert, P. Laroche, A. Eybert-Berard, L. Barret, Triggered lightning in New Mexico, J.G.R., vol. 89 D-2, 2511-2521, april 1984

/17/ B. Djebari, J. Hamelin, C. Leteinturier, L. Nicot, Electromagnetic pulses emitted by lightning. International aerospace conference on lightning and static electricity. OXFORD, march 1982

/18/ C. Leteinturier, J. Hamelin, M. Le Boulch, Tropical lightnings electromagnetic characterization COPT 81 experiment (Tropical deep Convection) 8th International Aerospace and ground conference on lightning and static electricity - FORT WORTH Texas USA - June 1983.

/19/ J. Hamelin, A. Meesters, Mesure des champs électromagnétiques créés par les décharges orageuses Colloque international sur la mesure en télécommunications URSI, octobre 1977

/20/ J. Hamelin, C. Leteinturier, la foudre. Première partie : mécanismes et impacts direct - L'écho des Recherches n° 108 - avril 1982.

/21/ M. Le Boulch, J. Hamelin, Rayonnement en ondes métriques et décimétriques des orages Annales des Télécommunications - 40 N° 5-6 Mai-Juin 1985

/22/ C. Leteinturier, J. Hamelin, Analyse expérimentale des caractéristiques électromagnétiques de décharges orageuses dans la bande 200 Hz - 20 MHz Annales des Télécommunications Mai-Juin 1984.

/23/ J. Hamelin, M. Le Boulch, C. Leteinturier, C. Weidman, Radiation emitted by leaders and intracloud lightning discharges processes and the fiels resulting at the ground. VII International WROCLAW Symposium on Electromagnetic Compatibility WROCLAW - juin 1984

/24/ C. Leteinturier, E.P. Krider, J.C. Willet, "Subnanosecond Structure of the Radiation Fields produced by lightning", 10 th Inst. Aerospace and ground Conf. on Lightning and Static Electricity (I.C.L.O.S.E.) 3 B6 PARIS, June 1985

/25/ D.E. Proctor "VHF Radio Pictures of cloud flashes". J.G.R. Vol 86 - N° 5 p. 4041/4071 - May 1981.

/26/ P.L. Rustan - M.A. Uman - D.G. Childers - W.H. Beasley - C.L. Lennon "Lightning source locations from VHF radiation data for a flash at Kenndy space center". J.G.R. vol 85 - N° 9 - p 4893-4903 - 1980

/27/ C.O. Hayenga - J.W. Warwick "Two-dimensional interferometric positions of VHF lightning sources". J.G.R. - vol 86 - N° 8 - p 7451-7462 - August 1981

/28/ C. Weidman - E.P. Krider "The radiation field waveforms produced by intracloud lightning discharge processes". J.G.R., vol 84, 3157-3164, 1979

/29/ W. Beasley - M.A. Uman - P.L. Rustan Jr "Electric fields preceding cloud-to-ground lightning flashes". J. G.R., vol 87, p 4883-4902, 1982

/30/ Le Boulch "Caractérisation et mécanismes du rayonnement VHF/UHF des décharges orageuses". Thèse Docteur-Ingénieur-University Clermont II - France - March 1984

/31/ K. Berger et al. "Parameter of lightning flashes", Electra n° 41 - 1975

/32/ C.D. Weidman and E.P. Krider, J. Geophys. Res., 83 (1978) 6239-6247

/33/ C.D. Weidman and E.P. Krider, in Preprints VII th International Conference Atmospheric Electricity, Am. Meteorol. Soc., June 3-8, 1984

/34/ E.P. Krider and C.D. Weidman, in Proceedings 8 th International Aerospace'& Static Electricity, Ft. Worth, TX, June 21-23 (1983) DOT/FAA/CT-83/25, 69.1-69.8

/35/ M.A. Uman and E.P. Krider, IEEE Trans. on EMC, EMC-24 (1982) 79-112

/36/ V.P. Idone and R.E. Orville, J. Geophys. Res., 87 (1982) 4903-4915

/37/ E. Garbagnati, F. Marinono, and G.B. Lopipero, Proceedings 16 th Conf. on Lightning Protection, Szeged, Hungary, July 1981

/38/ Vance E.F., Coupling to shielded cables, J. Wiley Ed., N.Y, 1978

/39/ Fontaine J.M., Umbert A., Djebari B. and Hamelin J., 4 th Symp. on E.M.C., Zurich 1981, Proceedings pp. 109-114

/40/ Blumer W. and Ianovici M., Annales Telecom., vol 39, n° 9-10, 1984, pp 405-412

/41/ Degauque P. and Zeddam A., Note Technique CNET, NT/LAA/ELR 465, 1985

/42/ Zeddam A. and Degauque P., Int. Symp. on E.M.C.; Tokyo 1984, Proceedings pp 925-929

/43/ Chang D.C. and Olsen R.G., Radio Science, 1975, n° 8-9, pp ,823-831

/44/ Kuester E.F., Chang D.C. and Olsen R.G., Radio Science, 1978, vol. 13, pp 605-623

/45/ Courbet G., Degauque P. and Gabillard R., Annales Telecom., vol. 37, n° 5-6, 1982, pp 225-237

/46/ Degauque P., Courbet G. and Heddebaut M., I.E.E.E. Trans. on E.M.C., vol. 25, n° 4, 1983, pp 422-427

/47/ Jecko B., Dafif O. and Guedira R., Annales Telcom. vol 39, N° 9-10, 1984, pp 397-403

/48/ Paladian F., Plumey J.P., Roubertou D. and Fontaine J., 6th Int. Symp. on E.M.C., Zurich 1985, Proceedings pp 279-282

/49/ Roubertou D., Fontaine J., Plumey J.P. and Zeddam A., 1984, Int. Symp. on E.M.C., Tokyo 1984, Proceedings pp 717-720

/50/ Demoulin B., Degauque P. and Gabillard R., I.E.E.E. Int. Symp. on E.M.C., San Diego 1979, Proceedings pp 19-26

/51/ Demoulin B., Degauque P. and Cauterman M., 4th Symp. on E.M.C., Zurich 1981, Proceedings pp 491-495.

/52/ Demoulin B., Etude de la pénétration des ondes électromagnétiques à travers des blindages, Thèse Doctorat, Lille, 1981

/53/ Demoulin B., Duvinage P. and Degauque P., 6th Symp. on E.M.C., Zurich 1985, Proceedings pp 521-524

/54/ Demoulin B. and Degauque P., 1984 Int. Symp. on E.M.C., Tokyo, Proceedings pp 936-939

/55/ Demoulin B., Degauque P., Cauterman M. and Gabillard R., I.E.E.E. Trans. on E.M.C., vol. E.M.C. 22, n° 3, 1980, pp 173-180

/56/ Paul C.R., I.E.E.E. Int. Symp. on E.M.C., San Diego 1979, Proceedings pp 268-273

/57/ Demoulin B., Duvinage P., Cornic P. and Degauque P., 5th Symp. on E.M.C., Zurich 1983, Proceedings pp 297-302

/58/ Demoulin B., Duvinage P. and Degauque P., Annales des Télécom., vol 39, n° 9-10, 1984, pp 433-445

/59/ Castenot A. and Jecko B., Annales des Télécom., vol 39, n° 9-10, 1984, pp 447-456

/60/ Heddebaut M., Degauque P. and Demoulin B., Annales des Télécom., vol 39, n° 9-10, 1984, pp 457-464

/61/ J. Ménez, B. Djebari, A. Zeddam, A. Caron, 5th Int. Symp. on E.M.C., Zurich 1983, Proceedings pp 185-190.

Nonlinear and Environmental Electromagnetics, edited by H. Kikuchi
Elsevier Science Publishers B.V., Amsterdam, 1985 – Printed in The Netherlands

VESTIGIAL RADIATION FROM INDUSTRIAL, SCIENTIFIC, AND MEDICAL RADIO-FREQUENCY EQUIPMENT

Ryszard G. Strużak
International Radio Consultative Committee CCIR
B.P. 820, CH-1211 Geneva 20, Switzerland

Abstract. Unintentional vestigial radiation from industrial, scientific, and medical radio-frequency equipment may cause interference to radio communications. The paper discusses the main aspects of this problem, and gives relevant data. It is addressed to all those interested in environmental electromagnetics, electromagnetic compatibility, and radio spectrum use.

1. INTRODUCTION

The electromagnetic environment consists of natural and man-made components: active ones, able to generate and radiate electromagnetic energy, and passive ones, absorbing that energy and transforming it into another form. Radio and TV transmitters are good examples of man-made radio frequency (RF) generators that intentionally produce electromagnetic emissions carrying inteligence for telecommunication purposes.

There are also other man-made radio frequency generators, used for non-communication purposes, in industry (I), science (S), medicine (M), and elsewhere. Generally, they are intended to transport and concentrate radio frequency energy in a restricted working area to produce physical, chemical and/or biological effects; for example to heat a product, to cause mechanical vibrations, or to ionise a gas. In technical jargon such applications are known as "ISM" applications. ISM or non-telecommunication applications of radio waves have been developed into a highly useful and reliable tool. Many modern processes have been improved, or even made possible, by ingenious applications of radio waves. ISM radio frequency generators offer immediate readiness to operation, easy and accurate dosage of the energy, rapid control and regulation, easy automation, small dimensions, high efficiency, clean and safe working conditions. With all those benefits, they became more and more popular. Table 1 gives a non-exhaustive list of possible non-telecommunication applications of radio waves.

Due to unavoidable imperfections in construction, production, and exploitation of ISM equipment, and due to the laws of nature, there is always an unintentional leakage of RF energy from this equipment. As a result, each

Table 1. **ISM APPLICATIONS** (Sample)

INDUSTRIAL
BUILDING OPERATIONS (CONCRETE DRYING & CRUSHING)
CHEMICAL INDUSTRY
CERAMIC AND GLASS INDUSTRIES
ELECTRONIC INDUSTRY
FOOD INDUSTRY
METALLURGICAL INDUSTRY
MINING INDUSTRY (CRUSHING ROCK)
PAPER INDUSTRY
PLASMA TECHNOLOGY APPLICATIONS
RUBBER INDUSTRY
TEXTILE INDUSTRY
WOOD INDUSTRY
SCIENTIFIC
ENERGY TRANSFER EXPERIMENTS
PARTICLE ACCELERATORS AND SEPARATORS
THERMONUCLEAR FUSION EXPERIMENTS
MEDICAL
MEDICAL TREATMENT (BONE FRACTURES TREATMENT, CANCER TREATMENT BY RF-INDUCED HYPERTHERMIA, MEDICAL DIATHERMY)
RF SURGICAL APPARATA
NUCLEAR RESONANCE IMAGING
OTHERS
DOMESTIC INDUCTION AND MICROWAVE OVENS
ROAD MAINTENANCE (ASPHALT MELTING)

ISM generator acts like an unintended radio transmitter producing signals with no intelligence.

The ISM generators cover a wide range of frequencies and powers. Frequencies may be as low as 50 Hz or as high as many GHz, and powers generated range from a hundred watts up to tens megawatts (Table 3). Even with a very small vestigial leakage, the amount of energy radiated into space may be sufficient to interfere with other systems using the same, adjacent, harmonic, or even sub-harmonic frequency bands. In particular, radiocommunication systems that use radio-frequency signals of very low energy (especially in the radioastronomy and in some of the Earth-exploration services) are extremaly vulnerable to interference from ISM generators.

With random distributions of ISM generators and victim receivers, other factors being constant, the number of interference cases should increase with increasing number of receivers and generators. With a great number of radiating generators the threat may be serious. At the present time, the total amount of ISM installations is estimated as 120 million or more on a world scale, see Tables 2a and 2b. It is estimated that the population of ISM generators constantly increases with a rate of 3% to 7% per year. With such a growth, the number of ISM generators expected for the year 2000 would be 2 to 4 times greater than now. Their undesired radiation, in connection with their number, and with their growth rate, cause significant problems for national and international radiocommunications.

The paper introduces the main problems associated with undesired vestigial radiation from ISM equipment and discusses potential interference to radiocommunications; possible interactions with other man-made systems, the biosphere, and the inanimate nature, are beyond its scope. Section 2 gives a short overview of ISM generators and some of their applications. Their radiation at the operating frequency is discussed in Section 3. Various mechanisms are outlined, and available measurement data are compiled here. Section 4 is devoted to radiation at spurious frequencies, that is at frequencies other than the operating frequency. A simple model is introduced here to explain some measurement results. The model may also be used for prediction purposes. A compilation of pertinent measurement data available is included. Propagation aspects are discussed in Section 5. Section 6 outlines some interference mechanisms possible. Known effects of ISM interference, including statistical data, are presented in Section 7. Section 8 discusses the problem of limitation of ISM radiation. The bibliography contains some sixty papers and books. International recommendations concerning the operating frequencies and radiation limits are given in the appendices.

2. ISM GENERATORS

All known features of radio waves are now employed for non-telecommunication purposes and there is a great variety of ISM generators, differing in their application, size, power, and frequency, see Tables 2a, 2b and 3. For instance, one can find a micro-metallurgical, 100 Watt, 0.8 MHz, induction generator, designated to melt small metallic elements used in dentistry . Its volume is about 0.03 cubic meter, it consists of one pulse thyratron, and it weighs about 10 kG. At the other extreme one can mention here the Solar Power Satellite System, projected (few years ago) for production of electrical energy at the geostationary satellite orbit and for

Table 2a. **ESTIMATED TOTAL NUMBER OF ISM EQUIPMENTS INSTALLED**

Country	Number of equipments installed per capita	per 1 km2	total	Source
Denmark	0.0008	0.1	4 000	[1]
Finland	0.002	0.03	10 000	[2]
Germany F.R.	0.004	1	250 000	[3]
Japan	0.1	30	12 000 000	[4]
United Kingdom	0.01	2	580 000	[5]
U.S.A.	0.07	2	15 000 000	[6]
	* * * * *			
World total	**0.02**	**0.9**	**120 000 000**	[7]

Adapted from the following sources:

[1] Doc.CCIR IWP 1/4-20 (Larsen/DK) August 1981
[2] Doc.CCIR IWP 1/4-8 (Nikkila/SF) April 1981
[3] Doc.CCIR IWP 1/4-16 (Lehning/FRG) June 1981
[4] Doc.CCIR IWP 1/4-13 (Arai/J) May 1981
[5] Doc.CCIR IWP 1/4-17 (Whitehouse/UK) June 1981
[6] Doc.CCIR IWP 1/4-11 (Wall/USA) May 1981
[7] Doc.CCIR IWP 1/4-18 (Spash/UIE) June 1981

Note: The distribution of ISM equipment is highly non-uniform.

Table 2b. **ESTIMATED NUMBER AND POWER OF ISM EQUIPMENT**

Application	Percentage of total number	power
Domestic	78 %	12 %
Industrial & Scientific	20 %	87 %
Medical	2 %	1 %
* * * * *		
Total number	100 % (580 000 units)	
Total power		100 % (2.5 GW)

Note: Data for a developed country with the G.N.P. of ca 9000 US$ per capita (U.K.)

Adapted from Doc.CCIR IWP 1/4-17 (Whitehouse/UK) June 1981

transmission of it to the Earth. Its space segment would handle 5.000.000.000 Watt at 2.45 GHz, would weigh about 50.000.000 kG, and would use 100.000 klystron tubes. It would be 10 km long, 5 km wide, and 0.5 km deep. Its earth segment would be elliptical in shape, roughly 13 km long and 10 km wide, and would consist of 10 billion half-wave dipoles and rectifiers (rectennas).

Despite the differences, in all cases the ISM equipment consists of four basic elements:

power supply unit,

RF generator/amplifier unit

RF applicator/antenna

auxiliary and control circuitry.

Electrical DC energy supplied is converted into radio-frequency energy in the generator/amplifier, and then injected by the applicator into the specimen (object of application). Here the energy of the radio waves directly interacts with matter, producing physical, chemical, and/or biological effects. The spatial distribution of energy in the specimen depends on the radio-frequency and on the applicator used, as well as on electromagnetic characteristics of the matter. This creates the possibility to control the energy concentration (and thus the physical, chemical, and/or biological effects) within the work piece, in a way which is not achievable by other methods.

A good example here is the RF cancer treatment in which the energy is concentrated into the affected organ of the body. Another example is the surface hardening of metals. Here, due to the skin effect, only a thin surface layer is processed and the metallic core remains elastic, without internal stresses.

Most ISM applications involve the production of heat. Radio-frequency heating receives a preference over other methods as it provides direct heating without external heat sources, and with minimum heat transfer. This guarantees minimum energy loss, minimum thermal inertia, and minimum heating time.

In electrically conducting materials, the electromagnetic field is induced by means of a special applicator-coil with radio-frequency current. In heating technology this is known as induction heating. RF currents of the order of hundreds of Amperes or more in the coil, and field-strengths of thousands of Amperes per meter in the specimen, are not uncommon. The depth of penetration of the field in conducting materials is a function of the material and decreases with the frequency. Thus, for production of heat at greater depth, lower frequencies must be used, and conversely, for heating only a thin layer of the work piece, higher frequencies are appropriate. Frequencies below about 10 kHz are, therefore, used for heating large work pieces. For smaller ones, or for partial heating of large parts, as for example surface hardening, higher frequencies are used, typically 300 kHz to 5 MHz. The unit power of induction heaters is about 1 kW to 700 kW or more, dependent on the mass of the specimen (Table 3).

Table 3. **TYPICAL FREQUENCIES AND POWERS USED CURRENTLY FOR ISM APPLICATIONS**

Nominal frequency, frequency variations, nominal power	Typical uses
Below 9 kHz 10 % up to 20 MW	Metallurgical industry: induction heating hardening, soldering, melting, refining. Medical applications.
9 kHz - 1 MHz 10 % 100 W - 1 MW	Ultrasonic cleaning and welding. Electro-erosion treatment. Induction heating and melting of metals. Power inverters. Medical applications.
1 MHz - 5 MHz 10 % 100 W - 100 kW	Induction heating. RF excited arc welders Sealing for packaging. Production of semiconductor materials. Medical applications.
5 MHz - 20 MHz 2,5 % 100 W - 300 kW	Dielectric heating. Wood drying and glueing. Plasma heating. Medical applications.
20 MHz - 100 MHz 1 % 100 W - 300 kW	Dielectric heating. Plastic welding Food processing. Plasma heating. Particle separation. Medical applications.
100 MHz - 500 MHz 0.5 % 100 W - 1 MW	Food processing. Medical applications. Particle accelerators.
500 MHz - 3 GHz 100 W - 100 kW	Food processing. Insect desinfestation Medical applications. Materials manufacture. Plasma heating. Particle acceleration.
Above 3 GHz up to 15 GW	Particle acceleration/separation. Plasma heating. Thermonuclear fusion experiments. Microwave-powered vehicles (project). Free-space power transmission by microwave beam (project). Satellite Solar Power Station (project)

In non-conducting, or poorly-conducting materials, the electromagnetic field is produced by special electrodes, essentially in the form of capacitor plates. Field-strengths of the order of tens kV/m or more in the specimen are not uncommon here. In heating technology this method is known as dielectric heating. It offers nearly uniform electromagnetic field distribution, and thus nearly uniform heating of the whole volume of the specimen. Note that most electrically non-conducting materials are also poor thermal conductors, and they cannot be uniformly heated by external sources. Frequencies and powers employed are determined by the loss characteristics of the material, required temperature increase, heating time, and the size of the work piece to be heated. Typical applications include food processing, wood glueing, furniture assembly, and plywood moulding. Typical frequencies are between 1 MHz and 100 MHz, and the power varies from about 100 watts to 100 kW or more (Table 3).

At higher frequencies, and especially at microwaves, the coil-type and capacitor-type applicators are replaced by (loaded) cavity resonators, waveguides, and special antenna systems tailored to a specific application. The most popular applications include microwave domestic ovens and physical instruments using interaction between radio waves and elementary particles, for instance particle accelerators and separators.

Another example can be found in Nuclear Magnetic Resonance (NMR) Imaging. Here short radio frequency pulses are used to stimulate atomic nuclei placed in a magnetic field. Then the nuclei re-emit some of the absorbed energy in the form of radio signals, carrying significant information about the nuclei. Operating frequencies are of the order of 10 MHz. NMR Spectroscopy is now a very powerful tool used for elucidating the structure of chemicals as well as the nature of static and dynamic interactions between nuclei in a wide variety of materials and biological systems (Pykett et al., 1982). RF generators are also suitable for plasma heating at the ion-cyclotron resonance frequencies (20 to 80 MHz), electron-cyclotron resonance frequencies (30 to 150 MHz) and the lower (1 to 4 GHz) hybrid resonance (Menes, 1985).

Various methods are used to generate the required power, the most common being alternators at low frequencies, and semiconductor, thyristor, as well as vacuum-tube-oscillators at the higher frequencies. Primitive equipment may be quite simple and may consist of a power-line transformer, rectifier, and oscillator with a suitable circuit to couple the radio-frequency energy to the workpiece via an applicator. At microwaves magnetron and klystron oscillators are used in both pulsed and continuous-wave modes. Several oscillators may be combined to produce the required RF power. In addition to the Solar Power Satellite System, we can mention here, for instance, the Stanford Linear Electron Accelerator which uses 200 pulsed klystrons distributed over a distance of about 3 km, each delivering peak power of more than 30 megawatts.

3. RADIATION AT THE OPERATING (FUNDAMENTAL) FREQUENCY

There is a number of separate mechanisms by which radio frequency power is leaked from ISM equipment:

a. radiation from the applicator and associated leads carrying radio frequency current,

b. radiation from the oscillator body/cabinet, along which vestigial RF surface current flows,

c. radiation from surrounding structures in which radio frequency currents are induced.

In general, the data published on the radiation from ISM equipment at the operating frequency is scarce.

3.1. Radiation from applicator

The amount of energy radiated from the applicator and associated leads depends on the particular arrangement of the devices and the work piece, which together act like an antenna. Because the system is not designed to serve as an intended antenna, its radiation efficiency is usually very low. Most often the dimensions of the applicator constitute only a small fraction of the wavelength at the operating frequency, and the radiation source may be approximated by a small loop and/or small dipole, known as elementary radiators in antenna theory. Although the efficiency of such small antennas is very low, the amount of radiated power may be considerable, if the nominal power is high. The radiation resistances of the elementary loop and dipole antennas, describing their radiation ability, are given in Table 5. For instance, the radiation resistance of the dipole equalling only 0.5% of the wavelength, amounts to about 0,02 Ohm. With 100 Amper RF current it gives 200 Watt of radiated power. The same radiation resistance (and the same power) one obtains, for example, with a loop antenna of 2% of the wavelength diameter and 2,5 turns (with the same RF current).

3.2. Radiation from cabinet and surrounding structures

Stray radio frequency fields in the vicinity of the equipment are associated with currents flowing over the surface of the body/cabinet and over the surrounding structures. The equipment acts as a complex antenna system consisting of coupled radiating surface elements, resonating at some unspecified frequencies.

3.3. Radiation from the power supply leads

For economic reasons some manufacturers do not include filtering in the mains and control leads. Often all power and control wires go close to radio frequency power circuits with no shielding. As a result a considerable amount of radio frequency energy is fed into these leads and is conducted outward at a distance and then reradiated. No detailed data are available concerning that component of ISM radiation, which may, nevertheless, be significant.

3.4. Directive radiation pattern

Radio waves, radiated by various elements of the system, coincide in phase in some directions and may be in opposite phases in other directions. As a result the directive radiation pattern may be very irregular in shape, with sharp maxima and deep minima, depending upon the frequency and specific situation of conductors in the vicinity of the generator. In practice it is not possible to take into account all details and usually it is assumed that the generator acts like an isotropic point-source of energy, or as an elementary antenna mentioned above.

3.5. Measurement results

Practical results of measurements of some ISM installations are compiled in Table 4. As it can be seen, the mean value of EIRP at the operating frequency may reach about 10 Watt, the standard deviation being 2 dB to 16 dB. If the normal distribution can be assumed (assumption not proved!), the levels exceeded with 10% probability can be determined: they are about -40 dB or 0.01% (induction heaters), and ca. -20 dB or 1% (medical diathermy and microwave ovens) in relation to the nominal power level. These data concern the low- and medium power installations. Estimations by other authors give the average value of EIRP as 13 W to 200 nW (dependent on the distance and formulas used to derive the power from the field strength measurements) and the standard deviation of 30 dB, with no mention of its relation to the nominal power (Harry and Hasler, 1980). It must be noted, however, that no correlation has been found between the nominal power and radiated power for the general population of ISM equipment.

Data on directive radiation patterns of ISM generators, especially on radiation at high elevation angles (relevant to aeronautical and satellite communications) are very scarce. In the horizontal plane, differences between maximum and minimum radiation may be as high as 20 dB (this is the average value from measurements of 36 inductive heaters, each from 4 directions: North, South, East and West: standard deviation is 12 dB; source: Doc. CCIR IWP 1/4(Whitehouse/UK) 41, Feb. 1983).

4. RADIATION AT SPURIOUS FREQUENCIES

The common feature of all ISM generators is a rich frequency spectrum of radiation. In addition to the fundamental frequency, spurious frequencies are also observed. The spurious radiation components include:

harmonics, occurring at integer multiples of the operating frequency,

local oscillator/synthesizer leakage (if any) through the circuitry,

parasitic oscillations, occurring mainly in output RF power stages. (Note that parasitic oscillations are not harmonically related to the operating frequency.)

intermodulation, spurious modulation products, and noise.

Table 4. **POWER RADIATED AT FUNDAMENTAL FREQUENCY: MEASUREMENT RESULTS**

APPLICATION FREQUENCY POWER	MEAN IN DB 1 W.	E.I.R.P. RELATIVE NOM.POW	STAND. DEV. (DB)	TOTAL NO.	SOURCE
INDUCTION HEATER 0.2 to 1.6 MHZ 8 to 45 kW	-17	-62	16	39	[1,2]
DIELECTRIC HEATER 27 MHZ	2	--	12	20	[1]
MEDICAL DIATHERMY 27 MHZ 1 KW	0	-30	9	2	[2,3]
MICROWAVE OVEN 0.45 GHZ 0.5 to 2 KW	11	-20	2	3	[4]

NOTE: Equivalent Isotropically Radiated Power (EIRP) computed from the field-strength (FS) and distance (D) measurements, using free-space propagation model: EIRP=(FS*D)**2/30

Adapted from the following sources:

[1] Doc. CISPR/B/WG1(Whitehouse/UK)8, Febr. 1983

[2] Skomal E.; Man-Made Radio Noise; pp. 135-172, Van Nostrand Reinhold Co., New York, 1978

[3] Sowton., Britton G.; Radio Interference P.5: Industrial, Scientific and Medical Apparatus and Radiating Receivers; The Post Office El. Eng. Journal, Vol. 51, p. 3, Oct. 1958, pp. 202-205

[4] Interference to Radioastronomy from Microwave Ovens Operating in the 2450 MHz Band - Rep. 854 Recommendations and Reports of the CCIR, 1982, Vol.II, pp. 546-548

4.1 Device nonlinearity effects

The major cause of harmonic radiation from ISM equipment is the waveform distortion in radio-frequency output power stages. Their active devices are non-linear, often operating in pulse conditions, like switches, in order to achieve the maximum efficiency. Ideally, in the "switch-on", and "switch-off", states no power would be absorbed in the switching device. With the zero- transit-time from the "on" state to "off" state, the power loss in the transit state would be nil too, and the energy efficiency would reach 100%.

Unfortunately, the price to be paid for the high efficiency is that the radio frequency current contains many harmonics. If the current has a form of very short pulses of constant amplitude (switching mode of operation), its frequency spectrum contains a number of harmonics with amplitude independent of the frequency. In more realistic cases (Raab, 1977; Sokal and Raab, 1977), harmonics decrease as N**B, N being the harmonic number and exponent B equalling (-1), (-2), or (-3), depending on the circuitry.

In all cases, even in the continuous mode of operation, and including magnetrons and klystrons at microwaves, the radio frequency current in the active device which is responsible for producing the power output is highly non-sinusoidal, and the amplitudes of the fundamental, second, third, etc., harmonics are comparable to the dc component. The level of harmonic power depends also upon the radio frequency impedance of the output circuits and the radiating elements at these harmonics, as well as upon the characteristics of filters that may be included to reduce unwanted radiation. In the literature one can find estimates of the typical levels of the second to the fifth harmonics (relative to the fundamental), for the communication transmitter output stages. They are -32 to -45 dB, -38 to -50 dB, -52 to -62 dB, and -58 to -70 dB, respectively (Thompson, 1981). These values may be assumed as representing the state-of-art in the communication technology; in non- communication applications rather more pessimistic values should be expected.

4.2. Load nonlinearity effects

Other spurious radiation sources are the nonlinearities in the load. Nonlinear load enhances the power spectrum radiated. Data are scarce on this issue. An increase of two orders of magnitude (20 to 40 dB, 4-th to 10-th harmonics) in the power radiated at specific harmonics was observed in practice when a resistive (linear) load was replaced by a plasma load (West, 1983).

4.3. Surrounding nonlinearities

Harmonics and intermodulation products are generated also by layers of oxide and other surface corrosion products that are exposed to high radio frequency fields. Oxide layers existing on conducting parts of the structure transform metal-metal junctions into metal-insulator-metal or metal-semiconductor-metal junctions which have non-linear characteristics. Radio frequency fields induce RF currents in metallic parts. The currents flow through the junctions and the non-linear characteristics cause harmonic and intermodulation products to be generated and re-radiated by the structure. Metal-to-metal junctions are common intermodulation product sources, especially when between dissimilar metals and when exposed to the atmosphere.

This interference mechanism was observed in practice with 50 W power at frequencies 100 MHz - 8 GHz, and with 500 W power in HF band. More than 30 harmonics generated by that mechanism were observed in communication applications (Watson, 1983). No data are available specifically concerning ISM equipment.

4.4. Spurious resonances

The components of the output circuits, and matching and filtering networks (if any), as well as coupled surounding structures, can combine to produce resonant conditions at specific frequencies. This may easily occur because of the physical size of the structures and components used in high-power circuits. If this occurs, the radiation at these frequencies may be accentuated to a high level. The same mechanism may be responsible for extra absorption of the radiation at anti-resonance frequencies. As a result the harmonic radiation may be highly irregular, however, no detailed data have been published on these effects.

4.5. The simplest model

In order to understand the process better, we assume here, for simplicity, that the ISM installation can be replaced by the (equivalent) elementary radiators (small loop or small dipole) and that the output spectrum can be approximated by the expression shown in Table 5a. The harmonic content of RF current is characterized here by two coefficients: A and B, where coefficient A represents an attenuation of harmonics independent of the frequency, and exponent B shows the frequency dependence of the harmonic level. The spectrum is changed by the radiation abilities of the radiator (due to the law of nature higher components are usually radiated more easily than lower ones): Table 5b gives numerical examples.

The model shows that the amount of power radiated at higher harmonics may be greater than that radiated at lower ones, including the fundamental frequency. This effect may occur if $B>-2$ for loop mode, and $B>-1$ for dipole mode radiation. Note that with $B=-2$ (loop) or $B=-1$ (dipole), the RF power radiated does not depend on frequency, but if $B<-2$ (loop) or $B<-1$ (dipole), the amount of the power radiated diminishes with frequency. These conclusions are restricted to the frequency range for which the radiating antenna is still small in comparison with the wavelength. The spurious resonances and anti- resonances may be taken into account, too, if the necessary data are available. The practical significance of such considerations seems, however, doubtful.

4.6. Measurement results

Table 6 compiles available measurement results of the harmonic radiation from ISM installations, separately for dielectric heaters, induction heaters and medical diathermy. Note that with the average value lower than the standard deviation multiplied by factor three, the EIRP observed at harmonics may be indeed greater than the power radiated at the fundamental frequency, as predicted theoretically in section 4.5. Note also that on average, power radiated at harmonics does not necessarily decrease with the frequency but may be rather constant within a very wide frequency band. Such a possibility has also been predicted by our simple model (Table 5a and 5b).

Table 5a. **POWER AND FIELD-STRENGTH RADIATED: LOOP MODE AND DIPOLE MODE**

No		Expression
1.	FREQUENCY [HZ]	
	FUNDAMENTAL	F(1)
	N-TH HARMONIC	F(1)*N
2.	RF CURRENT [A]	
	FUNDAMENTAL FREQUENCY:	I(1)
	N-TH HARMONIC (N>1):	I(N) = I(1)*A*N**B
3.	RADIATION RESISTANCE - LOOP MODE [OHM]	
	FUNDAMENTAL FREQUENCY	R(1) = 19000*T**2*(D/L)**4
	N-TH HARMONIC	R(N) = R(1)*N**4
4.	POWER RADIATED - LOOP MODE [W]	
	FUNDAMENTAL FREQ.	P(1) = I(1)**2*R(1)
	N-TH HARMONIC	P(N) = P(1)*[I(N)/I(1)]**2*[R(N)/R(1)]
		P(N) = P(1)*A**2*N**(4+2*B)
5.	TOTAL POWER RADIATED - LOOP MODE [W]	
		PT = P(1)+P(2)+...+P(N)+..
		PT = P(1)*{1+A**2*[2**(4+2B)+3**(4+2*B)+...+N**(4+2*B)+...]}
6.	RADIATION RESISTANCE - DIPOLE MODE [OHM]	
	FUNDAMENTAL FREQUENCY	R(1) = 80*PI**2*(D/L)**2
	N-TH HARMONIC	R(N) = R(1)*N**2
7.	POWER RADIATED - DIPOLE MODE [W]	
	FUNDAMENTAL FREQ.	P(1) = I(1)**2*R(1)
	N-TH HARMONIC	P(N) = P(1)*[I(N)/I(1)]**2*[R(N)/R(1)]
		P(N) = P(1)*A**2*N**(2+2*B)
8.	TOTAL POWER RADIATED - DIPOLE MODE [W]	
		PT = P(1)+P(2)+...+P(N)+..
		PT = P(1)*{1+A**2*[2**(2+2*B)+3**(2+2*B)+...+N**(2+2*B)+...]}
9.	FREE-SPACE FIELD-STRENGTH E [V/M] AT DISTANCE R [M]	
	FUNDAMENTAL FREQ.	E(1) = SQR[30*P(1)]/R
	N-TH HARMONIC	E(N) = E(1)*SQR[P(N)/P(1)]
	LOOP	E(N) = E(1)*A*N**(2+B)
	DIPOLE	E(N)'= E(1)*A*N**(1+B)
10.	HARMONIC CONTENT:	H = SQR{[I(2)**2+I(3)**2+...+I(N)**2+...]}/I(1)
	CURRENT	H(C) = A*SQR{2**(2*B)+3**(2*B)+...+N**(2*B)+...}
	FIELD-STRENGTH, LOOP	H(L) = A*SQR{2**(4+2*B)+...+N**(4+2*B)+...}
	FIELD-STRENGTH, DIP.	H(D) = A*SQR{2**(2+2*B)+...+N**(2+2*B)+...}

KEY: A, B - COEFFICIENTS , A<1, B<0
D - DIAMETER OF THE LOOP OR LENGTH OF THE DIPOLE [M]
L - WAVELENGTH [M]
T - NUMBER OF TURNS

Table 5b. **NUMERICAL EXAMPLE: POWER RADIATED AT HARMONICS, RELATED TO POWER RADIATED AT FUNDAMENTAL FREQUENCY, in dB. (LOOP AND DIPOLE MODES, DIFFERENT RF-CURRENT HARMONIC STRUCTURES)**

HARMONIC NUMBER N	B = 0		B = -1		B = -2		B = -3	
	LOOP	DIPOLE	LOOP	DIPOLE	LOOP	DIPOLE	LOOP	DIPOLE
A = 0.1								
1	0	0	0	0	0	0	0	0
2	-8	-14	-14	-20	-20	-26	-26	-32
3	-1	-10	-10	-20	-20	-30	-30	-39
4	+4	-8	-8	-20	-20	-32	-32	-44
5	+8	-6	-6	-20	-20	-34	-34	-48
6	+11	-4	-4	-20	-20	-36	-36	-51
7	+14	-3	-3	-20	-20	-37	-37	-54
8	+16	-2	-2	-20	-20	-38	-38	-56
9	+18	-1	-1	-20	-20	-39	-39	-58
10	+20	0	0	-20	-20	-40	-40	-60
A = 0.01								
1	0	0	0	0	0	0	0	0
2	-28	-34	-34	-40	-40	-46	-46	-52
3	-21	-30	-30	-40	-40	-50	-50	-59
4	-16	-28	-28	-40	-40	-52	-52	-64
5	-12	-26	-26	-40	-40	-54	-54	-68
6	-9	-24	-24	-40	-40	-56	-56	-71
7	-6	-23	-23	-40	-40	-57	-57	-74
8	-4	-22	-22	-40	-40	-58	-58	-76
9	-2	-21	-21	-40	-40	-59	-59	-78
10	0	-20	-20	-40	-40	-60	-60	-80
A = 0.001								
1	0	0	0	0	0	0	0	0
2	-48	-54	-54	-60	-60	-66	-66	-72
3	-41	-50	-50	-60	-60	-70	-70	-79
4	-36	-48	-48	-60	-60	-72	-72	-84
5	-32	-46	-46	-60	-60	-74	-74	-88
6	-29	-44	-44	-60	-60	-76	-76	-91
7	-26	-43	-43	-60	-60	-77	-77	-94
8	-24	-42	-42	-60	-60	-78	-78	-96
9	-22	-41	-41	-60	-60	-79	-79	-98
10	-20	-40	-40	-60	-60	-80	-80	-100

Expressions

10*log[P(N)/P(1)] = 20*log(A) + 20*(2+B)*log(N) {Loop}
= 20*log(A) + 20*(1+B)*log(N) {Dipole}

Table 6. **POWER RADIATED AT HARMONICS: MEASUREMENT RESULTS**

HARMONIC NUMBER	MEAN E.I.R.P. IN DB RELATIVE TO POWER RADIATED AT FUNDAMENTAL FREQUENCY		
	DIELECTRIC HEATERS	INDUCTIVE HEATERS	MEDICAL DIATHERMY
1	R E F E R E N C E	L E V E L	(0 dB)
2	-18	-56	-43
3	-19	-62	-40
4	-19	-48	-38
5	-26	-49	-46
6	-19	-44	-43
7	-20	-38	*
8	-27	-45	*
9	-26	-50	*
10	-19	-58	*
11	-21	-78	*
12	-25	-83	*
13	-20	-82	*
14	-20	-78	*
AVERAGE	-21	-60	-42

Notes:

1. Equivalent Isotropically Radiated Power (EIRP) computed from the field-strength (FS) and distance (D) measurements using free-space propagation model: EIRP=(FS*D) 2/30

2. Dielectric heaters: 13 to 80 MHz, 1.5 to 25 kW, about 420 installations. Standard deviation <30dB. Results adapted from Skomal 1978 and Czyzewski 1984.

3. Induction heaters: 0.35 to 0.5 MHz, 20 to 40 kW, 3 installations. Standard deviation ca. 20 dB. Results adapted from Skomal, 1978.

4. Diathermy apparata: 13 to 36 MHz, 0.1 to 1 kW, 2 installations. Standard deviation ca. 25 dB. Results adapted from Sowton 1958 and Skomal 1978.

* = no data available.

The number of significant harmonics may be very great in ISM generators. The highest harmonic number found in publications available was 582. This was an induction heater, 30 kW, 378 kHz, and the highest harmonic was measured at 220 MHz. Between the fundamental frequency and its 282-nd harmonic at 220 MHz there were 47 other harmonics with significant power. The field strength at about 10m distance measured at the highest harmonic was 40 uV/m (equivalent EIRP about 4 nW), 48 dB below the level of the second harmonic (Skomal, 1978). Thus, an ISM equipment acts as a number of unintentional radio transmitters, each radiating at a separate frequency.

No data are available on directive radiation pattern at spurious frequencies.

4.7. Noise and modulation spectrum

Due to economy reasons, ISM equipment is usually designed for minimum cost, and typically is reduced to the bare essentials for operation. Frequency stability and spectral purity of the power delivered to the load are not the main objectives. In almost every case the work piece is strongly coupled to the oscillator/amplifier, and since the electromagnetic characteristics of the material change during the work cycle, the magnitude, phase, and frequency, of the radiation, all may be unintentionally modulated due to these changes. Note that if the working current has a varying amplitude, phase and/or frequency, the harmonics also show similar variations.

Pulse-type operation of ISM equipment results directly in the pulse modulation of the field radiated. A modulation may also result from poor filtration of the power supply. In multiple-device installations, part of the output of one device may be coupled inadvertently into the other one and cause modulation, too. Therefore a great level of unintentional modulation, as well as noise from output devices may be expected from ISM equipment, much higher than in communication transmitters. Due to these mechanisms a significant portion of RF power may be radiated also in the frequency bands adjacent to the operational frequency of any ISM generator, as well as adjacent to its harmonics.

5. PROPAGATION ASPECTS

It is well-known that the power density of any radiation diminishes with the distance from its source, and different propagation models are used to explain the propagation phenomena or for prediction purposes. Often it is assumed that the ISM radiation can cause interference to radiocommunications only within limited and short distance from the ISM installation, of the order of 300 m (Europe), or 1 mile (about 1,8 km, U.S.A.). This is probably a result of mixing-up the measuring distances defined for standardisation purposes, with non-interfering distances that depend on several other factors in addition to the ISM radiating power.

As known, the service coverage by transmitting stations is via normal or standard modes of propagation. These modes are highly stable and assure required reliability of the service during, for example, 99% of the time. However, there are also non-standard, or anomalous, modes of propagation possible. These modes are often more efficient than, but not as reliable as, the standard modes, propagating beyond the normal coverage area. The anomalous modes are not sufficiently stable to provide reliable service fields, except perhaps very special applications, but may be responsible for long-distance interference from ISM equipment, during, for instance, 1% of the time. The occurrence of the anomalous modes generally depends upon unusual conditions in the propagation media.

5.1. VLF, LF, and MF bands

At very low (VLF), low (LF), and medium (MF), frequencies (3kHz to 3MHz) normal service coverage is provided by the highly stable ground-wave propagation. However, at VLF, whistler-mode propagation can provide signals as far as to the opposite hemisphere. At LF and MF, during night-time, reflections from the ionosphere can support propagation over very long ranges, of the order of 2000 km, without great attenuation. At HF (3MHz to 30 MHz), in addition to the ground-wave and sky-wave mode of propagation via ionospheric reflections, scattering from ionospheric irregularities, meteor trails, and highly efficient guided or ducted modes along ionospheric layers are also possible.

5.2. VHF, UHF, and SHF bands

At very-high (VHF), ultra-high (UHF), and super-high (SHF) frequencies (30MHz to 30GHz) service coverage is provided by line-of-sight (short paths), diffraction (intermediate paths) or by forward scatter (long paths) propagation modes. From about 30 to 150 MHz the interference mode may be due to scattering from ionospheric irregularities, sporadic E-layers, and meteor trails to distances that can readily exceed 1000 km. From about 30 to 1000 MHz surface and/or elevated tropospheric ducts may play a significant role. Data from radio amateurs show that transmissions over the distance of 2000 km or more are possible at very low power levels due to anomalous tropospheric propagation (144 and 432 MHz frequency bands). At higher frequencies, transhorizon interference fields may occur by diffraction, by tropospheric forward scatter, by hydrometeor (rain, snow, hail, ice clouds) forward-, side-, or back-scattering, and by tropospheric ducting.

5.3. High elevation angles

The abovementioned remarks concern victim receivers situated on or near the Earth surface. If they are elevated, as for example in aeronautical communications, the free-space propagation model is valid. If the victim receiver is above the ionospheric layers (satellite and space communications), the free-space propagation model applies for the 10 MHz to 37,5 GHz frequency band. This band, known as the "radio-window" is limited at lower frequencies by the shielding action of the ionosphere and by atmospheric absorption at higher frequencies. In practice the free-space propagation model is perhaps the most often used to assess the small-time-percentage interference effects from ISM radiation, independently of the receiver situation and radiated frequency.

6. INTERFERENCE MECHANISM

Effects of interfering signals depend upon the susceptibility of receiving equipment and upon the relationships among the power levels, types, modulations, and bandwidths of useful and interfering signals at the receiver. (The susceptibility is the characteristic of an electronic receiving equipment that permits undesirable responses when subjected to electromagnetic energy).

The following three types of interference may be generally differentiated in the frequency domain, in descending order of importance:

a - fundamental or co-channel, when the legitimate receiver pass-band coincides with the ISM operating frequency

b - second-order, when one or more ISM spurious frequencies coincide with the legitimate receiver pass-band (b1), or inversely, when the ISM operating frequency coincides with one or more receiver spurious responses (b2)

c - third-order, when one or more spurious frequencies of the ISM equipment coincide with one or more spurious responses of the receiving installation.

If the interfering signal is strong enough to drive an active device in the receiver into a non-linear condition, serious disturbances can occur. An instance of this is when a receiver is overloaded by a signal outside its legitimate (nominal) pass-band with consequent loss or degradation of the desired input. Blocking may occur as a result of receiver-generated intermodulation, when two or more signals applied to a non-linear circuit element result in the generation of a number of sum-and-difference products. Cross-modulation, a special case of intermodulation, occurs when the desired signal is modulated by an undesired signal. All these effects are most likely to occur in receiver front-ends, but may happen at other points in the receiving equipment as well.

In the case of impulsive interference, much depends on the receiver time-response, which, in turn, is determined by the bandwidth and characteristics of the detector and automatic gain control circuits, if any. Receivers using narrow band filters with steep skirts have transient responses which persist for a long time under impulsive interference conditions, and are thus vulnerable to this type of disturbance.

Some relief from interference may be gained by improving the receiver immunity, that is by making it less susceptible. This may sometimes be done by the use of selective filters and/or a directive antenna at the receiver input. However, when the interfering power falls within the legitimate pass-band of the receiver, and arrives from the same direction as the useful signal, no remedy is available, except a time-coordination between the ISM and receiving installation activities.

7. POTENTIAL vs ACTUAL INTERFERENCE EFFECTS

ISM generators have a great interference potential indeed. However, as concerns real cases of interference, we can note here only limited information. One of the most recent and most spectacular is that in China, due to the vestigial radiation from ISM generators about 25% of the TV coverage area was lost (Chen and Zhu, 1985). This means that 25% of the total maintenance costs of TV studios and TV transmitting network was lost, too.

At least one specific case of long-distance interference has been published, in which a 15 kW ISM generator located in Sydney disturbed essential radio services in New Zealand, about 2000 km away (Mondel, 1970).

This supports the theoretical considerations and measuremet data reported above, but is in disagreement with informations from other sources. In 1980 this author prepared and disseminated a questionnaire among the experts involved in a special working party (IWP 1/4) of the International Radio Consultative Committee (CCIR). Replies to the questionnaire indicated that the number of verified complaints on ISM-generated interference was surprisingly low: in some countries in which such information was available only about 1 per 800 ISM installations was the source of verified interference complaint, and this constituted only about 1% of all interference complaints registered. No long-distance interference cases were reported to the party. It should be noted, however, that not all interference cases can be resolved in practice, and a part of them may remain unregistered, unclassified, or falsely classified. It is also well-known that the number of interfering radiations observed is steadily increasing, but their sources most often could not be positively identified.

With the proliferation of ISM applications, and with the growth of the number of susceptible receiving systems, a corresponding increase in the volume of interference complaints should be expected (assuming other factors constant). With the ISM growth rate 3% to 7% per annum, a similar growth rate of ISM interference complaints might be expected. Actually, however, the number of identified ISM complaints in the most industrialised countries shows just the opposite trend: according to the statistics reported to the CCIR IWP 1/4, the number of ISM interference complaints was diminishing. The discrepancy between potential and actual interference effects has not been explained and needs further study. One of the possible reasons is that the increase of the interference potential is compensated by a parallel preventive activities in these countries.

According to the data submitted to the party, the majority of interference cases observed in practice, is of type **b2** (Section 6), with harmonics of the ISM operating frequency falling into the legitimate pass-band of receivers located nearby. This is in line with the data presented in section 4. Among the other major causes of the interference, the replies to the abovementioned questionnaire indicate the inadequate installation, maintenance or use of ISM equipment, inappropriate location of the equipment versus victim receiver installations, insufficient immunity of receiving systems, excessive harmonic radiation in specific frequency bands, and insufficient stability of the operating frequency (Doc. CCIR 1/186, October

1981). This may suggest that the receiver immunity and ISM radiation levels, achievable in practice, are not very well matched to the ITU Radio Regulations, or vice versa.

8. ISM RADIATION LIMITATION

8.1. Spectrum sharing

The frequency spectrum is a limited resource with dimensions of frequency, time, and space. It is extensively used for medical care, public safety (including law enforcement), navigation, management, information transmisssion, industrial and scientific processes, entertainment, etc. The demand for the spectrum is steadily increasing, and more and more difficulties are encountered in meeting the requirements. It is, however, plain that the spectrum should be shared between as many users as possible, without one user causing harmful interference to another, and without placing undue restrictions or undue economic penalties on the shareholders. What constitutes acceptable or unacceptable interference, and necessary or undue restrictions, is often a matter of subjective assessment. For the radio user, factors to be considered in determining interference protection criteria may include the consequences of interference occurring (annoyance related to interference to broadcast reception as against catastrophe for interference to safety of life services). In general, the cost of an alternate technology and of alternative means of communication may need to be taken into account.

Radiation limitation is a price to be paid by radiating equipment for sharing the spectrum. At the other end, the price for the receiving equipment is the limitation of its immunity. Both limits, radiation and immunity must complement one another. The development of the limits requires a cooperation between all parties interested. Ultimately, the cost of ensuring electromagnetic compatibility between ISM installations and radio services will fall on society at large. Ideally, the radiation and immunity limits should be chosen to minimize the the total social costs involved (cost of radiation suppression and immunity limitation, costs of alternative technology/communication, costs of preventive and curative activities necessary).

In the saturation state, when the total demand exceeds the resource available, a problem of preference arises: which requirements must be satisfied and which ones not. In making any reasonable decision here, the concept of a value of the service to the society may be introduced. Then the service of greater value would be preferred over that of lower value. But all spectrum uses contribute to the well-being of humanity, and it is very difficult to judge their relative value or importance to the society.

8.2. Social value

There is no general consent yet as concerns the social value of ISM applications and radio services. Similarly, there is no agreement as concerns the required degree of reliability and required degree of protection of many radio services. For aeronautical services some directions may result from general safety performance criteria. For example, the air-traffic systems which ensure separation between aircrafts flying "en route" are required to reduce the chance of mid-air collisions to a level not greater than 2 in 100.000.000. Clearly, the probability of harmful interference due to ISM radiation should be less, if there are no other navigational aids. For comparison, the probability that any particular person in the United States will die within a year as a result of an accident due to any cause is about 1 in 2000.

Generally, it is difficult to assess the social value of any given service and the required degree of protection against radio interference, on the basis of our current knowledge. Radioastronomy may serve as an example. It bases on the detection of extremely weak signals of natural origin against a background of noise and interference. The magnitude of radioastronomic signals is of the order of that which would be received from a 0.00001W transmitter on the geostationary orbit. A time factor is also of importance. With an observation of lunar occultation, for instance, it might be necessary to wait up to 19 years before the Moon returns to the same position in the sky so that the observation can be repeated. Who knows the social value of such observations and can estimate the loss due to interference ?

8.3. Statistical aspects

ISM generated interference is dependent on various factors, many of which are difficult not only to analyse, but often to take into account. Some of them are uncontrollable and of a random character. The probability of interference depends upon the propagation characteristics, the radiation patterns of the receiving antenna and of the ISM installation involved, the susceptibility of the receiving system, the time coincidence of both, ISM equipment and receiving system activity, the radiated power and spectrum signature of ISM installation, the intensity and the modulation of the wanted signal.

Generally, two factors determine the degree of interference suffered by radio services. One is the response of the receiving equipment to the type of interference generated by a particular type of ISM apparatus, as discussed above. This determines the ability of that equipment to function normally while in the presence of that type of exposure. The other factor is the probability that the exposure will occur. This probability is a function of the distributions of the radio equipment and of the ISM apparatus, in space, time, and frequency, and depends upon the operational procedures involved.

No radiocommunication system can be guaranteed to be 100% reliable. Therefore, for each radio service it is necessary to determine and agree a level of probability of interference occurring from ISM apparatus and from other sources of radiation. The lower the probability specified the greater the cost to that radio service, and to those industries using ISM apparatus. Thus the agreed level of probability will depend upon the social value of the service, i.e. upon economic, human, social and political consequences of harmful interference occurring.

8.4. ITU Radio Regulations

The first attempts to solve the problem of electromagnatic compatibility with radio services on the international scale go back 50 years or so, when the International Radio Consultative Committee (CCIR) was formed to study and recommend technical standards that should be adopted by all countries to ensure the rational use of the radio spectrum and the elimination of harmful interference.

Undoubtedly, the most significant international achievement since that time is the Radio Regulations, issued by the International Telecommunication Union ITU, which now includes the CCIR. The Radio Regulations, among others, designated frequency bands for ISM applications (Annex I). By the allocation of specific frequency bands to specific uses, the regulations eliminate interference mechanism type **a** (Section 6). Indeed, if the ISM equipment and receiving installations are ideal, that is without spurious emissions and responses, no interference would be observed. Imperfections of technically realizable equipment cause interference mechanisms type **b** and **c** to exist even if the Radio Regulations are strictly followed.

The World Administrative Radio Conference, Geneva 1979, reconsidered this problem and asked the CCIR to study, in collaboration with the International Electrotechnical Committee (IEC) and International Special Committee on Radio Interference (CISPR), the radiation from ISM equipment (Kirby, 1981). Inter- national recommendations are sought on any limits that might be imposed on radiation from ISM equipment, both outside and inside the frequency bands designated for its use in the Radio Regulations, in order to ensure adequate protection against interference to radiocommunication services. A special CCIR Interim Working Party (IWP 1/4) was set up for this purpose in 1980, under the chairmanship of this author and the work began with the questionnaire already mentioned. (Since 1985 the chairman of the party is Mr G. Lehning, Federal Rep. Germany.)

The replies to the CCIR IWP 1/4 questionnaire show that in some countries only a part of the ISM equipment really use the ITU-designated frequency bands. It is estimated that only about 30% of the equipment presently operating outside of ISM-designated bands could theoretically be moved to one of these bands. The main reason is that there are only few such bands, and there are no ISM-designated bands for some applications that, for physical reasons, must use specific frequencies not foreseen in the Radio Regulations. The replies showed also that only a part of the equipment nominally operating on the designated frequencies really meets the stipulated frequency stability requirements (Doc. CCIR 1/186, October 1981).

8.5. "Natural" radiation limits

The radiation from ISM equipment is limited by the very nature of ISM applications: there are no ISM equipments designated specially to radiate electromagnetic energy into space. Any amount of the energy irradiated outside its working area is lost. For that reason the outward radiation from ISM generators is limited at the design stage voluntarily, without any official regulation. Some experts estimate that the power radiated at the fundamental frequency may reach up to 3% (i.e. -15 dB) of the nominal power of ISM generator, with additional 10 dB shielding/attenuation effect of surrounding structures (Doc. CCIR IWP 1/4-9 (De Jong/NL) May 1981). With 10 kW nominal power it gives 30 W of the power radiated, which is equivalent to 300 mV/m free-space field-strength at 100 m distance or 30 mV/m at 1000 m distance.

Note that these values are of the same order of magnitude as the measurement results, see Table 4.

8.6. "Safety" limits

Radio-frequency power density in the vicinity of ISM equipment may reach very high values. Health and safety considerations have resulted in regulations on the non-ionizing radiation hazard to be now in force in several countries. These regulations require the radiation from ISM equipment to be limited to specific values in the working place of the operator. The national hazard limits vary from country to country, and no common internationally agreed limits are known to date. One of them requires the power density to be less than 1 mW/cm2 (Lerner, 1980). Assuming an equivalent isotropic point- source of radiation and 1 m distance it gives about 125 W EIRP. These numbers, however, may be misleading at lower frequencies, at which there are near- field, rather than far- or radiation-field conditions.

8.7. CISPR limits

Limits internationally agreed and based on interference considerations are those given in CISPR Publication No. 11 (Annex 2). They limit radiation from ISM equipment only outside the frequency bands designated (before the WARC, Geneva 1979) for ISM applications in the ITU Radio Regulations. The limits are expressed in terms of the field strength at the specific distances from the equipment and spread from 30 μV/m at 30 m to 250 μV/m at 300 m, depending on the frequency band and working conditions (Doc. CCIR IWP 1/4-7 {Jackson/CISPR} Dec. 1980). Assuming a plane wave and isotropic radiation source this is roughly equivalent to 30 nW and 200 μW EIRP, respectively. The 30 nW limit is suggested, among others, for the protection of specific safety radio services in particular areas. These values are 60 dB to 100 dB below the abovementioned radiation levels resulting from the efficiency and health hazard considerations. Radiation within the frequency bands designated (before WARC 79) for ISM applications is unlimited, according to the CISPR publication. Some kinds of equipment (e.g. RF excited arc welding equipment, surgical apparatuses) are not covered by the publication. The CISPR limits have been reached as a compromise, after a long process of studies and discussions; they have often been considered as pragmatic manufacturer specifications.

Answers to the CCIR IWP 1/4 questionnaire indicated that in some countries the CISPR limits are actually observed in practice, whereas in others it is estimated that only a small percentage of the ISM equipment installed conforms to the limits. It appears that in the majority of countries such information is not available at all.

9. CONCLUSIONS

ISM applications and communication applications of radio waves are both indispensable for society, now and in the foreseable future. Unfortunately they both use the same limited spectrum resource, and this leads to conflict. Complex economic, technical, and legal issues are involved here. The system approach is applicable with difficulty because separate elements, i.e. communication and ISM equipment, are usually within the competence of different parties, often with opposing interests. Unless it is recognized that any progress here necessarily involves compromise, it will be impossible to arrive at a globally acceptable solution.

In this paper we presented technical problems generated by vestigial radiation from ISM equipment. We compiled available data, discussed possible mechanisms of leakage and interference, and developed a simple model for spurious radiation. The paper also indicates that there are problems unsolved yet and that our present knowledge is limited. Further studies, and a continuous effort is necessary to guarantee adequate protection of radiocommunications against harmful interference due to unintended RF radiation from ISM equipment.

10. ACKNOWLEDGEMENT

The author wishes to thank Mr. Richard C. Kirby, Director of the CCIR and Mr. Richard E. Butler, Secretary-General of the ITU, for publication permission, as well as Mr. Richard Nickelson and Dr. Kevin A. Hughes of the CCIR Secretariat for valuable comments. Thanks are also addressed to Ms. Linda Mostert for help in the preparation of the manuscript for publication. The article is based largely on the paper presented at the URSI meeting (Strużak, URSI, 1984), and prepared while the author was with the Institute of Telecommunications and Wrocław Technical University, Poland.

The author gratefully acknowledges numerous discussions with the members of CCIR Interim Working Party IWP 1/4, and their kind cooperation during his chairmanship of the party. However, as the paper involves subjective judgements on matters on which there are reasonable difference of opinion, the ackowledgment does not imply the endorsement of the work by any of these persons or by any organization.

BIBLIOGRAPHY

1958

Sowton C.W., Britton G.A.; Radio Interference P.5: ISM Apparatus and Radiating receivers; The Post Office Electrical Engineers' Journal, Vol. 51, p.3, October, pp.202-205;

1970

Mondel R.H.: Electromagnetic Interference Caused by Industrial Users of Radio Frequency Generators; Proc. IREE Austral., August, pp. 246 - 250

1972

Paolini E., Picciolini L.: Actual Attenuation of UHF Interferences from ISM Apparatus Over Test Sites and Over Real Grounds: Proc. EMC Sympos. Wroclaw, Sept. 14-16, pp.15-27;

1973

Rusakiewicz W.: International and Polish Regulations Concerning the Limitation of Radio Interference Caused by Industrial HF Equipment - A Comparison (in Polish); Metaloznawstwo i Obrobka Cieplna, No. 5, pp. 52-57;

1974

Voss W.A.G. (Ed.): ISM Applications of Microwaves, Special Issue; Proc. IEEE, January;

1975

Garlan H.: US Approach to Equipment Approval to Achieve Electromagnetic Compatibility; Proc. EMC Sympos. Montreux, May 20-22, pp. 182-186;

Horner F.: Some Sources of Interference to Radio Astronomy; Proc. EMC Sympos., Montreux, May 20-22, pp. 329-332;

Meyer de Stadelhofen J.: Organization for the Protection of Radiocommunications by Planning Frequencies and Testing Electrical Equipment; Proc. EMC Sympos. Montreux, May 20-22, pp. 337-340;

Rusakiewicz W.: Attempts of Frequency Limitations for ISM Equipment in Poland and their Technical Aspects; Proc. EMC Sympos. Montreux, May 20-22, pp. 24-28;

Strużak R.G.: Urban Electromagnetic Environment: Facts, Models, Trends; Proc. EMC Sympos. Montreux, May 20-22, pp. 88-92;

Wilson T.L.: Dielectric Heating - A Useful Industrial Tool; Proc. EMC Sympos., Montreux, May 20-22, pp. 333-336;

1976

McLachlan A.S.: Radio Interference from RF Heating Equipment; The Radio and Electronic Engr., Vol. 46, No. 6, June, pp. 267-275;

Rusakiewicz W.: Observations on the Limits of Interference for ISM Radio Frequency Equipment; Proc. EMC Sympos. Wroclaw Sept. pp. 270-284;

1977

Raab F.H.: Idealized Operation of the Class E Tuned Power Amplifier; IEEE Transactions on Circuits and Systems, Vol. CAS-24, No. 12, Dec. 1977, pp.725-735;

Rusakiewicz W.: Measurements of Interference from RF Heating Equipment with Rapidly Changing Field-Strength and Operating Frequency; Proc. EMC Sympos. Montreux, June 28-30, pp. 420-430;

Schulz R.B.: A Review of Interference Criteria for Various Radio Services; Proc. EMC Sympos. Montreux, June 28-30, pp. 143-148;

Sokal N.O., Raab F.H.:Harmonic Output of Class-E RF Power Amplifiers and Load Coupling Network Design; IEEE Journal of Solid-State Circuits, February 1977, pp.86-88;

1978

Abramson Y, Gatova A., Kapitonov V.; Radio Interference from ISM Apparatus and Establishing their Limits (in Russian); Proc. EMC Sympos. Wroclaw Sept. 13-15, pp. 335-348;

Skomal E.: Man-Made Radio Noise; Van Nostrand Reinhold Co. New York;

Stumpers F.L.: EMC from Pragmatic to the Scientific Approach; Proc. EMC Sympos. Wroclaw, Sept. 13-15, pp. 3-15;

1979

Paolini E., Corona P.: Experimental Procedure to Determine Limits of ISM Interference Affecting Navigational ILS Equipment; Proc. EMC Sympos., Rotterdam, May 1-3, pp. 141-146;

1980

Harry R.J., Hasler R.J.: Measurement of the Immunity of Television Receivers to RF Interference; The Radio and Electronic Engineer, Vol. 50, No.1/2, January/February, pp.22-28;

Lerner E.J.; RF Radiation: Biological Effects: IEEE Spectrum, December, p. 51-59

Rusakiewicz W.: EM Field Produced by Frequency-Stabilized RF Plastic Welders; Proc. EMC Sympos., Wroclaw, Sept. 17-19, pp. 174-179;

1981

Dvorak T.J.: The Problem of Limits Interpretation in Type Approval Testing and Production Control; Proc. EMC Sympos., Boulder Co., August 18-20, pp. 264-268;

Kirby R.C.: Impact of WARC-79 on the Studies of the International Radio Consultative Committee (CCIR); IEEE Trans. Electromagnetic Compt., Vol. EMC-23, No. 3, August, pp. 174-178;

Thompson A.R.: Effects of a Satellite Power System on Ground-Based Radio and Radar Astronomy; Radio Science, Vol. 16, No. 1, January/February, pp.35-45;

1982

Garlan H., Knapp J.P., Wall L.A.: The USA Approach to Regulation of ISM Equipment; Proc. EMC Sympos. Wroclaw, June 22-24, pp. 537-544;

Pykett I.L. et al.: Principles of Nuclear Magnetic Imaging: Radiology, Vol. 143, April, pp.157-168;

Showers R.M.: EMC Control, its Status, and its Future; Proc. EMC Sympos. Wroclaw, June 22-24, pp. XXVII-XXX.

1983

Bascoulergue P., Sedaries R.: Mesures et Statistiques Etablies par Telediffusion de France sur les Appareils ISM; Actes de la Demi Journee ISM du 2eme Colloque National sur la compatibilite Electromagnetique, 3 June, pp. 38-66;

Bassole M.: Evolution de la Reglementation des Telecommunications Relative aux Brouillages Radioelectriques; Actes de la Demi Journee ISM du 2me Colloque National sur la Compatibilite Electromagnetique, Tregastel, 3 June, pp. 1-25;

Beck M.D.: Problemes Poses aux Constructeurs dans la Cadre de la Reglementation ISM; Actes de la Demi Journee ISM du Colloque National sur la Compatibilite Electromagnetique, Tregastel, 3 June pp. 93-98;

Chef M.: Pollution Radioelectrique et Securite de la Navigation Aerienne - La situation Internationale Actuelle; Actes de la Demi Journee ISM du Colloque National sur la Compatibilite Electromagnetique, Tregastel, 3 June pp. 86-92;

De Jong P.: EMC Standards and their Application in European Countries; Proc. EMC Sympos. Zurich, March 8-10, pp. 331-333;

Knapp J.P., Wall L.A.: Update on the Proposed Changes in the FCC Rules Governing
ISM Equipment; Proc. EMC Sympos. Arlington, August 23-25, p.249-253;

Le Luc M.: Brouillage des ILS; Actes de la Demi Journee du 2eme Colloque National sur la Compatibilite Electromagnetique, 3 June , pp. 84-85;

Nard G.: Simulation de Brouillages d'Appareils de Navigation par des ISM. Determination des Champs Susceptibles d'Entrainer des Brouillages Prejudiciables; Actes de la Demi Journee ISM du 2eme Colloque National sur la Compatibilite Electromagnetique, 3 June, pp. 67-83;

Nickum J.D.: Summary of the ISM Measurement Program at Ohio University; Proc. IEEE Internatl. Sympos. EMC, Washington DC, 23-25 August, pp. 274-276;

Schulz R.B.: A Procedure for Estimating the Risk of Interference; Proc. EMC Sympos. Zurich, March 8-10, pp. 313-318;

Struzak R.G. :ISM RFI Radiation Problem; Proc. EMC Sympos. Zurich, March 8-10, pp.343-348;

Struzak R.G.: Protection of Radiocommunications Against Interference From SM Equipment; Actes de la Demi Journee ISM du 2eme Colloque National sur la Compatibilite Electromagnetique, Tregastel, 3 June, pp.26-37;

Turnage H.C.: EMC Assurance Measurements of Industrial Facilities; Proc. IEEE Internatl. Sympos. EMC, Washington DC, August 23-25, pp. 265-269;

Watson A.W.D.; Improvements in the Suppression of External Nonlinearities (Rusty-Bolt-Effects) which affect Naval Radio Systems; Proc. IEEE Internatl. Sympos. EMC, Washington DC, Aug. 23-25, 1983, p. 157-160;

Weinberg D.I.: An Administrative Program to Limit Radio Frequency Interference from Industrial Plants and Laboratories; Proc. Internatl. IEEE Sympos. EMC, Washington DC, August 23-25, pp. 269-274;

West L.G.: Part 18, A Manufacturer's Viewpoint; Proc. IEEE Internatl. Sympos. EMC, Washington DC, August 23-25, pp.265-269;

Whitehouse A.C.D.: The Compatibility of Standards; Proc. EMC Sympos., Zurich, March 8-10, pp.309-312;

Wilson T.L.: RF Heating Equipment and International ISM Standards; Proc. Internatl. Sympos. EMC, Washington DC, August 23-25, pp. 258-265;

Yazar M.N.: New Directions for EMC Standards; Proc. EMC Sympos. Zurich March 8-10, pp. 325-330;

1984

Betzky O.V., Palatov K.I.: Spurious Oscillations in Microwave Electron devices; Proc. EMC Sympos. Tokyo, October 16-18, pp. 464-467;

Czyzewski S.: Measurement Results of RF Field-Strength in the Vicinity of Dielectric Heaters (in Polish); Bul. Inform. Techn. PIR, No. 1/84, pp.8-13;

Minozuma F.: Radio Noise Interference Activities in Japan; Proc. EMC Sympos. Tokyo, October 16-18, pp. 1-17;

Struzak R.G.: Protection of Radiocommunications Against Interference from ISM Equipment; Annales des Telecommunications, Vol. 39, No. 5-6, May-June, pp. 209-218;

Struzak R.G.: Radiation from Industrial, Scientific and Medical Radio Frequency Equipment; XXI General Assembly of the International Union of Radio Science URSI, Florence, August 28-September 5, Commission E4, pp.383;

1985

Audonne B., Cazzola R., Barale G.: Statistical Evaluation of the EMC Safety Margin at System Level; Proc. EMC Sympos. Zurich, March 5-7, pp. 339-245;

Chen Q., Zhu Y.C.; The Application and Development of EMC in China; Proc. EMC Sympos. Zurich, March 5-7, 1985, p.259-262;

De Jong A.:Statistical Aspects of Noise and Limits; Proc. EMC Sympos. Zurich, March 5-7, pp. 229-232;

Shupe R.E., Hornback N.B.: The Friendly Fields of RF; IEEE Spectrum, June pp. 64-69;

Smith B.L.: High-Power Electron Tubes Designed for Modern Microwave Systems; Microwave System News, Vol. 15, No. 7, June, pp.68-81;

Whitehouse A.C.D.: Radio Interference - the Probability Problem; Proc. EMC Sympos. Zurich, March 5-7, pp.233-238;

ANNEX I - Frequency bands designated for ISM applications in the ITU Radio Regulations

Frequency band	Centre frequency
6 765 - 6 795 kHz	6 780 kHz**
13 553 - 13 567 kHz	13 560 kHz*
26 957 - 27 283 kHz	27 120 kHz*
40.66 - 40.70 MHz	40.68 MHz*
433.05 - 434.79 MHz	433.92 MHz**
902 - 928 MHz	915 MHz*
2.4 - 2.5 GHz	2.45 GHz*
5.724 - 5.875 GHz	5.8 GHz*
24.0 - 24.05 GHz	24.125 GHz*
61.0 - 61.5 GHz	61.25 GHz**
122 - 123 GHz	122.5 GHz**
244 - 246 GHz	245 GHz**

* Radiocommunication services operating within this band must accept harmful interference which may be caused by ISM applications. Administrations shall take all necessary steps to ensure that radiation from ISM equipment is minimal and that, outside the bands designated for use by this equipment, is at a level that does not cause harmful interference to a radio navigation, or any other safety service, operating in accordance with the provisions of the Radio Regulations (RR 534, 546, 548, 707, 752, 806, 881, 1815).

** The use of this frequency band for ISM applications shall be subject to special authorization by the administration concerned, in agreement with other administrations whose radiocommunications might be affected (RR 524, 661, 911, 916, 922).

ANNEX II - CISPR limits of radiation from ISM equipment (not on a test site)

Frequency band MHz	Limit [uV/m] at a distance 30 m*	100 m*	300 m**
A : General			
0.15 - 0.285	-	50	50
0.285 - 0.49	-	250	250
0.49 - 1.605	-	50	50
1.605 - 3.95	-	250	250
3.95 - 30	-	50	50
30 - 470	30[3]	-	200
	500[4]	-	200
470 - 1 000	100[3]	-	200
	500[4]	-	200
B : Particular* **			
0.2835 - 0.5265	-	50	-
74.6 - 75.4	250	-	-
108 - 136	30	-	-
242.95 - 243.05	30	-	-
328.6 - 335.4	100	-	-
960 - 1 215	100	-	-

* From the boundary of user's premise.
** From the equipment.
*** For the protection of specific safety services in particular areas.

Notes:

1. Surgical diathermy apparatuses and RF excited arc welding equipment are excluded.
2. The CISPR limits do not concern the frequency bands designated for ISM applications listed in Annex I.
3. Compliance with these limits is required only for the TV channels in use at any time at the site.
4. Limits for use outside the TV channels in use at the time at the site.
5. These limits are under review by CISPR and may be subject to alteration.

Nonlinear and Environmental Electromagnetics, edited by H. Kikuchi
Elsevier Science Publishers B.V., Amsterdam, 1985 – Printed in The Netherlands

Behavioural Ontogeny in Rats Exposed to Extremely Low Frequency Electric Field

J. Behari, R. Mathur* and K.N. Sharma*
School of Environmental Sciences
Jawaharlal Nehru University
New Delhi-110067, INDIA

* Department of Physiology
University College of Medical Sciences
Ring Road, New Delhi-110016, INDIA

Abstract

The effect of chronic exposure of 50 Hz electric fields on rats is examined. The animals in two age groups are considered. One in the developing age (94-138 days) and other mature rats. The animals were kept in acrylic cages with metallic plates on two sides connected to a high voltage (5 kv/m) ac power supply (50 Hz). The control group of animals of same age and sex were kept in identical boxes without the field being on. A number of physiological parameters including, food and water intake, body weight gain and the activity of either group of rats are compared. Survival time of the rats and their behavioural pattern are also noted. It is found that the food intake of experimental group of animals is consistently less than the control ones. The body weight of experimental group also show a corresponding pattern. The motor activity of the experimental groups is found to be higher than the control ones. It is concluded that the effect on younger animals is more than the older ones and the changes observed are within the physiological limits.

I. Introduction

In recent years (1-4) there has been a growing interest in finding out the effects of electric and magnetic fields on biological systems. Efforts have been made to cover a wide range of frequencies and power levels, besides covering systems from unicellular organisms to humans. The mechanisms by which the electromagnetic fields could affect biological matter also vary and depend on the electric and magnetic properties of such matter. One important aspect of the interaction between the field and the biosystem is the amount of energy that could be transferred from the field to the system. At very high frequencies, the electromagnetic fields with relatively modest intensities are capable of producing immediate effects. At low frequencies the effects are much less known, particularly the mechanism at work. Since the frequencies of the physiological range of interest falls in this range there is increasing speculation if this would have any effect on the system in question.

Recent conflicting claims regarding the nature of effects produced, prompted the present investigators to undertake a quantitative study of the possible phenomena at work. The observations were so planned so as to reveal fairly long term effects. Also the intensity of exposure was chosen so as to fall in the range of usual exposure due to various electrical appliances in day to day use (5), so as to assess the quantum of effects.

II. Materials and Methods

Male and female rats of same age group were kept singly in each cage. The number of cages was equivalent to the number of animals. The animals were housed in two separate large acrylic cages placed between two parallel electrodes. The two metallic sheet electrodes were of the size of the cage side.

The animals were located in the middle between the two electrodes. The electrodes were connected to the high voltage transformer and the voltage adjusted to give an unperturbed electric field intensity of 5 kv/m between the sheet electrodes. A similar set up was used for the control animals but with no field applied. The sawdust in the cages were changed after every 48 hours. The temperature inside the laboratory was around 25°C during winter and summer. Lighting conditions were natural light from large windows plus fluorescent light during the day and complete darkness during the night. Animals were fed normal diet and the food and water was freely available. Food intake was measured every twenty four hours. Field was put off for one hour each day during cleaning and providing fresh food and water. The animals were exposed 23 hours/day, 7 days/week. Food and water intake as obtained for each day was averaged over for a period of 8-10 days for both experimental ($\bar{a}$) and control

($\bar{b}$) group animals.

Each animal was ear coded and they were put in the electric field in the respective cages. They were exposed to the field at various periods of their age. This was done for convenience and also to find out if the field would have a different effect on the rats exposed during their formative days. Experiments were extended on animals falling in two age groups: (94-138 days) and more than 150 days of age. Each group had ten animals taking not more than 4 at a time.

III. Results and Discussions

Table No.1 shows that the animals of the same age group and of the same sex develops apparently different food intake patterns after the field exposure begins. Typical set of data are presented here. Initially the control group of animals had larger food intake as compared to experimental ones (averaged over 8-10 days). Subsequently it becomes less, with the result that ($\bar{a}$ - $\bar{b}$) becomes negative. The food intake which was insignificantly different has become significantly different after a period of about three weeks. Water intake pattern also show a similar pattern though significance level fluctuates.

Another set of mature group of animals show less increase in their weight after exposing to the electric fields as compared to the control group (Table No.IV). In fact a slight decrease in animal weight is also recorded. The average food intake falls after the exposure starts but water intake continues to be higher (Table No.II).

It is further observed that the activity of the animals after exposure to the field is consistently higher than that before exposure. A typical value for the average value of the activity turns out to be 435 (arbitrary unit), which is lower than the experimental one (830). However, we do not observe the death of animal during the exposure period and also no abnormality in their behaviour was observed.

It may be inferred that the electric field at 50 Hz (5 kv/m) has more definite effect on young animals as compared to the mature ones, though the effects are within the physiological limits.

References

1. Speigel, R.J. 1977. IEEE Trans Biomed Eng. 24, pp 466-72.

2. Sheppard, A.R. and Eisenbud, M. 1977. In: Biological Effects of Electric and Magnetic fields of Extremely low frequency', New York, New York University Press.

3. Fam, W.Z. 1980. IEEE Trans Biomed. Engg. 27, pp 376-81.

4. Galvin, M.J. and RcRee, D.I. 1981. J. Appl Physiol. 50(5), pp 931-5.

5. Pressman, A.S. 1970. Electromagnetic Fields and Life, Plenum Press.

Table No. I

	Age days	Male	days	Food $\bar{a}$	$\bar{b}$	$\bar{a}-\bar{b}$	SD	P		Water $\bar{a}$	$\bar{b}$	$\bar{a}-\bar{b}$	SD	P
1.	94	F	10	15.3	14.65	0.65	3.1	NS	1.	27.0	23.0	4.0	4.8	NS
2.	94	F	10	9.3	14.6	-5.3	2.5	NS	2.	25.55	23.0	2.55	4.2	NS
3.	94	F	10	11.4	14.8	-3.4	1.8	NS	3.	23.5	46.4	-22.9	6.7	VS
1.	104	F	10	12.5	19.42	-6.92	2.0	VS	1.	30.0	34.5	-4.5	5.5	NS
2.	104	F	10	15.4	19.42	-4.02	1.9	NS	2.	30.0	34.5	-4.5	4.4	NS
3.	104	F	10	15.5	15.9	-0.40	1.2	NS	3.	34.0	36.5	-2.5	3.2	NS
1.	114	F	8	6.87	14.8	-7.93	1.1	VS	1.	30.6	46.4	-15.8	7.4	NS
2.	114	F	8	5.25	14.8	-9.55	1.0	VS	2.	26.87	46.4	-19.5	3.2	VS
3.	114	F	8	5.5	16.77	-11.27	1.8	VS	3.	21.66	40.0	-18.3	6.8	S

F = Female, S = Significant, NS = Not significant, VS = Very significant

Comparison of food and water intake pattern among control and experimental group of growing rats.

Table No.II : Mature Rats

		Food					Water				
Days	Sex	$\bar{a}$	$\bar{b}$	$\bar{a} - \bar{b}$	SD	P	$\bar{a}$	$\bar{b}$	$\bar{a} - \bar{b}$	SD	P
10	M	16.8	19.0	-2.2	1.9	NS	41.7	30.5	11.2	3.80	S
10	M	16.4	17.5	-1.1	1.90	NS	45.1	31.0	14.1	5.80	S
10	M	21.3	21.3	0.0	1.77	NS	50.5	34.0	16.5	2.90	VS
10	M	19.7	19.0	0.7	2.14	NS	43.0	30.5	12.5	4.0	S
10	M	15.3	17.5	-2.2	1.58	NS	36.5	31.0	5.5	3.35	NS
10	M	20.2	21.3	-1.1	1.72	NS	51.0	34.0	1.70	3.0	VS
10	M	17.3	19.0	-1.70	2.00	NS	54.4	30.5	23.9	3.30	VS
10	M	14.0	17.5	-3.5	1.50	S	54.4	31.0	23.4	5.00	VS
10	M	22.4	21.3	1.1	1.25	NS	41.6	34.0	7.6	4.30	NS
10	M	16.7	19.0	-2.3	2.00	NS	54.0	30.5	23.5	3.40	VS
10	M	13.6	17.5	-3.9	1.60	S	54.0	31.0	23.0	5.50	VS
10	M	21.2	21.3	-0.1	1.50	NS	45.0	34.0	14.0	4.90	S
6	M	15.8	19.0	-3.2	2.50	NS	56.1	30.5	25.6	3.60	VS
6	M	13.3	17.5	-4.2	1.90	S	46.6	31.0	15.6	6.70	S
6	M	21.6	21.3	0.3	1.60	NS	64.1	34.0	30.1	7.00	VS

M = Male, NS = Not significant, S = Significant, VS = Very significant

Comparison of food and water intake pattern among control and experimental group of mature rats.

Table No. III

Age (days)	Weight (gms) (Experimental)	(Control)
66	151	220
80	204	240
93	205	250
99	206	260
110	220	270
125	241	280
139	248	300

Typical weight gain data for young experimental rats (male)

Table No.IV

Duration of Measurements (days)	Weight (gms) Experimental	Control
10	230	260
20	245	280
30	284	290
40	208	305
50	303	315
60	298	310
70	288	315

Typical weight gain pattern of mature rats
($>$ 150 days of age)

V. PONDEROMOTIVE FORCE IN AN EM ENVIRONMENT

Nonlinear and Environmental Electromagnetics, edited by H. Kikuchi
Elsevier Science Publishers B.V., Amsterdam, 1985 – Printed in The Netherlands

A THEORY OF PONDEROMOTIVE FORCES ON A VLASOV PLASMA

H.AKAMA and M.NAMBU

College of General Education, Kyushu University,
Ropponmatsu, Fukuoka 810, JAPAN

Abstract A general formula of ponderomotive forces induced by electrostatic waves in a dispersive nonuniform plasma is derived. The ponderomotive forces are made up of three different kinds of forces: a force on resonant particle, a ponderomotive force in a narrow sense with its correction term and a force expressed as the Poisson bracket of the real part of the response function and wave energy density. For Langmuir waves, the wave kinetic equation gives rise to a cancellation between the first and the third force.

1. Introduction

During the past years, nonlinear forces in plasma physics, known as ponderomotive forces, have been intensively studied because they play an important part of radio frequency confinement, self-focussing, and anomalous transport in plasma physics. For a survey of early work, we refer to a book by Motz (1979).

The ponderomotive forces do not seem to have been given a soundly established theory. There are as many different avenues of approach as there are authors on the subject. The purpose of this article is to give a general formula of ponderomotive forces which include all of the results reported by the previous authors. Furthermore, the formula includes a few new terms due to inhomogeneities of the dielectric response function in space and time.

2. Ponderomotive force in a narrow sense

The α-component of the ponderomotive force in a narrow sense (F_α) is obtained from the quasi-static force acting on the dipole (Penfield and Haus, 1967). The time average of the force density exerted by the electrostatic wave upon the electron plasma reduces to

$$F_\alpha = \frac{1}{2} R_e \{P_\beta \, \partial/\partial x_\beta \, E_\alpha^* \}, \tag{1}$$

here P_β and E_α are the polarization and the electric field, respectively. The symbols R_e and * show the real part and the complex conjugate of the relevant term, respectively.

The complex polarization vector can be written as

$$P_\beta = - \varepsilon_0 \frac{\omega_p^2}{\omega^2} E_\beta , \tag{2}$$

where ω_p, ω and ε_0 are the electron plasma frequency, angular frequency of electrostatic wave and the permittivity of free space. If we substitute Eq.(2) to Eq.(1), we obtain

$$F_\alpha = - \frac{1}{4} \varepsilon_0 \frac{\omega_p^2}{\omega^2} \frac{\partial |E|^2}{\partial x_\alpha} . \tag{3}$$

The ponderomotive force in a narrow sense [Eq.(3)] was first obtained by Boot and R-Shersby-Harvie (1957), Gaponov and Miller (1958), and Weibel (1958). The rf confinement of plasmas is studied based on Eq.(3) (Hora, 1979). However, for hot plasmas, there are many additional ponderomotive forces which give the same order of magnitude as that of Eq.(3). In the next section, we show a general formula of ponderomotive forces for hot plasmas.

3. A general formula of ponderomotive forces

The α-component of the force exerted by an electrostatic wave on a unit volume of a plasma is given by (Akama and Nambu, 1981)

$$f_\alpha = \frac{1}{4} [\rho^*(\vec{k},\omega)\varepsilon_\alpha + c.c.] , \tag{4}$$

where $E_\alpha = \varepsilon_\alpha \exp[i(\vec{k}\cdot\vec{x} - \omega t)]$. Here, $\rho(\vec{k},\omega)$ is the Fourier component of the polarization charge density and is given by

$$\rho(\vec{k},\omega) = - ik_\beta [\varepsilon_0 \{K(\vec{k},\omega) - 1\}] (- ik_\beta \phi) , \tag{5}$$

where $K(\vec{k},\omega)$ and ϕ are the dielectric response function and scalar potential, respectively.

We now proceed to the case in which both the amplitudes of electrostatic waves and the dielectric function depend weakly on $\vec{x}$ and t. Then, we get

$$\varepsilon_\alpha = \varepsilon_\alpha^{(0)} + \varepsilon_\alpha^{(1)} = - ik_\alpha \phi - \frac{\partial \phi}{\partial x_\alpha} , \tag{6}$$

and

$$\rho(\vec{k},\omega) = \rho^{(0)}(\vec{k},\omega) + \rho^{(1)}(\vec{k},\omega) , \tag{7}$$

with $$\rho^{(0)}(\vec{k},\omega) = ik_\beta \varepsilon_0 \varepsilon_\beta^{(0)} - ik_\beta \varepsilon_0 K' \varepsilon_\beta^{(0)} + k_\beta \varepsilon_0 K'' \varepsilon_\beta^{(0)} , \tag{8}$$

$$\rho^{(1)}(\vec{k},\omega) = ik_\beta\varepsilon_0 K' \frac{\partial\phi}{\partial x_\beta} - 2ik_\beta\varepsilon_0 \frac{\partial\phi}{\partial x_\beta} + \frac{\partial}{\partial x_\beta}[\varepsilon_0 K'(ik_\beta\phi)]$$

$$+ ik_\beta\varepsilon_0 \left(\frac{1}{i}\frac{\partial K'}{\partial k_\gamma}\frac{\partial}{\partial x_\gamma} + i\frac{\partial K'}{\partial\omega}\frac{\partial}{\partial t} - \frac{1}{i}\frac{\partial K'}{\partial x_\gamma}\frac{\partial}{\partial k_\gamma} - i\frac{\partial K'}{\partial t}\frac{\partial}{\partial\omega}\right)(ik_\beta\phi), \tag{9}$$

here we assume that $|K''| < |K'|$, where K' and K" denote the real and imaginary parts of K, respectively. Accordingly, the α-component of ponderomotive forces reduces to

$$f_\alpha = \frac{1}{4}\{\rho^{(0)*}(\vec{k},\omega)\varepsilon_\alpha^{(0)} + c.c.\}$$

$$+ \frac{1}{4}\{\rho^{(0)*}(\vec{k},\omega)\varepsilon_\alpha^{(1)} + c.c.\}$$

$$+ \frac{1}{4}\{\rho^{(1)*}(\vec{k},\omega)\varepsilon_\alpha^{(0)} + c.c.\} . \tag{10}$$

On substituting $\varepsilon_\alpha^{(0)} = -ik_\alpha\phi$ and Eq.(8) into the first term of Eq.(10), we obtain

$$\frac{1}{4}\{2\varepsilon_0 K'' k_\alpha k^2|\phi|^2\} . \tag{11}$$

We substitute $\varepsilon_\alpha^{(1)} = -\frac{\partial\phi}{\partial x_\alpha}$ and Eq.(8) in the second term of Eq.(10). Then we get

$$\frac{1}{4}\{\varepsilon_0(K' - 1)k^2 \frac{\partial|\phi|^2}{\partial x_\alpha}\} . \tag{12}$$

In obtaining Eq.(12), we have omitted a minor term $i\varepsilon_0 K'' k^2(\partial\phi^*/\partial x_\alpha\, \phi - \partial\phi/\partial x_\alpha\, \phi^*)$. Next, we substitute Eq.(9) and $\varepsilon_\alpha^{(0)} = -ik_\alpha\phi$ into the third term of Eq.(10). Then, adding the result with Eqs.(11) and (12), we obtain the α-component of the ponderomotive forces as follows.

$$f_\alpha = \frac{1}{4}\{2k_\alpha\varepsilon_0 K'' k^2|\phi|^2\} + \frac{1}{4}\{\varepsilon_0(K'-1)k^2 \frac{\partial|\phi|^2}{\partial x_\alpha}\}$$

$$+ \frac{1}{4}\{2k_\alpha k_\beta\varepsilon_0(1-K')\frac{\partial|\phi|^2}{\partial x_\beta}\}$$

$$+ \frac{1}{4}\{k_\alpha k^2\varepsilon_0([K',|\phi|^2]_{\omega,t} - [K',|\phi|^2]_{k,x})\}, \tag{13}$$

where $[K',|\phi|^2]_{\omega,t}$ is the Poisson bracket defined by $[K',|\phi|^2]_{\omega,t} \equiv \partial K'/\partial\omega\times\partial|\phi|^2/\partial t - \partial K'/\partial t\times\partial|\phi|^2/\partial\omega$.

The first term of Eq.(13) is the dissipative forces (Akama and Nambu, 1981 and 1982) due to the resonant wave particle interaction (Landau resonance). The dissipative forces are the origin of the anomalous resistivity, (Sagdeev, 1979), the rf induced transport (Myra and D'Ippolito, 1984), and the anomalous radiation from a turbulent plasma (Nambu and Akama, 1985). The second term on the right-hand side of Eq.(13) stands for a ponderomotive force in a narrow sense [Eq.(3)]. The third term of Eq.(13) shows its correction due to nonresonant wave particle interaction. The last term is the force due to the adiabatic interaction between waves and particles. The last term includes four different effects. We thus come to the conclusion that these many kinds of forces have to be considered on an equal footing. Unfortunately, all of the previous authors except Kadomtsev (1965) consider only a part of Eq.(13) (Landau and Lifshitz, 1960; Cary and Kaufman, 1980; Barash and Karpman, 1983 and references therein).

4. Wave kinetic equation for inhomogeneous plasma

The extended dispersion relation for homogeneous plasma reduces to

$$ik_\beta \varepsilon_0 K(\, -ik_\beta \phi\,) = 0\ . \tag{14}$$

We obtain the general dispersion relation for inhomogeneous medium under the following interchanges

$$ik_\beta \rightarrow ik_\beta + \frac{\partial}{\partial x_\beta}\ ,$$

$$K \rightarrow K + \frac{\partial K}{\partial k_\gamma}\frac{1}{i}\frac{\partial}{\partial x_\gamma} - \frac{\partial K}{\partial x_\gamma}\frac{1}{i}\frac{\partial}{\partial k_\gamma} + \frac{\partial K}{\partial \omega}\, i\,\frac{\partial}{\partial t} - \frac{\partial K}{\partial t}\, i\,\frac{\partial}{\partial \omega}\ . \tag{15}$$

On substitution of Eq.(15) into Eq.(14), we obtain the extended dispersion relation to the lowest order

$$\begin{aligned} & ik_\beta \varepsilon_0 K(-ik_\beta\phi) + \varepsilon_0 \frac{\partial K}{\partial x_\beta}(-ik_\beta\phi) \\ & + \varepsilon_0 K(-ik_\beta)\frac{\partial \phi}{\partial x_\beta} + ik_\beta\varepsilon_0 K(-\frac{\partial\phi}{\partial x_\beta}) \\ & + k_\beta \varepsilon_0 [\frac{\partial K}{\partial k_\gamma}\frac{\partial}{\partial x_\gamma} - \frac{\partial K}{\partial x_\gamma}\frac{\partial}{\partial k_\gamma} - \frac{\partial K}{\partial \omega}\frac{\partial}{\partial t} + \frac{\partial K}{\partial t}\frac{\partial}{\partial \omega}](-ik_\beta\phi) \\ & = 0\ . \end{aligned} \tag{16}$$

We note the following identity,

$$\frac{\partial K}{\partial x_\gamma} \frac{\partial}{\partial k_\gamma} (-ik_\beta \phi) = - i \frac{\partial K}{\partial x_\beta} \phi - ik_\beta \frac{\partial K}{\partial x_\gamma} \frac{\partial \phi}{\partial k_\gamma} . \tag{17}$$

Substituting Eq.(17) into Eq.(16), we get

$$k^2 \varepsilon_0 K\phi - 2i\varepsilon_0 K \frac{\partial \phi}{\partial x_\beta} k_\beta + ik^2 \varepsilon_0 \{[K,\phi]_{\omega,t} - [K,\phi]_{k,x}\} = 0 , \tag{18}$$

where $[K,\phi]$ is the Poisson bracket. We substitute the relation $K = K' + iK''$ into Eq.(18), and multiply the result by ϕ^*, we obtain

$$\begin{aligned} & k^2 \varepsilon_0 K' |\phi|^2 + ik^2 \varepsilon_0 K'' |\phi|^2 \\ & + 2\varepsilon_0 K'' k_\beta \frac{\partial \phi}{\partial x_\beta} \phi^* - 2i\varepsilon_0 K' k_\beta \frac{\partial \phi}{\partial x_\beta} \phi^* \\ & + k^2 \varepsilon_0 \{[K'',\phi]_{k,x} - [K'',\phi]_{\omega,t}\} \phi^* \\ & + ik^2 \varepsilon_0 \{[K',\phi]_{\omega,t} - [K',\phi]_{k,x}\} \phi^* = 0 . \end{aligned} \tag{19}$$

Subtracting the complex conjugate of Eq.(19) from Eq.(19), we get

$$\begin{aligned} & 2k^2 \varepsilon_0 K'' |\phi|^2 - 2\varepsilon_0 K' k_\beta \frac{\partial |\phi|^2}{\partial x_\beta} \\ & + k^2 \varepsilon_0 \{[K',|\phi|^2]_{\omega,t} - [K',|\phi|^2]_{k,x}\} = 0 . \end{aligned} \tag{20}$$

Thus, we obtain the wave kinetic equation for the inhomogeneous plasma which corresponds to the generalization of the eikonal equation derived by Stix (1962).

One of the most conspicuous results comes out if we substitute Eq.(20) into the general formula of the ponderomotive forces [Eq.(13)] for one component plasma. For one component plasma, Eq.(13) reduces to

$$f_\alpha = \frac{1}{4} \{ \varepsilon_0 (K'-1) k^2 \frac{\partial |\phi|^2}{\partial x_\alpha} + 2k_\alpha k_\beta \varepsilon_0 \frac{\partial |\phi|^2}{\partial x_\beta} \} . \tag{21}$$

Note that, for Langmuir waves ($K' - 1 = - \omega_p^2/\omega^2$), the first term of Eq.(21) agrees with Eq.(3). Accordingly, the dissipative force disappears for Langmuir waves. However, for two component plasma (ion sound waves, ion cyclotron waves, MHD waves etc.), the dissipative force appears for electrons. The dissipative force is the origin of the anomalous diffusion (Petviashvili, 1963) the resistivity (Sagdeev, 1979), and the radiation (Nambu and Akama, 1985).

5. Discussion

A general formula of the ponderomotive force [Eq.(13)] has four terms on the right-hand side. As pointed out in Sec.3, the most of them can be found in the literature, but only separately. We now want to emphasize that these different terms must be treated on same footing. For Langmuir waves, the net force per unit volume on a plasma simply reduces to Eq.(21) due to interrelations between the ponderomotive forces [Eq.(20)]. In deriving Eq.(13), we have kept only the lowest order contribution. It is straightforward to consider the next order ponderomotive force (Bel'kov and Tsytovich, 1979; Kono et al., 1981) within the framework of our analysis. Finally, our results do not rely on Vlasov equation. Accordingly, Eq.(13) is applicable to both a plasma and plasma-like medium.

References

Akama, H. and Nambu, M.: 1981, Phys. Lett. 84A, 68.
Akama, H. and Nambu, M.: 1982, Physica 116A, 155.
Barash, Y. S. and Karpman, V. I.: 1983, Sov. Phys. JETP 58, 1139.
Bel'kov, S. A. and Tsytovich, V. N.: 1979, Sov. Phys. JETP 49, 656.
Boot, H. A. H. and R-Shersby-Harvie, R. B.: 1957, Nature 180, 1187.
Cary, J. R. and Kaufman, A. N.: 1980, Phys. Rev. 21A, 1660.
Gaponov, A. V. and Miller, M. A.: 1958, Sov. Phys. JETP 7, 168.
Hora, H.: 1979, Nonlinear Plasma Dynamics at Laser Irradiation, Lecture notes in physics, Springer-Verlag, Berlin.
Kadomtsev, B. B.: 1965, Plasma Turbulence, Academic Press, London, 28.
Kono, M., Skorić, M. M. and D. ter Haar: 1981, J. Plasma Phys. 26, 123.
Landau, L. D. and Lifshitz, E. M.: 1960, Electrodynamics of Continuous Media, Pergamon, Oxford.
Motz, H.: 1979, The Physics of Laser Fusion, Academic Press, New York.
Myra, J. R. and D'Ippolito, D. A.: 1984, Phys. Rev. Letters 53, 914.
Nambu, M. and Akama, H.: Nuovo Cimento (in press).
Penfield, Jr. P. and Haus, H. A.: 1967, Electrodynamics of Moving Media, The M.I.T. Press, Cambridge.
Petviashvili, V. I.: 1963, Zhur. Eksptl i Theoret. Fiz. 44, 1993.
Sagdeev, R. Z.: 1979, Rev. of Mod. Phys. 51, 1.
Stix, T. H.: 1962, The Theory of Plasma Waves, McGraw-Hill, New York.
Weibel, E. S.: 1958, J. Electron. Cont. 4, 435.

Nonlinear and Environmental Electromagnetics, edited by H. Kikuchi
Elsevier Science Publishers B.V., Amsterdam, 1985 – Printed in The Netherlands

PLASMA DENSITY DEPRESSION BY STRONG ELECTROMAGNETIC PULSE

H. Sugai

Department of Electrical Engineering, Nagoya University
Nagoya 464, Japan

Abstract

A laboratory experiment of the strong microwave pulse incident on a plasma demonstrates that a nonlinear interaction of electromagnetic wave with the plasma gives rise to the local depression of plasma density due to the radiation pressure. After turn-off of the microwave pulse, a large-amplitude oscillation of the plasma density is found, which is identified as an ion acoustic wave driven by the ponderomotive force.

1. INTRODUCTION

Recently, much attention has been given to the nonlinear interaction of electromagnetic waves with a plasma in various fields; laser fusion, radio-frequency heating and current drive in magnetic confinement fusion devices, as well as ionospheric plasma modification by microwave beam in a solar power transmission system. The nonlinear interactions are classified into two groups. One is a resonant type of wave-wave interactions(parametric instabilities) and wave-particle interactions(nonlinear Landau or cyclotron dampings). The other is a nonresonant type of interactions, such as collisional heating, wave mixing, and so on. Among the nonresonant interactions, the growing interest has been focussed to the ponderomotive force[1], the simplest collisionless effect of high-frequency waves on slow plasma motion.

Expressions for the ponderomotive force can be obtained by averaging the nonlinear terms in the fluid equations, or directly from a single particle motion. Recently, a quite general but compact expression has been derived[2] for the ponderomotive scalar and vector potentials, ϕ_p and $\vec{A}_p$, as

$$\phi_p = - \langle \vec{\xi} \cdot \vec{E} \rangle /2 \tag{1}$$

$$\vec{A}_p = - \langle \vec{\xi} \times \vec{B} \rangle /2 \tag{2}$$

where $\vec{\xi}$ denotes an excursion of particle due to rf electric and magnetic fields, $\vec{E}$ and $\vec{B}$, respectively. The ponderomotive force corresponds to the spatial gradient of the potentials given by Eqs.(1) and (2).

So far, the laboratory experiments on the ponderomotive scalar potential given by Eq.(1) has been reported concerning the electrostatic waves with $\vec{E}$ parallel to the wavenumber $\vec{k}$. For example, the first experiment of the ponderomotive force was reported on an electron plasma waves[2,3]. The observation showed that the plasma density is depressed by the ponderomotive force at the critical layer where the local value of electron plasma frequency ω_p coincides with the wave frequency ω. Subsequently, many observations have been reported on the ponderomotive force effects of electrostatic waves[4-10].

However, only a few experiments on the ponderomotive vector potential given by Eq.(2) have been reported on electromagnetic waves with $\vec{E}$ perpendicular to $\vec{k}$. The physical basis of the ponderomotive force is simply explained in terms of an electromagnetic radiation pressure. The nonlinear theories of electromagnetic waves in a plasma predicted interesting effects of the ponderomotive force depending on the wave frequency ω. Firstly, in a range of low-frequency($\omega<\omega_c$=electron cyclotron frequency$<\omega_p$), the ponderomotive force of whistler waves gives rise to the density hump($\delta N>0$) which in turn leads to a self-focusing of whistlers[11, 12]. However, the high-power laboratory experiments on self-focusing whistlers turned out to be unsuccessful[13,14] because of collisional heating. Secondly, in the case of overdense unmagnetized plasma($\omega_c<<\omega<\omega_p$), the ponderomotive force of electromagnetic wave results in the density depression($\delta N<0$), in which the wave is trapped to generate a filamentation instability[15-17]. Thirdly, the density dip due to the ponderomotive force is also predicted for the underdense unmagnetized plasma($\omega>\omega_p>>\omega_c$)[15].

In the underdense condition, a pressure balance between the plasma kinetic pressure and the electromagnetic radiation pressure leads to the expression[15] for the plasma density N as

$$N = N_o \exp(e\Psi_p/\kappa T_e) \qquad (3)$$

where the ponderomotive potential $\Psi_p=-eE^2/(2m_e\omega^2)$, and the electron temperature $T_e>>$the ion temperature T_i. This expression can be rewritten as

$$N/N_o = \exp[-(v_d/v_e)^2/2] \qquad (4)$$

where the quiver velocity $v_d=eE/m_e\omega$, the electron thermal velocity $v_e=(T_e/m_e)^{1/2}$, and E=the peak amplitude of the wave electric field

In this paper, the experimental verification of the ponderomotive vector potential in the underdense plasma is presented[18]. A well-defined laboratory experiment is performed with use of a plasma-filled waveguide in which an intense microwave pulse is injected.

2. EXPERIMENTAL SETUP AND METHODS

In order to compare the experiment with the one-dimentional ponderomotive force theory, we use an oversize(7 cm × 7 cm) rectangular waveguide which is comletely filled with a discharge plasma. Microwave pulse up to 10 kW and time width 3 μsec at frequency $\omega/2\pi$=9.37 GHz are launched into the plasma waveguide of 25 cm length, via a tapered section which is connected with a standard x-band waveguide. The end of the plasma-filled waveguide is terminated by a stainless steel fine mesh which is the anode for the discharge using an oxide-coated cathode of 10 cm diam. The plasma is produced in argon at a pressure of 3×10^{-4} Torr, under a weak axial magnetic field(B_o=150 G, $\omega_c<<\omega$). The experiment is performed for the underdense($\omega>\omega_p$) plasma with the density $N_o\simeq5\times10^{11}$ cm^{-3} and the electron temperature $T_e \simeq 4$ eV.

Almost all the incident power is reflected at the terminating position of mesh anode, and the standing wave is formed in the plasma waveguide(see the top trace in Fig.2). Two wire probes are utilized as both a microwave antenna and a Langmuire probe. One is movable in the z-direction along the axial slit of the waveguide, and the other traverses the axis in the x-direction perpendicular to the wave electric field in the waveguide.

3. EXPERIMENTAL RESULTS

(a) Time Evolution

In the absence of the plasma, axial and transverse measurements of electromagnetic fields cofirmed the standing wave of TE_{10} mode in the waveguide. As the plasma density is increased, the axial pich of the standing pattern is observed to become long. The measured axial wavenumber as a function of the plasma density coincides well with the calculated value[19] given by $k_z=(k_o^2\varepsilon-k_x^2)^{1/2}$ where the dielectric constant $\varepsilon=1-\omega_p^2/\omega^2$, the wavenumber in free space $k_o=\omega/c$, and the transverse wavenumber $k_x=\pi/7\ cm^{-1}$ for the TE_{10} mode.

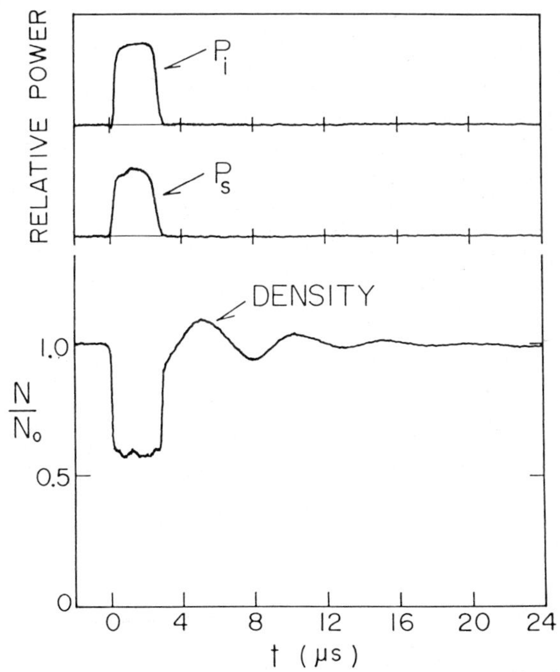

Fig. 1. Temporal evolution of plasma density N normalized by the initial density N_o, incident power P_i, and received power P_s. ω_p/ω=10 kW.

After turn-off of the microwave pulse, a damped slow oscillation of the plasma density is found as demonstrated in Fig. 1. Here the plasma density is approximate by the electron saturation current flowing to the Langmuir probe. This approximation is admissible except for $0<t<3$ μsec since the electron temperature is measured to be constant. As is well known, the Langmuire probe method is observed to be subject to errors due to the intense microwave field for $0<t<3$ μsec when the high-power microwave is turned on.

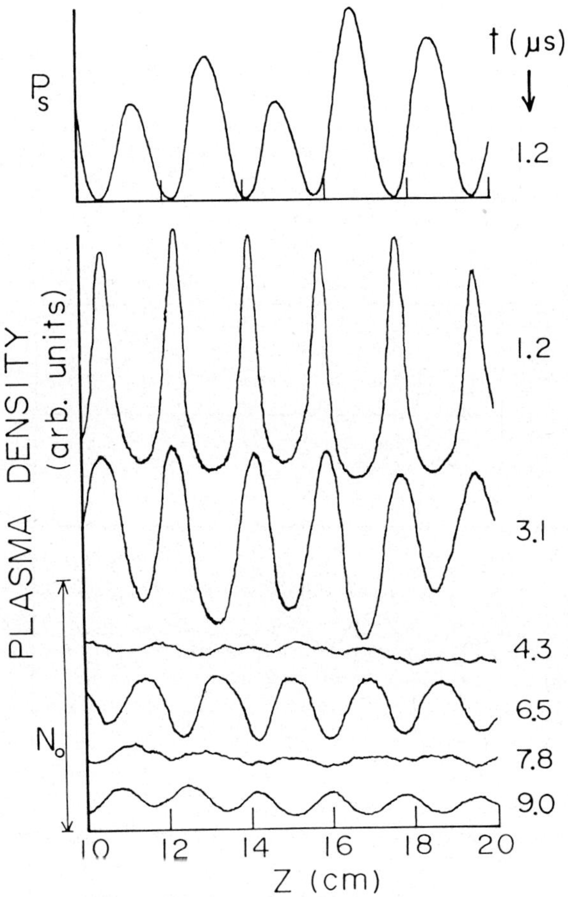

Fig. 2. Axial distribution of the received microwave power P_s and the density variations, with the time t after the turning on of the microwave power as a parameter. Zero level of the plasma density is shifted for each trace, and the magnitude of the initial density N_o is shown. z is the distance from the mesh anode, $\omega_p/\omega=0.60$, $P_i=8.8$ kW.

(b) Spatial Evolutions

Spatial evolution of the plasma density at different times t are shown in Fig. 2, together with the distribution of the received microwave power. The top trace shows the microwave field in the waveguide demonstrating the standing wave pattern. A little irregularity of the amplitude comes from the coarse movement of the motor-driven antenna.

We find two important results in Fig. 2. First, the spatial correlation is recognized between the microwave field intensity and the plasma density variations. Comparing the field distribution with the density distribution at t=3.1 μsec immediately after turn-off of the microwave power, we notice that the plasma density is depressed at the antinodes and raised at the nodes of the standing wave. That is, the wave radiation pressure depresses the plasma density with its spatial periodicity, i.e., with the pitch of half the wavelength of the guided wave.

Second, the plasma density modified by the radiation pressure begins to oscillate freely as a low-frequency standing wave for t>3 μsec. For example, the axial density distribution at t=3.1 μsec and that at t=6.5 μsec are 180^0 out of phase, and the nodes in the temporal behavior are found at t=4.3 μsec, 7.8 μsec, and so on.

(c) Ion Waves Driven by Ponderomotive Force

The time and space resolved measurements show that the axial and transverse wavelengths of the low-frequency oscillation are half of those of the microwave field. Provided that the low-frequency oscillation is an ion acoustic wave, the oscillation period is given by

$$2\pi/[(\kappa T_e/m_i)^{1/2}(4k_z^2 + 4k_x^2)^{1/2}].$$

The measured electron temperature of 4.5 eV and the observed wavelengths give the calculated period of 5.5 μsec, which is in good agreement with the observed value of 5.1 μsec in Fig. 1. Thus the low-frequency oscillation is identified to be the ion acoustic wave. In usual experiments, the ion acoustic wave is generated by a local exciter such as a grid, and it propagates as a travelling wave from the exciting source. In our experiment, the radiation pressure (ponderomotive vector potential) is regarded as a distributed exciter which is composed of an infinite number of local exciters. Thus the radiation pressure of standing microwave excites the standing ion acoustic wave.

(d) Density Depression

The magnitude of the density modification due to the radiation pressure can be estimated from the temporal decay of the ion acoustic wave. Assuming the exponential decay, we can picture the envelope of the wave amplitude in data such as in Fig. 1, to find the density at t=3 μsec.

In this way, the density depression at the antinode(t=3 μsec) was measured for different values of the incident power P_i. The density dip $\delta N(=N_o-N)$ normalized by the unperturbed value N_o is shown in Fig. 3. With increasing the microwave power, the plasma density is expelled linearly as much as 20 % by the ponderomotive force effect.

The microwwve electric field E in the plasma can be estimated from the measured incident power, and the plasma dielectric constant $\varepsilon=1 - \omega_p^2/\omega^2$. Thus, the data

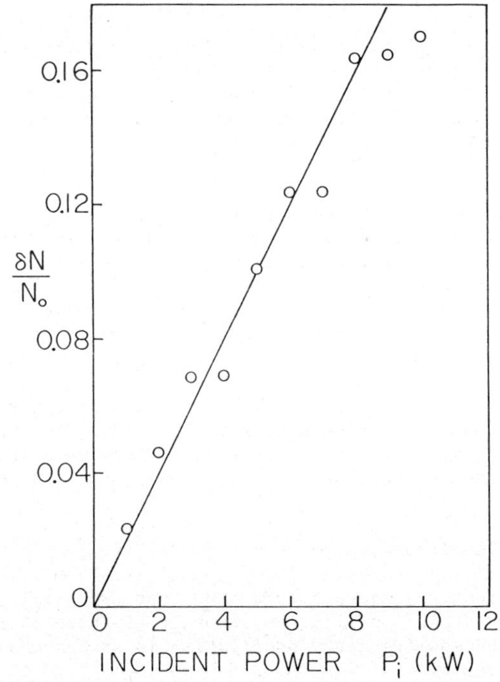

Fig. 3. Density depression δN normalized by the unperturbed density N_o as a function of incident power P_i.

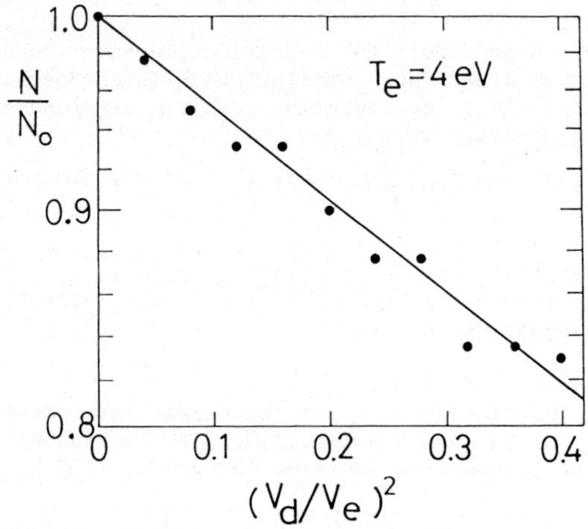

Fig. 4. Plasma density N normalized by the initial density N_o as a function of $(v_d/v_e)^2$. Experimental points and theoretical line.

shown in Fig. 3 were replotted as a function of the quiver velocity $V_d = eE/m_e\omega$. Figure 4 shows the experimental points of the plasma density as a function of v_d normalized by the electron thermal velocity v_e. The solid line indicates the ponderomotive force theory given by Eq.(4). The experimental points fit well to this theoretical lines as seen in Fig. 4. Thus, the observed density modification is explained by the poderomotive vector potential.

4. CONCLUSIONS

After a brief review of the poderomotive force effect, a laboratory experiment of the ponderomotive vector potential associated with a strong electromagnetic pulse is presented. A 9.37-GHz, 10-kW, pulsed microwave with 3 μsec duration was injected into the underdense plasma where the electron plasma frequency is less than the wave frequency.

During application of the intense electromagnetic pulse, the plasma density is observed to be depressed by the radiation pressure(ponderomotive vector potential). Furthermore, a relaxation oscillation of the plasma density is found after turn-off of the microwave pulse. This slow oscillation was identified as an ion acoustic wave driven by the ponderomotive force.

Finally, the poderomotive force effect is a general and fundamental process in the nonlinear wave-plasma interactions. This effect may take place in the wide areas of space or fusion plasmas as well as laboratory plasmas.

References

1. Landau, L.D. and Lifshitz, E.M., Electrodynamics of Continuous Media(Pergamon, New York, Adison-Wesley, Reading, Mass., 1960).
2. Kim, H.C., Stenzel, R.L., and Wong, A.Y., 1974, Phys. Rev. Lett. 33,pp.886.
3. Ikezi, H., Nishikawa, K., and Mima, K., 1974, J. Phys. Soc. Japan 37,pp.766.
4. Wong, A.Y. and Quon, B.H.,1975, Phys. Rev. Lett. 34, pp.1499.
5. Ikezi, H., Chang, R.H., and Stern, R.A., 1976, Phys. Rev. Lett. 36, pp.1047.
6. Stenzel, R.L. and Gekelman, W.,1977, Phys. Fluids 20, pp.108.
7. Sugai, H., and Takeda, S., 1980, Proc. Int. Conf. on Plasma Physics(Nagoya, 1980), Vol.I, pp.113.
8. Sugai, H. and Takeda, S., 1980, Relation between Laboratory and Space Plasmas, edited by H. Kikuchi(D. Reidel Publishing Company, 1981) pp.251.
9. Cho, T. and Tanaka, S., 1980, Phys. Rev. Lett. 45, pp.1403.
10. Sugai, H., 1983, Proc. 16th Int. Conf. on Phenomena in Ionized Gases(Duesseldorf, 1983), Invited Lectures, pp.266.
11. Taniuti, T. and Washimi, H., 1969, Phys. Rev. Lett. 22, pp.454.
12. Karpman, V.I. and Washimi, H., 1977, J. Plasma Physics 18, pp.173.
13. Sugai, H., Maruyama, M., Sato, M., and Takeda, S., 1978, Phys. Fluids 21,pp.690.
14. Sugai, H., Niki, H., Takeda, S., and Inutake, M., 1980, Phys. Fluids 23,pp.2134.
15. Kaw, P.K., Schmidt, G., and Wilcox, T., 1973, Phys. Fluids 16, pp.1522.
16. Drake, J.F., et al., 1974, Phys. Fluids 17, pp.778.
17. Langdon, A.B., and Lasinski, B.F., 1975, Phys. Rev. Lett. 34, pp.934.
18. Sugai, H., Maruyama, M., Sato, M., and Takeda, S., 1977, Phys. Lett. 64A,pp.295.
19. Allis, W.P., Buchsbaum, S.J., and Bers, A., 1963, Waves in Anisotropic Plasmas (MIT Press, Cambridge, MA, 1963), Ch.9 and 10.

VI. ENVIRONMENTAL EFFECTS OF NEMP

Nonlinear and Environmental Electromagnetics, edited by H. Kikuchi
Elsevier Science Publishers B.V., Amsterdam, 1985 – Printed in The Netherlands

"Some Aspects of the EMP/EMC-Activities in the Federal Republic of Germany"

Joachim F. Nedtwig and G. Geilke
AEG-TELEFUNKEN
EMC/EMP-Center
Sedanstreet 10
7900 Ulm, F.R.G.

1. Summary

First a definiton of the term EMP as a summary of all existing electromagnetic pulses is discussed. Then a survey of EMP/EMC-Activities in West-Germany is presented as follwos: EMP-Threat, EMC-Measuring Techniques, the Status of the EMP-Standardization (Lightning and NEMP), NEMP-Simulation Techniques and EMP-Protection Measures.

2. Introduction

The electronic industries in West-Germany estimate the electro-magnetic-pulses EMP, e.g. Lightning, Nuclear EMP, Electrostatic Discharge etc. as a severe threat. Numerous efforts in the following fields are made:
- Analyzing the threat data and coupling data
- Measurement techniques (EMC, NEMP, Lightning)
- EMC- and EMP-Standardization
- Protection measures for electronic systems and equipments

Although most activities were initiated from the military region, the know-how propagates more and more into the civil region.

3. Survey of the EMP-Activities in West-Germany

Differing from the USA in West-Germany we mostly define the term EMP not only as a nuclear EMP, but as a summary of all existing electromagnetic pulses. The block diagramm of Fig. 1 shows the most important EMP e.g. Lightning (L), Nuclear Electromagnetic Pulse (NEMP), Electrostatic Discharge (ESD), Radiation of Pulsed Power (RPP) and pulses from switch-gears (PS) of electric power stations.

The EMP-threat in West-Germany is presented in Fig. 2. Besides the natural EMP, the Lightning, this paper mainly deals with the NEMP.

As you know both parts of Germany declared after world war II not to produce nuclear weapons. Nevertheless, I don't know any other country in the world, where the super mights have installed as many thousands of nuclear weapons as in Germany. Regarding this point of threat we estimate the NEMP also to be a severe threat for us. However, the solution of these problems are a task for the politicians. The technicians are demanded to analyse the NEMP-effects and to provide protection measures.

After discovering the NEMP-phenomena in 1959 the NEMP was only a military domain. Today many electrical and electronical institutes as well as EMP-departments in the industry are interested in this phenomena.

In West-Germany we have two military universities, one in Hamburg, one in Munich, which are involved in EMP activities. Most studies are theoretically, analyzing EMP-coupling and interaction. Tests concerning EMP-measuring techniques and EMP-protection measures are carried out in cooperation with industrial companies.

We have no special institutes in West-Germany who are exclusively involved in the EMC/EMP-discipline. The reason for that is the inhomogeneity of the problems to be solved, e.g. the field distributions, the transfer of pulses along lines, the measurement technique of very fast pulses with high voltages. Therefore different institutes are interested in EMP-Problems to be solved, e.g.

- High-voltage-Institutes with
  EMP-Analysis,
  HV-Test with fast rise time,
  special EMP-Components for Lightning- and NEMP-Protection

- Theoretical Electrotechnique Institutes with
  field coupling, penetration and interactions,
  field distributions and propagations

- Radio Frequency and Communication Institutes
  transmission on lines,
  Coupling into antennas,
  measurement techniques and
  protection measures (line protections and shields)

- Institutes for Electronic Measurement Techniques
  EMP-Sensors, test methods for fields and for EMP-injection into cables, test methods for the assurance and qualification of protection measures.

Most EMC/EMP-Activities in industrial companies, e.g. in central EMC/EMP-departments, are supported by the Ministry of Defence and the Military Procurement Agency.

The following explanations contain informations about

- EMC-Measurement Technique
- Status of Standardization (NEMP and Lightning)
- EMP-coupling into lines and cables
- Computer-Programs for the computation of EMP-coupling
- NEMP-Measurement Technique for electronic equipments and components
- EMP-Protection measures (NEMP and Lightning)

4. EMC-Measurement Technique

The assurance of electromagnetic compatibility EMC for systems and equipments becomes more and more important. For EMC many instructions exist, for example MIL-STD's, VG-Standards, VDE-Standards, FTZ-Standards and others. After EMC-Analysis usually EMC-measurements are carried out at an internationally comparable status, see Fig. 4, 5 and 6.

5. Status of EMP-Standardization (NEMP and Lightning)

In order to uniform terms and definitions, procedures and analysis, program-management and measurement techniques and at least the protection measures we produce in West-Germany a new EMP-Standard, called VG-Norm 96 900 - 909 (Fig. 7). This EMP-Standard considers NEMF and Lightning. In the case of Lightning we use threat models as shown in Fig. 8, in the case of NEMP in Fig. 9.

The threat data are defined in the Standard VG 96 902, part 4. With the knowledge of EMP-coupling and penetration the number of indirect Lightning damages increased being registered by the insurance companies in West-Germany (Fig. 10).

The lightning threat data for systems and equipments which are working near the earth are given in Fig. 11 for direct hits, and in Fig. 12 for indirect influences by magnetic fields.

Fig. 13 shows the Standard-Lightning Pulse 10/350 in time domain. The rise time is 10 µs, maximum amplitude 100 kA for normal hits with 95 % probability distribution, and 350 µs half-width-value. The spectrum of this pulse is given in Fig. 14. For flying objects, as aircrafts and missiles, in VG 96 901 part 4, special threat data are demanded.

In the case of NEMP two classes of threat data are discerned, Standard-EXO-NEMP and Standard-ENDO-NEMP. This means data of the exo-atmospheric and of the endo-atmospheric (near the earth) detonation. Differences of these data can be made by the employer. Fig. 15 shows the Standard-EXO-NEMP in time domain, Fig. 16 in frequency domain. For the Standard-ENDO-NEMP electrical and magnetical fields are not in phase. Fig. 17 and 19 give the pulse courses in time domain, Fig. 18 and 20 in frequency domain.

Comparisons between the different curves can easily be made by choosing the same axe scaling.

6. EMP-Coupling

For the determination of EMP-coupling into lines, antennas and structures computer programs are useful (Fig. 21) and usually additional measurements are carried out. A simple but very effective model consists of an electronic equipment and a line positioned over metallic ground (Fig. 22).

The measurement results of a 2 m long and o,5 m high single wire line under the influence of one Standard-EXO-NEMP is the typical course of the damped sine wave in Fig. 23. In addition to that pulse response a unipolar double exponential curve is obtained by matched line impedance or long lines. In the case of short antennas (≤ 10 m) the EMP-Incoupling to the feet usually will be computed (Fig. 24, 25) and additionally measured beneath a NEMP-Field-Simulator (Fig. 26).

7. Pin-test for equipments

EMP-Tests of systems are not discussed in this paper. For the EMP-threat of individual equipments interactions in a system must be additionally regarded. Nevertheless we make in analogy to the EMC-Standars MIL-STD-461B in West-Germany a proposal for testing electronic equipment installed over metallic ground like ships, aircrafts, shelters etc. to support deliverers. The following pulse shapes and limit values are valid under the conditions of STANDARD-EXO-NEMP (Fig. 15, 16). The pin-test 1 is a test at the line inputs of electronic equipments with a damped sine wave, shown in Fig. 27. The pin-test 2 considers impedance matching or long lines and is a unipolar double exponential pulse test (Fig. 28).

The limit curves are given in Fig. 29 for the current from pin to earth and in Fig. 30 for the voltage from pin to earth, regarding maximal amplitudes. The ratio of voltage and current values gives a line impedance of 100 Ω between a single wire and the metallic ground (U_{max} = 1000 V, I_{max} = 10 A).

8. NEMP-Simulators for electronic equipments and components

In this paper large field simulators for systems are not discussed. Some experiences with a smaller size of field simulators like TEM-Cells (Fig. 31, 32, 33) or open wave-guides (Fig. 35, 36) are given. These simulators are very suitable for electronic companies to be used during and after development. The advantage of TEM-Cells is the very clean transfer behaviour (Fig. 34) in the empty measuring room. The disadvantage is due to the cut-off frequency for creating higher modes than the TEM-type.

The cut-off frequency is approximately given by the expression

$$f_c \approx \frac{250\ \text{MHz}}{a}$$

for a TEM-Cell with a square cross-section and the square length a. For investigations and test series it is more convenient to use an open wave guide of 3 m height for example (Fig. 35, 36). The pulse generator reaches the maximal amplitude of 100 - 300 kV in less than 5 ns at a line impedance of 100 Ω. For these type of open wave guides STANDARD-EXO-NEMP and STANDARD-ENDO-NEMP can be realised, in the single and in the multiple-pulse-mode.

9. EMP-Protetion (Lightning and NEMP)

EMP-protection measures can be classified into
- shields for rooms, cabinets and for lines
- line and antenna protection boxes.

A survey of EMP-protection measures is shown in Fig. 37. For lightning requirements there are well-known additional protection measures like rods and leads for high currents.

The possible ways of EMP-coupling and penetration into electronic cabinets are drawn in Fig. 38a and the protections with surge arrestors in Fig. 38b. Inputs with high EMP-voltages and -currents are fitted out with arrestors of high absorption energy.

The surge behaviour of typical protection elements is different, see Fig. 39. The real surge behaviour depends on the dynamic surge data and on the electrical connections (Fig. 40).

If the connecting wires of a protective element are too long, easily resonance circuits were built up with higher surge voltages (Fig. 41) which may damage the following electronic.

On the market different forms of gas arrestors are offered (Fig. 42, 43). The higher the surge absorption energy is the larger the size of the gas-arrestors.

Special electronic companies offer printed cards with complete protection circuits (Fig. 44). Fig. 45 shows an example of a protection circuit of a symmetrical signal line (telephone line).

The protection of power lines is very important too. In the case of mains protections the protection box must be installed at the input or outside of a building.

The use of gas-arrestors for mains protections must comply with the protection condititons for persons according to VDE-Standards. That means, that it usually is dangerous to lead lightning or NEMP currents to the surfaces of electronic cabinets, except in the case of double isolations. Further it is more efficient to separate distortions from cabinets at the entrances of cables into the buildings.

The principal block diagram of an one-phase power line is shown in Fig. 46, with a high absorption energy gas-arrestor at the input, a serial inductance as a low-pass-impedance and a low absorption element, e.g. a varistor. In the combination of surge protection circuits and line filters compatibility (resonance effects) must be taken care of. A gas-arrestor with high energy absorption for mains protection boxes is presented in Fig. 47 and 48. The schematic diagram of a complete 3-phase overvoltage protective mains box is given in Fig. 49 for transportable communication systems with protective insulation. The current through a firing spark gap shows Fig. 50 a and the mains voltage with short interrupts Fig. 50b. A 3-phase EMP-protection circuit without following filters against EMI is given in Fig. 51.

The principal concept for line protection circuits is also valid for antennas: high-current input gas-arrestor, serial inductance, low engergy fast surge arrestor (see Fig. 52, protection of an antenna). The parasitical input capacity must be controlled for guaranty of the receiving data. In many cases it is opportune to use coaxial arrestors offered at the market, Fig. 53. Fig. 54 shows the application possibilities for EMP-protection of communication shelters.

10. Conclusions

Lightnings and NEMP are severe threats. The EMC-discipline will be supplemented by EMP-activities. EMP-analisis are carried out in close connection with EMC-analises.

The EMP-protection technique in West-Germany has reached a high level. Every electronic is efficiently protectable. The question of EMP-protection today is how to optimize the cost-to-use ratio. We are involved to define new EMP-Standards for Lightning and NEMP, concerning systems, equipments and components.

11. Literature

/1/ Fleck, K. — Elektromagnetische Verträglichkeit (EMV) in der Praxis
VDE-Verlag GmbH
Seminar am 25. März 1982

/2/ Vance, E. — Coupling to Shielded Cables
John Wiley & Sons, Inc., 1978

/3/ VG-Norm — 96 900 - 96 909
Beuth Verlag, 5 Köln 1

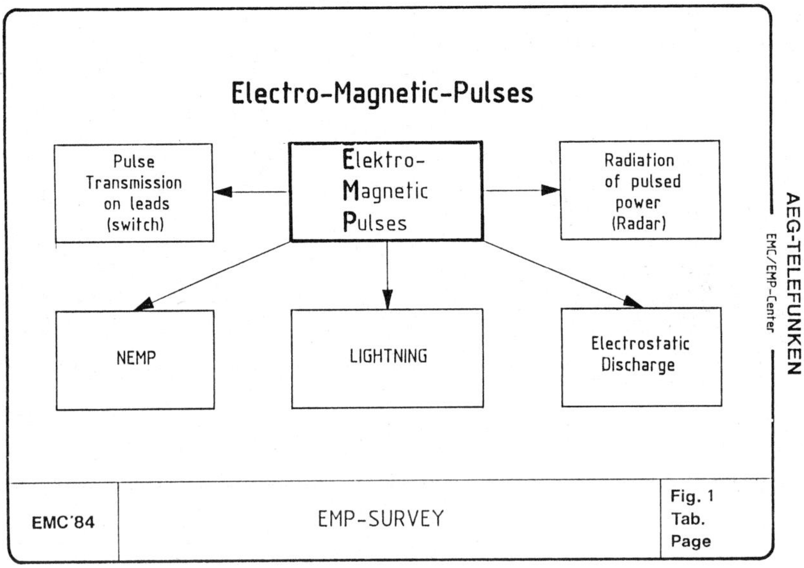
Electro-Magnetic-Pulses
Pulse Transmission on leads (switch)
Elektro-Magnetic Pulses
Radiation of pulsed power (Radar)
NEMP
LIGHTNING
Electrostatic Discharge
AEG-TELEFUNKEN
EMC/EMP-Center
EMC'84
EMP-SURVEY
Fig. 1
Tab.
Page

AEG-TELEFUNKEN
EMC/EMP-Center

EMP-threat situation in West-Germany

- **Lightnings**
  a real, daily possible event
- **NEMP's**
  we hope, that this event
  never happens,
  but we protect our electronic
- **others:**
  ESD, Transients on Leads,
  Radar, etc.

EMC'84	EMP-threat situation	Fig. 2 Tab. Page

A1E161 1084E

AEG-TELEFUNKEN
EMC/EMP-Center

Fields of EMP-activities in West-Germany

- military
  - Army
  - Airforce
  - Navy
- military universities
  - Hamburg
  - München (Munich)
- industrial
  - electronic companies
  - construction companies
- universities
  - high voltage institutes
  - high frequency institutes
- civil protection
  - power supply
  - PTT, Broadcast, TV
  - railways
  - computer centers
  - water supply
  - industrial companies
  - home protection
  - constructions, buildings

EMC'84	Fields of EMP-activities	Fig. 3 Tab. Page

A1E161 1084E

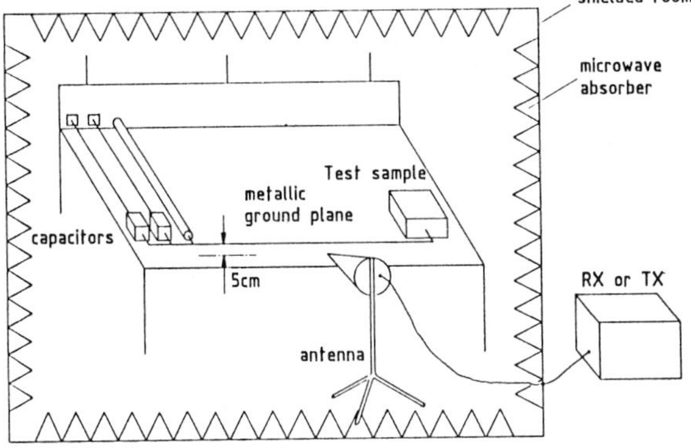

AEG-TELEFUNKEN
EMC/EMP-Center

EMC'84	TYPICAL TEST-SET-UP FOR "RADIATED EMISSION" AND "RADIATED SUSCEPTIBILITY" TESTS	Fig. 4 Tab. Page

AEG-TELEFUNKEN
EMC/EMP-Center

EMC'84	Radiated-Susceptibility-Test of commercial Cars in a shielded, anechoic chamber	Fig.5 Tab. Page

AEG-TELEFUNKEN
EMC/EMP-Center

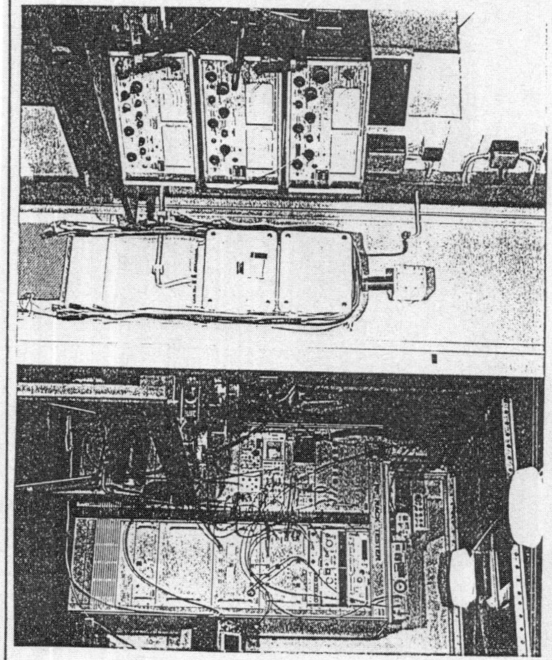

EMC'84	Typical EMC-Test-Set-Up in a shielded room	Fig. 6 Tab. Page

AEG-TELEFUNKEN
EMC/EMP-Center

LIGHTNING's

- Lightning-direct-strokes (LDS)
- Lightning-induced-EMP (LEMP)
- Nuclear-Lightnings (NL)

EMC'84	Lightning's	Fig. 8 Tab. Page

2 N E M P and L i g h t n i n g P r o t e c t i o n; s u r v e y

2.1 Schematic assembly of VG-standards for NEMP and Lightning Protection

-DRAFT-

VG 96 900-909

NEMP - and LIGHTNING-
STANDARD of Federal Republic
of Germany

Fig. 7

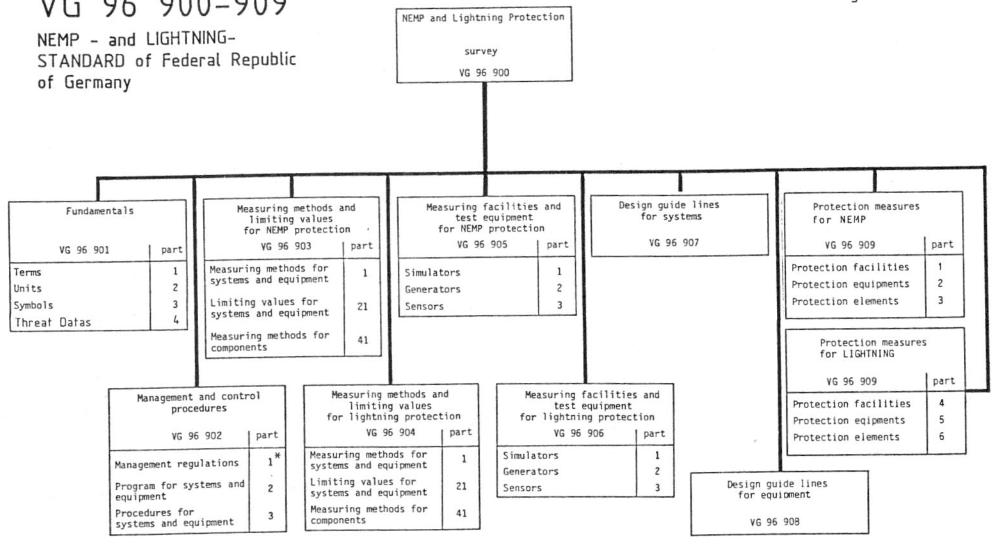

*This standard is published
Status February 1982

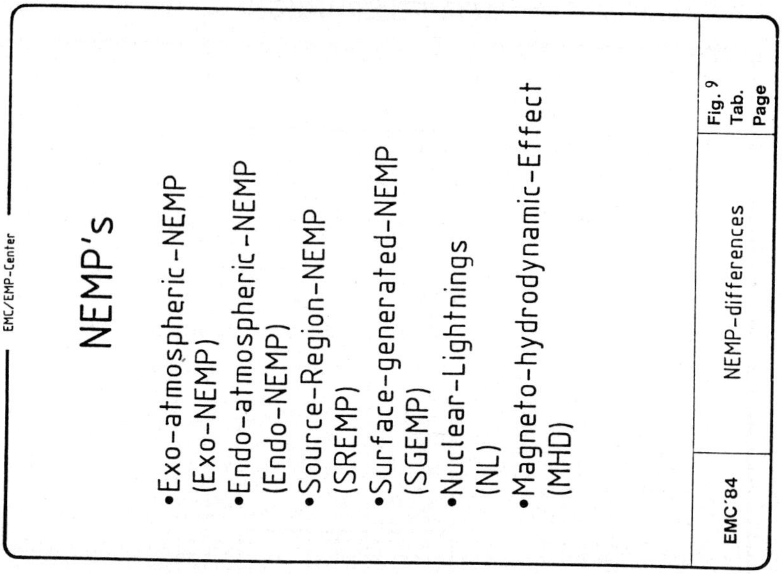

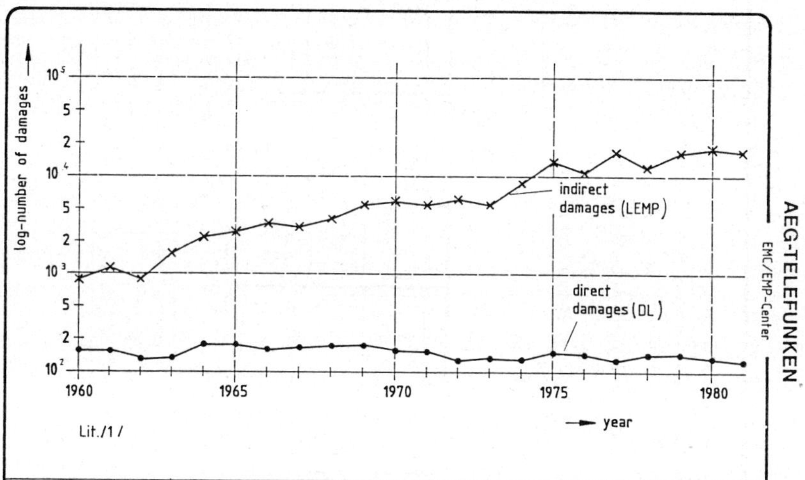

EMC'84	Lightning damages in West-Germany registered by insurance agencies	Fig. 10 Tab. Page

AEG-TELEFUNKEN
EMC/EMP-Center

threat-class	$\hat{i}_{max}$	Δi/Δt for 0,25 μs	Δi/Δt for 10 μs	Q_{surge} in 1 ms	$Q_{Longtime}$ in 250 ms	W/R
	kA	kA/μs	kA/μs	As	As	MJ/Ω bzw. $(kA)^2 \cdot s$
normal	100	100	10	50	100	2,5
high	200	200	20	100	200	10

EMC'84 | Primary LIGHTNING-Datas of German-STANDARD VG 96901, part 4 | Fig.11 Tab. Page

A1E161 10.84 E

AEG-TELEFUNKEN
EMC/EMP-Center

threat-class	H_{max}	$\hat{H}_{max}$ for 0,25 μs	H_{max}/T_1 for 10 μs
	kA/m	$\frac{kA/m}{\mu s}$	$\frac{kA/m}{\mu s}$
normal	$\frac{16}{s/m}$	$\frac{16}{s/m}$	$\frac{1,6}{s/m}$
high	$\frac{32}{s/m}$	$\frac{32}{s/m}$	$\frac{1,6}{s/m}$

$$H(t) = \frac{I}{2\pi s};$$

$$E(t) \approx 500 kV/m$$

$$0 \leq s \leq 100 m$$

(s = distance)

German Standard VG 96901, part 4

EMC'84 | Fieldstrength derived from primary LIGHTNING Datas | Fig. 12 Tab. Page

A1E161 10.84 E

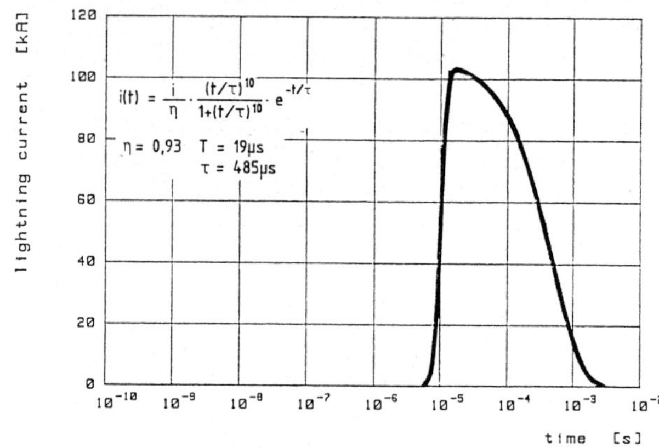

I(t) = 100*(1/.93)*[(t/1.9E-5)^10/(1+(t/1.9E-5)^10)]*EXP(-t/4.85E-4) [kA]

AEG-TELEFUNKEN
EMC/EMP-Center

EMC'84	100 kA-LIGHTNING 10/350	Fig. 13 Tab. Page

A1E161 10.84 E

A1E161 11.83 E

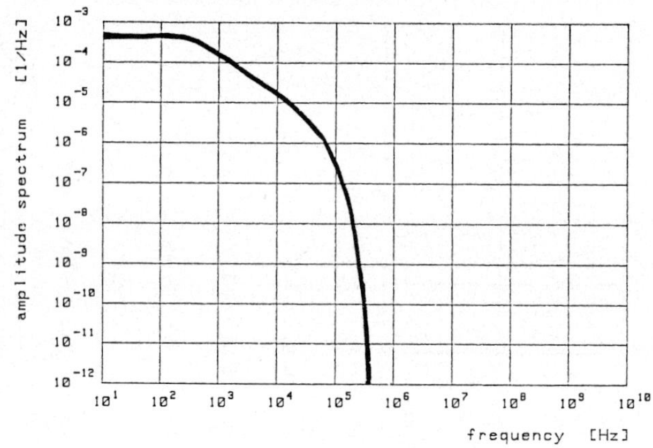

Peak Value normalized :
I(t) = (1/.93)*[(t/19.E-6)^10/(1+(t/19.E-6)^10)]*EXP(-t/485.E-6)

AEG-TELEFUNKEN
EMV/EMP-Zentrum

10/84	LIGHTNING 10/350 Amplitude Spectrum of the Current	Abb. 14 Tab. Seite

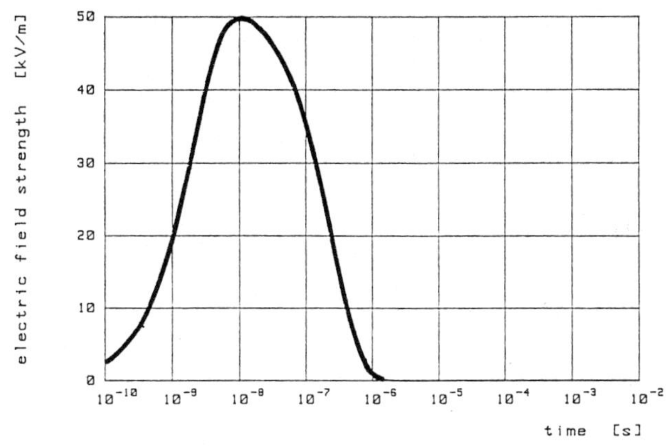

E(t) = 50*[1.05*EXP(-4.E+6*t)-1.05*EXP(-4.76E+8*t)] [kV/m]

EMC'84	Standard-EXO-NEMP Electric Field Strength	Fig. 15 Tab. Page

AEG-TELEFUNKEN
EMC/EMP-Center

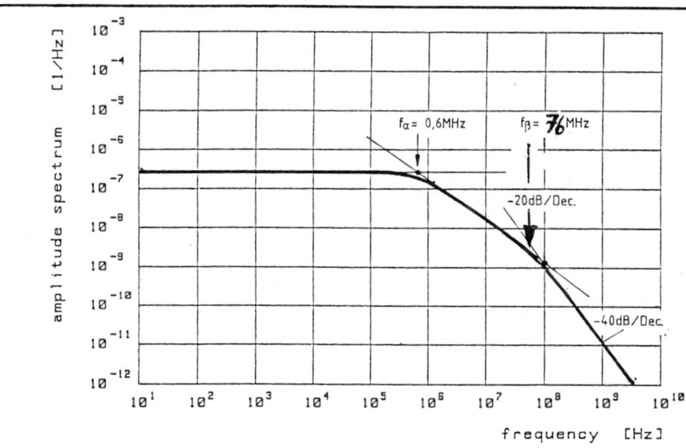

Peak Value normalized :

E(t) = 1.05*EXP(-4.E+6*t)-1.05*EXP(-4.76E+8*t)

EMC'84	Standard-EXO-NEMP (Difference of 2 Exponentials) Amplitude Spectrum	Fig. 16 Tab. Page

AEG-TELEFUNKEN
EMC/EMP-Center

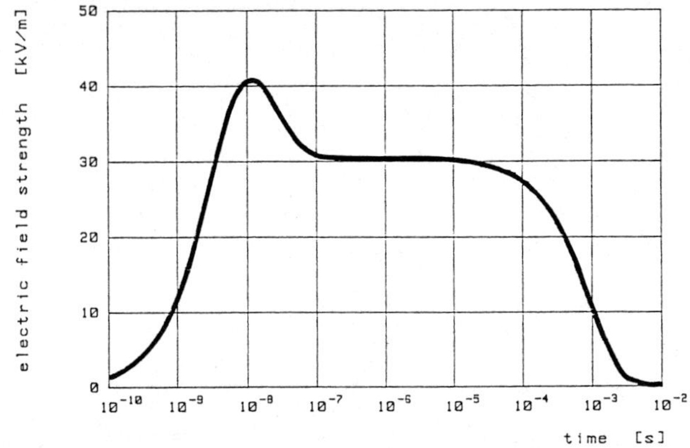

E(t) - 40*[.475*EXP(-4.E+7*t)-1.235*EXP(-3.E+8*t)+.76*EXP(-1.E+3*t)] [kV/m]

EMC'84	Standard-ENDO-NEMP Electric Field Strength	Fig. 17 Tab. Page

AEG-TELEFUNKEN
EMC/EMP-Center

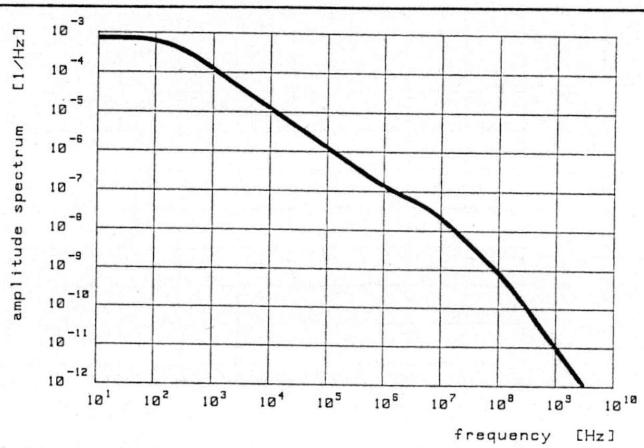

Peak Value normalized :

E(t) - .475*EXP(-4.E+7*t)-1.235*EXP(-3.E+8*t)+.76*EXP(-1.E+3*t)

EMC'84	Standard-ENDO-NEMP (Difference of 3 Exponentials) Amplitude Spectrum of the Electric Field	Fig.18 Tab. Page

AEG-TELEFUNKEN
EMC/EMP-Center

A1E161 084 E

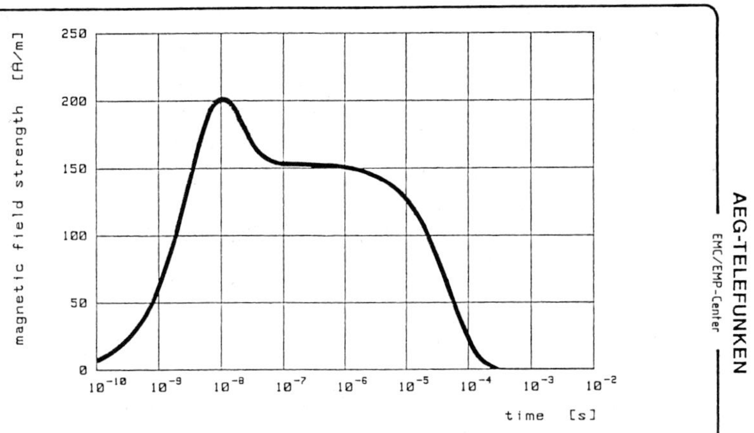

H(t) = 200*[.55*EXP(-6.E+7*t)-1.32*EXP(-3.E+8*t)+.77*EXP(-2.E+4*t)] [A/m]

EMC'84	Standard-ENDO-NEMP Magnetic Field Strength	Fig.19 Tab. Page

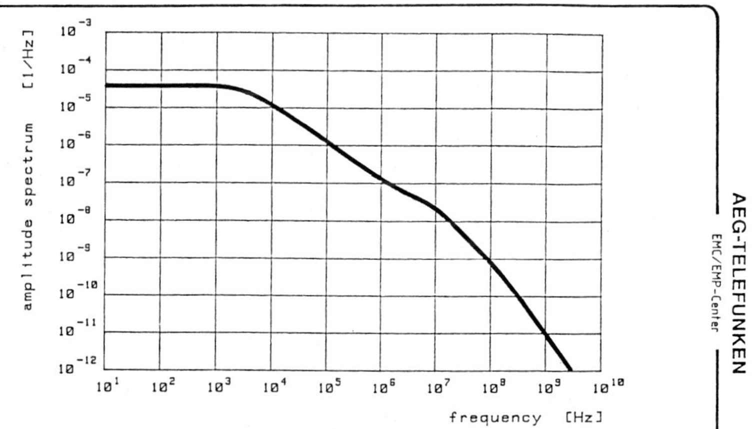

Peak Value normalized :
H(t) = .55*EXP(-6.E+7*t)-1.32*EXP(-3.E+8*t)+.77*EXP(-2.E+4*t)

EMC'84	Standard-ENDO-NEMP (Difference of 3 Exponentials) Amplitude Spectrum of the Magnetic Field	Fig. 20 Tab. Page

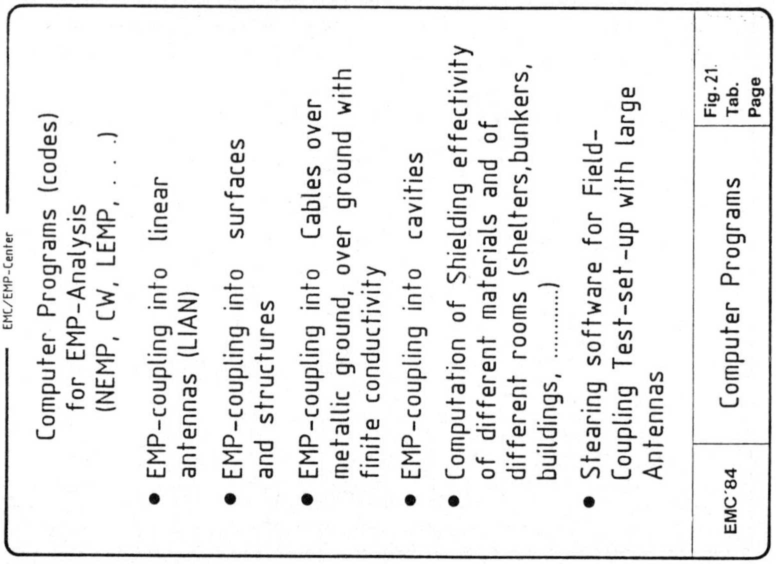
AEG-TELEFUNKEN
EMC/EMP-Center
Computer Programs (codes)
for EMP-Analysis
(NEMP, CW, LEMP, . . .)
● EMP-coupling into linear antennas (LIAN)
● EMP-coupling into surfaces and structures
● EMP-coupling into Cables over metallic ground, over ground with finite conductivity
● EMP-coupling into cavities
● Computation of Shielding effectivity of different materials and of different rooms (shelters, bunkers, buildings, ...)
● Stearing software for Field-Coupling Test-set-up with large Antennas
EMC'84
Computer Programs
Fig. 21
Tab.
Page

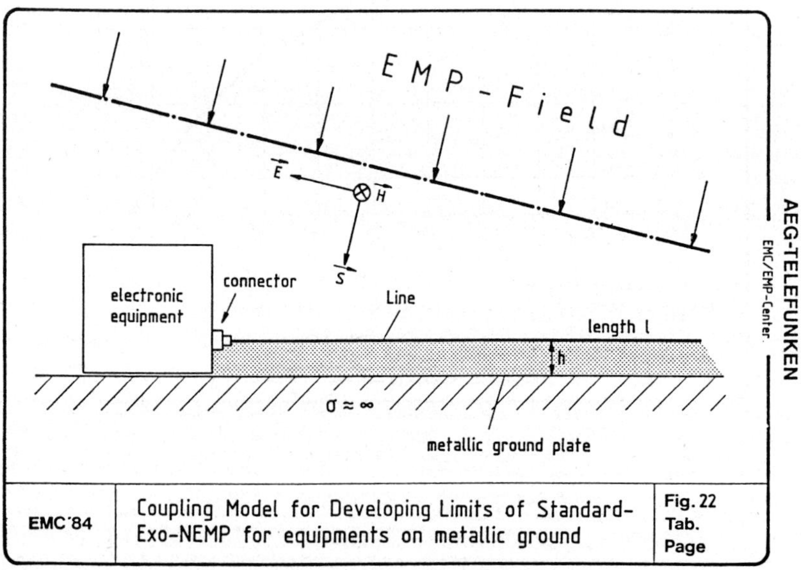
EMP-Field
E
H
S
electronic equipment
connector
Line
length l
h
σ ≈ ∞
metallic ground plate
AEG-TELEFUNKEN
EMC/EMP-Center
EMC'84
Coupling Model for Developing Limits of Standard-Exo-NEMP for equipments on metallic ground
Fig. 22
Tab.
Page

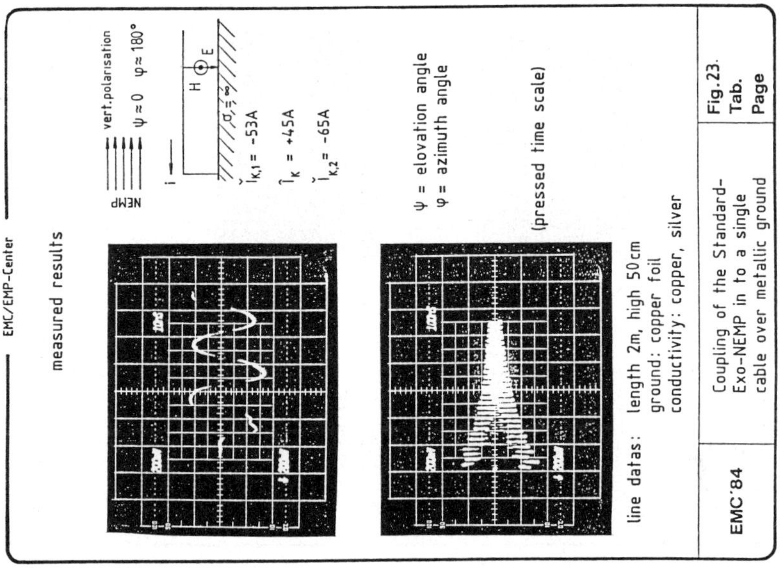

EMC'84 — Coupling of the Standard-Exo-NEMP in to a single cable over metallic ground — Fig.23 Tab. Page

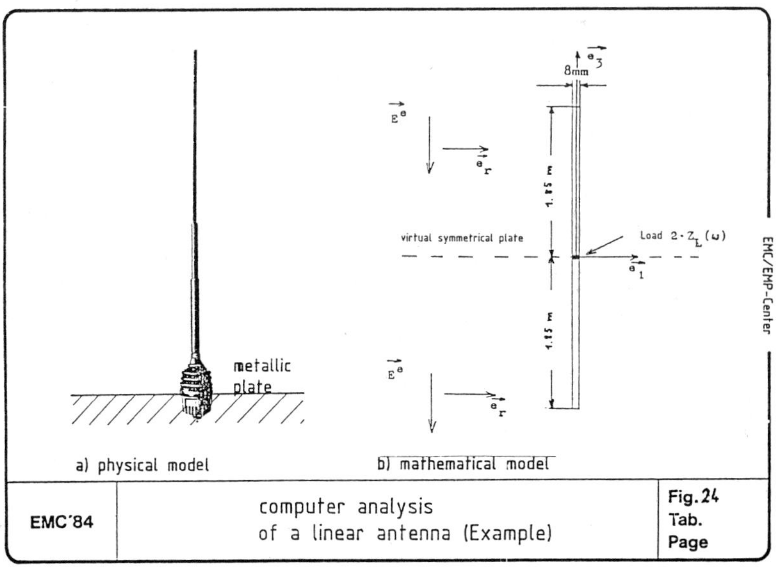

EMC'84 — computer analysis of a linear antenna (Example) — Fig.24 Tab. Page

AEG-TELEFUNKEN

EMC/EMP-Center

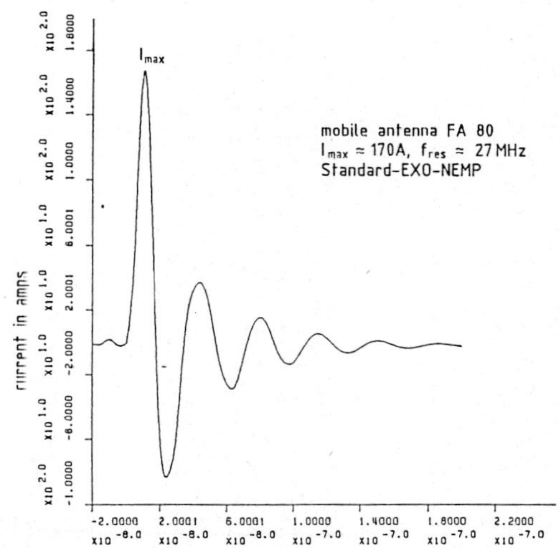

EMC'84	Computed EMP-data of antenna FA 80	Fig. 25 Tab. Page

A1E161 10.84 E

AEG-TELEFUNKEN

EMC/EMP-Center

a) measured current
I_{max} = 180A
E = 50kV/m
h = 180cm
R_L = 50Ω

I = 50A/Div

b) measured voltage
U_{max} = 9kV
E = 50kV/m
h = 180cm
R_L = 50Ω

U = 2kV/Div

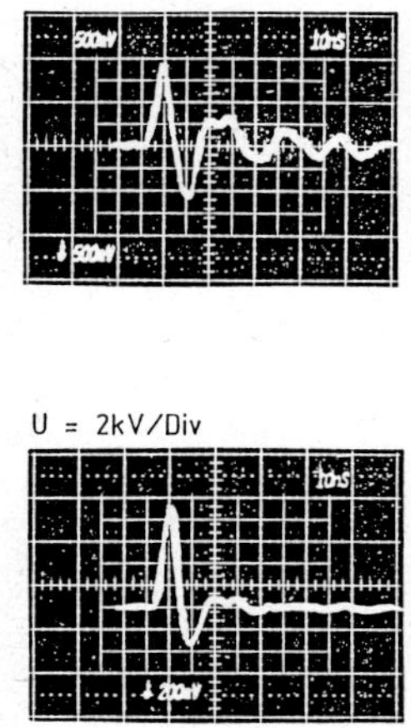

EMC'84	measured EMP-data of an antenna	Fig. 26 Tab. Page

A1E161 10.84 E

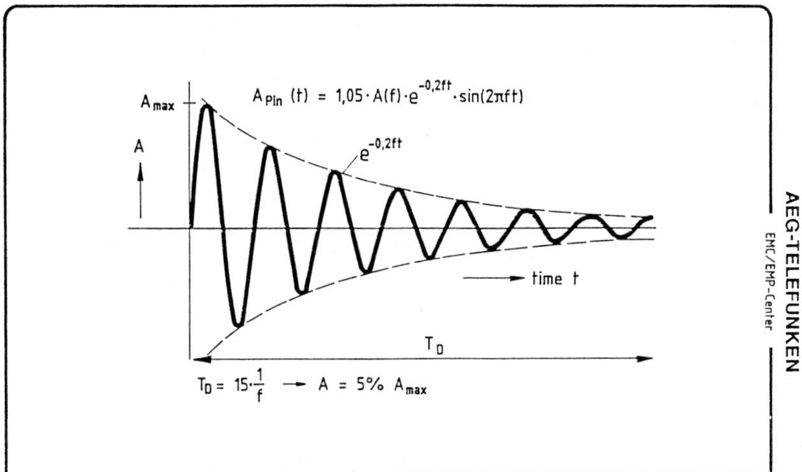

EMC'84	Testimpulse 1 at the PIN's of an electronic equipment (Exo-NEMP)	Fig. 27 Tab. Page

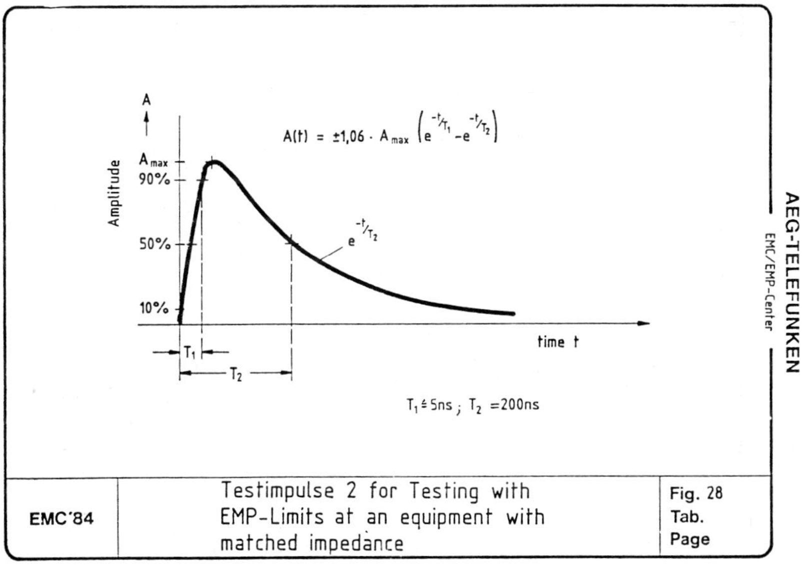

EMC'84	Testimpulse 2 for Testing with EMP-Limits at an equipment with matched impedance	Fig. 28 Tab. Page

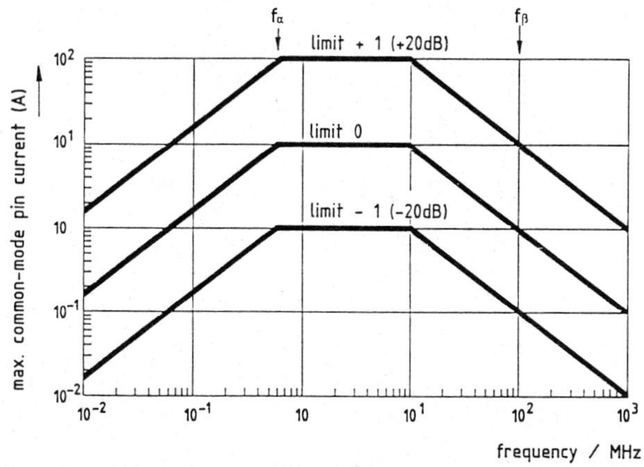

AEG-TELEFUNKEN
EMC/EMP-Center

EMC'84	Proposal for Standardization Exo-NEMP Interface-PIN-Current-test at electronic equipment	Fig. 29 Tab. Page

A1E161 10.84 E

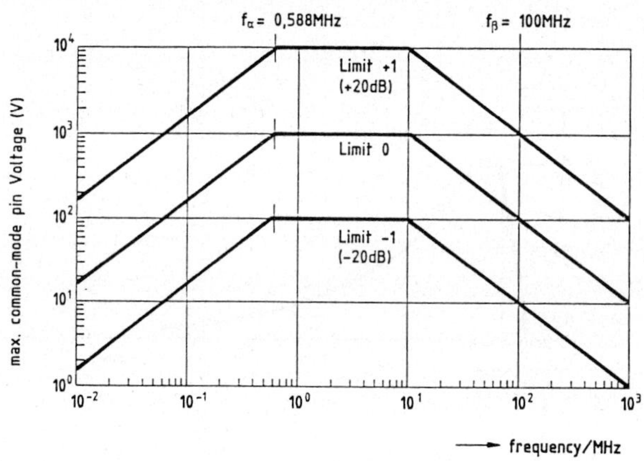

AEG-TELEFUNKEN
EMC/EMP-Center

EMC'84	Proposal for Standardization Exo-NEMP Interface-PIN-Voltage-test at electronic equipments	Fig. 30 Tab. Page

A1E161 10.84 E

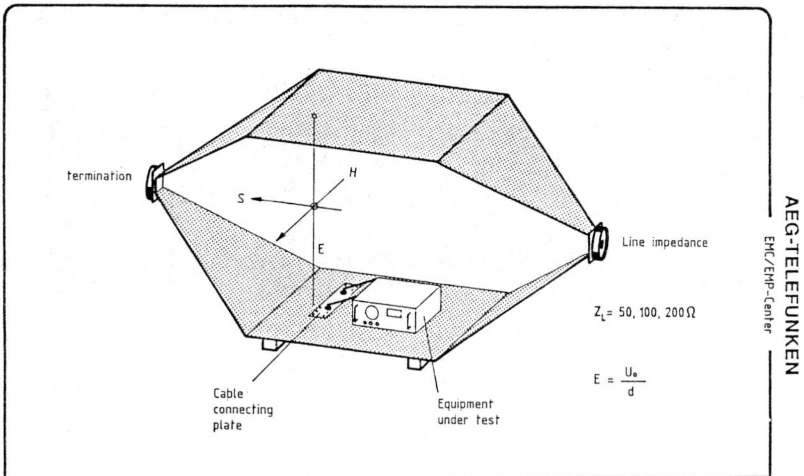

EMC'84 — Principle structure of open TEM-Cell — Fig. 31 Tab. Page

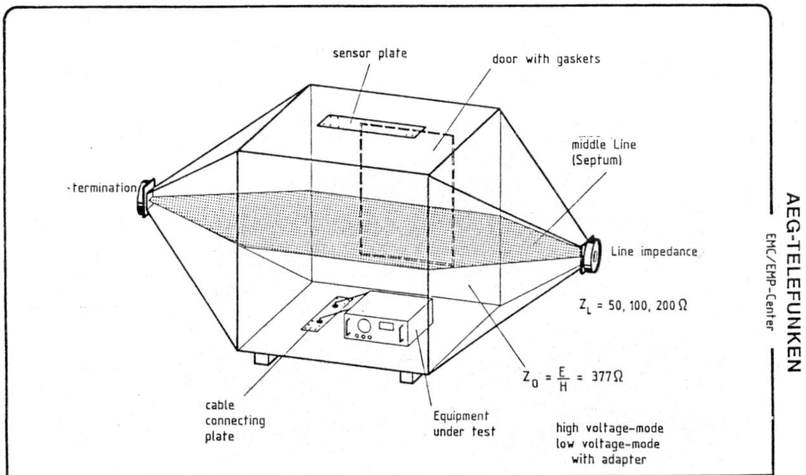

EMC'84 — Principle structure of shielded TEM-Cell — Fig. 32 Tab. Page

AEG-TELEFUNKEN
EMC/EMP-Center

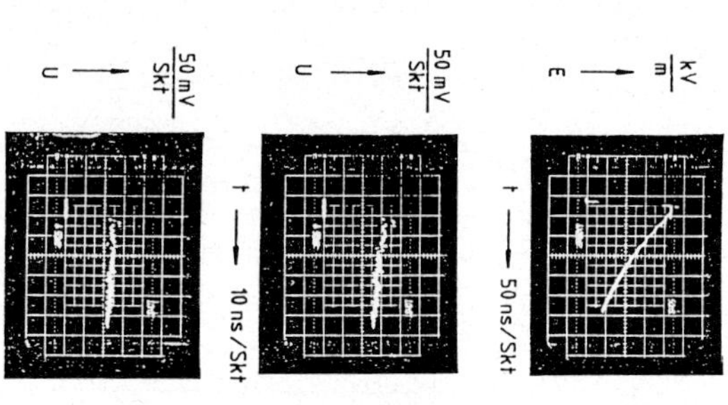

E-Field
measured with
E-field sensor
T_1 = 3ns
T_2 = 280ns

$$\frac{E}{\left(\frac{kV}{m}\right)} = 0{,}23 \frac{U}{(mV)}$$

E_{max} = 75kV/m

(stretched time scale)

H-Field
measured with
H-Field sensor

$$\frac{H}{\left(\frac{A}{m}\right)} = 0{,}8 \cdot \frac{U}{(mV)}$$

H_{max}=200A/m

EMC'84	E- and H-Field measured in a TEM-Cell (Standard-EXO-NEMP)	Fig. 34 Tab. Page

AEG-TELEFUNKEN
EMC/EMP-Center

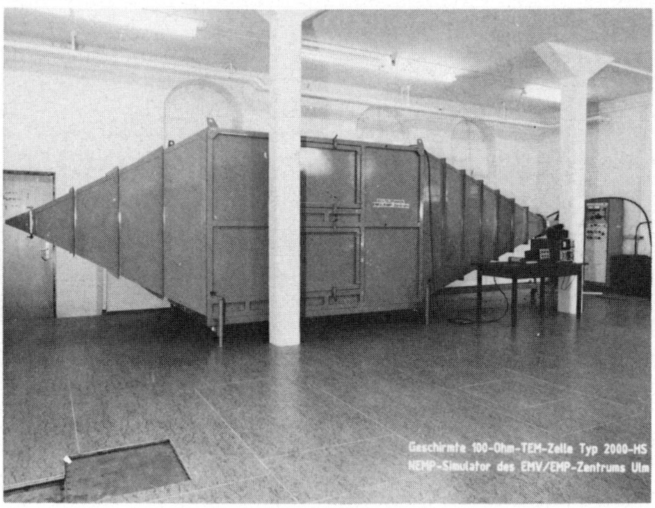

EMC'84	Self-constructed TEM-Cell Z_L= 100Ω , High: 2m , Breadth: 2m Length: 8,5m	Fig.33 Tab. Page

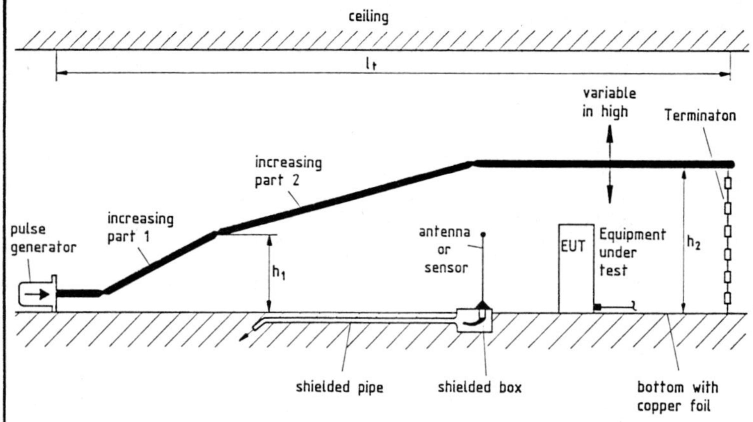

AEG-TELEFUNKEN
EMC/EMP-Center

EMC'84	Side view of the Open-Waveguide-Field-Simulator for Exo-NEMP	Fig. 35 Tab. Page

A1E161 10.84 E

AEG-TELEFUNKEN
EMC/EMP-Center

EMC'84	Self-constructed NEMP-Field-Simulator with Vertical Polarisation	Fig. 36 Tab. Page

AEG-TELEFUNKEN

EMC/EMP-Center

EMP-Protection facilities

- Shielding:
  of rooms, shelters and cables
  (20, 40, 60, 80, 100 dB-
  attenuation-steps)

- Protection devices:
  - mains termination boxes
  - antenna protection boxes
  - two-wire protection cards for audio-frequency lines
  - coaxial arrestors
  - filters

- Protection elements:
  Arrestors, Varistors, Diodes,
  Inductance, Capacitors, Resistors

EMC'84	NEMP- and Lightning Protection -Survey	Fig. 37, Tab. Page

A1E161 10.84 E

AEG-TELEFUNKEN

EMC/EMP-Center

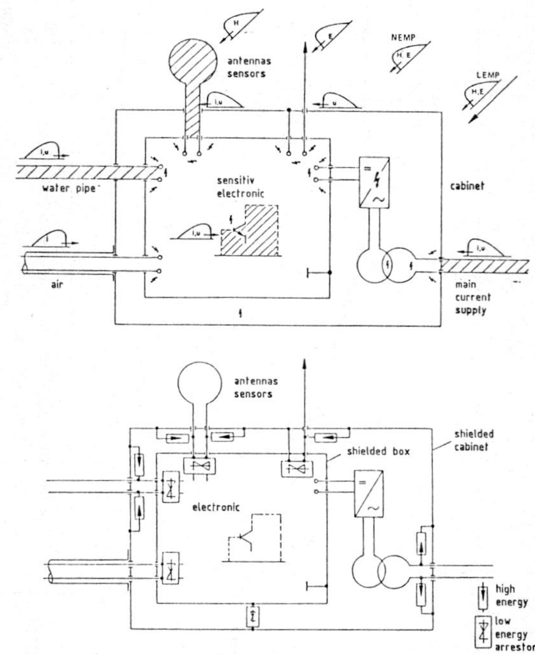

EMC'84	possible high-voltage breakdowns and protection with arrestors	Fig. 38 Tab. Page

A1E161 10.84 E

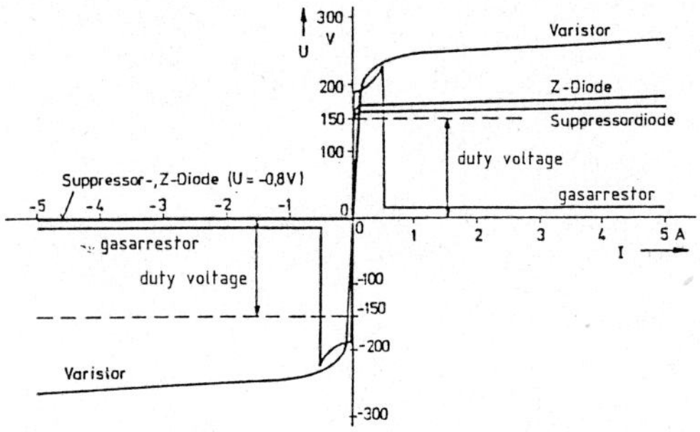

AEG-TELEFUNKEN
EMC/EMP-Center

EMC'84	Suppressing behaviour of typical protection elements	Fig. 39 Tab. Page

A1E161 10.84 E

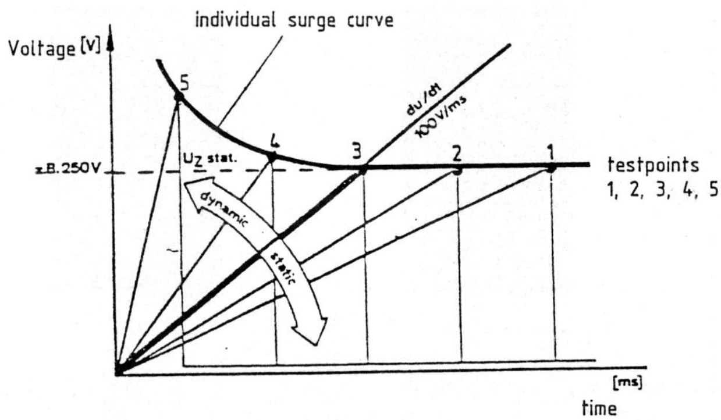

AEG-TELEFUNKEN
EMC/EMP-Center

EMC'84	Characteristic surge-voltage-curve of a protection element or equipment	Fig. 40 Tab. Page

A1E161 10.84 E

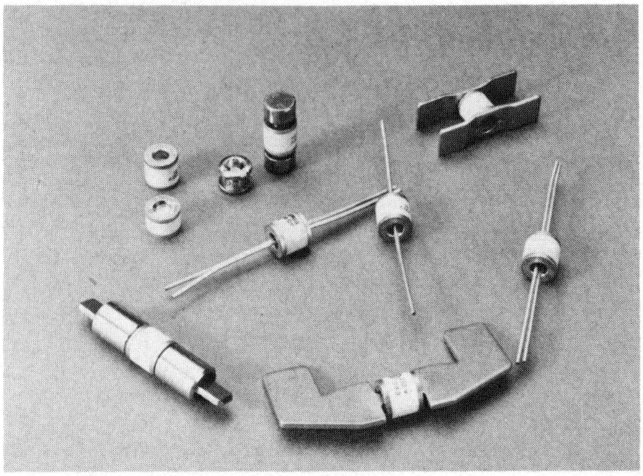

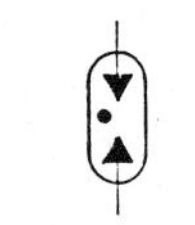

AEG-TELEFUNKEN
EMC/EMP-Center

EMC'84	different forms of gas-arrestors for middle currents	Fig. 42 Tab. Page

AEG-TELEFUNKEN
EMC/EMP-Center

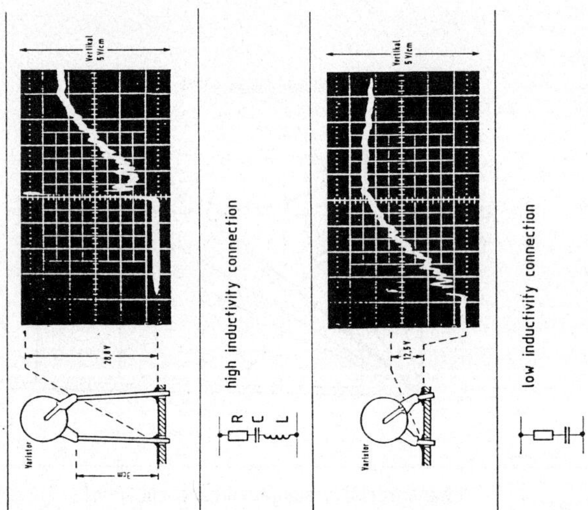

EMC'84	short leads of sup-pressor elements - good protection effects	Fig. 41 Tab. Page

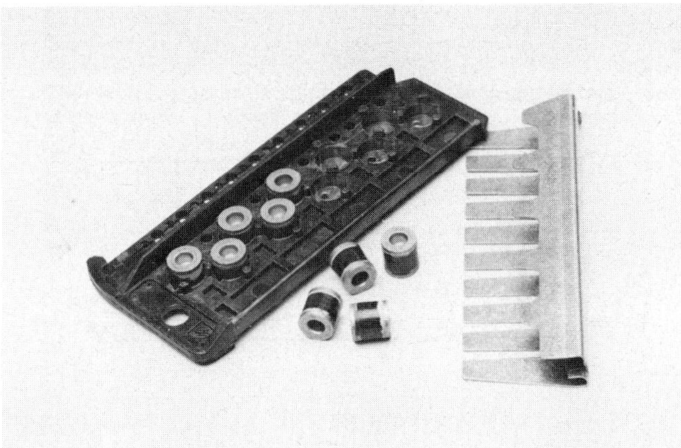

AEG-TELEFUNKEN
EMC/EMP-Center

Lit.16

EMC'84	Gasarrestors at the end of Communication Lines to protect Telefon-Centers	Fig. 43 Tab. Page

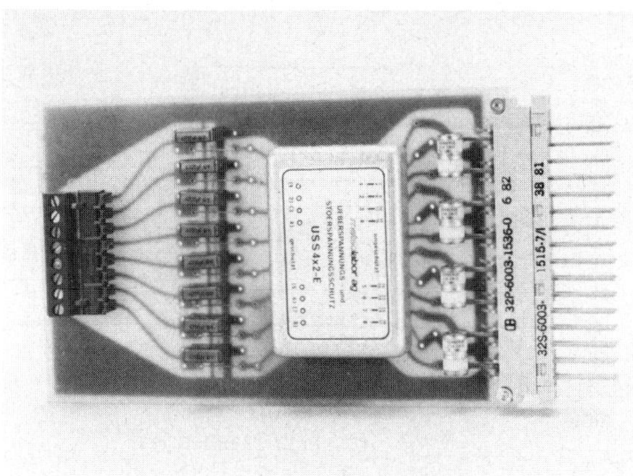

AEG-TELEFUNKEN
EMC/EMP-Center

Lit.17

EMC'84	Europa-card with protection for 8-unbalanced signal lines	Fig. 44 Tab. Page

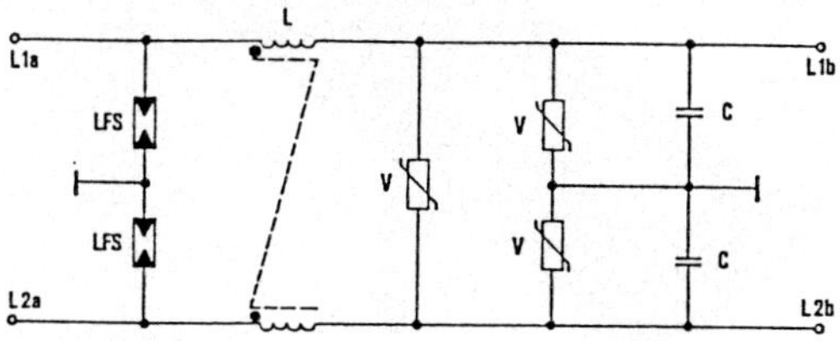

LFS = quenching gap

Lit.141

EMC'84	Schematic Diagram of the EMP-Protection for audio-frequency lines	Fig. 45 Tab. Page

AEG-TELEFUNKEN
EMC/EMP-Center

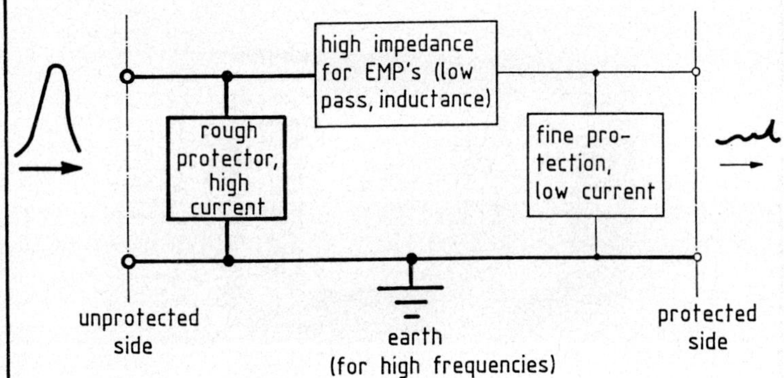

EMC'84	Principle device for Line-Protection against NEMP and Lightning	Fig. 46 Tab. Page

AEG-TELEFUNKEN
EMC/EMP-Center

AEG-TELEFUNKEN
EMC/EMP-Center

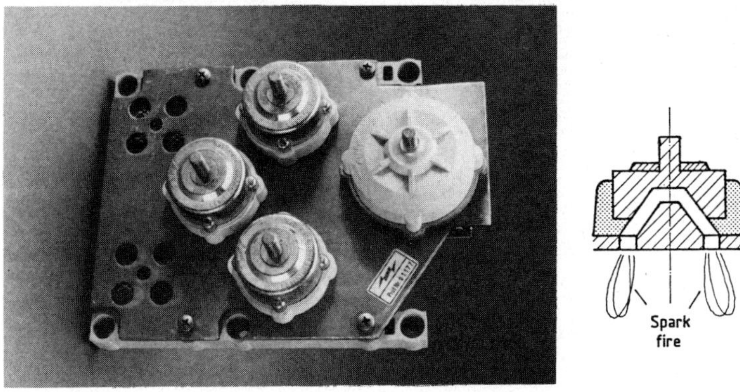

Lit. 151

EMC'84	High current quenching gaps for main current protection	Fig. 47 Tab. Page

AEG-TELEFUNKEN
EMC/EMP-Center

Mains Termination unit

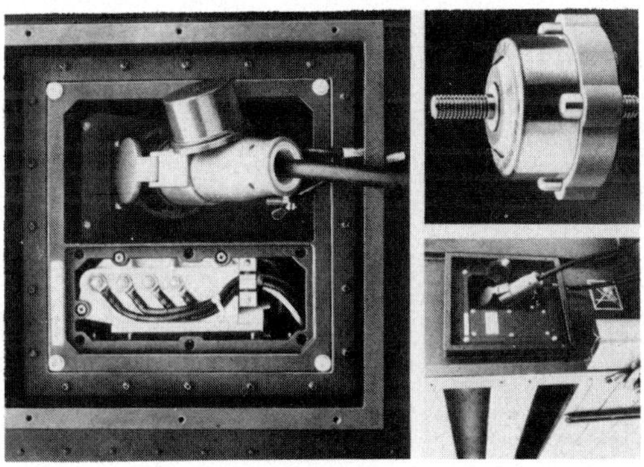

build in a shelter

High current Spark Gap

Lit. 151

EMC'84	Mains Termination Box for Lightning-, NEMP- and EMI-Profection	Fig. 48 Tab. Page

1 Input Connector
2 High-Current Spark Gap HSFS
3 Quenching Gaps LFS
4 Spike Chokes
5 Housing of the Mains Connection
6 Primary Fuses
7 Turn-On Current Limiting
8 Basic Insulation
9 Additional Insulation
10 Double or Reinforced Insulation
11 Isolating Transformer
12 Varistors
13 Feed-Through Filters
14 Housing of the Isolating Transformer
15 Enclosure of the Transportable Communication System
16 Station Ground
17 Voltage-Compensation Line

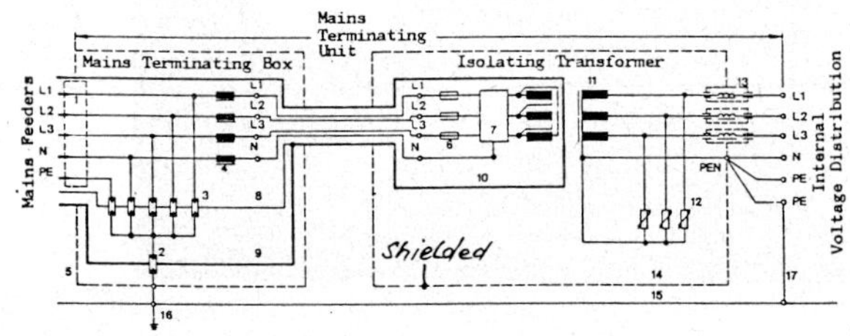

Lit.141

AEG-TELEFUNKEN
EMC/EMP-Center

EMC'84	Schematic Diagram of an Overvoltage-Protected Mains Connection for a Transportable Communication System with Protective Insulation in the Input Circuit	Fig. 49 Tab. Page

AEG-TELEFUNKEN
EMC/EMP-Center

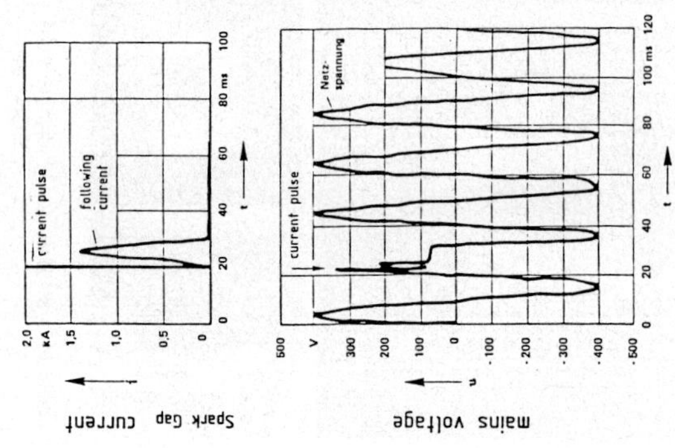

Lit.[1]

EMC'84	Current-and Voltage-Behaviour behind the mains termination box	Fig. 50 Tab. Page

AEG-TELEFUNKEN
EMC/EMP-Center

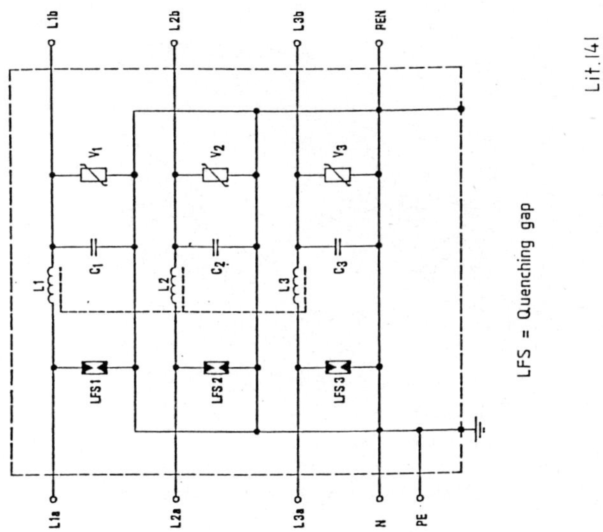

EMC'84 | - Three-Phase EMP-Protection for 63A Mains | Fig. 51 Tab. Page

AEG-TELEFUNKEN
EMC/EMP-Center

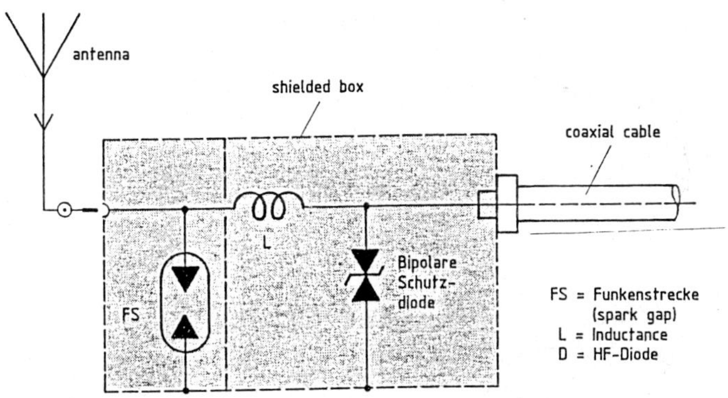

EMC'84 | Typical antenna protection circuit against Lightning and NEMP | Fig. 52 Tab. Page

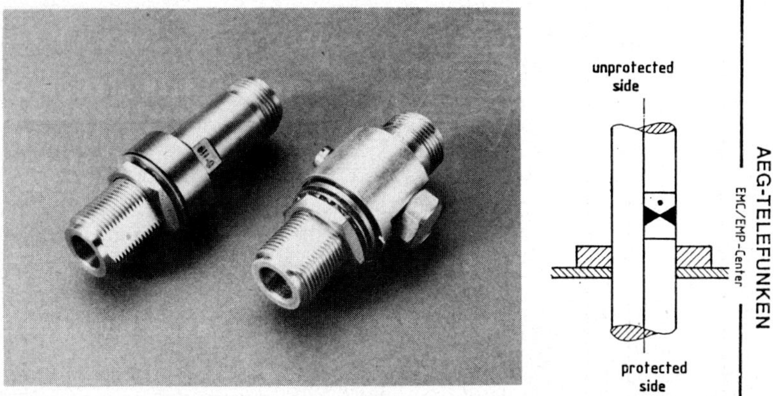

EMC'84	coaxial feedthrough with a gas-arrestor (EMP-protector)	Fig.53 Tab. Page

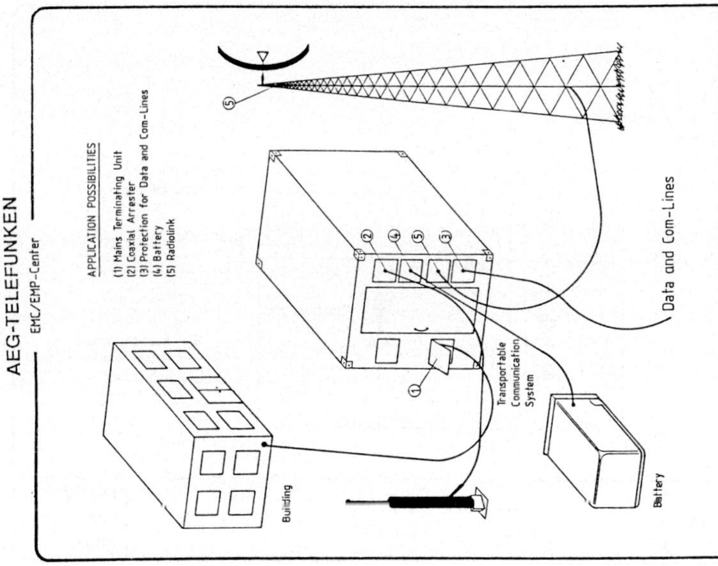

EMC'84	Application Possibilities for Lightning-, NEMP- and EMI-Protection	Fig.54 Tab. Page

Nonlinear and Environmental Electromagnetics, edited by H. Kikuchi
Elsevier Science Publishers B.V., Amsterdam, 1985 – Printed in The Netherlands

EMC AND EMP ACTIVITIES IN SWITZERLAND

M. IANOVICI

Laboratory of Electrical Networks
Swiss Federal Institute of Technology
Lausanne, Switzerland

Abstract : An overview of EMC and in particular EMP activities in Switzerland is presented. The EMC problem is regarded from the point of view of the industrial activity and of the research in the industry and the universities. The teaching aspect is also presented. Some specific domains concerned by EMC problems are discussed. Concerning EMP an activity exists in the Army Laboratories, Civil Defense organizations, Universities and in the industry. The main studies and experimental work taking place in these different places are briefly discussed and the principal results are presented. Emphasis is put in particular on problems concerning the response of transmission cables to an EMP and tranfer impedance determination.

1. INTRODUCTION

The increasing number of sensitive electrical and electronic installations working in the vicinity of power circuits along with the use of more and more sensitive components have made the electromagnetic compatibility problems of great actuality in all industrial countries.

Among the 3 millions of wage earners from Switzerland, only 7% are working in agriculture, while 39% are employed in industry, trades and construction and 54% in service operations. This shows the important place occupied by the industry and services in the swiss economy. The first one is using modern methods for the control of the production process and the computed aided design. The services are using management by computer (banks, financial services) and must be provided with reliable communications means.

A significant increase of communications devices can also be noticed in the every day life as seen in the graph from fig. 1 showing the development of this field in Switzerland during the last 30 years with an extrapolation in the future |1|.

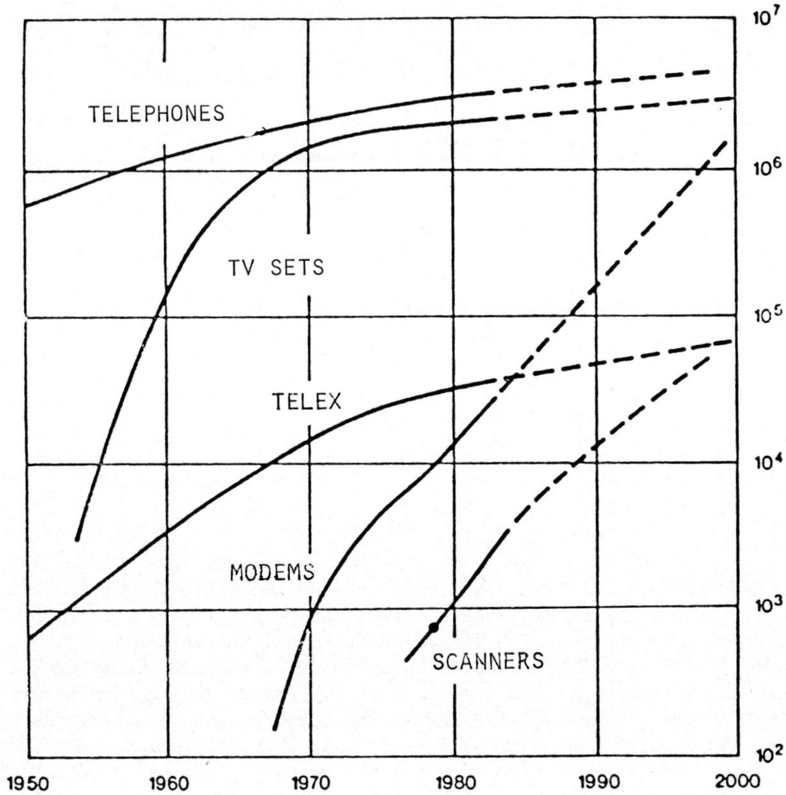

Fig. 1 - Increase of the number of information transmission devices in Switzerland

In order to insure a high reliability of sensitive devices working in an electromagnetic poluted environment, measures have to be taken and methods to compatibilize systems have to be developed.

A particular aspect of the EMC are the EMP problems. Due to its large spacial extension the effect of a high altitude nuclear electromagnetic pulse (HAEMP) can cover the surface of a continent and therefore even a neutral country like Switzerland has to consider the possibility of beeing concerned by it and to take protection measures.

The aim of this paper is to present some aspects of the development of these two fields in Switzerland.

2. EMC ACTIVITIES

2.1 History

The first problems caused by disturbances and also the fight against them have appeared in the early 30's together with the development of radio broadcasting.

The CISPR (Comité International Spécial des Perturbations Radioélectriques) is founded and Switzerland is from the begining an active member.

A semi-private company Pro Radio is charged by the Swiss Post and Telecommunications to perform interference measurements and noise suppression.

The first standards in this field appear in Switzerland in the early 40's.

After the war period during which the main activity takes place in the army services in order to insure reliable telephonic connexions and to collect informations from allied broadcasting disturbed by the german services, a new development of this field begins in the 70's with the development of the computer techniques.

2.2 To-day's situation in Switzerland in the EMC field

2.2.1 Types of activity

Several types of activity in EMC can be distinguished :

1) Practical activity in the industry
2) Research in universities and industry
3) Teaching
4) Special application fields

In what follows a brief description of the activity in Switzerland in these four direction will be given.

2.2.2 Industrial activity

A large practical activity in the EMC takes place in the Swiss industry. The aim of it is to solve in a practical way all compatibility problems appearing in systems and installations of high technology built up by different companies. The approach often based on the years of experience of specialized engineers and technicians is changing in the last years more attention beeing given to the use of a methodology based on models and numerical calculations.

The different aspects of this activity are the following :

1) EMC consulting and engineering field in which companies are active.
2) Production of EMC protection, i.e. components as filters, suppressors, feed-through capacitors, shielded cables and shielded enclosures (Faraday cages).
3) EMC simulation, consisting of a production of various types of disturbance generators in order to simulate : lightning strokes, EMP, ESD (electrostatic discharges) and bursts |2,3|.

As a result of an increasing number of computers of different types from the big units for scientific work to the PC (personnal computers) penetrating all aspects of human activity many EMC problems must be found and solved by the consulting companies. The first problem is that of disturbances comming from the mains consisting of : voltage fluctuations (from 1 s to several hours), voltage clips (20 ms - 1 s), harmonics, micro-disconnections and over-voltages. As an example, fig. 2 shows the statistics of switching over-voltages occuring in s swiss distribution network |4|.

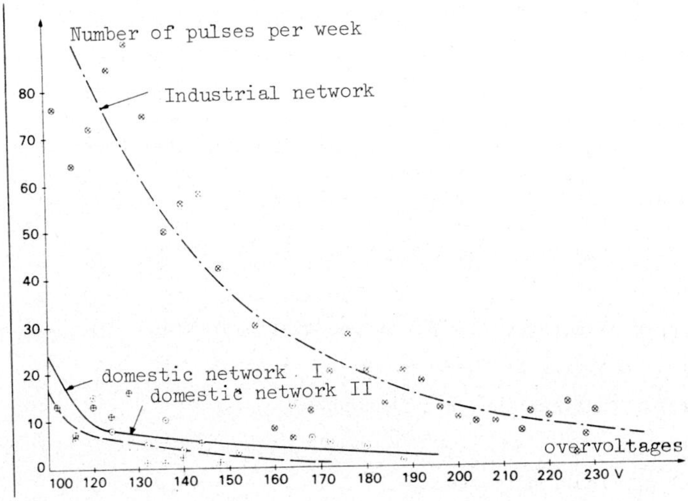

Fig. 2 - Overvoltage statistics in domestic and industrial networks in Switzerland

Another possible disturbance source is the radiated fields comming from trains, radars, talkie-walkies, static converter supplies, experimental physics or hospital equipment (Nuclear Magnetic Resonance) |5|.

As an exampke of a problem solved by a swiss engineering company in collaboration with a university is the study for the protection of a big computer center situated in the proximity of a 400 kV line and of an airport.

The proposed agenda for the EMC study was the following :

- calculation of induced current in transmission cables between two computer units, using a computer code. The disturbance can occur in case of a short-circuit in the HV network ;
- calculation of the potential distribution around the grounding of the nearest tower of the HV line (current due to a lightning stroke) ;
- measurement of the potential distribution around the grounding of the tower using a reduced current simulation ;

- measurements of radiated electromagnetic fields from the HV line (corona effect) and from the airport radars.

2.2.3 Research in EMC

The research in the EMC field takes place either :

1) In the two federal institutes of technology from Switzerland.

2) In some industries in connexion with :

  - a production of sensitive devices which must be protected against disturbances ;
  - production of EMC test installations.

The practical activity in the industry was discussed under point 2.2.2. It is obvious that in order to achieve new and reliable products and to introduce new technologies in the manufacturing process the companies must have a research activity. Often, especially in the case of small industries the research is performed in collaboration with the industries.

A typical example is the following problem. A company producing telephonic switchboards was obliged to harden its products against the effect of ESD for a voltage level of 15 kV instead of 9 kV as previously, the standard beeing changed in the meantime. A research performed in collaboration with the Federal Institute of Technology from Lausanne has permitted to find a practical solution which is now used for a series production.

A research activity centred on specific problems takes place in the federal institutes of technology.

At the Swiss Federal Institute of Technology from Zürich (original name in german : ETHZ), the EMC studies are performed in the Institute for Communication Technology. Several problems are investigated |6-12|.

- generation of homogeneous fields for immunity measurements ;
- theoretical and experimental investigations of the noise immunity of spread spectrum ;
- communications both in conducted and radiated channels.

These studies are applied to solve the problem of mobile communications and chordless telephon.

Every two years the Institute for Communication Technology of the ETHZ organizes under the presidence of professor P. Leuthold and with Dr. T. Dvorak as organising chairman the International Symposium and Technical Exhibition on EMC. This meeting which is the largest in the world in the field was organised for the first time in 1975. This year 850 participants from 26 countries attended the symposium and about 120 paper covering various fields of EMC were presented.

At he Swiss Institute of Technology from Lausanne the research in EMC is performed in the Laboratory of Electrical Networks. The main subjects developed for more than 10 years are :

- methods to calculate the coupling between electromagnetic fields and conductors |13| ;

- transfer impedance calculation and measurements |14| ;
- transient phenomena in electrical networks.

The first two points were mainly considered in connexion with Nuclear Electromagnetic Pulse (NEMP) and Lightning Electromagnetic Pulse (LEMP) effects and will be discussed in detail in the next section.

Concerning the last point transient behaviour of power networks was studied in the case of lightning or switching overvoltages. These investigations were applied to insulation coordination studies |15|.

2.2.4 EMC Teaching

In Switzerland EMC teaching in the universities has began in form of dedicated lectures in 1983. It is obvious that elements of EMC have been given long time before in different courses as telecommunication, electromagnetics, high voltage engineering (measurements in high voltage laboratories), protection of power networks, etc. However it is only in 1983 that a special course on EMC in the form of a postgraduate course was organised at the Swiss Institute of Technology from Lausanne and has been attended by 30 students. This type of teaching will be organized in Lausanne every two years.

This year (1985) an EMC course of one semester, for undergraduate students has also begun at the electrical department of the Swiss Institute of Technology from Lausanne.

Another form of teaching are workshops organized by different companies with the aim to present specific products or methods to potential customers. Such a workshop on EMC problems which took place in 1980, had 400 participants in Zürich (german speaking part) and 100 in the french part of Switzerland, coming from different industries, universities, research centers as the European Center for Nuclear Research (CERN), and state administrations and organizations (Swiss Post and Telecommunication) |16|.

2.2.5 EMC activities applied to specific fields

Two institutions can be considered as having an activity applied to specific fields.

First are the Swiss Army Laboratories the activity of which is mainly directed toward EMP hardening. This aspect will be discussed in detail in next section.

The second organization is the Swiss Post and Telecommunication which laboratories are concentrating on :

- measurements of interference and noise ;
- EMC design ;
- elaboration of standards concerning disturbances and protection against noise.

3. EMP ACTIVITIES IN SWITZERLAND

3.1 General

The activity in the EMP field can have various aspects among which one can mention the following :

- numerical calculations
- experimental tests using simulators
- protective measures
- elaboration of standards.

Such activities can take place :

- in Army institutions and laboratories
- in Civil Defense organizations
- in universities and research centers
- in the industry.

In the last two they are in general performed in connection with the EMC activity.

In Switzerland EMP studies take place in all four types of institutions. These studies are mainly directed in the following directions :

- sensitive electronics protection
- transmission cables hardening
- effects on transport and distribution power networks.

The first two points have been studied in different laboratories and industies. Very little attention has been given up to now to effects on power networks. The first studies have begun last year at the Federal Institute of Technology from Lausanne.

3.2 EMP activity of swiss army laboratories

All activities mentionned in what follows are a summary of released publications from the army laboratories.

3.2.1 The swiss army laboratory from Spiez

In this institution a theoretical activity based on numerical calculations is combined with practical measurements using simulators. The main direction of the studies performed by this group are the following :

1) Physical phenomena of HAEMP using numerical modelling with computer codes.
2) Electromangetic field interaction with structures and transmission lines using different computer codes.

In order to study the interaction of an electromagnetic wave with an object or with an antenna the laboratory uses several codes among which the Numerical Electromagnetic Code (NEC) developed by Poggio and Burke |17|, which calculates in the frequency domain the field generated by an antenna or response of an object to an electromagnetic field.

The coupling between an electromagnetic wave and a line is investigated with the aid two codes EMPLIGN and EMPCOAX developed at the Federal Institute of Technology from Lausanne, which will be described under point 3.3 and also with programs comming from other laboratories from abroad.

3) EMP simulation

This type of activity is used to test objects and systems in order to verify their hardening performances. From 1978 to 1982 the laboratory has worked with a guided wave simulator called GESIBAN with the following main characteristics |18| :

- supply : a 450 kV, 600 J generator
- symmetrical type antenna 2,5 m height, 6 m wide and 5 m deep with a total length of 20 m terminated on a characteristic impedance of 90 ohms (fig. 3).

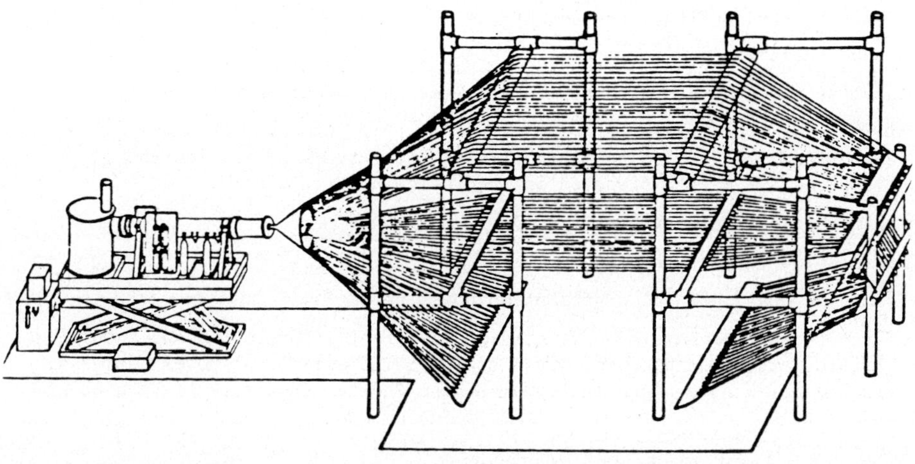

Fig. 3 - The GESIBAN simulator

To-day a simulator called HEISS 1 derived from the first one is operational the main difference being the fact that it has an asymmetrical type of antenna but with slightly larger dimensions.

The measurement instrumentation at disposal in the laboratory consists of :

- electric and magnetic field probes ;
- current and voltage probes ;
- optical fiber transmission between the probes and the data acquisition systems ;
- transient digitizers and memory oscilloscopes for data acquisition.

Other simulation possibilities are used in the laboratory as for instance :

- current injection in systems ;
- radiation in the frequency domain on systems ;
- continuous wave injection ;
- scale modelling.

The scale modelling installation developed under contract at the Federal Institute of Technology from Lausanne consists of a plateform with a conical antenna supplied by a 1 kV generator producing pulses with less than 200 ps time of rise. It can be used for tests of objects at reduced scale to estimate for instance the surface current distribution.

4) Transfer impedance measurements

The laboratory uses two types of installations for these measurements :

a) A continuous wave equipment using the quintaxial method (fig. 4) |19|.

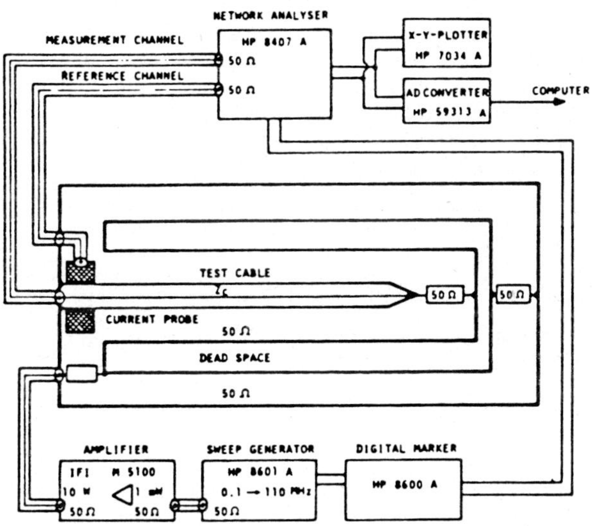

Fig. 4 - The quintaxial transfer impedance measurements installation

b) An impulse current installation (fig. 5). The advantage of this method is that a test at a real scale permits to simulate a nonlinear behaviour for instance the saturation effect due to a shield of ferromagnetic material |20|.

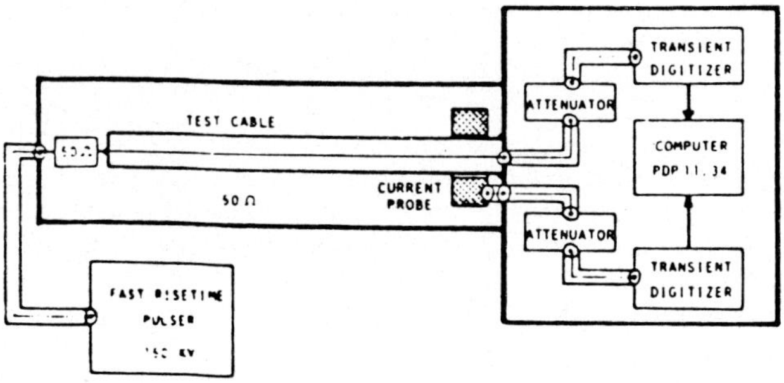

Fig. 5 - The transfer impedance impulse current measurement installation

3.2.2 The Research Institute for Protective Materials from Zürich |21|.

This institute was the first in Switzerland where the basic phenomena of the nuclear electromagnetic pulse has been studied at the begining of the 70's |22|. Computer codes for the calculation of the response of cables to an EMP and for the study of the interaction of plane electromagnetic waves with multilayer structures have been developed in the last years |23,24|.

A work performed in collaboration with the Federal Institute of Technology from Lausanne concerns a numerical method for the calculation of the propagation constant of buried conductors in a multilayer soil |25|.

The institute has also a teaching activity and lectures on EMP are given at the Federal Institute of Technology from Zürich on the following subjects :

- the physical phenomena ;
- the free-field effect ;
- interactions with structures ;
- simulation principles ;
- protection measures

3.3 Civil Defense activity on EMP

Concerning civil defense problems a concept with respect to EMP has been eleborated. Its aim is to protect electric power supplies and the transmission of information to the population. As a total protection cannot be coceived the decision was to achieve a selective defense. This goal should be achieved using :

- shielded cables for the transmission of information ;
- metallic enclosures for switchboards ;
- suppressors ;

- radio broadcasting stations and receivers of special construction.

3.4 EMP activity in universities

A research activity in EMP takes place in the Federal Institute of Technology from Lausanne since 1978. At that time a contract with a cable industry started the investigations on the subject. The object of the contract was a study on improvement possibilities of the shielding of coaxial cables protected against NEMP. Wave-conductor coupling models have been imagined, a calculation method was developed and two computer codes (EMPLIGN and EMPCOAX) were written |26,27|. These two codes ares calculating :

- EMPLIGN : the current induced by an EMP in an aerial or buried conductor (for instance the shield of a coaxial cable).
- EMPCOAX : the voltage induced by an EMP between the sheath of a coaxial cable and the internal conductor.

The calculations performed with the aid of these two codes have permitted to define how to optimize the cable shielding |28|. Figure 6 shows the variation of the transfer impedance for three shielding structures of equal thickness. A result obtained using EMPCOAX shows in fig. 7 the reduction of the induced voltage if a structure of three sheets (AL-STEEL-AL) is used |26,27|.

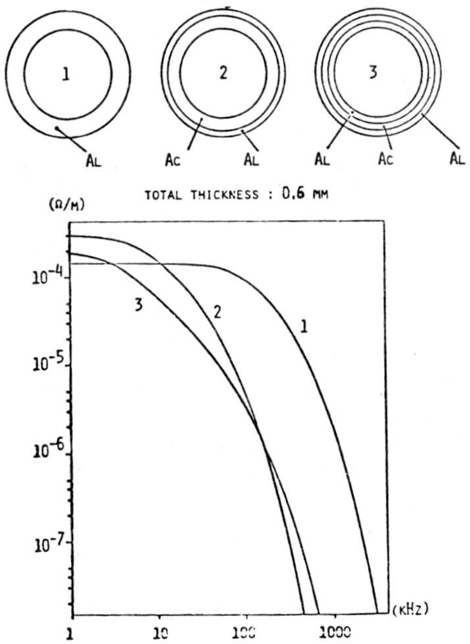

Fig. 6 - Transfer impedance for three shielding structures of equal thickness

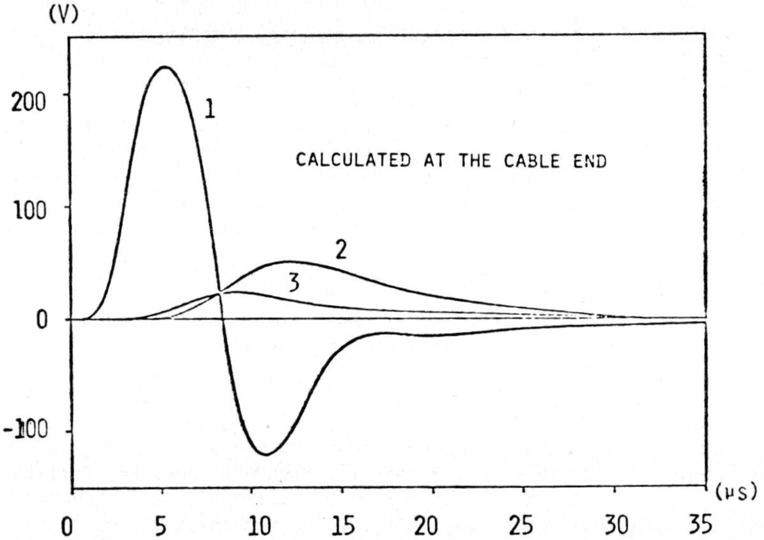

Fig. 7 - Induced voltage on the inner conductor for three shielding structures of equal thickness (0,6 mm)
1 : 1 sheet (AL)
2 : 2 sheets (AL-STEEL)
3 : 3 sheets (AL-STEEL-AL)

These first calculation have shown :

- the necessity of testing the calculation results by comparison with measurements ;
- the important place occupied by the transfer impedance in all coupling studies for transmission cables.

Therefore during the following years the activity was focussed on :

- testing of the calculation method ;
- transfer impedance measurements methods.

However two more subjects have been investigated in this period, i.e. EMP effects on objects and EMP effects on power lines.

3.4.1 Testing of the calculation method

Two ways have been found interesting in order to test the numerical calculation method developed at the Federal Institute of Technology from Lausanne.

The first one consists of current injections in the cable sheath. This aim of this kind of simulation is two-fold :

1) To measure the effect of a current having the same order of magnitude and same shape as a current induced by an EMP in the cable sheath on the internal conductor. In this way it is possible to obtain (assuming that the current in the sheath is known) the response of a long cable to an EMP. Such long cables which length is larger than a few hundred meters cannot be tested in a simulator.

2) To verify the calculation method by comparing measurement with results obtained using the transmission line approximation. The mathematical procedure used to compute the internal voltage from the sheath current is the same as the one used for calculating the coupling between a plane wave and a conductor both using the transmission line approximation.

The results obtained with the current injections have shown a very good agreement between calculated and measured data (fig. 8).

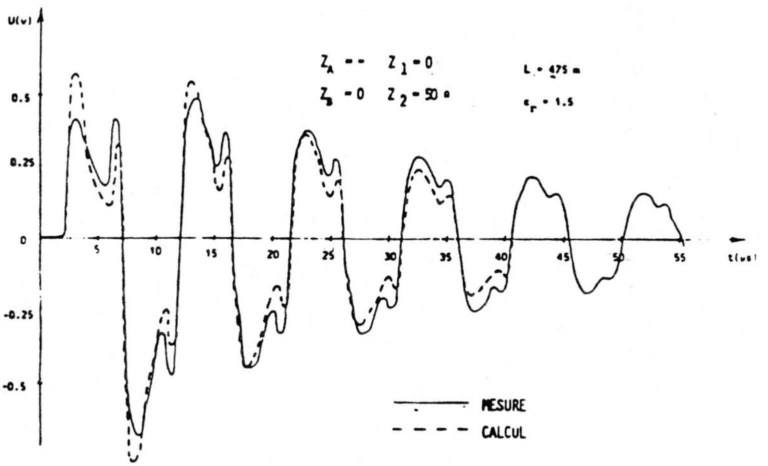

Fig. 8 - Comparison between calculated and measured response of a cable to an current injected in the sheath

A second method of verification was to use the electromagnetic field of a lightning stroke. In this case the shape of a lightning EMP is rather different from that of a NEMP. However the transmission line approximation can also be used to compute the effect of a LEMP on an aerial line placed in the proximity of the lightning stroke fall point. The first results obtained with this simulation have been reported | 29 | and work is still in progress.

3.4.2 Transfer impedance measurements

A test system using impulse current has been built up at the Federal Institute of Technology from Lausanne. This installation similar to the one existing at the Swiss Army Laboratory from Spiez (fig. 5) permits the determination of the transfer impedance at a real scale of the injected current. The studies in progress now at Lausanne concern the following points of interest :

- transfer impedance of braided cables ;
- multiconductor cables ;
- behaviour of ferromagnetic sheaths.

3.4.3 EMP effects on objects

This problem was investigated under contract with the Swiss Army Laboratory from Spiez. A scale modelling installation (fig. 9) has been constructed and delivered to Spiez where it is now operated |30|.

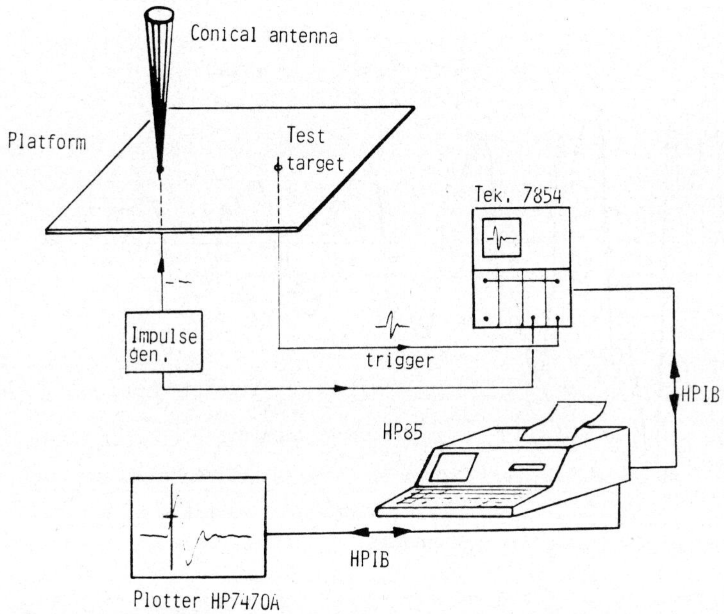

Fig. 9 - Scale modelling installation

3.4.4 EMP effects on power networks

Very little work has been done up to now in this field. An approach to the problem has been defined and preliminary studies performed |31|. As the modelling of a transformer in transient state is difficult, the main problem is the behaviour of power transformers to a pulse of current propagating on a power line.

3.5 EMP activity in the industry

In the EMP field the approach of the industry is similar to that discussed for EMC problems. The swiss industry produces installations NEMP hardened, components for the protection against the effect of the EMP and test systems for EMP simulation. An example of such a test installation is a mobile EMP-system with high flexibility developed for testing electronic instruments. Uniform field strength of 50-400 kV/m depending on the size of the antenna in a volume of 3-25 m^3 can be produced with this system |32|. Another impulse generator shown in fig. 10 produces rectangular shaped voltages and currents |33|.

Fig. 10 EMP generator producing rectangular shaped voltages and currents

EMP calculation pertaining to specifie problems for instance field penetration through circular apertures are also performed in some industries having a research activity |34|.

CONCLUSION

It is obvious that in a few pages an activity as rich and as various as the EMC activity of a country cannot be presented in an exhaustive way. We hope however, that the main features have been mentioned, the principal directions have been emphasized and that the few examples chosen to illustrate the different problems will give an idea about this field which is to-day in full development in Switzerland. In our opinion however, it is not only the quantity of problems, solutions and practical measures which grows up, but a different approach can be observed in the last years. From a domain in which people were looking mostly for empirical solutions based on years of experience, EMC becomes now a field in which numerical calculations based on electromagnetic theories are more and more used.

This tendency should give the possibility in the future to optimize the solutions and to arrive at the desirable situation in which EMC becomes part of the design of a product and not a later intervention when it has been seen that interferences cause malfunction or failure of the system.

REFERENCES

|1| P. Leuthold : Grundlagen der Elektromagnetischen Vertraeglichkeit, in "Compatibilité Electromagnétique". Editors : M. Ianovici and J.-J. Morf, Presses Polytechniques Romandes, Lausanne, 1983.

|2| A. Rodewald, H. Kunkel, M. Lutz : Interfence generated by switching operations and its simulation. Proc. Int. Symp. on EMC, EMC'84, Tokyo, Oct. 16-18, 1984, paper 16 PD1.

|3| H.A. Kunz, P. Kull, B. Amstutz : Reproductible simulation of on interference burst pulse. Proc. Int. Symp. on EMC, EMC'84, Tokyo, Oct. 16-18, 1984, paper 16 PD2.

|4| M. Aguet : Informatique et compatibilité électromagnétique. Output, No 6, 1982.

|5| H. Sauvain : Protection contre le rayonnement. Output, No 10, 1984.

|6| P. de Bruyne, P. Leuthold : Room to room communications via waterpipes with modulated ultrasounds. Abstr. Int. Symp. on underwater Acoustics, Tel-Aviv, 1981, p. 11.

|7| T. Dvorak, H. Ochsner : Low tension power line as a fast digital data transmission channel. 4th Int. Symp. on EMC, Zürich, March 10-12, 1981, paper 1A1.

|8| T. Dvorak : Compatibility of spead spectrum signals with narrow band FM receivers in VHF mobile networks. IEEE Trans on EMC - vol. EMC-21, No 2, May 1979, p. 131-136.

|9| T. Dvorak : Kompatibilität von Spread Spectrum Systeme und Konventionellen Funkdiensten Nachrichten Techn. Kolloquium, 1981-1982 Bandspreitstechnik, Universität Bern, s. 91-101.

|10| H.-P. Baer : Effect of hard limiting in a PN Spread-spectrum system. 4th Int. Symp. on EMC, Zürich, March 10-12, 1981, paper 2A2.

|11| G. Meyer : The TEM measuring line - A critical overview. 4th Int. Symp. on EMC, Zürich, March 10-12, 1981, paper 74M5.

|12| G. Meyer : Devices for EMC measurement, European Electronics, No 2, 1982, pp. 22-28.

|13| M. Ianovici : NEMP effects on transmission cables. Energia Elettrica (in press).

|14| Ph. Blech : Calcul et mesure de l'impédance de transfert de câbles coaxiaux. Energia Elettrica (in press).

|15| M. Aguet, Ph. Blech, M. Ianovici : Coordination probabilistique de l'isolement des réseaux électriques à haute tension. Bull. ASE/UCS, t.68 (1977), No 24, p. 1297-1303.

|16| EMC-Seminar 80, Probleme der Entstörung und des Schutzes von elektrischen Geräten und Systemen. Fabrimex AG, Kirchweg 5, 8032 Zürich.

|17| G.J. Burke, A.J. Poggio, J.C. Logan, J.W. Rockway : Numerical Electromagnetics code - A Program for Antenna System Analysis. Proc. 3rd Symp. Techn. Exh. on EMC, paper 18DC, Rotterdam, May, 1979.

|18| T. Rüdy, J. Bertuchoz, B. Wamister : NEMP simulation and tests in Switzerland. 3rd Int. Symp. on EMC, Rotterdam, May 1-3, 1979, paper 39G6.

|19| T. Rüdy, J. Bertuchoz, B. Wamister : Estimation of NEMP induced voltages in coaxial cables with the help of transfer impedance. 4th Int. Symp. on EMC, Zürich, March 10-12, 1981, paper 45H7.

|20| B. Brändli, W. Jenni, J. Bertuchoz : High current pulse cable testing system. 5th Int. Symp. on EMC, Zürich, March 8-10, 1983, paper 34G7.

|21| J. Gut : The Swiss EMP Concept of general Defense. FMB 84-6, Zürich, December 1984.

|22| K. Appert, W. Jöhl : Der elektromagnetische Impuls von Nuclearexplosionen in grosser Höhe. FMB 76-17, Zürich, Dezember 1976.

|23| W. Blumer : EMP response estimation in shielded cables. FMB 79-4, Zürich, Juni 1979.

|24| W. Blumer : NEMP-Wechselwirkung mit Mehrschichstrukturen. FMB 84-1, Zürich März 1984.

|25| W. Blumer, M. Ianovici : Calculation of the propagation constant of buried insulated conductors. 5th Int. Symp. on EMC, Zürich, March 8-10, 1983, paper 10C3.

|26| M. Aguet, M. Ianovici, C.C. Lin : Transient Electromagnetic field coupling to long shielded cables. IEEE Trans. on EMC, vol. EMC-22, No 4, 1980, p. 276-282.

|27| M. Aguet, M. Ianovici, C.C. Lin, F. Fornerod : Comportement des lignes et des câbles de transmission soumis au rayonnement électromagnétique provenant d'une explosion nucléaire (NEMP). BUll. ASE/UCS, t. 71, No 17, 1980, p. 911-918.

|28| F. Fornerod, C.C. Lin : Recherche d'optimisation de construction pour les armures des câbles protégés contre les perturbations. NEMP, Bull. ASE/UCS, t.71, No 17, 1980, p. 919-923.

|29| A. Caron, B. Djebari, A. Zeddam, Ph. Blech, Y. Dijamatovic, M. Ianovici : Validation of EMP calculation methods using the response of an aerial cable to a lightning stroke. 6th Int. Symp. on EMC, Zürich. March 5-7, 1985, paper 72L8.

|30| E. Bergerie, Y. Dijamatovic, M. Ianovici : Installation d'essais pour la modélisation à échelle réduite des impulsions électromagnétiques. Ann. Télécom., t.39, No 5-6, Mai-Juin 1984, p. 191-197.

|31| M. Ianovici, F.M. Tesche : EMP effects on a HV networks. Int. Symp. on EMC, EMC'84, Tokyo, Oct. 16-18, 1984, paper 17AD4.

|32| K. Feser, M. Modrusan, K.H. Gonschorek, H. Singer : Mobile EMP-system with high flexibility. 4th Int. Symp. on EMC, Zürich, March 10-12, 1981, paper 25E7.

|33| M. Modrusan, E. Gockenbach, O. Frey : A new impulse generator for impulses with short rise times. Int. Aerospace and Ground Conf. on Lightning and static Electricity, 1984, Orlando, USA.

|34| N. Ari, D. Hansen, H. Schär : Electromagnetic Pulse (EMP) penetration through circular apertures in the frequency domain. Proc. IEEE, vol. 73, No 2, February 1985, p. 368-369.

Nonlinear and Environmental Electromagnetics, edited by H. Kikuchi
Elsevier Science Publishers B.V., Amsterdam, 1985 – Printed in The Netherlands

ELECTROMAGNETIC SURGE PROBLEMS ON TELECOMMS PLANT

R. Tharby
Head of Local and Junction, Network Operations, British Telecom

1. The 3 principal problems which arise are:

   - Physical Damage
   - Software Corruption
   - Service Degradation

   taking each in turn.

2. Physical Damage

This may be catastrophic in degree occurring from exo-atmospheric NEMP discharges which cause radiation surges up to 50Kv/m with rise-times of the order of nano-seconds and could lead to widespread damage to telecomms lines and terminal equipment.

It can also have a more localised form, e.g. as a result of lightning discharges directly or near to telecomms line plant. For this reason lines are often fitted with surge arrestors to minimise the damage. International standards have been prepared to ensure that protection requirements are well-understood and surge measurement programmes have been instituted in many countries to assess the probability of these surges arising.

Localised surges are also caused by faults on nearby power or electric railway systems, and protective measures are taken when they exceed acceptable limits. Electrostatic discharges near to subscribers terminal equipment may damage vulnerable components unless suitable design standards are followed.

3. Software Corruption

Many telecomms terminals incorporate memories or microprocessors in which control information is stored and on which the correct operation of the equipment depends. Such software may be corrupted by electromagnetic interference arising from surge voltages due to lightning or electrostatic discharge. The correct design of this equipment can minimise the probability of this corruption occurring. The levels that can arise are well-established and shielding methods can be used to minimise their effects.

4. Service Degradation

This may have very local effects, eg where equipment is disturbed by the operation of electrical equipment nearby. It may also have a more general nature in a telecommunications network where impulsive surges or noise is induced from neighbouring circuits which are being switched by electromagnetic sources or where adjacent high-amplitude digital signals are being transmitted.

Either of these effects may give rise to errors on modern digital systems of a level beyond that which users will tolerate. This matter is assuming major importance as national telecomms systems are being converted from analogue to digital operation. The performance of digital circuits may be measured in terms of error rates, in terms of circuit availability, or in error-free seconds. The degradation is highlighted by the many circuit quality monitors now being used by subscribers.

Nonlinear and Environmental Electromagnetics, edited by H. Kikuchi
Elsevier Science Publishers B.V., Amsterdam, 1985 – Printed in The Netherlands

Hardening of Telecommunication Networks against Electromagnetic Pulses

Manuel W. Wik

The Defence Materiel Administration of Sweden
Electronics Directorate
Stockholm

Nuclear explosions at high altitudes generate strong electromagnetic pulses (EMP), which can induce large currents and voltages, for example in power and telecommunication networks over a whole continent simultaneously.
The author describes the increasing EMP threat and why it concerns telecommunications administrations and the electronics industry. He also describes the generation mechanisms and pulse waveform. The EMP impact on telecommunication facilities is discussed, together with hardening and testing methods.

Printed with permission
of Telefon AB LM Ericsson

Nuclear explosions can generate very strong electromagnetic pulses (EMP, or nuclear EMP = NEMP). Such pulses can in certain cases dominate over other more well-known effects, such as nuclear and thermal radiation, blast and shock. For high altitude explosions (above 30 km) EMP is practically the only effect apparent at ground level, and the following description is mainly concerned with this case.

A nuclear EMP can roughly be compared to the electromagnetic field very close to a lightning stroke, but the field can extend much further and cover whole countries. It also has a broader frequency spectrum covering the whole radio frequency communication band. Currents and voltages are induced in metallic objects. Large aerial or buried power and telecommunication networks can absorb considerable amounts of energy and be damaged, but even short radio antennas and other lines can be sufficient to cause serious damage to the connected equipment. Consequently telecommunications administrations and electronics designers – civil as well as military – have every reason to consider the problems posed by EMP.

Fig. 1
Diagram showing the increasing EMP threat posed by the development of nuclear weapons technology and electronics

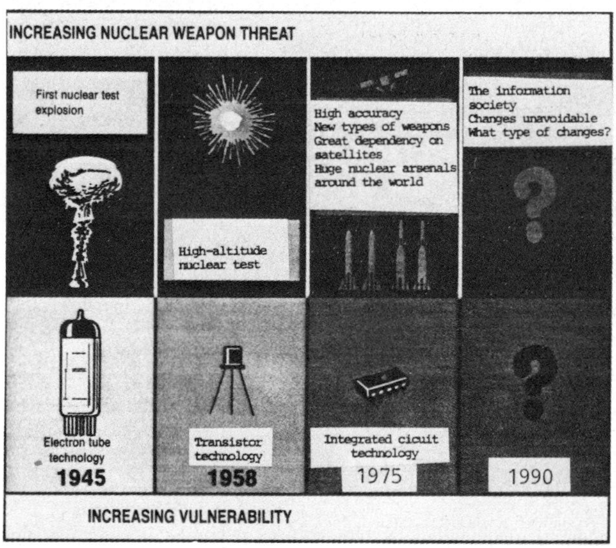

The threat

The continuous development of nuclear technology over the years has resulted in new generations of nuclear devices and weapon systems. Outer space has become increasingly important as satellites having important tasks in communication and surveillance have been launched. Intercontinental ballistic missiles (ICBM) with nuclear warheads have been programmed for high trajectories, and so have other weapons intended to combat ICBMs. All in all the probability of nuclear explosions at high altitudes (tens to hundreds of km) has increased considerably.

In 1963 the limited test ban treaty came into force, which prohibits atmospheric nuclear test explosions. Before that time both the US and the Soviet Union had started to carry out test explosions at high altitudes. It was then discovered that the nuclear EMP effect, which was already known from explosions closer to the ground, occurred with great intensity over enormous areas at ground level. The information was classified as secret, and after the test ban agreement the superpowers were confined to theoretical calculations and simulations. However, some information leaked out, such as the following newspaper report about a US high-altitude test over the Johnston Island in the Pacific Ocean in 1962:

– The quiet predawn in Honolulu was shattered by the simultaneous pealing of hundreds of burglar alarms. At the same time circuit-breakers on the power lines started blowing like pop-corn. Not a cloud in the sky, so lightning could not be blamed. The power company failed to trace any gigantic electrical surge able to blow out virtually the entire systems simultaneously. The mystery was solved later – then promptly sealed under a "top secret" stamp. The culprit: A high altitude nuclear test burst more than 500 miles away.

More than twenty years have passed since then, and society has developed and changed in such a way that the nuclear EMP threat has grown. The development in electronics has gradually led from transistors to large and very large-scale integrated circuits (LSI, VLSI) that work with low signal and power levels in the range 10^{-3} to 10^{-8} W per transistor. Susceptibility thresholds are increasing and the pulses and energy surges need not be very high in order to cause damage (junction burnout), only of the order of 10^{-3} to 10^{-6} J.

At the same time society has become much more dependent on electronics. For example, telecommunication networks, high-voltage power supply networks, railway networks, air traffic, water supplies and process industries are all controlled and regulated by equipment which usually contains susceptible semiconductor components. The increasing anxiety about nuclear EMP has recently been mirrored in the increase in newspaper articles and TV reports on the subject.

The greater vulnerability makes EMP attacks more attractive. It is quite possible that special EMP weapons already exist. An attacker could find it profitable to start military operations with an EMP attack. This would create such chaos that subsequent measures would be more effective and counter-offensives made more difficult. A high-altitude explosion can be triggered at a large distance from the borders of the country to be attacked and still have a great impact on that country. The country attacked might not even realize that a nuclear explosion had occurred, let alone who set it off. With a high-altitude explosion (at above 30 km) effects other than EMP could be negligible at ground level, and perhaps be impossible to detect without special equipment.

Large sums have been invested in EMP protection. Ten years ago the US was spending over 250 M US$ a year on EMP protection and testing. In October 1981 president Reagan stated that the US defense project with the highest priority was not the MX system or the B1 bombers but "to strengthen and rebuild our communications and control systems –

Fig. 2
An example of the impact area of EMP from a high-altitude nuclear explosion over Europe

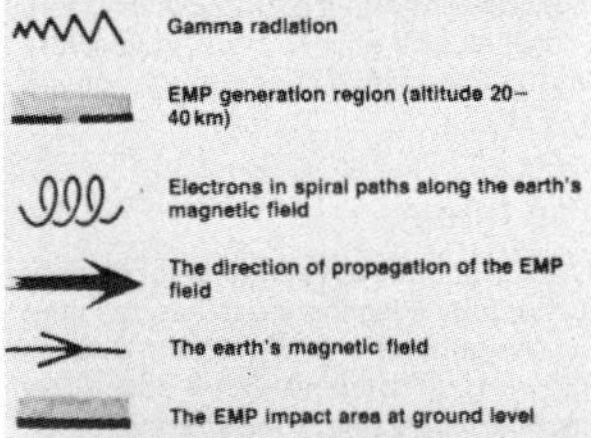

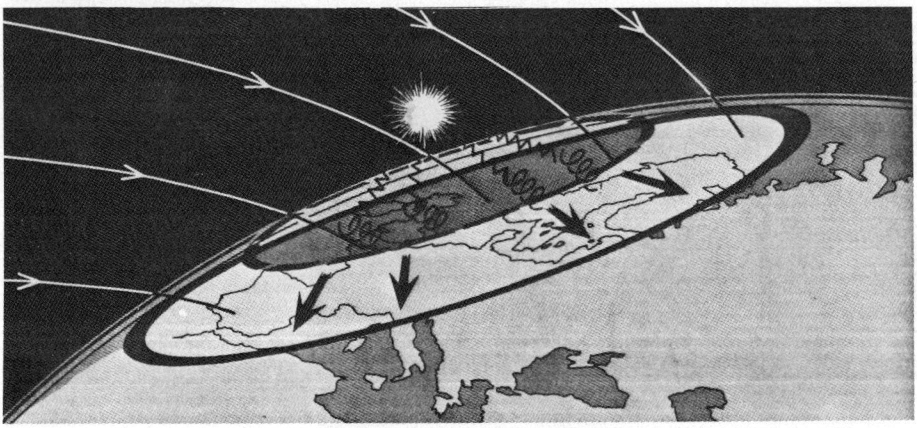

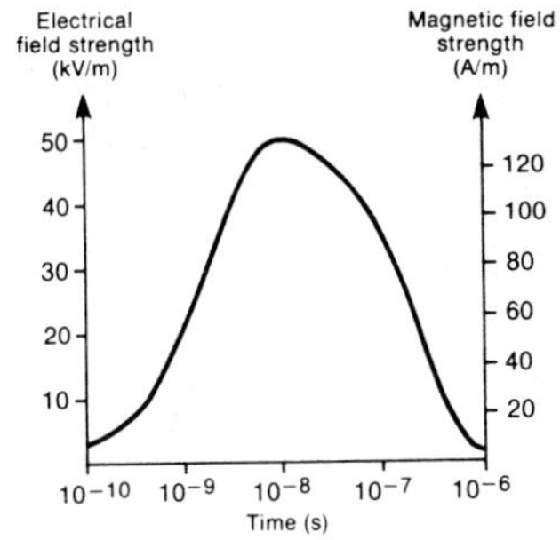

Fig. 3
The electrical and magnetic EMP field from a high-altitude explosion, plane wave and horizontal polarization. The time dependency is generalized

a much neglected factor in our strategic deterrent". 20 billion (10^9) US$ was earmarked for this project, a large part of which is intended for EMP hardening. In the Soviet Union EMP protection has even been provided for facilities of less than top priority for about 20 years.

Telecommunication networks that still function after high-altitude explosions are vital if diplomatic and political communication between nations is to be possible so as to avoid the military escalation of a limited nuclear confrontation into a total nuclear war. In the case of a nuclear war elsewhere a country needs a working telecommunication network, partly in order to be able to collect and process weather and nuclear radiation reports. It must be possible to inform the population of the country about the situation if chaos and exposure to nuclear radiation are to be limited. This could save many lives. Without a secure telecommunication network it would also be very difficult to rebuild and reconnect the country's electric power system before chaos arises and spreads. Disruption of electric power would seriously affect pumping of water and fuel supplies, and many other vital functions.

The generation of nuclear EMP

From a nuclear explosion an intense gamma radiation pulse propagates at the speed of light. The gamma radiation consists of photons (radiation quanta) which collide with air molecules or other matter and release free electrons. The so-called Compton electrons leave positively-charged ions behind and move on in the direction of propagation. They are slowed down by collisions, and each Compton electron thereby generates tens of thousands of secondary electron-ion pairs. The charge separation causes a strong electric field and the charging movements produce a current.

For explosions close to the ground, conditions are strongly asymmetrical. There is a net electron current with a strong upward component from ground zero. Electrons travelling outward in the air from the burst return toward the burst point through higher-conductivity ground and ionized air paths. The current loops generate very large azimuthal magnetic fields. For exo-atmospheric explosions Compton electrons are not generated until the gamma rays that travel downward reach the denser atmospheric layers, at a height of 20 to 40 km above ground. This is known as the deposition layer. The electrons encounter the earth's magnetic field and are deflected, producing a transverse electric current. This results in electromagnetic radiation which is directed radially out from the explosion point, adding in phase. This EMP has a tremendous coverage. In the case of large explosion yield, the higher frequency part of EMP extends to the horizon. The lower frequencies will follow along a duct between the earth and the bottom of the ionosphere as well as along the surface of the earth far behind the horizon. For example, with a height of burst of 100 km, large parts of Europe could be covered (impact radius approximately 1200 km along the surface of the earth; 400 km altitude would give a radius of approximately 2200 km). The EMP effect would be noticeable even outside this area.

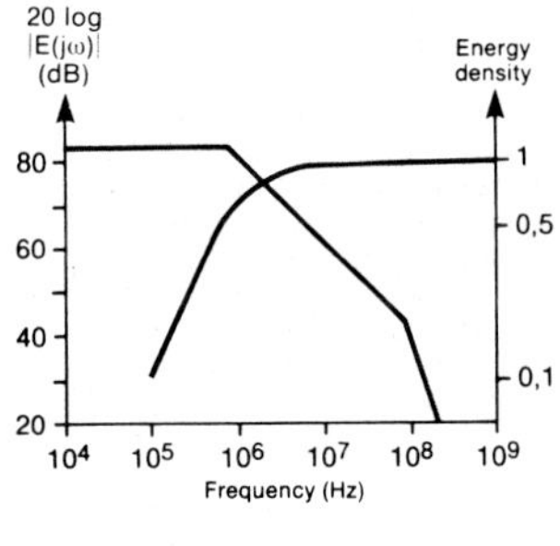

Fig. 4
The spectrum of the electrical field and the corresponding normalized cumulative energy density of EMP from a high-altitude explosion

Magnitude of the EMP

The EMP from a high-altitude explosion consists of a plane wave (E/H=377 ohms), which is propagated along the line of sight from the explosion point. The E and H fields are perpendicular to each other and to the direction of propagation. A part of the pulse is reflected by the ground, and the rest is propagated into the ground and is gradually attenuated. The strength and polarization of the field vary within the impact area due to various factors. Vulnerability and

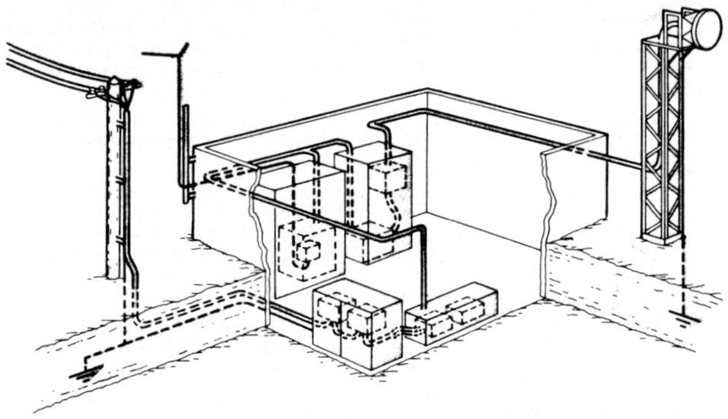

Fig. 5
Simplified picture of a communication installation with internal and external cabling

hardening calculations are often based on an electromagnetic field in free space before reflection, in accordance with fig. 3. The high amplitude (50 kV/m) and the short rise time (5 ns) are of particular importance for the effects. For comparison purposes it may be mentioned that a field amplitude of approximately 200 V/m is obtained near radar stations and approximately 0.01 – 0.1 V/m in urban areas. A generalized, time-dependent E field is often given as

$$E(t) = 5.25 \times 10^4 \times (e^{-4 \times 10^6 t} - e^{-4.76 \times 10^8 t})$$

Fig. 4 shows the frequency spectrum of the pulse. The energy density in this case is 0.9 J/m^2, and 99.9 % of the total energy lies below 100 MHz.

The effect on telecommunication facilities

Generally speaking the electromagnetic field induces currents and voltages in all sorts of conducting objects, which act as antennas, intentionally or unintentionally. The induced energy can find its way to connected objects where it is dissipated as heat, in some cases in combination with flash-overs. In a widespread network, pulses will be able to destroy or interfere with connected devices almost simultaneously in a number of places. The conductors can be compared to magnifying glasses which gather solar energy and focus it on points where heat is concentrated to such an extent that it is destructive. EMP energy is transmitted primarily by means of electromagnetic coupling (radiated field) and secondarily through galvanic coupling, inductive coupling ($u = M \times di/dt$) and capacitive coupling ($i = C \times du/dt$). In view of the circumstances all paths must be considered as risks. The current and voltage rise rates are considerable.

In order to give a more detailed picture of the EMP effects a communication facility with a microwave radio tower, fig. 5, is used as an example. This model is not necessarily typical, but is used to make the description more concrete.

The large external conductors are the power mains, telephone cables, antenna cables, waveguides and antennas. In addition there are large metal structures such as microwave radio towers, grounding conductors, fences, reinforcing netting in concrete walls, as well as air conditioning, water and sewage installations in the building. The building also contains internal cabling with electric power and telephone lines and cable ducts, racks and power and telecommunication equipment.

This model can be represented by a topological diagram as shown in fig. 6. The external environment includes air and ground.

Fig. 6
A topological diagram

Volume	External boundary	Attenuation	Total
V_1	Air	0	0
V_2	Concrete	20 dB	20 dB
V_3	Shielded cabinet	20 dB	40 dB
V_4	Shielded unit	20 dB	60 dB

Transient protection
Shielding boundary
Penetrating conductor

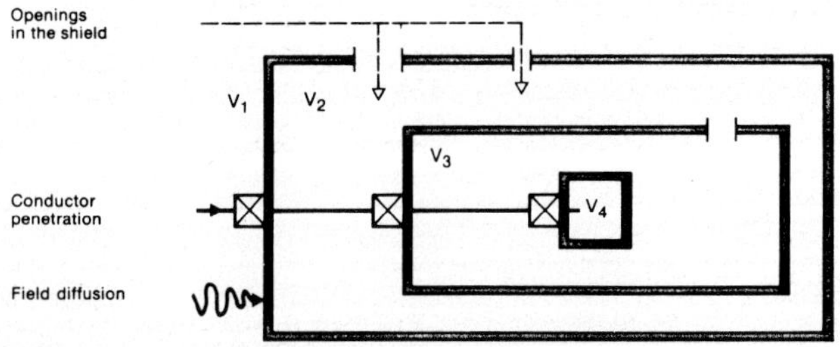

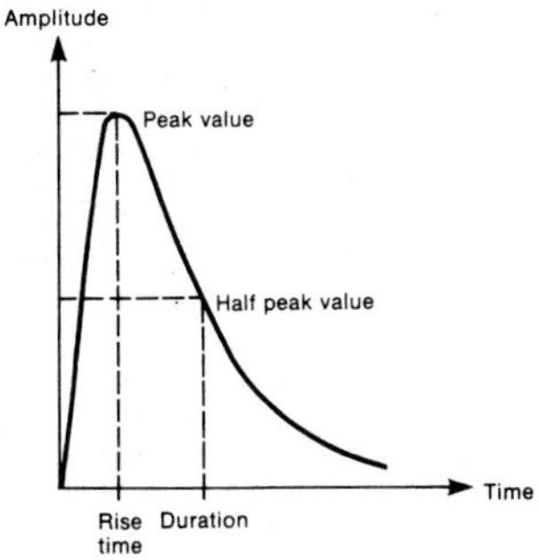

Fig. 7
Double exponential pulse waveform from induced currents in aerial and buried cables caused by EMP from a high-altitude explosion. If the lines are long, the pulses can have long decay (not included in the table below)

		Aerial line	Buried cable
Peak current	kA	5	1.5
Rise time	μs	0.1	0.1–0.5
Time to half value	μs	1	1–5
Rate of rise	kA/μs	50	15

Fig. 8
Short-circuit current for a 3 m long monopole antenna excited by an EMP field in accordance with fig. 3

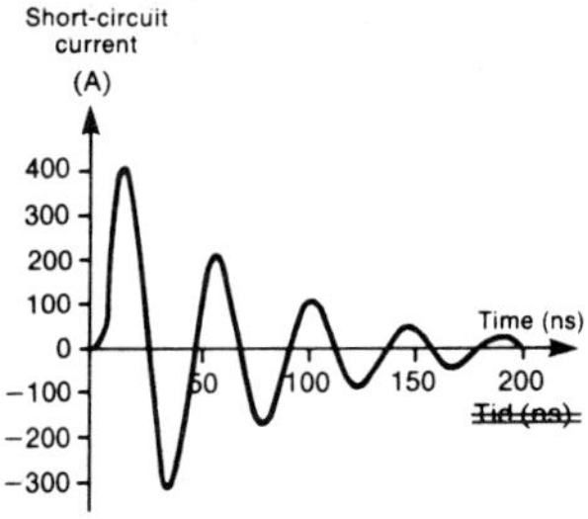

Overhead power lines and telephone cables are severely exposed to EMP. The voltage to ground can be between 10 and 1000 kV and is limited primarily by the insulation resistance. The current is limited by the characteristic impedance of the conductors, which can be a few hundred ohms. At discontinuities, for example a transition from an overhead to a buried cable or to a transformer connection, strong pulse reflections occur. Buried cables obtain a certain amount of protection from the ground and cable sheath, and the induction is therefore less than for overhead lines. Fig. 7 gives some examples of EMP values for overhead and buried cables. The pulses have a double exponential waveform. Antennas and other short conductors that are exposed to EMP respond with an oscillating current and voltage, determined by the natural resonant frequency of the antenna or conductor. The oscillation has an exponential decay. Fig. 8 shows an example of a 3 m monopole antenna. Fig. 9 gives the voltage, current and energy as a function of the resonant frequency for excitation of conductors that can be considered as monopole antennas. The microwave radio tower in fig. 5 also acts as a large antenna, and peak currents of the same magnitude as for overhead lines can be induced. The long overhead lines and the microwave radio tower thus give rise to the largest induced currents, and buried lines to somewhat lower currents.

The conductors carrying the induced currents penetrate the building. The EMP field can also penetrate the building, through openings and, with some attenuation through the roof and walls. This field causes induction in the internal cabling. Concrete walls can attenuate the field by between 5 and 35 dB depending on their construction, the size of the building, frequency etc. If the building is equipped with a Faraday cage made of steel sheet the field attenuation can be some 50 to 100 dB. The higher frequencies are attenuated more than the lower, which makes the internal pulse rise time longer. However, the resultant attenuation is strongly dependent on how the shield penetrations are implemented.

The residual field can induce currents of the order of 1 to 20 A on internal cabling. But large currents entering via a lead-through can induce higher values in other cables which run parallel. The currents in the internal cabling can be a damped oscillation of between 1 and 10 MHz and with a time constant of between 1 and 10 μs, cf. fig. 8.

Reduction of induced pulses on conductors largely depends on whether the conductors have been equipped with some form of transient protection (surge arrester, gas tube spark gap, filter etc.), and on the design of the potential equalization. Special power conversion equipment (e.g. rectifiers and d.c./d.c. converters with accumulators) can in favourable cases provide attenuation of some 35 to 70 dB. Specially screened isolating transformers, in combination with filters, can provide even higher attenuation. If good protection is to be ensured, power supply lines must be shielded or at least installed at a considerable distance from other installations. It must also be ensured that other conductors do not form loops that can be closed by flash-overs.

The internal cabling terminates in various kinds of equipment cabinets. At best these can provide electromagnetic shielding, but whether the shielding is effective depends to a great extent on the cabling and how it is terminated. Inside the cabinet there is the equipment wiring, which leads to printed board assemblies and similar modules. In the best designs these are screened and the wires are terminated satisfactorily from an electromagnetic point of view, but often this is not a rigorous requirement. The equipment power supply can in favourable cases provide good supplementary attenuation of external transients. Finally the wiring on the printed board assemblies, in its turn, is terminated in components. However if the latter consist of integrated circuits there is a still lower level of wiring and components.

Knowledge of the threshold levels of the equipments and components for functional damage and operational interference is necessary in order to be able to assess the possible effects of EMP. The induced EMP energy must be compared with the energy threshold failure level of each component in order to esti-

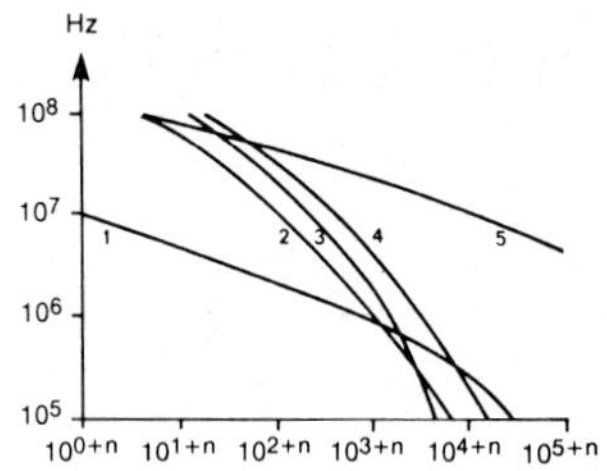

Fig. 9
Estimated voltage, current and energy as a function of the resonant frequency ($f = c/4l$, where c is the speed of light and l is the line length) when conductors that can be considered as monopole antennas are excited (field in accordance with fig. 3)

Curve		Abscissa, n =
1	EMP energy dissipated in a 50-ohm load (J)	0
2	Short-circuit current (A, peak value)	1
3	Open-circuit voltage (V, peak value)	3
4	Voltage across a 50-ohm load (V, peak value)	2
5	EMP energy dissipated in a 50-ohm load (J)	−4

For example, for a conductor with a resonant frequency of 100 MHz:
- **the peak short-circuit current is 50 A**
- **the peak open-circuit voltage is 15 kV**
- **the peak voltage across a 50-ohm load is 2 kV**
- **the EMP energy dissipation in a 50-ohm load is 0.5 mJ**

The fraction of this energy that reaches sensitive semiconductor components in the connected equipment can be sufficient to cause damage.

mate the probability of damage. The charge or voltage can also be the limiting factor for thin oxide layers in semiconductors and electrolytes. In addition EMP energy can trigger dissipation of the system's own energy in inadmissible places, thus causing serious secondary damage. For short pulse widths, failure usually depends on the energy content of the pulse. Table 10 gives typical minimum threshold energy levels causing permanent damage. In semiconductor devices failure modes are typically junction burnout, oxide punchthrough and metallization burnout. With short pulses (less than approximately 1 to 0.1 μs) breakdown occurs in the semiconductor at a constant pulse energy regardless of the pulse duration, because the energy does not have time to disperse. Local melting occurs, particularly at any manufacturing defects. Such defects may be due to poor quality, which has not been detected during routine production control. Good quality and good control can give at least a tenfold improvement of the threshold energy compared with the values for low-quality circuits.

Although semiconductor components are particularly susceptible to transients, damage must also be expected in other components such as resistors (particularly metal oxide), capacitors (particularly tantalum electrolytic), relays, and indicator instruments. During the first nuclear tests in the middle of the 1940s many measuring instruments suffered EMP damage. Since then EMP simulations in communication facilities have resulted in false alarms, subscriber services outages, program interruptions in computers and certain permanent damage to the systems. However, it is difficult to make EMP simulations correctly on such large objects as communication installations so as to simulate high-altitude explosions correctly.

Among the conclusions from EMP simulations are:
- Permanent damage or impaired performance may be caused, particularly for active components (especially high-frequency transistors, integrated circuits and microwave diodes), passive components (particularly those with very low power or voltage ratings or precision components), semiconductor diodes and silicon controlled rectifiers (especially in power supply units connected to the mains) and insulated high-frequency and power cables (especially if they operate close to their maximum power and voltage levels, or are exposed to humidity or abrasion).
- Functional interference and status faults may be caused in low power or high speed digital processing systems, in memory units, control and alarm systems and in subsystems employing long integration or recycling times for synchronization, data acquisition or signal processing.

Hardening methods

A system that has to survive in an EMP environment must not undergo such disturbances or permanent damage as lead to functional failure. In order to reduce the effects of EMP on a system it will generally be necessary, either to limit EMP exposure or to increase the EMP susceptibility threshold. This can be achieved in various ways, most of which are well known and well proven e.g.:
- Electromagnetic shielding
- Potential equalization
- Isolation
- System delimiting (topological and/or functional)
- Well-designed cabling and wiring
- Transient protection
- Filters
- Availability of repair facilities and spares
- Use of more robust electrical components
- Use of non-electrical functions (e.g. mechanical, pneumatic, hydraulic, optical, acoustic, thermal)
- Disconnection of cables that are for the moment not absolutely necessary
- Redundancy.

Several of the measures derive from methods of protection against lightning, static electricity, network transients and high frequency interference, as used in electromagnetic compatibility (EMC) technology. Hardening must be effective both for radiation and for

Component	Typical threshold energy levels (mJ)
Microwave diodes	$10^{-4}-10^{-2}$
Digital IC circuit	$10^{-3}-1$
Low power transistors	$10^{-2}-1$
Switch diodes	$10^{-2}-1$
Electroexplosive devices	$10^{-2}-1$
Tantalum electrolytic capacitors	$5\cdot10^{-2}-$
Metal oxide resistors	$10^{-1}-10$
High power transistors	$10^{-1}-10$
Silicon-controlled rectifiers	$10^{-1}-10$
Zener diodes	$10^{-1}-10^{2}$
Rectifier diodes	$10^{-1}-10^{2}$
Metal film restistors	$5\cdot10^{-1}-10$
Carbon film resistors	$5\cdot10^{-1}-50$
Electron tubes	$1-10^{4}$
Fuel-air mixture	3
Relays	$10-10^{3}$
Carbon composition resistors	$50-$
Wire-wound resistors	$10^{3}-$

Table 10
Typical minimum threshold energy levels (mJ) to cause permanent damage for pulse durations ≤ 1 μs. Threshold energies for temporary functional interference are 10 to 100 times smaller than the values for permanent damage

Fig. 11
Relative transfer impedance as a function of frequency for different cables. The d.c. resistance of the sheath, R_0, is often 1–10 mohm/m

1 Aluminium tape wound with a helix angle of 25°
2 Power cable EKFR (10 x 1.5)
3 Braided coaxial cable RG 11
4 Mains cables (EKLR or FKLR)
5 1.4 mm thick homogeneous lead pipe
6 1 mm thick homogeneous aluminium pipe
7 Steel tape between two homogeneous aluminium pipes

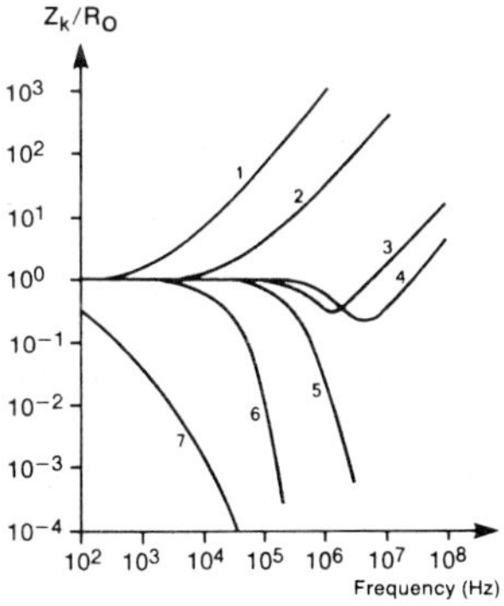

conduction effects. Good EMC practice can be regarded as a step towards EMP hardening, but many methods must be applied more rigorously and systematically than is otherwise customary.

The principles of protection are quite simple. The problems arise in applying them without incurring excessive costs. Since it is difficult to predict, calculate and verify the effects of EMP exactly, it is often easier to choose better protection than may be absolutely necessary. Detailed analyses and simulations can turn out considerably more costly. A compromise consisting of a combination of a reasonable number of analyses, tests and protective measures is usually to be recommended.

The protection level should be related to the required probability of survival. For example, a survival probability of 99% for the worst possible EMP case for a large complex system would require an enormous investment. With more reasonable probability requirements and with protection planning included right from the start of the planning of a new system, the protection costs can be reasonable and quite acceptable. However, when modernizing old systems the costs can be unrealistic unless the measures are limited primarily to protecting certain parts, or raising the survival probability to a more moderate level.

EMP protection methods can be exemplified with the aid of the communication installation described above. The choice of action for a given overall result must be based on an assessment of the cost effectiveness of the various measures in relation to each other. The hardening alternative chosen must be documented in a project plan which gives instructions in the greatest possible detail, based on a topological decription of the object to be hardened, figs. 5 and 6. This work has to be systematic and disciplined. There is a much wider choice of action if the communication site and the associated telecommunication network are in the early planning stage. It might then even be possible to choose non-metallic optical fibre cables and microwave radio connections that work in high frequency bands. Aerial lines should be avoided if possible.

We face a rapid changeover to optical fibre cables in the telecommunication networks during the next few years. This provides a unique opportunity for planning an exclusive network with EMP protection, superposed on the ordinary network. The cables can be equipped with extra fibre pairs (single mode), which are run to subscribers of major importance for the administration and maintenance of the country. The optical-electrical conversion should take place as close to the exchanges, telephones, computers, video terminals etc. as possible. All such associated electrical functions must be equipped with EMP protection. The power supplies for cable repeaters, exchanges and subscriber terminals must be separate and protected. If the suggested EMP-protected network is to be implemented in a reasonable way it must be included in the planning for the optical fibre expansion right from the start. A new EMP-protected network for top priority subscribers will then be obtained at a relatively marginal cost as the ordinary telecommunication network is extended. The new protected network can be commercialized and justified for top priority subscribers for a number of reasons, such as very low congestion in crises and conflicts, very high capacity and flexibility as regards transmission of different types of information (including wideband services), and good protection against listening-in and disturbances (lightning, power failures etc.).

The power distribution lines connected to the EMP-protected network must be equipped with large primary surge arresters at the point where transition is made to buried cable. This point should be remote from the communication facilities, and the connecting cable should be shielded if possible. The external power line system should be isolated from the installation by means of, say, rectifiers, accumulators and d.c./d.c. converters connected up for maximum transient attenuation, or by motor-generator transmission on an insulating shaft. The installation should be equipped with a standby power generator which starts up automatically if a mains failure occurs. The electrical installation must be carried out with great care, since external transients must be attenuated as much as possible by surge arresters, filters and other devices before they can reach standby power systems and finally internal loads.

The building could perhaps be constructed so that it provides electromagnetic shielding. This can be done by using small-mesh reinforcement and ensuring that the bars are in contact with each other wherever possible (may give 20 to 30 dB), or by supplementing the basic structure with metal sheets which are all welded or permanently joined in any other way (may give 50 to 80 dB).

The layout of external and internal wiring can also be modified during the planning stage. External power lines and telecommunication cables and grounding conductors can then be planned so that they are all run into the building at the same point. At this common entry point, cable shields and grounding conductors should be connected to the outside of a common inlet plate, which in its turn would be connected to the metal framework of the building. This arrangement diverts some of the currents induced in the lines to the outside of the installation, and only a minor part would penetrate inside.

At the inlet plate, transient protectors for the power and telecommunication lines should be installed with the shortest possible connections to the plate. The protective effect of even top quality overvoltage protectors is ruined if their connections are not as short as possible. The length of the connecting lead gives an additional voltage $u = L \times di/dt$, which can be of the order of 1 kV/cm. The transient protectors give potential equalization between different conductors and between the power and telecommunication networks, and thus reduce the risk of flash-overs in the internal installation.

If possible, cables should be shielded or run in shielded ducts or conduits. In some countries cables are run in iron conduits with threaded joints. A measure of the shielding effectiveness is the so-called transfer impedance. It is defined as the voltage drop, per unit length along the cable, along the inside of the cable shield in relation to the current on the outside of the shield. Fig. 11 shows typical values as a function of the frequency. Braided or in other ways transparent shields give considerably less protection against rapid transients which contain high frequencies than homogeneous (solid-walled) shields. Special sheath designs have been developed for EMP-hardened telecommunication cables. As regards the internal cabling it may be easier to run the cables together in special shielded cable ducts. However, such common cable runs must not include any conductors that can bring in large EMP currents from outside into the installation. Neither should the cabling be run near openings in the building shield, for example doors.

In shielded facilities the feed-through for air, light, fuel, water and sewage should be arranged with special care, e.g. by using waveguids. These are often sectioned and given a honeycomb structure, which makes them short and makes the overall attenuation the same as the attenuation of each individual cell.

The grounding or potential equalization in a telecommunication installation must be designed in accordance with

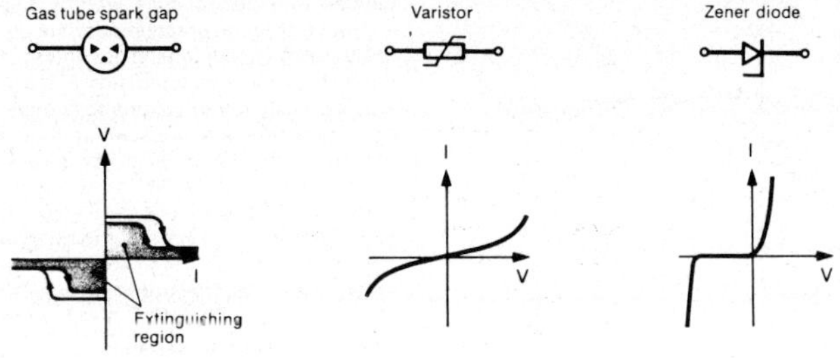

Fig. 12
Common components for transient protection

electrical safety regulations, and also in such a way that the grounding does not become a source of disturbances to the electronic equipment, or a conductor of EMP and lightning transients to susceptible points. In order to avoid damage from lightning or from earth faults in the power system the Swedish Telecommunications Administration now designs its installations so that external power and telecommunication cables are brought in as close to each other as possible. The power and telecommunication terminal grounding strips are joined together at the point of entry, either direct or via gas-type arresters, and the cables are equipped with overvoltage protection. If possible the internal cabling is arranged so that the power and telecommunication cables cannot form loops.

The protective measures are repeated at each new topological barrier, fig. 6. Some type of transient protection is usually required at equipment cabinets, and sometimes also at internal modules. The protective measures must be coordinated, which can be difficult. The type of transient protection varies from the outer barrier (primary protection) to the subsequent barriers (secondary protection), but the connection principles are the same. The final protectors must be placed as close as possible to the devices they are to protect. Protection components, such as gas-type arresters, varistors, zener diodes and filters, are often combined. Resistor fuses and protection diodes are sometimes used on printed boards. Fig. 12 shows some protections against transients. Circuits with differential coupling can suppress EMP induction by the order of 100 times (40 dB) as compared with circuits which are sensitive to common mode voltages.

The protection levels of the transient protectors and of the shielding barriers must be effective both for conducted and for radiated effects. It might be more economical to arrange protection in several topological steps of approximately 20 to 30 dB each. However good an electromagnetic shield is, its protective ability is largely lost if a cable is led through it without special protection being arranged at the point of entry.

The amplitude limiting given by overvoltage protectors is a function of time, whereas that given by filters is a function of frequency. Transient protectors can consist of one of these types or a combination of both. Power filters become larger and more expensive with increased power level. It is therefore more economical to use filtered power only for very sensitive equipment, and to connect units with lower sensitivity or priority to less well-protected power lines. This implies that conductors having different protection levels must be run in such a way that the mutual induction is negligible.

If a transmission equipment only uses a relatively narrow frequency band, it may be economical to install filters that attenuate the signal outside the operating range. At high frequencies, where a quarter wave corresponds to a reasonable length, a quarter-wave shunt can be used. It consists of a short-circuited transmission line connected in parallel with the ordinary transmission line. The

		Gas tube spark gap	Varistor	Zener diode
Ability to withstand energy with 1 ms pulse	J	10^2-10^3	$1-10^3$	$10^{-2}-10^1$
Ability to withstand current with 1 ms pulse	A	10^4	10^2-10^3	1–500
Leakage current	A	$10^{-9}-10^{-12}$	10^{-3}	$5 \cdot 10^{-3}-10^{-6}$
Response time	s	$10^{-9}-10^{-6}$	10^{-9}	10^{-9}
Capacitance	F	$10^{-11}-10^{-12}$	$10^{-10}-10^{-8}$	$10^{-10}-10^{-8}$
Voltage range	V	$75-3 \times 10^4$	20–1 500	2–300
Polarity		bipolar	bipolar	unipolar
Failure mode		open or short-circuit	degradation and short-circuit	short-circuit
Operation mode		short-circuit	voltage clamping	voltage clamping
Ability to extinguish		*)	yes	yes

*) Can require combination with a non-linear resistor

Table for fig. 12
Comparison chart of common transient protection devices. These are typical values; deviations can occur

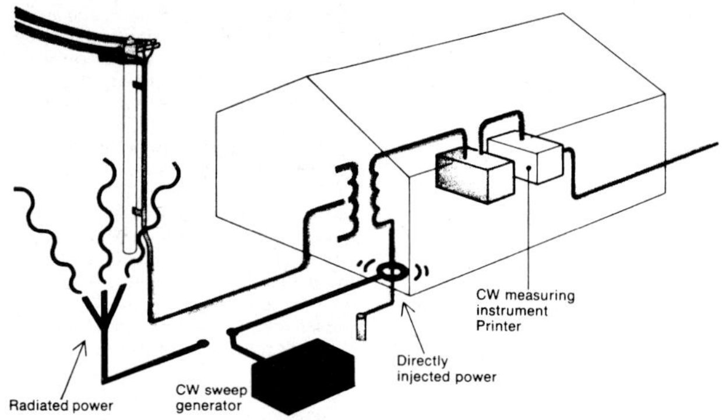

Fig 13
A telephone exchange can be tested by means of low-level simulation with continuous wave (CW)

short circuit acts as an open circuit for frequencies corresponding to an odd number of quarter wavelengths. For other and lower frequencies it acts as a shunt.

In order to simplify the introduction of EMP hardening, detailed requirement specifications are needed to form the basis for a standardized range of protective devices. It will then also be easier to plan and estimate the cost of hardening plant or systems. Although a multitude of protective devices are now available, there is a lack of requirement specifications which could lead to homogeneous and standardized protection systems. This important omission must be remedied soon if EMP protection is to be more widely undertaken.

An EMP protection standard for radiated and conducted effects is being prepared. The proposal is similar to the current military EMC standard specifications. A number of different classes of radiated EMP environments, starting at 50 kV/m and divided into steps of 20 dB, are defined in the proposal. Cable-associated EMP environment is defined for a number of classes of double exponential and damped sinusoidal currents respectively. Test procedures for the various classes are also being prepared.

Most protection methods described here also provide protection against lightning. The handbook "Practical Methods for Electromagnetic Interference Control" published by Ericsson's Networks Department contains detailed instructions for protection against lightning strokes and EMP.[7]

Hardening verification

It must be possible to verify the hardening level in connection with installation and commissioning, and later periodically during service life. Protectors can deteriorate with time for environmental reasons (e.g. wear or corrosion of contacts) as well as maintenance and modification reasons (e.g. installation of new cables, which are run straight through a Faraday screen without any protection at the entrance).

The operating staff must be trained and have access to instructions and spare parts in case the EMP threat becomes a reality. The protective arrangements must be simple, clear and readily accessible. Training is also needed in order that the personnel should not incapacitate any protection by improper action. Test equipment needed for routine tests could be built into the equipment. The shielding effectiveness can, for example, be tested regularly with the aid of permanent measuring loops around the main body of the building. Surge devices can be used to test the transient protection. Fig. 13 shows one way of testing a communication installation at the time of commissioning.

The nuclear power nations, particularly the US, have extensive programs for EMP testing and simulations. Much work is also being carried out in Europe, for example in the UK, France and West Germany. A number of nations that do not have nuclear weapons also have EMP hardening programs, for example Sweden and Switzerland. Practical measurements are supplemented by theoretical calculations. A number of computer programs have been developed to aid complicated calculations of EMP environments, propagation penetration, induction response in different structures from macro (e.g. countrywide networks) to micro (e.g. amplifier stages, integrated circuits, simple components), and for respective threshold levels for interference of function and permanent damage (for final comparisons). Statistics programs have also been developed for determining the accuracy and reliability etc. of the data.

The classes of EMP tests include:

- low-level mapping of currents induced in subsystems that have not been activated
- high-level current injection
- high-level exposure of operational subsystems to pulse-shaped electromagnetic fields.

Simulators are available for the free radiation of pulse or CW (continuous wave) fields. Current injection test equipment for pulse and CW is also available. Since the systems to be tested are usually too large to be accommodated within the test volume of the simulators, it is often necessary to carry out extrapolations and supplementary tests, for example a combination of radiation and current injection, or scale tests.

References

1. Mindel, I. N.: *DNA EMP Awareness, Course Notes, Third Edition.* ITT Res. Inst., Chicago, Illinois, October 1977, AD – A058367.
2. Ricketts, L. W. et al.: *EMP Radiation and Protective Techiques,* John Wiley & Sons, Inc., 1976.
3. Bell Laboratories, Loop Transmission Div.: *EMP Engineering and Design Principles.* Whippany, New Yersey, 500 – 052, 2 – 76.
4. Air Force Weapons Laboratory, Air Force Systems Command.: *EMP Interaction.* Kirtland AFB, New Mexico 87117 USA, AFWL–TR–80–402, December 1980.
5. Cooley, W. W. et al.: *DNA PREMPT Program, Communication Facility EMP Response – Prediction Techniques.* Boeing Aerospace Co, Seattle, WA 98124, USA EMC Conf. Proc. 1977, Seattle.
6. Broad, W. J.: *The Chaos Factor,* Science, Jan-Feb. 1983, pp. 41–49.
7. *Practical Methods for Electromagnetic Interference Control.* Instructions published by LM Ericsson, Network Department, Stockholm 1984.

SUPPLEMENT - URSI COMMISSION E ACTIVITY DURING 1982-1985

Nonlinear and Environmental Electromagnetics, edited by H. Kikuchi
Elsevier Science Publishers B.V., Amsterdam, 1985 – Printed in The Netherlands

URSI FACTUAL STATEMENT ON NUCLEAR ELECTROMAGNETIC PULSE (EMP) AND ASSOCIATED EFFECTS

URSI NUCLEAR EMP AD HOC COMMITTEE:

Wik, M.W. (Chairman, Sweden),
Stone, W.R. (Secretary, USA),
Gjessing, D.T. (Norway),
Lefeuvre, F. (France),
Lundbom, P-O (Sweden),
Migulin, V. (USSR),
Schwartz, S. (USA),
Stumpers, F.L. (The Netherlands).

PREFACE AND ACKNOWLEDGMENTS

This factual statement on nuclear EMP and associated effects was prepared by URSI (INTERNATIONAL UNION OF RADIO SCIENCE) during the URSI XXIst General Assembly in Florence August 1984. The statement was requested by SCOPE - ENUWAR (SCIENTIFIC COMMITTEE ON PROBLEMS OF THE ENVIRONMENT - ENVIRONMENTAL CONSEQUENCES OF NUCLEAR WAR) within ICSU (INTERNATIONAL COUNCIL OF SCIENTIFIC UNIONS) and should be unemotional, non political, authoritative and readily understandable.

The statement was unanimously approved (in principle) by the URSI Council meeting on September 6th, 1984.

Due to the limited time available during the URSI General Assembly some editorial work including some remarks at the council meeting was left to be done.

The committee is grateful to those URSI delegates who have taken active part in the preparation of this factual statement during the URSI General Assembly in Florence.

Allow me to add one personal reflection. The unanimous acceptance of the statement and the fear what would happen if communications were to be disrupted also shows that URSI delegates all around the world are striving towards future better understanding among people.

Manuel W. Wik

DISCLAIMER

All material in this document is based on what is available in the open literature. Participation in the preparation of this document does not constitute more than an affirmation that the material in the document is consistent with what is in the open scientific literature. Such participation and the information in this document neither confirms nor denies the validity of such information.

TABLE OF CONTENTS:

1. INTRODUCTION

Communications, electronic control and electric power systems are fundamental to modern society. Without them there would be risk of widespread chaos if outages were to last for many hours or days. Nuclear explosions can create blast, ground shock, thermal and nuclear radiation and electromagnetic pulse (EMP) effects. In this document emphasis is given to widely distributed environmental consequences on electric and electronic systems due to EMP rather than to the more local effects. EMP from exoatmospheric or high altitude bursts and associated electromagnetic effects, Transient-Radiation Effects (TRE), and radio propagation changes are described. Further information and technical and scientific reports can be found in the open literature, and even more has been classified for military and other reasons.

2. HIGH ALTITUDE EMP

2.1 Phenomenology

Early Time EMP Environment

High altitude nuclear explosions can cause disruption of communications and electric power systems even though other nuclear effects such as shock, heat, and radioactive fallout are not present at ground level.

The electromagnetic pulse generated by a single high altitude explosion could have a devastating effect on communications and power systems over the whole area of the earth, that can be viewed from the burst location, covering a country or a continent, and cause more or less widespread communication and/or power blackouts.

The explosion emits gamma rays within times of the order of nanoseconds. The gamma rays that travel downward will reach denser atmospheric layers at a height of 20-40 km and at this pancake-shaped zone, known as the deposition region, the gamma rays strip electrons from air molecules. This process is called the Compton effect. The free electrons initially move in an average radial direction (from the detonation point). In the presence of the earth's magnetic field these electrons are deflected to give a transverse component to the current, which in turn produces a radially directed (and therefore propagating toward the earth's surface) high-amplitude pulse of electromagnetic energy.

This EMP interacts with all sorts of metallic conductors which by design or by accident act as resonant antennas. The energy induced in conductors will find its way to connected objects where it is dissipated as heat, in some cases in combination with flashhovers. The electric field-strength of the pulse can be billions of times greater than those normally used for radio communications, e.g. some 10^{11} times larger than radio fields that could be received by an FM radio receiver. The rise-time of the pulse is less than 10 nanoseconds and its duration is of the order of a few hundreds nanoseconds. The rise-time and duration of the EMP field are in general shorter than those of lightning and of transients normally occurring in power systems, but for severe lightning strikes rise-times may be comparable. The longer duration of lightning strikes gives larger power in the lower part of the frequency spectrum. Probability of harmful effects on human beings, due to illumination by nuclear EMP, is small.

In a widespread network, pulses might be able to destroy or interfere with connected devices almost simultaneously in a number of places. Lightning and power transient levels are thought to exceed those for EMP for many situations. However, the simultaneous appearance of EMP- induced pulses in most parts of complex networks is a completely new situation that under some circumstances can affect network stability and performance in a serious way. There may be additional, quite significant effects on synchronous (e.g. power) networks. Instability and accumulated stress could also be the result of several EMP's.

Late Time EMP Environment

A nuclear explosion also causes a disruption of the earth's magnetic field. This can result in a second type of damaging electromagnetic pulse of much lower magnitude but longer duration (hundreds of seconds). This is known as the magneto-hydrodynamic EMP (MHD-EMP), in some way comparable to auroral geomagnetic storms. The MHD-EMP fields can disrupt very long landlines and submarine cables, including telephone cables and power distribution cables.

2.2 Environmental Consequences

Society has entered the information age and is more dependent on electronic systems that work with components that are very susceptible to excessive electric currents and voltages. For example, telecommunication networks, high voltage power supply networks, railway networks, air traffic, water supplies and industrial processes are all controlled and regulated by equipment which usually contains semiconductor components susceptible to EMP effects.

High altitude nuclear explosions could thus "knock out" radio receivers and other communication, electronic and electric equipment temporarily, if not permanently, by overloading and damaging the front ends, various cable terminations and internal circuitry, both by coming through intentional antennas and other paths that are unintended. However, whether or not this occurs depends strongly on the design and configuration of the system.

EMP and associated effects will vary considerably due to a number of factors, including the following: burst location and observation points; type of weapon; yield; number of bursts; type of exposed system and its topology; direction and length of exposed conductors; susceptibility of subsystems; and protective measures taken.

Several questions can be raised concerning the impact of high altitude EMP and associated effects on communications and power systems in case of nuclear exchange. Can people still use their radios? Do the nationwide telecommunication networks still function so that information about the radioactive fallout situation and weather forecasts can be distributed? Do the communication systems needed to re-establish electric power, water and food supplies and other vital requirements work? Can electric power for light, heat, water, and gasoline pumps be obtained? Which transportation systems can be used? Do the hotlines between the political leaders still function, so that the conflicts can be solved politically and not militarily? Are the command, control and communication systems intact so that defence systems are under control?

It is clear that current landline and satellite communication systems and domestic power grids may be at a serious risk. As an example there have been various recent incidents in several countries where regional and national power networks have failed totally because a few elements of the system failed and a cascade effect resulted. Any practical assessment must anticipate massive failures of communication systems, power supplies and electronic equipment. The effects of this on the civilian population could cause chaos, and the possible disruption of communication networks could have major significance for the development of a nuclear exchange.

Electric power networks often use nationwide overhead power lines that act as unintenional antennas and collect EMP energy. Many power stations are computer controlled and involve components that are very sensitive to electromagnetic transient overstress. However, the most serious situation are possible long time outages that could arise if damage occurs to any unique high voltage equipment, such as some larger, specially designed transformers. If there is not enough redundancy and flexibility in the power system, this might result in a power outage until new equipment has been manufactured and installed.

2.3 Protection

Today EMP analysis, testing, and protective design is a discipline in which there is much scientific work done. Yet, additional research on many aspects of EMP is necessary. For decision-makers and for those responsible for civil defense, telecommunications, and electric power administrations awareness of these effects is a must if protective measures are to be taken in order to reduce system susceptibility.

Protection against EMP commonly includes shielding, filtering, transient protection using surge arresters and other more specialized techniques. There are generally associated benefits (e.g. reliability and electromagnetic compatibility). However, the protection (usually termed "hardening") of a power or telephone network is vastly different from the hardening of a radio receiver. As an example, a rapid changeover to optical fibre cables is occurring in telecommunication networks. This provides a unique opportunity for planning exclusive EMP-protected communication channels superposed on the ordinary networks. Nevertheless, one must then also be prepared to invest the additional financial means for these less vulnerable communication networks.

It is unlikely that widespread complex systems can be completely protected against EMP. However, just as a given level of reliability can be attained for a given effort and cost in design and engineering, so is the degree of hardening achievable dependent on the amount one is willing to spend on the hardening process.

It is vital to be aware that hardening can be accomplished much more economically in almost every case if it is brought in at the planning and design stages, rather than if an attempt is made to add it at some later time.

2.4 Testing and Assessments

The understanding of EMP simulation and system testing is important to achieving protection and to measuring the degree of protection achieved. The complexity of the phenomena and the inability to test them with actual EMPs make fully verified assessments impossible although sophisticated analysis and testing techniques exist and are widely used.

It is important to maintain a sound scientific basis for the assessments of the effects. This implies that there shall be a certain objectivity in any assessment. In particular, it is important that highly qualified scientists and engineers and their organisations which have nothing to gain by hiding deficiencies be selected for this task. As in any good scientific discipline the results of such assessments must be subjected to peer review and must be fundamentally based on experimental data (testing), if one is to have confidence in the accuracy of the results.

3. OTHER RELATED EFFECTS

3.1 Low Altitude EMP

Low altitude nuclear explosions create blast, ground shock, thermal radiation, prompt nuclear radiation, fallout and electromagnetic pulse (EMP) effects. Electric power networks and communication networks are vulnerable to most of these effects. It is possible that in some cases because of EMP, failures will extend far outside the region where most objects are damaged by the usual blast and thermal effects. However, at least for public systems this EMP effect does not appear to be nearly as significant as the widespread effects of high altitude EMP effects.

3.2 Radio Wave Propagation And Long-Term Radiation Effects

Depending upon the scenario, nuclear bursts can also modify the atmosphere and ionosphere and affect radio wave propagation at all frequencies from a few Hertz through tens of Gigahertz.

They also create intense radio-frequency noise. The bursts increase the electron and ion densities, and there are large- and small-scale structural modifications, and long-term chemical changes. The actual propagation disruption patterns depend on many factors, and are very complicated. The propagation effects can cause blackouts for at least several hours, especially for shortwave communications. Questions also remain about whether satellite signals would be able to penetrate the ionospheric disruptions from high-altitude bursts.

An additional radiation effect that can disturb terrestrial radio wave propagation and communication satellite functioning is the Argus effect. By this phenomenon charged particles spiral around the paths of the earth's magnetic field.

The nuclear explosions can also produce artificial electron belts that can persist for months. The ionizing dose to electronic components in space installations can reduce their performance and lifetime. Other effects could result from different scenarios.

3.3 Transient-Radiation Effects (TRE)

Transient-Radiation Effects (TRE) are primarily caused by the prompt gamma, neutron and X-radiation. Effects on electronics are normally named Transient-Radiation Effects on Electronics (TREE). Transient and sometimes permanent changes in the performance of semiconductor components and other sensitive components and material can occur. These effects are of special concern in space for communication satellites where the atmosphere is not present to filter out the radiation environment.

3.4 System Generated EMP (SGEMP)

In outer space, communication satellites can be exposed to gamma and X-rays at considerable distances from a nuclear burst. The interaction of this radiation with the system also produces what is termed system generated EMP (SGEMP). This SGEMP can be a significant threat to many satellite systems.

4. CONCLUSIONS

Nuclear EMP is one of the effects of nuclear weapons. The awareness of the high altitude EMP threat in different scenarios and the protection problems should be widely spread and taken seriously. This brief document touches only on a small portion of the known effects of EMP and associated effects. It is intended to make workers outside this field aware of:

1. The existence of nuclear EMP phenomena.

2. The fact that high altitude EMP effects can occur even though other nuclear effects such as shock, heat, and radioactive fallout are not present at ground level.

3. The fact that these effects can cover the whole area of the earth, that can be viewed from the burst location, e.g. a country or even a continent.

4. The potentially serious nature of the effects of high altitude bursts to telecommunications and power systems.

5. That in consequence, the possible disruption of communication networks could have a major significance for the development of a nuclear exchange.

6. That the possible disruption of power supply in large areas could cause chaos and lead to the collapse of the infrastructure of modern society if outages were to last for a longer time.

7. The various serious effects on satellite communications.

8. The techniques for protection against EMP and associated effects.

9. The necessity for valid testing and assessments of system performance in EMP environment.

The fact that nuclear EMP high-altitude explosions could have a major significance in the initial stage of a conflict is a very serious problem and can lead to a destabilization of the present deterrent balance in the world.

It is hoped that the information presented in this statement and its scientific basis will emphasize the need to recognize the seriousness of nuclear EMP and associated effects and the importance of telecommunications in crises management.

----- x -----

EXAMPLES OF REFERENCES

Scientific:

IEEE Transactions on Antennas and Propagation, Vol. AP-26, No 1, January 1978

Bowhill, S.A.: Review of Radio Science 1981-1983. URSI, Brussels, Belgium, 1984, pp. E-9, E-10

EMP INTERACTION: Principles, Techniques and Reference Data, Dec. 1980, US Air Force Weapons Laboratory, AFWL-TR-80-402

Technical:

Sherman, R. et al: EMP Engineering and Design Principles. Bell Telephone Laboratories, Whippany, New Jersey, 1975

Glasstone, S. and Dolan, P.J.: The Effects of Nuclear Weapons. Prepared and published by the US Dept. of Defence and the US Dept. of Energy, Third Edition, 1977

McGrath, E. and Nelson, G.P.: Command, Control and Communications: Vulnerability and Mitigation. Technical Training Corporation, State of the Art Seminars Division London 15-16 December, 1983

Wik, M.W.: Hardening of Telecommunication Networks against Electromagnetic Pulses. Ericsson Review, No. 1, 1984

General:

Broad, W.J.: The Chaos Factor. Science, Jan-Feb 1983, pp 41-49

Breuer, R. and Lechleitner, H.: Der Lautlose Schlag, Meyster Verlag Gmbh, Munchen, 1982

Voyennaya mysl, USSR 1968. For translation and discussion see Douglass Jr., J.D. Soviet Strategy for Nuclear War (Hoover Institution Press, Stanford, Calif., USA 19779)

Steinbruner, J.: Launch under Attack, Scientific American, January 1984, Nr. 1, p. 23-33.

------ x -----

Honoured by the task given by the URSI President, Professor William E. Gordon.

Respectfully submitted:

MANUEL W WIK,
Swedish National
Committee of URSI,
Commission E.
(Chairman ad hoc EMP Committee)

ROSS STONE,
US National
Committee of URSI.
(Secretary ad hoc EMP Committee)

PER-OLOF LUNDBOM,
Swedish National
Committee of URSI.
(Chairman Commission A)

Nonlinear and Environmental Electromagnetics, edited by H. Kikuchi
Elsevier Science Publishers B.V., Amsterdam, 1985 – Printed in The Netherlands

OVERVIEW OF TRIGGERED LIGHTNING EMP *

H. Kikuchi
Nihon University, College of Science and Technology
Tokyo 101, Japan

Introduction

As one of man-made or artificial impacts on natural electrical phenomena, there is a growing interest in artificially triggered lightning and its electromagnetic pulse (LEMP) in comparison with natural lightning and nuclear electromagnetic pulse (NEMP) from phenomena and EMC aspects. The present review is intended to outline the problem of triggered lightning EMP briefly and to highlight the state of the art as a base of comparative studies of natural and triggered lightning phenomena and EMC. The topics covered are:
(1) Tall-structure triggered lightning
(2) Rocket-triggered lightning.

The present session entitled "Triggered Lightning EMP" consists of four papers (by Kikuchi, The St. Privat d'Allier research group, Podgorski et al., Takeuchi et al.) to be introduced hereafter among a general review of topics covered.

Definition of Triggered Lightning

Natural cloud-to-ground lightning is usually initiated by a step-leader moving downward toward the ground. In contrast, in many strikes to towers and rockets, the leader is observed to travel upward from the top of the tower or rocket into the cloud. These upward-going leaders are characterized by upward branching of the channel into the cloud. When the leader is initiated from the structure or spacecraft, the lightning is termed to be triggered by them. That is, no lightning would occur in the absence of the structure or spacecraft.

Tall-Structure Triggered Lightning

According to a comprehensive review of triggered lightning by Clifford and Kasemir [1], the first documentation of lightning triggered by a tall structure was reported by McEachron on lightning strokes to the Empire State Building in New York [2]. More extensive research on Mount San Salvador in Switzerland was reported by Berger in 1977 on lightning discharges to the 70-m-high tower on the top of the mountain (314 m above sea level) [3]. Since 1976 lightning measurements of the 553 m CN Tower in Toronto have been conducted by a Canadian group to take the opportunity to examine the effects of the tallest free-standing

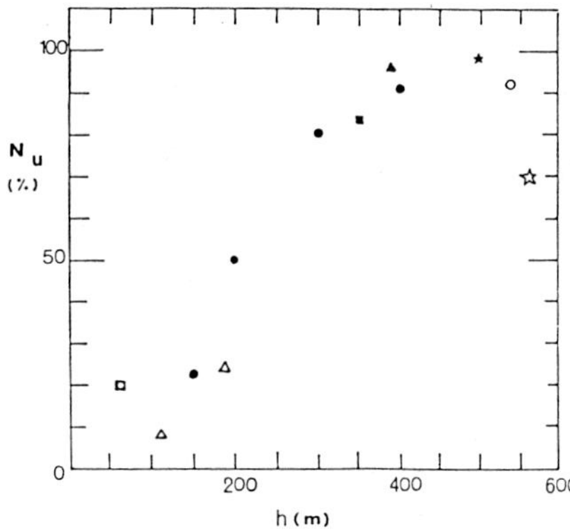

Fig. 1. Height dependence of the proportion of triggered lightning (upward leader). ☆ CN Tower [4,5,6], ◦ Golin et al. [7], * Garbagnati et al. [8], ▲ McEachron [2], ■ Berger [3], □ Erikson [9], ● Pierce [10], △ McCann [11].

structure in the world [4].

Incorporating lightning measurements made for various other tall structures, Fig. 1 represents the structure height dependence of the proportion of triggered lightning (upward leader) and indicates the increase in lightning flashes initiated by the structure with increasing height, reaching a maximum near 400-500 m [5]. A decreasing trend of the percentage beyond 500 m may be due to local lightning climatology or more specifically a new discharge channel formed by an EHD wind flow for downward leaders [6]. The reviewer's new model of electric cusp or reconnection might suggest this possibility, since electric reconnection can produce a strong EHD wind that will transport the ion clouds downward, forming a discharge channel [12].

In the past a transmission line (TEM mode) approach has extensively been used for modeling of towers. During the last few years measurements of lightning currents conducted by many researchers suggest the existence of rise times as short as 100 ns in the lightning induced EMP [4,13]. This indicates that for tall towers the wavelength of the spectral components of the fastest rise time generated by lightning may be shorter than the height of the modelled tower. Taking this into account, Podgorsky reports the implementation (for the first time) of a frequency domain computer code based on the solution of Maxwell's equations to the time domain modelling of the lightning-CN Tower interaction [14]. The ability of the frequency domain computer code to accept the ground and metal losses and tower footing resistance leads to a model that realistically predicts the details of the measured lightning waveforms as follows: prediction of the fastest rise time that can propagate along the tower; calculation of the tower impedance, prediction of current and EM field distribution. Such implementation may constitute a basic theoretical tool in lightning studies.

A very tall 200 m chimney located at a steam power station in Hokuriku, Japan provides another opportunity of studying triggered lightning, as reported by Takeuchi et al. [15]. During the winter in 1981-82, four lightning flashes hit the chimney. Three of them were identified to be triggered flashes with one positive and two negative polarity from electric field measurements by slow and fast antennas. The rise time of one positive flash is approximately 400 μs and nearly coincides with that of a natural positive return stroke. The durations of two leaders in the positive flash are approximately 10 ms. One of two negative flashes is composed of two return strokes and the other is of single return stroke. Their rise times are all estimated to be a few μs and also concide with that of a natural negative return stroke. The durations of three leader strokes preceding negative return strokes are all estimated to be approximately 200 μs that seems to be less than that of a natural leader stroke.

Some of remarkable features of tall-structure triggered lightning are summarized, particularly based upon CN Tower observations, as follows [6]:

1. A considerable number of positive flashes are expected in tall structures.
2. A large number of lightning strikes occur after large activity of electric field fluctuations with a strong electric field.
3. Lightning flashes to tall structures do not always need a strong electric field.

Rocket-Triggered Lightning

The first experiment of triggering lightning artificially was carried out successfully by Newman and coworkers as early as 1967, using wire-trailing rockets [16,17]. The wire length was usually about 100 m when the triggered lightning strike (upward leader discharge) was initiated.

Since 1973, a French group at Saint-Privat-d'Allier has developed the rocket-trailing technique for triggered lightning discharges [18,19]. During the eight-year period from 1973 to 1980 inclusive, 169 wire-trailing rockets were fired and 94 lightning flashes were triggered [20]. Although 56 percent of all the firings performed have triggered lightning flashes, the success ratio reached 93 percent in 1978. Out of 94 lightning flashes obtained, 31 were persistent current discharges of long durations (hundreds of milliseconds) and low amplitudes (a few kiloamperes) (type D) and 63 lightning flashes contained one or several strong current pulses (type C).

The ground electric field lies between 6 and 17 kV/m with an average of 10 kV/m for 56 triggered lightning flashes. The field polarity is generally negative (in the cloud) in 55 out of 58 cases cited above. For a positive field (positive cloud charges), the lightning flash is initiated by negative upward leaders developing from the rocket toward the cloud, requiring a field somewhat higher than that of negative polarity for triggered discharges. The height reached by the wire varies from 50 to 530 m for 71 triggered strikes.

A study of photographic film and electro-optical circuit documents reveals the following findings:

1. The triggered lightning flashes are ascending leaders that display most often many branches.
2. The lightning flash is initiated by an ascending leader attaining a speed between 2×10^4 and some 10^5 m/s and forming a continuous current of a few hundred amperes.
3. When this current is extinguished, one observes a brief and intense illumination that corresponds to a large-amplitude impulse with a rapid edge rise on the current recordings. The various branches do not light up simultaneously but successively and often at several instants. The speed of the dart leader (a luminous front that descends from the cloud just before the return stroke) measured by a streak camera is approximately 2 to 10×10^6 m/s. The speed of the return stroke that corresponds to current pulse propagation as measured by the electro-optical circuit lies between 0.5 and 1.8×10^8 m/s.
4. The total duration of the phenomenon can attain or even exceed one second. The lateral extention of the branches can reach 4 to 5 km and treeing shape is typical of an ascending lightning flash.

With regard to triggered lightning emissions, the study was initially confined to the most energetic emissions in the 150 Hz to 20 MHz band partly from the capability of equipments developed. More recently, new installations of VHF/UHF antennas and receivers and a fast time resolution system have largely extended the receiving capability to the 10 MHz to 1 GHz band. This enables us to obtain a great deal of information that cannot be extracted from photographs or current records and particularly to study the microstructure of VHF /UHF lightning emissions in relation to its discharge processes. Using these tools, new results obtained during the TRIP 82 experimental campaign in New Mexico and also the Saint-Privat-d'Allier experimental campaign in 1983 in France are as follows [21,22,23]:

1. Entire phenomena of triggered lightning discharge can be divided into two phases preceding and following the first return stroke. VHF/UHF activity in the initial phase starts well before the return stroke and is divided into three groups.
2. The first group of VHF/UHF activity with a duration of some tens of milliseconds is characterized by a sequence (or sequences) of closely spaced impulses at intervals of

tens to hundreds of microseconds. Comparing RF records with photographs, these emissions seem to be produced by upward moving leaders.

3. A few tens of milliseconds later, the second burst of emissions composed of isolated impulses or impulse packets is observed at intervals of about one millisecond for about 30 ms. Such a single intense RF burst is most likely produced by quasi-instantaneous 'partial discharges' visible on photographs typically at altitudes of 1 to 2.5 km.
4. The third RF burst occurs just before the first return stroke and consists of rapid sequences of impulsive emissions at intervals less than 3 μs lasting for about one millisecond. These emissions are thought to be produced by a dart leader propagating downward rapidly through a previously ionized channel.
5. The feasibility of triggered lightning is greatly increased by favourable processes and preparations in its initial phase. The formation of ionized channels is initiated by the development of an ascendant step-leader from the tip of the rocket into the cloud. VHF/UHF radiation of 'broken' spectrum is produced with a structure of impulse sequences in synchronization with the steps of the leaders. Subsequently, the broken spectrum of radiation is observed with a structure of impulse packets caused by partial discharges or 'intracloud' rapid leaders, probably in ionized branches which have penetrated the cloud. Only subsequently does the second descendant dart-leader develop which is more rapid, generating impulse sequences of broken type and completes the formation of an ionized channel for a subsequent return stroke.
6. The first return stroke is identified by the only abrupt impulse of intense VHF/UHF radiation originated from submicrosecond impulses as well as an abrupt change in the LF electric field.
7. In the development phase, VHF/UHF radiation is mainly organized in the form of impulse packets at intervals of several tens of milliseconds with a duration of about a hundred microseconds for each pulse and is attributed to dart leaders, K processes and junction phenomena that follow the ascension of the return stroke in the cloud. Except in the case of dart leaders, they are accompanied by charge neutralization. The spectra of dart-leader emissions are of the 'broken' type and their sources are located in an ionized channel. In contrast, the spectra of charge neutralization phases are of 'smooth' type and their sources are located in the cloud volume.

Since 1977, triggered-lightning experiments by wire-trailing rockets have been carried out in Hokuriku, a coastal region of the Sea of Japan. During the five-year period from 1977 to 1981, 97 rockets (48 during winters) were fired and 32 lightning flashes were triggered all during winters, although 7 firings during summers were not successful [24]. The success rate reached 66.7 percent in winter in contrast to no success in summer. This is because the Hokuriku area provides us with an excellent natural environment in terms of the cloud configuration and charge distribution for triggered lightning but only during the winter because of the following:

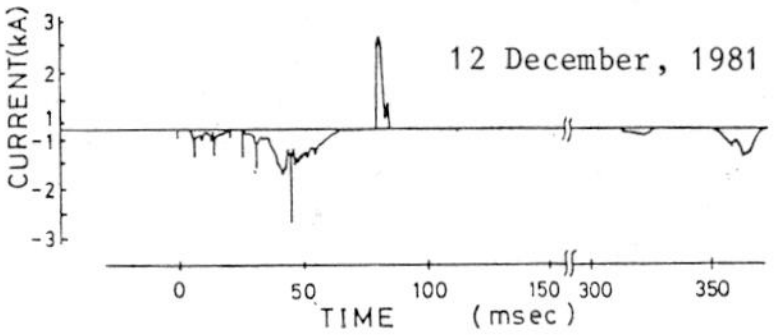

Fig. 2. A bipolar lightning flash.

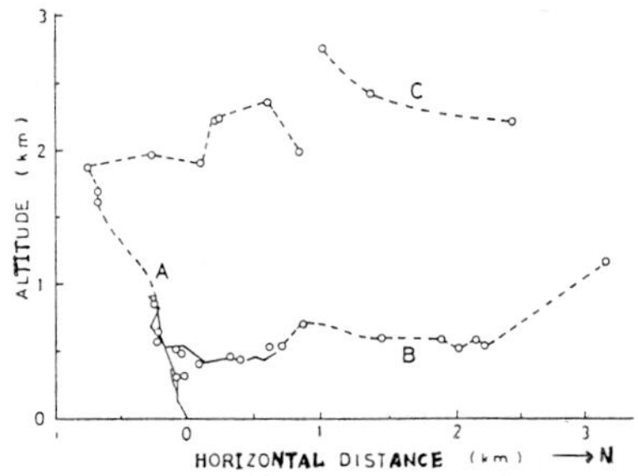

Fig. 3. Projection of lightning channels in a vertical north-south plane.

1. A cold north-west wind coming down from Siberia intersects a warm coastal current penetrating from the Pacific Ocean. As a result, there are many chances for thundercloud formation.
2. The cloud shape and charge distribution are extended more horizontally at lower heights.

The height reached by the wire varies between 37 and 326 m with an average of 142 m for 32 triggered flashes. An analysis of current, electric field, optical and acoustic measurements reveals the following:

1. Soon after launching a rocket, a predischarge current is initiated by corona discharge with a magnitude of a few microamperes and increases up to a few tens of milliamperes. When the rocket reaches a height of about 100 m, RF current pulses start flowing and being superposed on the continuous base current of several milliamperes. After the wire is vaporized by a large current pulse near 100 A, the main discharge occurs and its peak current ranges from 0.1 to 24 kA with a mean value of 15 kA. The duration of the discharge varies between 19 and 760 ms with an average of 420 ms.
2. During the ascent of a rocket, the electric field at ground decreases due to the induced charges at the tip of the rocket and space charges produced by a predischarge corona or streamer current.
3. Lightning traces produced from movie and still photographs reveal that upward leaders propagate from the tip of the rocket toward the cloud with a speed of 10^5 m/s and with a current of several tens of amperes, often developing into tree-like many branches.
4. Bipolar lightning flashes were observed with both positive and negative strokes for four flashes, for example as shown in Fig. 2. It has been found from both acoustic and photographic observations that the first and second strokes were produced by a vertical discharge into a negative thundercloud and a

horizontal discharge from a positive cloud through a previously ionized channel respectively, since the first and second strokes correspond to A and B as shown in Fig. 3: the projection of lightning channels in a vertical north-south plane.

5. An abrupt variation of the electric field at the moment of a return stroke was over 50 kV/m on the ground at a distance of 20 m from a discharge point and the horizontal electric field was over 10 kV/m at the discharge point itself.

Based on the observation that the cloud shape and charge distribution are extended more horizontally in winter in contrast to their vertical extent in summer, a new model of triggered lightning can be introduced in terms of a novel concept of electric cusp and reconnection, as proposed by the present reviewer [12]. The model is eventually represented in terms of an equivalent circuit and is particularly suited to explain bipolar and multiple strokes. In this connection, it should be remembered that multiple subsequent strokes were observed only when triggering occurred before the wire reached an altitude of 250 m for the case of French experiments and of 195 m for Japanese experiments. If the flash triggered at higher altitudes, the resulting discharge often exhibited only a continuing-current component, as might be inferred from this model. However, it is tentative and preliminary in the sense that the model only indicates the primary importance of geometrical and electrical configuration and is not going into any ionization physics, might be, of secondary importance.

Relation between Triggered and Natural Lightning

Natural cloud-to-ground lightnings, where tall structures or rockets are not involved, are initiated by downward stepped-leaders from the clouds to ground, although the formation of ionized channels by intracloud stepped-leaders and the consequent redistribution of charges in the cloud may be needed before the development of downward leaders. Once the leader contacts ground, the channel is rapidly discharged. Consequently, measurements of natural cloud-to-ground flashes made on the ground are always characterized by one or more discrete current pulses of high amplitude with an occasional continuing current following after one or more of the subsequent return-stroke pulses.

In contrast, a pronounced characteristic of rocket (or tall-structure) triggered lightning is the universal presence of a continuing-current component [1,17,19]. Many times, only a continuing-current discharge is experienced, but at other times hybrid flashes (continuous current plus discrete strokes) are observed. In this connection, it should be recalled that high-altitude rocket-triggering produced only continuing-current discharge. In any case, triggered-lightning is initiated by upward stepped-leaders that create an ionized channel between a cloud and the tip of a rocket or structure. Subsequently, dart leaders propagate downward rapidly from the cloud to ground through the thermalized path, further establishing a sufficiently ionized channel immediately before a return stroke. Accordingly, there are two processes for the formation of an ionized channel by upward stepped-leaders and by downward dart-leaders in the initial phase in contrast to only one process by downward stepped-leaders for the case of natural lightning.

It has been reported by French and Japanese groups that certain triggered lightning flashes did not follow the wire and struck the ground or other various objects near the predicted impact point [19,20,24]. Such an abnormal lightning flash displays a similarity with the first stroke of a naturally descending lightning flash in discharge characteristics.

In general, rocket-triggered lightning flashes display a marked similarity with structure-triggered flashes that strike metallic tall structures. However, a comparative study of various characteristics such as the rise time, duration and number of strokes should be done in order to clarify triggering-discharge mechanisms for both cases.

References

1. Clifford, D.W. and Kasemir, H.W., IEEE Trans. EMC-24, 112 (1982).
2. McEachron, K.B., J. Franklin Inst. 227, 149 (1939).
3. Berger, K., 'Lightning', Vol.1, ed. R.H. Golde (Academic Press, New York, 1977).
4. McComb, T.R. et al., Can. Elec. Eng. J. 5, 3 (1980).
5. Chang, J.-S., Proc. EMC Symp. Zürich (1983) p.17.
6. Chang, J.-S., Proc. EMC Symp. Wrocław (1982) p.467.
7. Golin, B.N. et al., Elektrichestvo, 8, 19 (1977).
8. Garbagnati, E. et al., Lightning Protection Conf. Szeged, Dec. (1980).
9. Erikson, A.J., Trans. SA Inst. Elect. Eng., Aug. (1978).
10. Pierce, E.T., Annual Meeting of Amer. Asso. for Advancement of Science, Phila. (1971).
11. McCann, G.D., Trans. AIEE, 63, 1157 (1941).
12. Kikuchi, H.,P. EMC Symp. Zürich ('85) p.47.
13. Berger, K. et al., Electra, 80, 23 (1975).
14. Podgorski, A.S.,EMC Zürich Symp.('85) p.57.
15. Takeuchi, T. et al., ibid., p.61.
16. Newman, M.M., 'Problems of Atmosphric and Space Electricity', ed. Coroniti (Elsevier, 1965), p.482.
17. Newman, M.M. et al., J. Geophys. Res. 72, 4761 (1967).
18. Gary, C. et al., Rev. Gén. Elect. 84, 44 (1975).
19. Fieux, R.P. et al., IEEE Trans. PAS-97, 725 (1978).
20. Saint-Privat-d'Allier Res. Group, Rev. Gén. Elect. 91, 561 (1982).
21. Leteinturier, C. et al., Atmospheric Electricity Conf. Albany, Jun. (1984).
22. Le Boulch, M. et al., ibid.
23. Hamelin, J. et al., EMC Wrocław Symp. Jun. (1984) p.1049.
24. Horii, K., Mem. Faculty Eng., Nagoya Univ. 34, 77 (1982).

* Introductory review at the session on triggered lightning EMP in the EMC Zürich Symposium, March, 1985.

Nonlinear and Environmental Electromagnetics, edited by H. Kikuchi
Elsevier Science Publishers B.V., Amsterdam, 1985 – Printed in The Netherlands

URSI COMMISSION E ACTIVITY IN JAPAN DURING 1982-1984

Haruji Ishikawa
RIA, Nagaya Univ., Toyokawa, Aichi 442 Japan

ABSTRACT : Following the idea agreed by URSI Commission E at Helsinki in 1978 to support 1984 International Symposium on Electromagnetic Compatibilitiy in Tokyo, the Japanese URSI Comm. E Committee collaborated with EMC'84 TOKYO Symposium having a URSI session principally in the domain of natural noise topics. We meet regularly once a year at Science Coucil of Japan, Roppongi, Tokyo to see the current scientific activity in and around our Japanese URSI E scientific community and review every three years, covering the period between successive URSI General Assemblies, the published research papers by our colleagues in and around our scientific community of Japanese Comm. E Committee of URSI in Science Council of Japan. Shown in the text are the review results mainly for 1982-1984.

E. I. Natural Electromagnetic Noise

E.I.1 EM Radiation Characteristics from Lightnings and Thunderstorm Activities

Electric field changes due to close lightning return strokes were recorded during winter thunderstorms in the Hokuriku area, Japan. Oscillating bipolar field changes were sometimes recorded together with ordinary unipolar field changes as recorded in summer thunderstorms. In order to understand the reason for the appearance of an oscillating field change, the return stroke current was calculated along a lightning channel with a transmission line model and the field change due to the current was computed at several distances. The calculation shows that the electric field due to nearby lightning (up to 30 km distance) oscillates because of the relatively small volume of the winter thundercloud [Takagi, et al., 1983].

Characterization of the frequency spectra was discussed of the return stroke field changes due to close lightnings in winter thunderstorms. It can be stated as follows: (1) The predominant frequency of oscillating electric field changes is about 16 kHz. (2) The spectra-character in VLF region depend on the distance while no such a tendency can be seen in summer lightning spectra [Nokubo, et al. 1983].

The multistation electric field change measurements were carried out in Nov.-Dec., 1979 in the vicinity of Unoke, Hokuriku area. The results of the analysis of six ground flashes were presented. It was found that the field changes for each return stroke and continuing current were relatively well fitted by the point charge model. As the preliminary results for the study of six ground flashes, the following items were pointed out:

(1) The number of strokes per flash was mostly one.
(2) All the flashes contained continuing currents.
(3) The temperature heights of the charge centers for negative ground strokes were 16° Centigrade on the average, while those for positive ground strokes were found in two different values, i.e. -22° and -2° Centigrade.
(4) The continuing currents of the short duration sustained the current of the order of 100 A for the initial 10 ms portion, regardless the polarity of the flash. The continuing currents of the relatively long duration associated with positive ground flashes sustained the current of the order of a kiloampare for the initial 20 ms portion [Funaki, et al., 1981].

Velocities of lightning leader and return strokes were measured with the optical detectors in two ground flashes (flash G-20 and G-22) initiated by upward streamers from a tall chimney. Dart-type leaders are confirmed both on the light signals and on the electric field changes. The leader propagating downward with a speed of 9x(E+6) m/s is noted in the flash G-20. 17 impulsive light signals are found to precede a return stroke signal in the development of upward streamers. On the three negative return strokes following downward dart type leaders in G-20 and G-22, the direction of the propagation is found to be upward and the speed to range from 1.2x(E+8) m/s to 2.3x(E+8) m/s [Nakano, et al., 1983].

Ogawa contributed the chapter of lightning current in the Handbook of Atmospherics and summarized the recent results of lightning research published mostly in 1970's and presented a general picture of lightning without neglecting the essential mechanism and features of lightning [Ogawa, 1982].

Ikeda, et al. collected the later dated lightning parameter data obtained by Japanese scientists interested in the field of lightning sciences. The data were selected from some groups which had sufficient number of the data and were presented at the scientific meeting of Society of Atmospheric Electricity of Japan in order to be discussed by the members. Then some new data were added to the first collection, thus more adequate lightning parameters in Japan were believed being compiled [Ikeda, et al., 1981].

Lightning charges, locations, and currents have been determined for 12 flashes from four winter storms observed on the Hokuriku coast of Japan during December 1977 through January 1978. Additional data are available from a total of eight winter storms. The heights and magnitudes of the charges in strokes-to-ground were calculated from simultaneous measurements of electric field changes made at seven stations covering an area of about 150 square kms. Discharges lowering positive charge to earth often exhibit large continuing currents (>10,000 A) for periods up to 10 ms. One positive discharge involved a peak current value of the order of 100,000 A and a charge that exceeded 300 C after 4 ms. Negative continuing current strokes are generally an order of magnitude smaller, about 4,000 A, involving charges of 100 C or less. The positive lightning discharges are located higher than the negative charges of the same storm, continuing a "normal" bipolar system similar to the charge configuration found in summer thunderstorms. For eight storms observed during this study, 26 out of total of 63 strokes-to-ground is a consequence of the wind shear. The shear provides a horizontal displacement between the charges that helps to ensure that an initiating positive streamer will continue down to ground rather than into the negatively charged region that would normally be located directly below it. The data suggest that positive strokes-to-ground would appear at a threshold shear value in the cloud layer of about 1.5 m/s/km [Takeuti, et al., 1982].

Experiment of artificial lightning triggered with a rocket pulling up the grounded wire has been carried out since 1977 in cooperation with the following members; I. Miyachi, K. Horii, Y. Kito, H. Akiyama, H. Yamashita, Y. Higashiyama (Nagoya Univ.), S. Muto, K. Nakamura (Nagoya Inst. of Tech.), G. Ikeda, S. Sumi (Chubu Inst. of Tech.), S. Aiba (Kanazawa Inst. of Tech.), R. Ishibashi (Kanazawa Univ.), H. Sakurano, (Ishikawa Tech. College) and N. Kitagawa (Saitama Univ.). The purpose of this experiment are to investigate the mechanism of the lightning discharge including predischarge phenomena, and to develop the technique protecting from the hazards by lightning discharge. Though the lightning has not been triggered in summers, 32 lightnings were triggered in winters at Hokuriku, the north coast of Japan. The peak value and the duration of discharge current, the electric field intensity at ground, the altitude of rocket, and others were measured, and the relations among them were discussed. The mechanism of lightning discharge was made clear from the optical observation and the measurement of current. The long gap discharges between the bobbin hanging in air and the ground were successfully performed three times, and the hairpin effect of mannequin and the safety in a motorcar for the lightning discharge were confirmed. The surge voltages induced on the test line of power distribution were measured and the shielding effect of the overhead ground wire to the test line was confirmed. And lastly the subjects to be investigated were described with the conclusions they obtained [Horii, 1982].

Strong corona discharges may be produced at ground levels in the intense electric field below the winter thunderclouds of low cloud-bottoms (less than 1km) which is one of the striking evidences found in winter thunderstorm at Hokuriku. Ions produced by corona along the ground level would be rearranged in the lower most atmosphere with strong electric field between thundercloud bottom and ground, thus producing a strong space charge layer over the ground. In the intent measuring the possible space charge layer in the lower most atmosphere over the ground, an electric field sensor with dual shutter field mills to be put on board a small model helicopter was instrumented for the following winter thunderstorm campaign at Hokuriku [Akiyama, et al., 1984].

Experiment of triggered lightning was conducted with 45 successful discharges in winters 1977-1984 at Hokuriku in Japan. The measurements were made on electric field intensity, discharge current including predischarge, light emission, thunder, surge voltage produced on power line, high speed movie and still pictures. Using the comprehensive data thus obtained, the mechanism of the triggered lightning was discussed in detail and the electric field intensity at the ground, the surge voltage on test power line, being induced by triggered lighning, were investigated showing the experimental evidences [Horii, 1984].

The final striking discharges to a power line or to the ground were observed from the near site in the experiment of triggered lightning with rocket in the winter season in Japan. The leader strokes propagated downwards set apart from the grounded wire trailed by the rocket, with branching and bending at every stepping point. The final stepping distnace was related to the peak discharge current comparing with the curves found in the literatures [Horii, et al., 1985].

The digital recording system used to measure a static electric field was developed and a lightning current wave thus obtained with it was described [Goto, et al., 1983].

To measure accurately lightning current waveforms having a long duration, a digital recording system which are controlled by a simple micro-computer board is developed. The current waveforms obtained with this are tentatively classified into five types [Goto, et al., 1983].

For the purpose of investigating lightning phenomena of winter thunderstorms, the lightning currents to an isolated high tower were measured by magnetic links, a magnetic tape recording device and an oscilloscope with a recording camera since the winter of 1976/77. Up to date, 40 thunderstorms in which lightning flashes struck the tower, were observed, and lightning current data were obtained by magnetic links and other devices. Of the 32 flashes, 56% were negative polarity, 31% were positive and unknown one were 13% because of their small current. The resultant oscillograms of current wave-form suggested the existence of lightning-currents having the very long duration. The still photographs of the flash to the tower showed that the most flashes may be of upward stroke. During one day or more when a trough passed over the region, several thunderstorms developed periodically and lasted usually for 2 or 3 hours. The electric fields at the ground were sometimes more than ± 20 kV/m, and they were highly variable in both magnitude and polarity [Goto, et al., 1981].

Ground flash density in Japanese winter thunderstorm area was estimated based on the registration of the CIGRE 500Hz lightning flash counter. It had been known that the thunderstorm days in winter at this area are not so less compared with summer, but in this study, it was clarified that the extremely low ground flash density per thunderstorm day is one of the remarkable features of winter thunderstorm. The extremely high flash incidence to tall structures despite the small ground flash density is another characteristic of winter thunderstorm. This indicates that the small value of ground flash density in winter can not be disregarded from the viewpoint of lightning protection of tall structures such as power transmission lines [Ishii, et al., 1981].

Kawamata et al.(1984) presented some results of the near field change signal due to the progressing leader which includes straight downward movement with refracting somewhere along the way backward or forward the observing site. They are obtained by calculation and represented as the graphic method which have the normalized scales of $n\Delta x/H$ on the axis of abscissa and of $\sum \Delta En 4\pi\varepsilon_0 D/2\rho$ on the vertical axis. On the other hand some oscillographic records have been obtained in the electric field which are of much value to compare with the results obtained by calculation [Kawamata, et al., 1984].

Kuwabara, et al.(1983) discussed indirect lightning surges at subscriber line ends, with the line being composed of overhead and underground sections. Lightning surge waves are calculated at both subscriber and telephone office ends. The relation between lightning surge waves at subscriber and telephone office ends is clarified. In addition, the lightning surge dependence on overhead cable sheath-earth resistance is also clarified. The analytical result presented in this paper is useful for lightning surge protection designing for telecommunication equipment [Kuwabara, et al., 1983].

Lightning surge inducing mechanism and calculated lightning surge waves were newly described. Lightning surges are induced by horizontal electric fields and vertical electric fields, when ground conductivity is finite. New equations which express lightning surge can be derived by using the above inducing mechanism. Lightning surge voltage dependences on cable length, earth resistance and on ground conductivity are obtained from both actual and calculated values. Lightning surge voltage distribution, front time distribution and time to half value distribution values are also clarified [Koga, et al., 1981].

In order to enable good protection against lightning surges, equipment should be tested by a reasonable test surge wave. Two test surge waves, a surge wave between paired wires in a cable and the ground, and a surge wave between two wires in the pair are needed. Two test surge waves are determined by equipment trouble probability distribution for surge energy and surge voltage, by using lightning surge data. An economical optimum design method for lightning surge-proof equipment is also determined [Koga, et al., 1981].

Kuwabara et el.(1983) discussed the design method of lightning surge test circuit for telecommunication equipment. The test circuit which represents telecommunication line and distributed inducing lightning surge on this line is obtained from network theory. The equivalent impedance of the test circuit is determined by the maximum lightning surge energy condition. The lightning surge test wave form is obtained from the distributions of lightning surge parameters based on equipment trouble occurrences due to lightning surge. The test circuit can be used for the equipment containing linear or nonlinear resistance, and also to estimate the reliability of the equipment [Kuwabara, et al., 1983].

In order to improve the fixing accuracy of atmospheric sources in South-East Asia along the base line of the direction finding (DF) network in Japan, the simultaneous observations of atmospherics at ELF/VLF were carried out at Toyokawa (34°50'N, 137°22'E), Japan and at Bangkok (13°44'N, 100°30'E), Thailand in the two autumnal seasons of 1979 to 1980. First, the coincidence of ELF atmospherics (5-800 Hz) received simultaneously at both the stations was identified by using NWC signal (22.3 kHz) as a time reference, then the fixing of atmospheric sources was, with high accuracy, made by means of the measurement of the propagation time difference between both the stations supplemented by the DF at Sakushima (TD method), in the region along the base lines where the effective triangulation DF was not usually realized. The direction finder was equipped at Bangkok during the 1980 campaign, in order to reconfirm the effectiveness of the TD method for the locating of atmospherics in South-East Asia where the accurate locating of atmospherics was predicted by the DFs at Sakushima and Bangkok because of the intersection at large angles of the direction lines measured at both the stations [Iwai, et al., 1981].

In order to estimate the accuracy in fixing atmospheric sources in South-East Asia by means of the triangulation and the two-station's direction findings (DF), Iwai, et al.(1982) made use of the measurements of difference in propagation time at Toyokawa and Bangkok in Thailand. As a result, it was found that atmospheric sources in South-East Asia can be located by the triangulation DF network with a rather small error. When a specific atmospheric source is fixed with its cocked-hat* area being less than 1x(E+4) square-km, the error in estimating the distance from Sakushima is less than $\pm$15%. The corresponding error is less than $\pm$20% for the fixed atmospheric source with its cocked-hat area of less than 5x(E+5) square-km. Furthermore, the location of atmospheric sources in South-East Asia by means of only two stations among the DF network is made within an error in distance of less than $\pm$22%. But one has to make an approp-

* "cocked-hat" is named after the triangle whose vertexes are made by the intersection of the direction lines at three DF stations.

riate choice on the combination of the stations either Moshiri or Kagoshima to be used along with Sakushima depending on the bearing angle measured at Sakushima [Iwai, et al., 1982].

E.I.2 Atmospherics Source Location and Thunderstorm Activity Distribution

Observation of global distribution of atmospherics, especially lightning discharges has been made by means of four channel HF receivers on board the Ionosphere Sounding Satellite-b. The operational frequencies of the receivers are set in the guard band of the standard frequency transmission at 2.5, 5, 10, and 25 MHz so as to avoid interference from terrestrial radio transmitters. The count of lightning discharges which were observed was accumulated in each geographic bin (10° x10° in both longitude and latitude) and normalized by the observation time for the area. Based on the data obtained during the period from June, 1978 to May, 1980, the active regions of lightning were mapped out in the form of global distribution divided into every four hours in universal time, 22-02, 02-06, 06-10, 10-14, 14-18, and 18-22 hr of a day of the four seasons. Estimated global lightning frequency was found to be 64 per second for the northern spring, 55 per second for the summer, 80 per second for the fall, and 54 per second for the winter. The peak lightning frequency occurs at 10° to 20° S during the northern winter and at 10° to 20° N during the summer. The diurnal variation of global lightning frequency has a peak at 12 hr UT [Kotaki, et al., 1981].

Based on global maps of thunderstorm activity observed by the Ionosphere Satellite-b, atmospheric radio noise levels over the world were calculated. Global distribution maps of atmospheric radio noise were made for frequencies of 2.5, 5, 10, and 20 MHz. for every four hours in universal time during autumn. The calculated results agree better with worldwide measurements than those by CCIR Report No. 322, and moreover, some characteristics of the atmospheric radio noise can be well explained by the present method [Kotaki, et al., 1983].

A new VHF (74 MHz) direction finding ('DF') system using Rocke-type Adcock antenna was developed to locate nearby lightnings with a bearing error less than 2 . It was found from the DF results that the location of lightnings were distributed within the region of a causative thundercloud, at ranges less than 120 km [Yamashita, et al., 1983].

E.I.3 Statistic Characteristics of Atmospherics

The impulse amplitude distributions (IADs) of impulsive noise received at a narrow-band receiver's input are theoretically derived by postulating physically motivated combinations of spatial distributions of sources and propagations. Good agreements are found between theoretical IADs and measured envelope crossing rate distributions (CRDs) for atmospheric radio noise. Derivation of envelope amplitude probability distributions (APDs) is considered for the input of the derived IADs. Particularly, if the input is represented only by a single power function of the amplitude, derived APDs are found to agree completely with a well known and important category of APDs [Nakai, 1983].

A model of atmospheric radio noise produced by a near thunderstorm is presented to give a physical interpretation to the properties of the fields in terms of the peak amplitude distributions of received impulses at the input of a narrow band receiver. In connection with the model, propagation over a short distance from any discharge inside the thunderstorm and the statistics of varying peak currents of return strokes and K changes are examined, and two different expressions, one expressing a lognormal distribution only for return strokes and the other a combination of two different lognormal distributions for the coexistence of the return strokes and K changes, are presented, both of which are available in calculating the respective part of the peak amplitude distribution. The emphasis is on the variation in the shape of the distribution with the distance to the thunderstorm. Comparisons have been made between calculated distributions and envelope crossing rate distributions (CRD's) at the output of the receiver. Comparisons have been found to be good for a frequency of 3 kHz in the high amplitude range of measured CRD's, which is nearly equal to the corresponding peak amplitude distribution at the input of the reveiver. Further application of the model is shown to be appropriate to higher frequencies, 50 kHz and 90 kHz, on the basis of the observed behavior of CRD's [Nakai, 1983].

E.I.4 Instrumentation for Whistlers, Hiss, and Chorus Measurements and Observation of Radio Noise in Space

A new whistler direction finder is developed and described and this has the following essential features: (1) The wave incidence situation is distinguished among the three cases, (i) vertical incidence with nearly circular polarization, (ii) oblique incidence with elliptical polarization, or (iii) horizontal incidence with nearly linear polarization. (2) The direction of arrival and polarization are deduced by the equation suitable for each wave incidence situation and yielding reduced polarization error. (3) The directional measurement is done at two frequencies for each whistler signals. (4) A method of recording and displaying the directional data is highly simplified so that it enables the measurement for a long period. (5) A real time whistler analyzer providing the occurrence rate and dispersion in real time is used, as an auxiliary equipment, to choose the times when to start and end the measurement. The classification of the wave incidence situation is based on the outputs from eight detectors, that is, six combinations of correlation among two horizontal magnetic field components at a point frequency, the corresponding phase-lagged (90°) components and the vertical electric field components, and two other signals yielding the nature of the polarization ellipse projected onto the ground. The overall system is described in detail in the present paper, which includes the principle of the direction finding measurement, antenna-receiver system, signal analyzer, data recording system, switching circuit of the filter frequency, sonagraphic display of the directional data and finally a real time whistler analyzer. Some results on the direction finding for whistlers observed at Moshiri (geomag. lat. 34.5°N) are presented to demonstrate the effectiveness of the present direction finder [Okada, et al., 1981].

The propagation characteristics of medium-latitude whistlers are studied on the basis of measurements with a new direction finder at Moshiri (geomag. lat. 34.5°N, L=1.6) during January and February 1978. Detailed analysis are made for two selected events: one with the highest daytime activity on January 22 and the other with the highest daytime activity on February 20. Daytime whistlers are of multi-flash or of an isolated type, and the dispersion remains constant during the event. The measured polarization is, on the average, close to exactly circular, indicating a negligible effect of multi-paths in the earth ionosphere wave-guide, implying that the ionospheric exit points are determined accurately. The characteristics on the direction finding results are compared with the properties expected for ducted and nonducted prolongitudinal mode propagation and it is concluded that the whistlers may be attributed to ducted propagation.

Just a single duct is present during the event. The dimension at the lower boundary of the ionosphere, illuminated by the duct which probably terminated at the F-2, is found to be 100 km. The small apparent movement of the exit regions may be due to a temporal fluctuation of the propagation conditions between the duct endpoint and the lower boundary of the ionosphere. For the night-time whistlers there are four distinct dispersions, but the discussion here is concentrated on the main dispersion which remained constant during the event. The observed polarization is less circular than for the case of day-time whistlers for the measured incident angle <60°. For larger incident angles, 60°< i <75°, one finds a great variability in the measured polarization. These indicate that multipath propagation may have a significant effect on the propagation in the waveguide at night. Correspondingly, there is a great scatter in the estimated exit points during each time interval. The exit region also shows a large apparent drift. This may be interpretated in terms of a joint effect of the presence of four ducts at the nearly same L shell and the temporal variation of the activity of the lightning sources at the feet of the four ducts in the conjugate zone. The scale of night-time ducts is difficult to determine due to the great scatter of the data [Hayakawa, et al., 1981].

In order to study the propagation mechanism of low-latitude whistlers, their ionospheric exit points have been determined by triangulation of the goniometer-polarimeter systems at the two stations of Kagoshima (geomagnetic coordinates 20°24'N, 197°53'E, L=1.22) and Ohgata (22°06'N, 199°51'E, L=1.25) supplemented by the routine observation at Sakushima (24°08'N, 203°19'E, L=1.28). Detailed analyses were made for January 23, 1978, when the whistler activity was most enhanced during one month's measurement. The difference in temporal variations of the occurrence rate at the three stations and the temporal variation of the single-valued dispersion enabled one to subdivide the daytime whistlers into three characteristic time intervals and the nighttime ones into two intervals. The characteristics during the whole intervals were found to be satisfactorily accounted for in terms of ducted propagation. The azimuth measurements have indicated that there was present a discontinuity in the characteristics between each two intervals, indicative of the different duct group for different time intervals, in other words, indicative of the identity and persistency of ducts for each period. Then the "apparent lifetime" of ducts of 1-2 hours was obtained. During the last interval for the daytime whistlers, a few ducts were found to be aligned on a certain latitude, which showed a temporal movement, or first increased and then decreased. The simultaneous presence of a few ducts at the same latitude was also seen at night [Hayakawa, et al., 1981].

A newly developed direction finding (DF) technique for auroral hiss based on the measurement of time differences of wave arrival was carried out in 1978 at Syowa Station (geomag. lat. -70.4°), Antarctica and its two slave unmanned observing points located at about 20 km distances from Syowa. The auroral hiss signals (0.3-100 kHz) received at the two spaced points were transmitted to Syowa by a wide-band telemeter of 2 GHz. The arrival time difference of auroral hiss between Syowa and each spaced point was automatically determined by cross-correlating the waveforms of the received signals and then the incident and azimuthal angles were measured with an accuracy of about 10°. It has been found that the new DF technique can determine localized exit regions at the ionospheric level which show rapid temporal movements. A comparison of the DF results with ground-based auroral data has shown that impulsive type auroral hiss with a wide-band frequency range has not emerged from the whole region of a bright aurora but from some localized regions of bright electron auroras at the ionospheric level, and that the arrival directions of auroral hiss change rapidly in accordance with the auroral movements [Nishino, et al., 1981].

Simultaneous spaced measurements of medium-latitude VLF/ELF emissions were carried out during the three northern winters from 1976 to 1979. The experiment was making use of two different kinds of direction-finding systems (a field-analysis method and a goniometric network) at two stations in Europe, namely Brorfelde in Denmark (L=2.9) and Chambron-la-Foret in France (L=1.9), this enabled one to locate the ionospheric exit regions of emissions over a wide range of L-values up to and beyond 4.0, the average plasmapause location. In order to study the time delay in the temporal evolution of VLF emissions or the longitudinal drift of the emission, observations from the Moshiri Observatory in Japan, widely separated in longitude, were also used. The overall system of the VLF equipment installed at the three stations was described. Then the VLF/ELF data of good quality obtained during the final year's campaign (Nov. 1978-Feb. 1979) were given. By making use of the direction-finding data, the observed emissions were classified into several categories, and some early results for some of the emissions were presented [Hayakawa, et al., 1981].

In order to measure precisely the dispersion of whistlers, the rho-theta transformation method and the templet matching method were tested. It has been shown from computer simulations that both the methods are effective for precise measurements of whistler dispersion and the latter is better to realize an apparatus for real-time measurements of whistler dispersion. On the basis of templet matching method, an improved real time whistler analyzer was realized and the principle of the dispersion analysis and the performance of the improved apparatus were presented [Tsuzuki, et al., 1981].

E.I.5 Source Activity of Radio Noise in Space

Latitudinal structures of plasmaspheric ELF hiss observed in the topside ionosphere have been statistically investigated by using ELF electric field spectra from ISIS-1 and -2 received at Kashima, Japan from 1973 to 1977. Most ISIS passes for ELF/VLF observations in mid- and low-latitudes include ELF hiss at frequencies below a few kHz. But sometimes a partial or complete lack of ELF hiss is noted along an ISIS pass. The ELF hiss is classified into two types: steady ELF hiss whose upper frequency limit is approximately constant with altitude and ELF hiss whose upper frequency limit increases with increasing latitude. The two types of ELF hiss occur in medium or quiet geomagnetic activities. Spectral shape and bandwidth of ELF hiss observed in the topside ionosphere are very similar to those of the plasmaspheric hiss or inner zone hiss observed by the OGO-5 and OGO-6. The occurrence rate of steady ELF hiss is about 0.3 near the geomagnetic equator and decreases rapidly with increasing latitude around L=3. Hence it seems likely that ELF hiss is generated by cyclotron resonant instability with electrons of several ten keV in the equatorial outer plasmasphere beyond L=3. Thirty seven percents of ELF hiss events at Kashima station occurred during storm times and sixty three percents of them occurred in non-storm or quiet periods, while Tsurutani et al. (1975) observed ninety two percents of 38 ELF events during storm times from the OGO-6 search coil magnetometer data. This discrepancy may result from the data difference between the ISIS electric field and the OGO magnetic field observations in the same frequency range. The electric field of ELF hiss becomes very weak every ten seconds of one half of the satellite spin period in mid- and low-latitudes but not near the geomagnetic equator. Ray tracing results suggest that waves of ELF hiss generated in the equatorial plasmasphere propagate down towards the low-latitude ionosphere bouncing back and forth between LHR reflection points in the both plasmaspheric hemispheres [Ondo, et al., 1981].

Latitudinal variations of narrow-band 5 kHz hisses have been statistically obtained by using VLF electric data received from the ISIS-1 and -2 at Syowa station, Antarctica and Kashima station, Japan, in order to study an origin of the narrow-band 5 kHz hisses which are often observed on the ground at mid- and low-latitudes. The results show that the narrow-band 5 kHz hiss occurs most frequently at geometrically invariant latitudes from 55° to 63°, that are roughly the plasmapause latitudes at various geomagnetic activities, both in the northern and southern hemispheres. The narrow-band 5 kHz hiss seems to be generated by the cyclotron instabilities of several keV to a few keV electrons for the most feasible electron density of 10 to 1000 per cubic cms in the vicinity of the equatorial plasmapause since the hotter electrons with energy of 10 to 100 keV are dominant just outside the plasmapause. This will be the origin of the narrow-band 5 kHz hiss observed frequently at mid- and low-latitudes [Ondo, et al., 1981].

Properties of VLF sources and auroral hisses have been studied by using ISIS VLF electric field data (50Hz-30kHz) received at Syowa Station, Antarctica. Source altitudes of VLF sources are obtained from frequency-time spectra of VLF sources by assuming the limiting model. The source distances of VLF sources are less than about 800 km below the satellite. VLF sources occur around the high- and low-latitude ends of the auroral hiss region. The active auroral hiss and the stable auroral hiss are classified according to the variability of the narrow-band intensity data processed from ISIS VLF tapes. The frequent occurrence region of the former agrees well with the usual auroral zone, while the latter occurs at invariant latitude from about 70 to about 80 degrees both on the day and night sides. The properties of the stable aurora have some similarities to those of the precipitating low energy electrons, the so called polar rain, in occurrence region and intensity. The narrow-band 5 kHz hiss occurs often at invariant latitudes of $61° \pm 5°$, which are close to the average plasmapause latitude, during the daytime and nighttime [Ondo, et al., 1981].

Latitudianal variations of chorus band frequency have been investigated to examine a model of the magnetospheric chorus obtained from OGO-1 and OGO-3 VLF observation by using field data (50 Hz-30 kHz) from ISIS-1 and ISIS-2 received at Syowa station, Antarctica in 1977 and 1978. The upper and lower limit frequencies of the dayside chorus decrease with L-value (or invariant latitude) and their latitudinal variations roughly agree with a latitudinal variation of one half of the equatorial gyrofrequency. This result supports the model of dayside magnetospheric chorus in which the chorus generated near the equatorial plane propagates along the field lines in the ducted mode or propagates inwards from the original field line in the nonducted mode. Nightside chorus were observed in disturbed periods by ISIS satellites as well as the nightside magnetospheric chorus. The upper and lower limit frequencies of choruses observed in the early morning, late evening and night time do not show any significant variation with L-value (invariant latitude). This may be due to different generation mechanism of the nightside chorus from the dayside one. Furthermore, a new case is found in that chorus band frequencies in latitudes beyond the plasmapause are higher than those in latitudes inside the plasmapause. Such a case has never been reported for observations of the magnetospheric chorus [Ondo, et al., 1982].

Using the fixed-frequency receivers on Ionosphere Sounding Satellite (ISS-b) the average galactic radio noise have been measured of 5, 10, and 25 MHz. Since the receiving antenna used for satellite observation has generally such poor directivity, it is difficult to distinguish the direction of coming radio waves. Since ISS-b has a circular orbit of about 1000 km altitude, the strength of the galactic noise from the center and anti-center can be distinguished by the reason that the galactic center is occulted by the earth at a certain position on an orbit. The average brightness temperatures of the galactic radio noise at frequencies of 5, 10, and 25 MHz have the values of (1.71+0.48)x(E+5)° K, and (3.47±1.07)x(E+4)°K, respectively. The radiated power coming from the direction of the galactic center is higher than that of the anti-center by 2.2 to 3.5 dB at high frequency band [Kotaki, et al., 1981].

E. II Man-made Electromagnetic Noise

E.II.1 Shinkansen Radio Noise

Measurements of impulsive noise radiation from individual bullet trains on the new Tokaido Line (called Shinkansen in Japanese) have been performed with the APD-CRD measuring apparatus using dipole antennas of vertical and horizontal types. Both an APD and a CRD have been simultaneously measured in a narrow-band at two frequencies of 50 and 100 MHz for noise samples taken in succession in time for individual bullet trains. Based on the analysis of these measured APDs and CRDs, the following remarkable results have been obtained: APDs and CRDs together with the parameters derived from individual APDs, average RMS, anti-log of amplitudes, and statistical parameter, V can be grouped into three different environments of train speed and acceleration and deceleration. A model for the impulsive noise was presented, saying that any elementary section of overhead electric wire radiates impulses with amplitudes varying randomly during the time interval, where it makes contact with any pantographs on the top of vehicles. Good fits are shown between time variations of values of the statistical parameter, V obtained as individual trains pass in front of the measuring site and those of a certain related field calculated based on this model [Nakai, et al., 1983].

E.II.2 City Radio Noise

Some data were shown of the radio noise due to a motor bicycle, which may be one of the component of the radio city noise. First, some noise waveforms which were picked up by a standard dipole antenna were shown, second, the INAD (impulse noise amplitude distribution) of the source were given, together with the radiation pattern, third, the INAD dependency on distances was given to lead to the INAD during the bicycle run. Finaly, the INAD measured beside a road was shown and the correspondence to the quasi peak indication was discussed [Echigo, et al., 1983].

Impulsive Noise Amplitude Distributions (INAD) were measured under the almost same condition as the previous measurements in 3 years ago, which was presented in the past technical report. The data showed the result that INAD were almost same as the earlier one, therefore the noise level was not changed so much for these years. Adding them, data in both polarizations (vertical and horizontal) were compared. Averaged curves revealed no differences between them, but each 1 min. interval data showed that vertical polarizataion data gave higher level for the cars. 5% values of coincidental quasi peak measurements and INAD levels which gave 100 pulse/min were compared and a straight correlation was shown [Echigo, et al., 1982].

E.II.3 Ignition Noise

The noise produced by the mechanical switching and gap breakdown was investigated. A brief view was given on discharges as a noise source, and some EM wave observations were made. The APD's were measured of the induced noise due to an inductive load circuit switching in which circuit voltage, current and inductance were changed respectively. Finally some theoretical and experimental studies were shown on the frequency spectrum of the noise by using sophisticated model systems. Analyses of noise due to mechanical switching and gap breakdown were found to be likely very complicated and difficult to obtain a substantial measure to reduce it. However, the problem was solved by taking a step-by-step way. As a result, it was shown that such a noise can be calculated in a macroscopic way such as a frequency spectrum, and also can be verified in the experiment to some extent. Thus the noise source should be analyzed in order to solve the EMC problem, and since the mechanical relay switching and gap breakdown are both very intense noise sources, the results shown in this investigation can effectively be used and developed for noise control in future [Takagi, 1983].

A simple method was presented of evaluating the electromagnetic ignition noise caused by the gap breakdown of a plug. A simple expression of the previously developed spectral evaluation was presented. By using above method, the effectiveness of the low breakdown field plug was discussed theoretically and experimentally. Then it was verified that the effective control of the ignition noise could be achieved by reducing the breakdown field strength [Fujiwara, et al., 1982].

E.II.4 Power Line Radiation

Observation of the power line radiation (PLR) had been requested to reveal the relationship between the PLR and VLF emission. The PLR intensity over Japanese Islands was observed by four times of balloon and rocket experiments in 1978 and 1979. From the results of balloon experiments for the measurements of horizontal PLR distribution on the Pacific Ocean at Sanriku Balloon Range, Tohoku, Japan, the horizontal attenuation rate of PLR was found to be approximately 1.3 dB/100km, and from the results of rocket experiments for vertical PLR intensity profile over Kagoshima Rocket Range, Kyushu, Japan, the attenuation rate in the ionosphere to be approximately 1.1 dB/100km [Yoshino,et al., 1981].

E. III <u>Interference</u>

E.III.1 EM Interference around a Shielded Cable

A consideration was given on an Intra-System EMI based on an example which showed interference with low level circuit produced by a high level shielded line. Equivalent circuits were introduced and analyzed to understand the mechanism of EMI. These results of analyses are thought to be very helpful to know EMI reduction techniques [Echigo, 1983].

E.III.2 Cycle Slipping on Omega Radio Waves

Phase lags (mostly, one cycle) of Omega signal (13.6 kHz) from Hawaii were often seen a few hours before the sunrise at Kagoshima, and those results were statistically reported by Taguchi, et al., (1982). A numerical analysis for this phenomena was thus presented. The east-west propagation below an anisotropic ionosphere with a transverse static field was applied. It resulted that if the ionospheric coefficient of the first order mode in the daytime is located in the fourth quadrant, the phenomena under consideration can be realized at Kagoshima before sunrise [Yamashita, et al., 1981].

Omega signals (13.6 kHz) from Hawaii were received at Kagoshima in Japan. As the cycle slipping (i.e., the signal phase retard by 1 or 2 cycles) was often seen during sunrise, a numerical work was previously done with use of the east-west mode propagation under an anisotropic and homogeneous ionosphere. To follow the previous investigation the cycle advance which occurred during sunset (i.e., signal phase advance by 1 or 2 cycles) was analyzed from the consideration of reciprocity between day-to-night propagation geometries. It was concluded that if the complex ionospheric reflection coefficient for the first order mode in the day-time is located in the fourth quadrant, these anomalous phenomena can be realized at sunset and sunrise [Yamashita, et al., 1982].

E. IV New Radio Noise Topics

E.IV.1 EM Emission Associated with Earthquakes

Recent observations in the USSR appear to suggest that wide-band electromagnetic radiation occurs just prior to earthquakes. To apply this phenomenon to earthquake prediction, measurements have been carried out at the Sugadaira Space Radiowave Observatory in Japan under the USSR-Japan Cooperation Program in 1980. The recorded noise level at 81 kHz is comparatively quiet throughout the day and night. However, about one-half hour before the main shock of a magnitude 7 earthquake at 07/33 UT on March 31, 1980, the instrument recorded an anomalous amplitude increase to 15 dB higher than the normal level. VLF data recorded synoptically at Sugadaira suggest that unusual impulsive radiation at frequencies below 1.5 kHz also occurred shortly before the earthquake. Similar 81-kHz emissions were observed prior to magnitude 5 and 6 earthquakes on September 25, 1980, and January 28, 1981 [Yoshino, et al., 1982].

Experimental studies carried out in various seismoactive zones of Soviet Union proved the manifestation of electromagnetic radiation in a wide range of radio frequencies during the preparation of the earthquake. Radiation anomalies related to the seismic activity have some typical characteristics that help to distinguish them from the background atmospheric and industrial noise. At the present time a large number of anomalous manifestations of EMI have been recorded, however, the study of dependence of emergence time, intensity, radiation spectra on such earthquakes parameters as magnitude, depth of focal region, epicenter location, and on accompanying geophysical parameters, for instance, the condition of the ionospheric waveguide, and geological structure of the region is just beginning. Collecting the relevant data for the most seismoactive regions of the globe seems most interesting. This was the reason for scientific-technical collaboration between the Schmidt's Inst. of Phys. of the

Earth (Academy of Science of USSR), and Univ. Electrocommunications, Tokyo, and Kyoto Univ. [Gokhberg, et al., 1982].

Simple observation systems to record natural electromagnetic noise have been developed. AGC signals from the radio receiver are amplified and filtered by a low-pass circuit to record a long period variations of the noise level, and filtered with a high-pass circuit to count-up the number of impulsive noises. The specific frequency of 9.5 kHz was used for the observation of VLF noises. For LF and HF continuous observations, 164 kHz and 20.007 MHz were used respectively. These frequencies have been chosen based on the results of surveying the artificial signal distribution in a wide frequency range. The integrated signals of each frequency are recorded on the strip chart paper with a speed of 25 mm/h. The counted number of pulses are converted into the analogue voltage to be recorded on the same chart paper. VLF flash type noises are clearly recorded by the latter method. Some examples of VLF imminent precursors have been obtained by these systems. They show that the signal level increases gradually before the occurrence of earthquake, and neither co-seismic nor post-seismic signals are recorded. ELF natural noises have been continuously recorded by the pen recorder with a speed of 4 mm/sec. The sensor of this system is a ball antenna installed on the roof of the institute. The continuous record have been obtained since the beginning of February, 1982. The Urakawa-oki earthquake of M7.3 occurred in Hokkaido on March 21, 1982. The focal depth of the earthquake was about 30 km and a part of after-shock region was located in the land part of the coastal region. Anomalous ELF signals were clearly recorded before and after this main shock. The ELF signals appeared about one day before the main shock. The signal intensity increased gradually and then rapidly decreased until the occurrence of the first fore-shock. The similar ELF signals appeared again after the occurrence of the main shock, and the maximum after-shock occurred at the peak of this ELF radiation [Oike, et al., 1982].

Schumann resonances have been examined in relation to occurrence of near earthquakes within 200 km from Kyoto city during the period February 1967 through June 1968. There were examples associated with the land earthquakes of magnitude of more than 3.0 and three with the sea earthquakes. The resonance amplitudes and frequencies are found to be increased before the earthquakes which occurred in land but not in the sea [Maki, et al., 1983].

References

[E.I.1]

Takagi, N., and T. Takeuti, Oscillating bipolar electric field changes due to close lightning return strokes, Radio Sci. 18, 391-398 (1983)

Nokubo, T., Z. Kawasaki, T. Takeuti, and T. Nakai, Spectral analysis of positive ground flashes in winter thunderstorms, Res. Letters Atmosph. Elect., Soc. A.E., Japan, 3, 33-40 (1983)

Funaki, K., N. Kitaqawa, M. Nakano, and T. Takeuti, Multistation measurements of lightning discharges to ground in Hokuriku winter thunderstorms, Res. Letters Atmosph. Elect., Soc. A.E., Japan, 19-26 (1981)

Nakano, M., T. Takeuti, Z. Kawasaki, and N. Takagi, Leader and return stroke velocity from a tall chimney, J. Met. Soc., Japan, 61, 339-345 (1983)

Ogawa, T., The lightning current, Handbook of Atmospherics, Edited by H. Volland, CRC Press, Inc. Boca Raton Florida (1982)

Ikeda, G., and S. Sumi, Lightning parameters obtained in Japan, Research Letters Atmosph. Elect., Soc. A.E. Japan, 1, 41-44 (1981)[in Japanese]

Takeuti, T., M. Brook, M. Nakano, and P. Krehbiel, The electrical structure of the Hokuriku winter thunderstorms, J. G. R. 87, C2, 1207-1215 (1982)

Horii, K., Experiment of artificial lightning triggered with rocket, Memoirs of Faculty of Engineering, Nagoya Univ., 34, 77-111 (1982)

Akiyama, H., H. Yamashita, K. Horii, Electric field and space charge below the winter thunderclouds in Hokuriku, Trans. IEE of Japan 104, No.3/4, 68 (1984)

Horii, K., Change of electric field intensity at ground and surge voltage induced on distribution power line in the experiment of triggered lightning with rocket, Trans. IECE of Japan, EMCJ, 25-30 (1984)

Horii, K., H. Sakurano, Observation on final jump of the discharge in the experiment of artificially triggered lightning, IEEE, PES Meeting, 85 WM 205-D, 1-6 (1985)

Goto, Y., F. Naito, and K. Narita, Automatic digital recording system for winter thunderstorm observation, National Convention Record, IEE, Japan, 985 (1983)

Goto, Y., K. Narita, and F. Naito, The digital recording system for lightning currents and some results obtained for winter thunderstorms, Res. Letter Atmosph. Elect., Soc. A.E. Japan, 3, 27-32 (1983)

Goto, Y., F. Naito, and K. Narita, The observation of lightning discharge in winter thuderstorms at Niigata coast, Japan, Proc. Int. Aerospace Conf. Lightning and Static Electricity, London, A-3 (1982)

Ishii, M., T. Kawamura, and J. Hojyo, Ground flash density in winter thunder storm, Res. Letters Atmosph. Elect., Soc. Atmosph. Elect., Japan, 1, 105-108 (1981)

Kishijima, I., H. Mitani, A. Inoue, Y. Ichihara, and M. Igarashi, Measurement of lightning stroke current at Kashiwazaki-Kariha in winter season (Part 1) Res. Report of CRIEPI, 180070, (1981)

Nakahori, K., T. Egawa, and H. Mitani, Characteristics of winter lightning current in Hokuriku district, IEEE Trans. on Power Apparatus and Systems, PAS-101, 4407-4412 (1982)

Kawamata, S., Y. Edano, and H. Kaminaga, A circuit model of lightning discharge to a transmission line., Proc. Int. Symp. EMC, Wroclaw, Poland (1984)

Kuwabara, N., H. Koga, and T. Motomitsu, Indirect lightning surges at subscriber line ends, Rev. Elect. Comm. Lab. 31, 539-549 (1983)

Koga, H., T. Motomitsu, M. Taguchi, and H. Yamaguchi, Lightning surge wave characteristics induced in transmission lines, Trans. IECE Japan, E64-B, 7, p. 627 (1981)

Koga, H., T. Motomitsu, and M. Taguchi, Lightning surge waves induced in transmission lines, Rev. Elect. Commun. Lab. 29, 797-809 (1981)

Koga, H., T. Motomitsu, and M., Taguchi, Test surge wave determination methods for telecommunication equipment, Rev. Elect. Commun. Lab. 29, 810-817 (1981)

Kuwabara, N., H. Koga, T. Ideguchi, and T. Motomitsu, Design method of lightning surge test circuit for telecommunication equipment. Trans. IECE, Japan, J66-B, 900-907 (1983)

Kuwabara, N., H. Koga, T. Motomitsu, and T. Ideguchi, Discussion of the design method of lightning surge test circuit, EMCJ 82-20 (1982)

Iwai, A., M. Kashiwagi, M. Nishino, Y. Katoh, and A. Kngpol, Fixing of the sources of atmospherics using the measurement of the arrival time differences of ELF atmospherics between Toyokawa and Bangkok, Proc. Res. Inst. Atmosphrcs, Nagoya Univ., 28, 13-29 (1981)

Iwai, A., M. Kashiwagi, and M. Nishino, On the accuracy of direction finding methods for atmospheric sources in South-East Asia, Proc. Res. Inst. Atmosphrcs, Nagoya Univ., 29, 35-46 (1982)

[E.I.2]

Kotaki, M., I. Kuriki, C. Katoh., and H. Sugiuchi, Global distribution of thunderstorm activity observed with ISS-b, J. Radio Res. Lab. 28, 49-71 (1981)

Kotaki, M., and C. Katoh, Global distribution of atmospheric radio noise derived from distribution of lightning activity, J. Radio Res. Lab. 30, 35-57 (1983)

World Distribution of Thunderstorm Activity obtained from Ionospheric Sounding Satellite-b, Observations June 1978 to May 1980, Radio Res. Lab. (1981)

Kotaki, M., I. Kuriki, C. Katoh, and H. Sugiuchi, Effects of the ionosphere on the radio horizon in HF band on the Earth-Space-Paths, J. Radio Res. Lab. 28, No. 125 (1981)

Yamashita, H., A. Iwai, M. Katoh, and T. Katoh, VHF direction finder for locating lightnings at close ranges, Proc. Res. Inst. Atmosphrcs, Nagoya Univ. 30, 15-24 (1983)

[E.I.3]

Nakai, T., Amplitude-Probability and Crossing-Rate distributions of atmospheric radio noise, Trans. IECE, Japan, E66, 290-297 (1983)

Nakai, T., Modeling of atmospheric radio noise near thunderstorms, Radio Sci., 18,2, 187-198 (1983)

[E.I.4]

Okada, T., A. Iwai, and M. Hayakawa, A new whistler direction finder, JATP, 43, 679-691 (1981)

Hayakawa, H., T. Okada, and A. Iwai, Direction findings of medium-latitude whistlers and their propagation characteristics, JGR, 86, 6939-6946 (1981)

Hayakawa, M., Y. Tanaka, T. Okada, and A. Iwai, Goniometric direction finding for low-latitude whistlers and their propagation mechanism, JGR, 86, 6781-6793 (1981)

Nishino, M., Y. Tanaka, A. Iwai, and T. Hirasawa, A new direction finding technique for auroral VLF hiss based on the measurement of time differences of arrival at three spaced observing points, Planet. Space Sci. 29, 365-375 (1981)

Hayakawa, M., Y. Tanaka, A. Iwai, J. Ohtsu, L.R.O. Storey, C. Beghin and T. S. Jorgensen, Simultaneous spaced direction-finding measurements of medium-latitude VLF/ELF emissions., Planet. Space Sci. 29, 505-520 (1981)

Tsuzuki, A., T. Okada, and A. Iwai, An improved real-time whistler analyzer using a micro-computer system., Proc. Res. Inst. Atmosphr., Nagoya Univ., 28, 31-48 (1981)

[E.I.5]

Ondo, T., Y. Nakamura, S. Watanabe, K. Aikyo, and T. Murakami, Plasmaspheric ELF hiss observed by ISIS satellites, J. Radio Res. Lab. 29, 159-181 (1982)

Ondo, T., Y. Nakamura, S. Watanabe, and T. Murakami, Narrow-band 5 kHz hiss observed in the vicinity of the plasmapause, Planet. Space Scci., 29, 65-72 (1981)

Ondo, T., Y. Nakamura, and T. Murakami, Characteristics of VLF saucers and auroral hiss from ISIS satellites received at Syowa Station, Antarctica, Proc. Third Symp. on Cordin. Obs. Ionosph. and Magnetosph. in Polar Region, Nat. Inst. Polar Res., Tokyo (1981)

Ondo, T., Y. Nakamura, S. Watanabe, and T. Murakami, Latitudinal variation of chorus fequency observed by the ISIS satellites, Proc. Fourth Symp. Cordin. Obs. Ionosph. and Magnetosph. in Polar Regions, Nat. Inst. Polar Res. 58-69 (1982)

Kotaki, M., I. Kuriki, C. Katoh, and H. Sugiuchi, Cosmic radio noise at high frequencies as observed with ISS-b, J. Radio Res. Lab., 28, 35-48 (1981)

[E.II.1]
Nakai, T., and Z. Kawasaki, On impulsive noise from Shinkansen., IEEE Trans. EMC-25, 396-404 (1983)

[E.II.2]
Echigo, H., N. Handa, T. Takagi, and R. Sato, An analysis of radio noise measured in a city area, Trans. IECE Japan, EMCJ 83-27 (1983)
Echigo, H., N. Handa, T. Takagi, and R. Sato, On the statistical measurement of impulsive noise in a city, Trans. IECE Japan, EMCJ 82-79 (1982)

[E.II.3]
Takagi, T., An overview of the noise sources due to switching and discharge, Proc. Sixth EMC Symp., Wroclaw (1982)
Takagi, T., Theoretical and experimental study of switching and ignition interference in motor vehicles, Fifth Symp. & Tech. Exhibit. EMC Zuerich, (1983)
Minegishi, S., T. Omori, H. Echigo, and R. Sato, Fundamental research on the noise generation from switching contact circuit, Proc. Sixth EMC Symp., Wroclaw (1982)
Fujiwara, O., Y. Amemiya, Suppression of ignition noise caused by plug gap breakdown, Proc. Sixth EMC Symp., Wroclaw (1982)
Fujiwara, O., Y. Amemiya, Calculation of ignition noise level caused by plug gap breakdown, IEEE Trans. EMC-24 (1982)

[E.II.4]
Yoshino, T., and I. Tomizawa, Rocket and balloon observations of power line radiation over Japanese Islands, Proc. Fourth EMC Symp., Zuerich, 4, 525-530 (1981)

[E.III.1]
Echigo, H., Electromagnetic interference around a shielded cable, Record of Electrical & Communication Engineering Conversation, Tohoku Univ. 52, 1 (1983)

[E.III.2]
Yamashita, M., and K. Taguchi, Numerical analysis of cycle slipping with Omega Hawaii (13.6 kHz) as received at Kagoshima, 65th Meeting, Nihon Kokai Gakkai, Oct. (1981)
Yamashita, M., and K. Tagushi, Numerical analysis of cycle advance with Omega Hawaii (13.6 kHz) as received at Kagoshima., 67th Meeting, Nihon Kokai Gakkai, Oct. (1982)
Taguchi, K., and M. Yamashita, Numerical consideration of occurrence of cycle slippings with Omega Hawaii (13.6 kHz) as received at Kagoshima, Trans. IECE, Japan, J65-B 281-287 (1982)

[E.IV.1]
Yoshino, T., I. Tomizawa, M. B. Gokhberg, and V. A. Morgounov, Experimental measurement of electromagnetic emissions possibly related to earthquakes in Japan, J. G. R. 87, 7824-7828 (1982)
Gokhberg, M.B., T.V.A. Morgunov, and T. Ogawa, Results of recording operative electromagnetic earthquake precursor in Japan., Izvestiya, Earth Phys. 18, 144-146 (1982)
Oike, K., and T. Ogawa, Observations of electromagnetic radiation related with the occurrence of earthquakes., Annual Report Disaster Prevention Inst., Kyoto Univ., No. 25, B-1, 89-100 (1982)
Maki, K., and T. Ogawa, ELF emission associated with earthquakes., Res. Letter Atmosph. Elect., Soc. A.E., Japan, 3, 41-44 (1983)

Ogawa, T., K. Oike, and T. Miura, Problems of EM radiation associated with earthquakes and a few results of rock experiments, Soc. Earthquake, Japan, Spet. Meeting (1983)

Nonlinear and Environmental Electromagnetics, edited by H. Kikuchi
Elsevier Science Publishers B.V., Amsterdam, 1985 – Printed in The Netherlands

URSI COMMISSION E ACTIVITY DURING 1982-1984

F. Louis. H. M. Stumpers
Chairman, URSI Commission E
Philips Research Laboratories, Eindhoven, the Netherlands (retired)
Royal Academy of Arts and Sciences, Amsterdam, the Netherlands,
Universities of Nymegen and Utrecht, the Netherlands

URSI Commission E activity during 1982-1984 is reviewed and highlighted, mainly based upon URSI Sponsored Symposia and Workshops.

EMC Symposium Wrocław 1982.

All preparations had been made for this symposium, and many texts of papers were already available, when, due to the state of war declared by the authorities, it became impossible to organize it, at the time foreseen, June 1982. It was decided to publish the avilable texts in two books of 750 pages in total under the usual name, Sixth International Wrocław Symposium on Electromagnetic Compatibility. The keynote-address: Electromagnetic Compatibility Control, its Status and Future by Professor Showers, Chairman of C.I.S.P.R. mentioned a possible revision of the basic measurement techniques in order to provide more meaningful limits, that are at the same time implementable from a regulatory point of veiw. However, much more work is required before adequate and economic interference control measures are available which have general acceptance.
74 other papers were published. Professor K. Rawer gave a survey of COSPAR's experience with detrimental activities in space. Plasma depletion by releases from spacecraft, and high power beams for Solar Power satellites, got much attention. Professor H. Kikuchi favored the use of a pulsed or bunched electron beam for space vehicle neutralization, over a pure continuous DC beam.
Pevnitsky, Chernomordik, Pawelec, and Betsky derived the interference effects of different processes from models. Takagi and his co-workers gave six papers on switching and discharge noise sources, based on Japanese investigations. New developments in antennas were discussed by six authors from U.S.S.R., Katulski, Katzenelenbaum, Jordanova, and so on. Several papers discussed radio propagation. Among them Bienkowski gave interesting results from the Orbiting Satellite Carrying Amateur Radio, OSCAR in 432, 144, and 48 MHz bands. Nonlinear resonance interaction of electromagnetic radiation with the polar medium was discussed by Gaiduk, and so on. Problems with earthing and shielding were discussed respectively by Castro Varela and Lattarulo. Eight authors discussed problems of frequency assignment and limitation of radiated power. Six papers reviewed mobile radio communication in fading. Boronichev treated the immunity of television and sound broadcast receivers, and Venskauskas developed guide lines for noise squelch circuits for voice signal radioreceivers. A session on Wire Communication was organized by Dr. Lorke. Disturbances induced by a DC traction line (Brunelli and Bombelli) and protective measures (Gotze, John, and

Schuppler) belonged to this theme. A session on Lightning would have had papers by Chang (Effects on a very high tower), Bruhin (Shielding by reinforced steel in concrete) and Sowa (Induced voltages in transmission cables). Four papers treated power line harmonics. Miscelleous noise sources had four papers: Garlan, and so on (I.S.M.), Vaculikova, and so on (Static power converters), Jermendy (High voltage transmission), Petrov (Broadband filter parameters for interference suppression). Among the measurement problems discussed, we mention: Bem and Zernicki (Radiation patterns) Novak (Computer aided envelope analysis), Hadrian (Dipole moments), Ivanova (Ignition noise), Glaba (Switching noise), Nano and Borsero, Rabiej (V and Delta Networks). The effect of electromagnetic radiation on biological tissues is a difficult problem, to which ten papers were devoted.

EMC Symposium Zürich 1983.

The Zürich Symposium of March 1983 had 629 participants, 103 papers, 29 exhibitors, 5 Workshops and 3 Technical excursions. Six papers got a citation. First prize winners were Goedbloed, Riemens and Stienstra for their paper: "Increasing the RFI immunity of amplifiers with negative feedback". They presented a theoretical background for the correct application of interference suppression capacitors in amplifiers with negative feedback. Results of experiments confirmed the theory. Second prize went to Dalby: "Linear antenna near field decoupling using a radial transmission line". The technique involves use of a radial-transmission-line equivalent of a $\lambda/4$ choke, mounted at the base of the monopole or stub-antenna. Stacked chokes have also been used successfully. Third prize was given to Demoulin, Duvinage, Comic and Degauque for "Penetration through an interruption of the shield of a coaxial cable". Based on the theory of an annular slot on an infintely long coaxial cable (Wait), the configuration of a slotted line parallel to a ground plane and of a cable inside a hollow metallic tube are discussed and experimentally confirmed. Bullough and Cotterill were cited for "Ariel 4 observations of power line harmonic radiation over North America and its effect on the magnetosphere". Their studies confirmed that VLF emissions in the North American zone are primarily due to power line radiation. Varakin was cited for "Electromagnetic compatibility of cellular communication systems with pseudo-noise signals". He showed that the cellular mobile communication system uses the spectrum more efficiently than traditional centralized systems. Max and Shaw got their citation for "Distributed low-pass filters for EMI filtering". Mutual enhancement of dielectric losses and reflections enables one to design distributed low-pass filters with reasonable low cut-off frequencies and high attenuation well into the upper GHz region. A prototype was built. It clearly indicated the superiority of the distributed filter.

There were 15 full and three half sessions. We mention Struzak's "Urban mf radio noise survey in Poland"; "Kikuchi's Charge neutralization problem for a space satellite"; Chang's: "Impact of lightning on a tall structure; Middleton: "Threshold detection in non-gaussian environment"; Koepke: "New methods for determining the emission characteristics of an unknown interference source"; Gallon: "On the measurement of soil permittivity"; Toler "Overview of biological effects of radiation"; Baum: "Review of hybrid and equivalent electric dipole EMP simulators"; Menez et al.: "Electromagnetic field coupling to telecommunication networks"; Takagi: "Theoretical and experimental study of switching and ignition interference in motor vehicles"; Jones: "The conceptual problem of ground planes"; Borsero and Nano: "Some considerations about interference voltages and limits"; Szentkuti, Peurala: "New views on the absorbing clamp in the testing of cable screening effectiveness"; Crawford: "Evaluation of shielded enclosures without and with RF anechoic material"; Whalen: "EMI in microelectronics".

"Power Line Radiation and its Coupling to the Ionosphere and Magnetosphere

H. Kikuchi, editor. Book Reidel, Dordrecht, Holland 1983. Also Space Science Reviews vol. 35, nos. 1 and 2.

Most of the original material in this book was first published during the Wroclaw EMC Symposium of 1980, where Professor Kikuchi organized the relevant session, but later it was found to be interesting to scientist outside the EMC Community, and with the agreement of the Wrocław Symposium Council this republication took place. Several special papers were added: H.G. Booker, "A simplified theory of ELF propagation in the earth-ionosphere transmission line"; H. Kikuchi, "Power line transmission and radiation"; A.B. Behroozi-Toosi and H.G. Booker, "Application of a simplified theory of ELF propagation to a simplified worldwide model of the ionosphere"; K. Bullough, "Satellite observations of power-line harmonic radiation". Nine papers were republished from Wrocław 1980.

EMC Symposium Wrocław 1984.

This Symposium with 107 papers (published in two books of 1070 pages) and over 200 participants was quite successful. That there were 105 participants from Poland alone shows for the interest the subject of EMC aroused in the country. The session on lightning, organized by Prof. Lundquist and me, was in the line of U.R.S.I., and so was Prof. Kikuchi's session on Active line models. In the opening plenary session, I gave an address on: Electromagnetic Compatibility and the International Union of Radio Science. I stressed the interest of U.R.S.I., not only of Commission E, but also of Commissions A,B,C,F,G and H in the subjects one usually treats in EMC Symposia. In the plan of that session Professor Seidler was to have spoken on "Dynamic Compatibility in Systems with State Identification Subsystem." Unfortunately an illness prevented his participation. His aim was to show that, if information of several transmitters and receivers is availabe, the performance of the system can be much improved. "A global satellite study of the 17.8 KHz NAA transmitter signal" by Davies, Strangeways, and Bullough gave evidence of cyclotron resonance amplification of the NAA signal with energetic electrons. Larkins and Likhter reviewed "The intensity of electromagnetic noise in the plasmasphere of the earth" at frequencies of 0.5, 1.6 and 2.5 KHz. Matthews, Omura, and Matsumoto reviewed power line harmonic radiation, based on a computer simulation of particle trapping by a coherent whistler mode signal. Bashkuev, Haptanov and Tsydypov reported on research in Central Asia on diurnal variations of spectral density in the 14-120 Hz range.

The importance of reproducible test results, e.g. by using metallic ground planes, was stressed by Gehrmann. Suckenik addressed the way in which interference measurements are defined in MIL-STD-461, and found the properties of accuracy, consistency and predictiveness to be lacking. He suggested some improvements. Klose described electronic equipment and systems for radiomonitoring in the domain of 10 KHz to 1 GHz, and even up to 18 GHz. Computer aided design of low-power TV network for efficient use of spectrum space was discussed by Waszkis. Other aspects of computer aided frequency assignment were discussed by Carmassi and Tomati; Rubinshtein, Kaganer and Bessonov; Sznajzycht. Active distributed parameter lines with ground return and with three conductors (cable) were treated by Kikuchi and by Matsuda, respectively. Ahermouche, Fontaine and Plumey presented a study of a vertical antenna penetrating the interface between a free space and a lossy half space. Chang continued his study of lightning beams on tall structures with an RC circuit model. Another circuit model of lightning discharge to a transmission line was discussed by Kawamata, Edano and Kaminaga. Lundquist and Cooray discussed Modelling of lightning. Sowa: Transient induced voltages in communication and control system in a large industrial building, struck by lightning. Scuka: EMI of lightning with digital communication systems. Bruhin: Shielding effectiveness of reinforced concrete against e.m. waves. Hamelin, Le Boulch, Le Teinturier and Weidmann: Radiation emitted by leaders and intracloud lightning discharge processes and the fields resulting at the ground.

Professor Mikolajczyk organised a session on EMC and Biology and gave Introductory Remarks. Progress in the healing of bone fractures was studied by Chang, Liu and Lin (Tai-wan). Cerri and De Leo treated waveguide applicators for use in microwave hyperthermia. For mobile radiocommunication four papers discussed cellular structures: Varakin, Pronichev, Tkachev and a joint paper of Varakin, Trubin, Pronitchev. Koushtuev and Trofimov used vocoders and spread spectrum equipment in mobile systems. Nine papers were availble on propagation problems. We mention the validation of the fourth power law by Grybkowski and Wojnar. Lorke organized two sessions on EMC and Wire Communications, cosponsored by CCITT Study Group V. He gave the general introduction. Professor David Middleton represented the IEEE EMC Society at this Symposium. His lecture on "Threshold signal detection for EMC in non-uniform, non-gaussian EMI fields". He spoke in his usual enthusiastic way, adding new results he found after his text was printed. In the session on EMC of Radio Systems, the 12 GHz band for television got the attention of Bykhovsky. Digital Secam was studied by Karwowska-Lamparska and Kedzierski and FM Color Television by Kalashnikov, Mordukhovich and Nemirowsky. A paper by Chang on the impact of electromagnetic field and pulse on the boiling of conducting and nonconducting fluids was given in my session on lightning, since it treated possible effects on nuclear reactors of lightning and NEMP. EMC and the Radio Amateur Service was the subject of a session organized by Cichon. He and Schwarzbeck studied the immunity of home entertainment devices and video recorders to rf fields. Johnson gave a written contribution on EMC and the New Zealand radio amateur.

Workshop: Lasting effects of transients on equipment performance. June 7, 1983

A booklet on this workshop was reproduced by Uppsala University. It consisted of a general introduction on the task of the working group by its chairman Dr. Scuka. Introductory remarks on the theme of the working group by the Commission E Chairman, Prof. Lundquist. A review paper on "Damage, caused by lightning, EMP and Geomagnetic Storms. A paper by Prof. Wiesinger on "Lightning hazards to low voltage power and communication systems and test procedures". Finally two papers by Dr. Scuka: "Deterioration of surge protection performance of metal oxide varistors (MOV)". "Deterioration of transmission characteristics of optical fibre cables caused by lasting effects of electrical transients". The large room of the auditorium was well filled, and a number of questions had to be answered.

Nonlinear and Environmental Electromagnetics, edited by H. Kikuchi
Elsevier Science Publishers B.V., Amsterdam, 1985 – Printed in The Netherlands

SUMMARY - EDITOR'S NOTE AND ACKNOWLEDGEMENTS

H. Kikuchi
Nihon University, College of Science and Technology, Tokyo 101, Japan

There is a growing interest in the problem of electromagnetic pulse (EMP) in relation to the impact of natural noise or lightning on man-made systems, the advent of nuclear electromagnetic pulse (NEMP) and other unconventional, nonlinear and environmental electromagnetics. For large amplitude waves or pulses with short duration, propagation media or transmission lines often exhibit nonlinear characters, leading to a variety of novel phenomena. Although a number of different approaches were devised for individual nonlinear phenomena, attempts to form a unified or clear view of nonlinear waves and oscillations have rather recently been made only for the past decade. In the field of electromagnetic compatibility (EMC), for example, transmission lines and installations exposed to an external electromagnetic environment due to man-made or natural sources such as lightning or nuclear electromagnetic pulse (LEMP or NEMP) are subject to damage, degradation or impairment that could be more severe by nonlinear effects of induction and interference. Another example is possible nonlinear or ponderomotive force effects of microwave beams from solar power satellites on ionospheric and magnetospheric plasmas. These are only a few examples among many problems in the field of EMC/EMP in which nonlinear effects may play a decisive role. Unfortunately, however, conventional environmental electromagnetics usually restricts itself to wave propagation in a linear medium equivalent to a small signal theory in electrodynamics or plasmadynamics. The main purpose of this book is to incorporate new studies and approaches for nonlinear and/or atmospheric effects into the category of environmental electromagnetics in order to make them feasible for a variety of basic problems in the field of EMC/EMP.

For the last several years, the EMC community is rapidly growing, since the EMC/EMP problems require interdisciplinary and multidisciplinary studies over a broad area of electrical, space science and technology, particularly when natural noise of terrestrial or even planetary origin is involved in a large scale, affecting our environment or interfering man-made systems. Because of this and because the URSI community is active along this line, the Workshop was intended for the first time, as a scientific meeting of the URSI Commission E, to bring together a broad area of scientists and engineers with different backgrounds who are interested in nonlinear and/or electromagnetic pulse aspects of environmental electromagnetics of man-made and natural origin from both theoretical and experimental points of view and to provide a forum for exchange of information among them, highlighting the state of the art in the EMC/EMP field.

The workshop was held at Gakushi-Kaikan (University Alumni Association) Kanda in Tokyo, Japan on 19 October, 1984 and consisted of five regular sessions and one special session with approximately sixty participants from twelve countries: Canada, D.R. Germany, England, France, F.R. Germany, India, Iran, Japan, Poland, Sweden, Switzerland, and U.S.A.

The book contains all the papers presented at the Workshop and covers the following topics:

(1) Atmospheric EMP and VLF Phenomena
(2) Nonlinear Waves, Quasi-Particles and Solitons
(3) Nonlinear and Chaotic Lines
(4) Environmental Electromagnetics
(5) Ponderomotive Force in an EM Environment
(6) Environmental Effects of NEMP.

The last topic was brought into a special session organized along a statement made at the URSI Plenary Assembly in Florence, 26 August-6 September, 1984. The final version of the URSI Factual Statement on Nuclear Electromagnetic Pulse and Associated Effects is included in Supplement with two additional URSI Commission E Review Reports and one review article at the URSI-Sponsored EMC Zürich Symposium.

In this book entitled "Nonlinear and Environmental Electromagnetics", the editor has thus attempted to highlight new topics of environmental electromagnetics in a broad sense from electromagnetic pulse aspects, incorporating both fields of natural/atmospheric and man-made EMP, in particular paying attention to nonlinear effects of strong electromagnetic pulses in terms of NEMP (nonlinear electromagmetic pulse), although NEMP is normally used for nuclear electromagnetic pulse as already mentioned. Since one of the basic problems of nuclear electromagnetic pulse is probably its behaviour as a "nonlinear electromagnetic pulse", both problems are related to each other in that sense. So that both aspects of NEMP are covered for the first time in this book, also incorporating atmopspheric and man-made EMP (AEMP and MEMP) as a common problem from the point of view of strong or large amplitude electromagnetic pulses.

Since topics are rather novel and are covered over a wide range of area from fundamentals to applications, it has been suggested that some summarized remarks may be useful and helpful for audiences and readers with a variety of backgrounds to better understand the purpose, discussions, relations among various articles and results of the Workshop, namely the contents of the Proceedings.

1. Atmospheric EMP and VLF Phenomena

In the introductory and tutorial paper celebrating the eighty-second anniversary of his birth, Professor Kimpara, the founder of atmospheric research in Japan, offers a comprehensive review of half a century's observations of atmospherics in Japan. His extensive and correlative survey covers thunderstorms, typhoons, polar fronts, jet streams, SEA, SID, sudden phase shift of VLF waves, SEA phenomena due to nuclear explosion, waveform of atmospherics and distribution of atmospherics in Aisa. Of particular interest are his correlative studies of atmospherics in an interdisciplinary and multidisciplinary manner with ionospheric anomalous phenomena, meteorology and even nuclear explosion as early as 1958, suggesting many interesting problems to be done in the future. Among them, for instance, nonlinear effects on the waveform of atmospheric EMP would be one of the problems directly relevant to the following chapter. In addition, a new introduction of electrohydrodynamics (EHD) in thunderstorms, typhoons and associated atmospheric phenomena is suggested in particular from the basic and theoretical points of view.

Hayakawa, Tanaka, Sazhin and Okada provide in their paper an interpretation of dawn-side mid-latitude VLF emissions in terms of quasi-linear electron cyclotron instability. Observed VLF results are particularly interesting to the editor, since they are surprisingly similar to our ELF results of short-period Pc-1 micropulsations in terms of generation regions in L-geomagnetic latitude (inside the plasmapause) and local time sector (pre-midnight and dawn) and increasing frequency drift with local time (Kikuchi, Nature, Lond. 229, 79 (1971), Kikuchi and Taylor, J. Geophys. Res. 77, 131 (1972), Kikuchi, J. Atmos. Terr. Phys. 38, 1127 (1976)) in spite of differences in mode, polarization and frequency. While a frequency range of VLF emissions can be in many cases satisfactorily understood in terms of instability of corresponding linear waves, this is not always true for emission amplitudes. In some cases, we have thus to take into account nonlinear effects. Along this line, the authors go on to a quasi-linear theory of electron cyclotron instability in order to explain observed VLF data. Such a direction is relevant to the main purpose of the Workshop.

A field-analysis direction finding system has extensively been used as a powerfull means of estimating absolute whistler intensities at Yamaoka (geomag. lat. 25°N) and Moshiri (geomag. lat. 35°N). By using this, Hayakawa, Okada, Tanaka and Ohta report on day-night and latitudinal variations of whistler intensities. While the intensity ratio of two stations is simply attributed to the latitudinal effect of the ionospheric absorption loss, the day-night asymmetry of intensities is complicated by additional effects of thunderstorm activity in the near and conjugate regions.

While VLF riser emissions triggered naturally or artificially have been observed in the magnetosphere, Ondoh reports on new types of giant riser emissions and VLF hiss with multiple rising-frequency structures, both having a frequency rise from 9 to 18 kHz in 2.5 seconds, which were observed by ISIS satellites at invariant latitudes between 47° and 60°N during the recovery phase of magnetic substorms. The giant riser emissions are different from normal riser emissions in structure and frequency: a frequency range of giant risers between 8 and 20 kHz is higher than that of normal risers between 5 and 10 kHz. Comparing to chorus emissions, however, the rise time, frequency range and frequency rise of both giant and normal riser emissions are much larger than those of chorus emissions. Giant and normal riser emissions are thought to be generated by precipitating ring current protons due to transverse proton cyclotron instabilities. In addition, groups of normal riser emissions whose frequency rise from 5 to 10 kHz in 1.5 seconds were repeatedly observed in a period of 26 - 27 seconds by ISIS-2 at invariant latitudes between 56.5° and 61.9°N during quiet days. Pc-3 micropulsations were also observed simultaneously at Memambetsu (geomag. lat. 34.1°N) with a periodic occurrence of normal riser emissions which is explained in terms of a periodic modulation of riser-emission growth rate resulting from particle precipitations modulated by Pc-3 micropulsations. These observations are a manifestation of nonlinear phenomena in general, arising from wave-particle and/or wave-wave interactions and an introduction of a simple nonlinear model relevant to observations would be an interesting problem from the point of view of nonlinear electromagnetics. Since all the observational results in this paper have been obtained at Kashima station in Japan from a series of Canadian satellites, Alouette-2, ISIS-1 and ISIS-2, this forms a connection with the next paper on review of Canadian activities.

Canadian studies of atmospheric EMP and of EM environment measurements were reviewed by Professor Pavlâsek over a broad range of activities, including studies of: the fundamental behaviour of naturally occurring sources of EMP - atmospheric EMP such as thunderstorms, geomagnetic storms, aurorae, natural ELF-VLF EMP; particular forms of humanly contrived sources (other than NEMP) such as power line radiation; detection, measurement and monitoring; development of sensors, monitoring systems, data management evaluation and interpretation; study of effect on and response of "victims" of EMP; analytical/numerical, data processing techniques for analysis and evaluation. Activities were also reported on "real life" and laboratory measurement, monitoring, simulation, analysis and prediction of the ambient EM environment in high density metropolitan core areas, both outdoors and indoors. Among a variety of these activities, a couple of specific studies should be mentioned a little bit in detail. There is a growing interest in a topic of tall-structure triggered lightning in comparison with usual natural and rocket triggered lightning. The 553 m CN Tower in Toronto is the tallest free-standing structure in the world and therefore it offers a rare and unique opportunity to study lightning EMP. Since 1976 lightning measurements have been conducted on the CN Tower under the auspices of the CN Tower Lightning Studies Steering Committee. An average of 38 flashes per year to the CN Tower has testified to the significance of the effects of the tower on lightning discharge and local climatology and convenience of the study of lightning characteristics. Some of results obtained are included in a review entitled "Overview of Triggered Lightning EMP" in the preceding chapter "SUPPLEMENT". Another specific topic is concerned with "Absorber Lined Chambers for EMP/EMC Measurements". Professor Pavlâsek provides a review of the evolution, methods of measurement and computational design of ALC for EM emission and susceptibility measurement purposes with future development and application trends (Proc. 1984 EMC Tokyo Symposium, p. 599) and in this connection he poses a question on the "Reference Ground Plane" for EMP/EMC measurements, saying that its concept, while meaningful at low frequencies, may have questionable utility at higher frequencies at which ground plane dimensions are comparable to or much greater than the wavelength (Proc. 1984 EMC Tokyo Symposium, p. 417). Further complication may arise at high frequencies for the purpose of simulation or site characterization from the fact that the actual ground is not a good conductor but becomes rather a semiconductor or a dielectric with increasing frequency (Kikuchi, Electrotech. J. Japan, 2, 73 (1956)).

2. Nonlinear Waves, Quasi-Particles and Solitons

In the introductory and tutorial paper celebrating the seventieth anniversary of his birth, Professor Marcuvitz offers a review of his comprehensive studies of "Quasiparticle View of Wave Propagation". Waves frequently have the form of finite wavepackets and a question is how such wavetrains are distorted on propagation through linear, nonlinear or turbulent media. Professor Marcuvitz describes a prescription for how a wavepacket may be viewed as a system of "quasiparticles" whose dynamics permits one both to visualize and approximate the average space-time propagation properties of a wavepacket in diverse media. Quasiparticles are distinguished by their "energy" Hamiltonian, which is a function of their position x and momentum k. One is thus led to a simple physical interpretation in terms of quasiparticles, indicative of a wave-particle duality that is not completely unfamiliar in classical wave dynamics where at least at linear levels it is often expressed in "ray-optic" terms. Time dependent quasiparticle distributions can be described in terms of variables of three kinds: kinetic (momentum and space dependent); fluid-dynamic (space dependent); a localized macroparticle (just time dependent). In the kinetic description, the motion of a system of quasiparticles is governed by the trajectory equations in phase space and their density in phase space is described in the form of the kinetic equation. The fluid dynamic description is obtained from the k moments of the distribution function of quasiparticles and a macroquasiparticle description is recovered from the k and x moments of their distribution function. The particle-wave duality implied in the analysis is discussed both from a quantum mechanical and classical standpoint. Several examples of the use of the quasiparticle method are presented analytically and computationally.

Attempts are made by Kikuchi and Hirota to treat a variety of problems in nonlinear electromagnetics in terms of quasi-particles, solitons and/or shocks and to apply such a concept to physical interpretations for several examples of numerical solutions or temporal evolution of the waveform described by the Korteweg-de Vries-Burgers and nonlinear Schrödinger equations subject to an initial condition. First of all, a new quasi-particle concept as described in the preceding paper is introduced into a nonlinear dispersive and dissipative system and is formulated in the Lagrangian coordinate system for the KdV-Burgers and nonlinear Schrödinger equations with reference to charged particle beams and space charge waves, distributed parameter lines, electrodynamics in the earth's plasmasphere, modulated waves, etc. While both stationary and nonstationary solutions are considered, numerical results for the latter case ensure analytical but approximate solutions obtained otherwise in fairly good agreement. Emphasis is placed on the existence of recurrence phenomena for both cases of the KdV and nonlinear Schrödinger equations that has been proved by numerical solutions as well as analytical but approximate solutions. Applications of the present approach to specific problems of environmental electromagnetics or EMP in a broad sense will be a largely open task of interest in the future. Part of the results obtained in this paper forms a close relation to the next one.

Noguchi reports on "Solitons in One and Two Dimensional Transmission Systems" and considers a transmission line composed of the series inductance with a uniform stray capacitance, which introduces a dispersive effect, and the parallel nonlinear capacitance for the former case. For low frequencies, the line equation is reduced to the Korteweg-de Vries or the modified KdV equation. Observations of the formation and recurrence phenomena of KdV, compressive and rarefactive solitons are displayed as well as those of overtaking and head-on collisions of two solitons. For sufficiently high frequencies and strong dispersion, the line equation is reduced to the nonlinear Schrödinger equation. Observations of the modulational instability and the formation of envelope and dark solitons are also displayed in good agreement with the theory described there and in the preceding paper. In a two-dimensional transmission system, the two-dimensional KdV equation, sometimes called the Kadomtsev-Petviashvili (KP) equation, is derived and a new soliton creation by an irregular collision between two solitons is indicated from an exact three-soliton solution with resonances and numerical experiments.

Nakamura describes his extensive study of "Ion Acoustic Solitons in a Multi-Component Plasma with Negative Ions". First, the theory is extended so as to include negative ions and some numerical simulations such as collisions of KdV and modified KdV solitons are presented. Wave characteristics are distinguished into three classes by a critical value of the negative ion concentration. When the concentration of negative ions is smaller than the critical value, compressive solitons are formed from a positive initial pulse and a negative pulse evolves into a wave train. On the contrary, when the concentration is larger than the critical, rarefactive solitons develop from a negative initial pulse and an initial positive pulse of small but finite amplitude evolves into a wave train, while a positive pulse of large amplitude becomes a solitary wave whose characteristics agree with a prediction by means of the pseudo-potential method. At the critical concentration, both compressive and rarefactive solitons are observed. All the observations presented are convincing and some of them may be interesting in relation to the corresponding observations in a transmission line presented in the preceding paper.

3. Nonlinear and Chaotic Lines

In the introductory and tutorial paper celebrating the eightieth anniversary of his birth, Professor Goto, the initiator of nonlinear and logical circuit theories in Japan, offers a review of his thorough studies of "Undamped Electric Oscillations and Electrical Instability in Transmission Lines", incorporating a new version of the extended theory into his earlier reports. Although this work was initiated as early as 1927 to elucidate anomalous electric oscillations that occurred in actual power lines, some applications of these undamped oscillations to a small size of simplified circuits led to several inventions such as frequency changers or down converters due to subharmonic generation, some of which were used later for the so-called "parametron" and also as a power supply for the telephone-bell and in particular for the bell to announce the opening of the Diet for a long time. In both cases of power line systems and laboratory sets, these anomalous oscillations occur due to the magnetic saturation characteristics of transformers and theoretical work in this connection is still rare in spite of the fundamental importance and occasional occurrence of related phenomena in power line systems, for which the theory should be available directly. In this sense, his work is not only of historical interest but also still stands out even at the present time, for instance with immediate application to the probem of power line re-radiation caused by geomagnetic storms that has recently been observed in a Canadian power line system (Kikuchi (ed), Power Line Radiation and Its Coupling to the Ionosphere and Magnetosphere (Reidel, Dordrecht/Boston/London, 1983), p. 38, p. 203). Then, storm-induced currents under the ground flow into a power line system through grounded lines and transformers and give rise to the increase in transformer-core saturation. Cnnsequently, new power line harmonic radiation is created and this may interact with the magnetosphere, affecting chorus activity and electron slot precipitation.

The Josephson junction line consists of two superconductor strips separated by a thin insulating (junction) layer. When the layer is thin, it permits the coupling of two superconducting states and causes a tunnel effect, the so-called "Josephson effect". Maginu discusses dynamical behaviour of this line with various oscillations described by the damped sine-Gordon equation, covering "Chaotic Oscillations". The spatially homogeneous periodic oscillation is stable when the length L of the junction line is small. However, when L exceeds a certain threshold length L_0, it becomes unstable and a spatially inhomogeneous periodic oscillation bifurcates from the homogeneous oscillation and is also amplified by the homogeneous oscillation as a result of parametric resonance. The amplitude of the inhomogeneous oscillation mode increases with increasing L. When it becomes sufficiently large, another period-doubling bifurcation takes place and consequently a more complicated inhomogeneous oscillation mode bifurcates. A chaotic oscillation is thus observed in the junction line after a sequence of this sort of period-doubling bifurcations. These phenomena can be simulated qualitatively by using a coupled pendulum model. The period-doubling bifurcation from a spatially homogeneous rotating states to a spatially inhomogeneous rotating state is observed when the torque exceeds a certain threshold. This model also exhibits a chaotic motion when the torque is taken in a suitable range.

4. Environmental Electromagnetics

Under the title of "Effect of the Lightning Electromagnetic Pulse on a Telecommunication Network", Degauque, Hamelin and Demoulin reviews research activities in France comprehensively. The paper consists of three parts: 1. electromagnetic field radiated by lightning; 2. coupling to cables; 3. COPELIA project. Part 1 presents the results obtained from both natural and triggered lightning discharges, including triggered lightning research in all aspects of techniques, phenomena and applications, based upon the past 13 years' activities. Three kinds of triggering techniques and three types of triggered discharges are summarized with comparison of natural and triggered lightning. Characteristics of intracloud discharge and return stroke radiation with rapid variations in lightning fields have been measured extensively over a wide range of frequencies up to a VHF or MHz range. Analysis of fine structures thus obtained has enabled us to clarify ionization and discharge processes significantly. This forms a close connection with an article, "Overview of Triggered Lightning EMP" in the preceding chapter, "SUPPLEMENT". Part 2 is concerned with the coupling of lightning to wires, cables (coaxial and multiconductors shielded) and antennas situated above or under the ground in terms of the transfer impedance, current and voltage distributions. The effects of various shielding and shield discontinuities are also reviewed. Finally, Part 3 outlines the new "COPELIA project" which is going on towards the evaluation of electrical surges on the subscriber's lines. Some preliminary results of lightning-induced pulse records, LF common mode voltage records and histograms of the peak amplitude values are included.

Electromagnetic environment consists - among others - of a great number of man-made sources of EM emission. Power lines and radio & TV transmitters are examples of such sources. Professor Stružak offered a review of his comprehensive survey of "Unintentional Radiation from Industrial, Scientific & Medical (ISM) Radio Frequency Generators (that are used for non-telecommunication applications) and Its Influence on Radiocommunications". Total amount of such generators exceeds 100,000,000 in the world scale, according to some estimations. Due to unavoidable imperfections, they radiate RF energy into space. This creates serious threat to radiocommunication services. The review contains main problems associated with non-telecommunication applications of RF equipments and outlines major factors, mechanisms, results and difficulties on the basis of his original model proposed to evaluate the probability of interference and to explain the role of various factors involved (Stružak, Ann. Telecommun. <u>39</u>, 209 (1984); Proc. 1983 EMC Zürich Symp., p. 343). This simple model sounds very interesting and allows us to estimate the fraction of the transmitter coverage area lost due to unacceptable interference. The topic results from current CCIR studies carried out by its Interim Working Party 1/4 for which he is serving as a chairman.

The subject of the interaction of electromagnetic fields with biological systems is traditionally in the domain of URSI Commission A, "Metrology" and is also of current interest to many scientists and engineers in the field of EMC. Behari, Mathur and Sharma report on the topic of "Behavioural Ontogeny in Rats Exposed to Extremely Low Frequency Electric Field". The animals of either sex were subjected to a field of 5 kV/m with a frequency of 50 Hz for a period of two to three months. Male and female rats of two age groups (developing age: 94-138 days and mature age: more than 150 days) were kept singly in each cage and each of groups was housed in a separate large acrylic cage. Besides the experimental group, the control group of animals of the same age and sex were kept in identical boxes without the field being on. The food intake of experimental group of animals is consistently less than that of the control one with the correspondinging effect on the weight gain pattern. In addition, the effect on younger animals is more than on the older ones, although the changes observed are within the physiological limits.

5. Ponderomotive Force in an EM Environment

The ponderomotive force is the nonlinear electromagnetic force acting on a single charged particle or body such as a fluid and a plasma and is defined by its average force on a time scale large compared to 1/f = T (f: frequency), being derived from the "quasi-potential" which is also referred to as the "ponderomotive potential". For the case of a single particle, its smooth guiding motion is derivable from the quasi-potential. For the case of a plasma, the ponderomotive potential takes a form of the radiation pressure that balances the plasma pressure in a state of equilibrium. For further details, refer to the book by Professor H. Motz entitled "The Physics of Laser Fusion" (Academic Press, London/New York/San Francisco, 1979), Chapter 6 that describes a comprehensive overview of the ponderomotive force in both physical and mathematical terms. When intense electromagnetic beams or pulses penetrate into a plasma, they produce various nonlinear effects, among which the ponderomotive force effect plays a significant role in nonresonant processes. For the past decade, the URSI has paied much attention to the environmental effects of microwave beams from solar power satellites and their consequencies for radio wave propagation. For this question, the ponderomotive force effect occupies one of the problems of fundamental importance. The same is true for the environmental effects of nuclear EMP and their consequences for telecommunications, for which the URSI has issued a factual statement, as seen in the preceding chapter, "SUPPLEMENT". Thus, the ponderomotive force effect is a general, fundamental and important process in nonlinear wave-plasma interactions in both laboratory and space plasmas.

Akama and Nambu elaborates a general formula for ponderomotive forces induced by electrostatic waves in a dispersive nonuniform plasma, incorporating all the results reported so far in different ways and even intoducing a few new terms due to inhomogeneities of the dielectric response function in space and time.

The ponderomotive force induced by an incident strong electromagnetic pulse is expected to depress the plasma density in order that the radiation pressure balances the plasma pressure. Sugai testifies to this prediction by a laboratory experiment after a brief review of the ponderomotive force effect. A 9.37-GHz, 10-kW, pulsed microwave with 3 μsec duration was injected into the underdense plasma where the electron plasma frequency is less than the wave frequency. During application of the intense electromagnetic pulse, the plasma density is observed to be depressed by the radiation pressure (ponderomotive potential). In addition, a damped slow oscillation of the plasma density is found after turn-off of the microwave pulse. This relaxation oscillation was identified as an ion acoustic wave driven by the ponderomotive force.

6. Environmental Effects of NEMP

During the XXIst General Assembly of the URSI in Florence, 26 August - 6 September, 1984, a factual statement on nuclear EMP and associated effects was prepared and soon after issued by URSI, as seen in the preceding chapter, "SUPPLEMENT". Along this line, the present chapter consists of four papers that highlight the state of the art from purely scientific and technical points of view, also incorporating the related field of EMC.

Following a brief discussion of a definition of the term EMP as a summary of all existing electromagnetic pulses, Nedtwig and Geilke reports on EMP/EMC activities in West-Germany in the following terms: analyses of threat and coupling data (lightning, nuclear EMP, electrostatic discharge etc.); measurement techniques (EMC, NEMP, lightning); EMC- and EMP-standardization; NEMP simulation techniques; protection measures for electronic systems and equipments. For a smaller size of NEMP simulators, TEM-cells and open waveguides have been developed. The advantage of TEM-cells is the very clean transfer characteristics in the empty measuring room. The disadvantage is the existence of the cut-off frequency for excited higher modes rather than the TEM mode. For investigations and a series of tests, the open-waveguide simulator (3 m high) is more convenient. The pulse generator reaches a maximum amplitude of 100-300 kV in less than 5 ns for a line impedance of 100 ohms. For this type of open waveguides, STANDARD-EXO-NEMP and STANDARD-ENDO-NEMP have been developed in a single and in a multiple-pulse mode operation. In general, their work is directed not only towards nuclear EMP but also towards all existing EMP in a multidisciplinary way, as indicated in an extensive survey with 54 figures which are most illustrative and instructive.

Ianovici presents a comprehensive overview of EMC and in particular EMP activities in Switzerland, including industrial activity, university and laboratory research and the teaching aspect. An EMP simulator of a guided wave type called GESIBAN was in operation from 1978 to 1982 to test objects and systems and to verify their hardening performances. It consists of a 450 kV, 600 J generator and a balanced wave transmission line terminated on a characteristic impedance of 90 ohms with a symmetrical type of antenna, 2.5 m high, 6 m wide, 5 m deep and a total length of 20 m. Today a modified simulator with an asymmetrical type of antenna but with slightly larger dimensions called HEISS 1 is operational. The Swiss industry produces many test installations for EMP simulation. An example is a mobile EMP-system with high flexibility, producing a uniform field strength of 50-400 kV/m in a volume of 3-25 m, depending on the size of the antenna. Another example is an impulse generator producing rectangular shaped currents and voltages. Tranfer impedance measurements are made by two types of installations. One is a continuous wave equipment using the quintaxial method, while the other is an impulse current installation. To estimate currents and voltages induced by EMP in an areal or buried conductor, a calculation method has been developed with two computer codes at the Federal Institute of Technology in Lausanne. This has enabled to define how to optimize the cable shielding for various shielding structures (for instance, a structure of three sheets: AL-STEEL-AL). Excellent agreement has been obtained between calculated and measured response of a cable to an current injected in the sheath, having the same order of magnitude and shape as a current induced by EMP. A study of EMP effects on objects is underway by using a scale modelling installation.

Tharby offers an overview of "Electromagnetic Surge Problems on Telecomms Plant" in terms of physical damage, software corruption and service degradation, mainly based on experiences on lightning surges but taking into account exo-atmospheric NEMP discharges which cause radiation surges up to 50 kV/m with rise-times of the order of nano-seconds. There is an essential difference between NEMP and conventional EMP such as lightning EMP and electrostatic discharges. While NEMP could lead to widespread damage to telecomms lines and terminal equipments, conventional EMP can have a more localized form directly or near to telecomms line plant. This may lead to considerable differences in protective measures.

As already mentioned, the URSI with Mr. Manuel Wik in the chair has issued a factual statement on nuclear EMP and associated effects. In the present paper, Mr. Wik describes a comprehensive overview of NEMP problems, covering the EMP threat, generation and magnitude of NEMP, effects on telecommunication facility, hardening and testing methods. Normally it is stated that the geographical coverage of exoatmospheric NEMP is limited by the tangent to the earth from the burst point. However, EM waves propagate further away within the earth - ionospheric waveguide as ground waves and waves reflected in the ionosphere and ground, as indicated from the detection of nuclear explosions. The fields can evidently strong enough to cause malfunction and possible damage further away than the tangent radius. The Teak explosion over Johnston Island on 1 August, 1958 at a height of 77 km had a tangent radius less than 1,000 km. In spite of this, the explosion burned out radios in the Wallis Islands 3,400 km away and caused trouble at Hawaii at a distance of 1,600 km from Johnston Island. These observations call for further studies of EM spectra at large distances.

Supplement - URSI Commission E Activity during 1982-1985

It has been suggested that the inclusion of some URSI Commission E activity in the Proceedings might be of general interest and of use to audiences, particularly among the URSI and EMC/EMP communities. Along this line, the editor has prepared a supplemental chapter of URSI Commission E activity during 1982-1985, consisting of four articles:

1. URSI Factual Statement on Nuclear Electromagnetic Pulse (EMP) and Associated Effects / URSI Nuclear EMP Ad Hoc Committee
2. Overview of Triggered Lightning EMP / H. Kikuchi
3. URSI Commission E Activity in Japan during 1982-1984 / H. Ishikawa
4. URSI Commission E Activity during 1982-1984 / F. L. H. M. Stumpers.

The first one is a final version of the factual statement on NEMP that was accepted unanimously by the URSI Executive Council and is closely related to the preceding chapter of "Environmental Effects of NEMP". The second is a review article at the URSI-Sponsored EMC Zürich Symposium, March 5-7, 1985 which was delivered tutorially but only briefly at the opening of the session on triggered lightning EMP. The third one is a revised version of the national URSI Commission E report. The last one is Professor Stumpers' report presented at the URSI Commission E Meeting in Florence in August, 1984, mainly based upon URSI-Sponsored European Symposia, since the report on Commission E activity in Review of Science is already available as an open literature.

Finishing up the editor's note, he would like to add a few more messages to audiences. Because of rather unconventional interdisciplinary topics, a wide range of scientists and engineers with different backgrounds have participated in the Workshop and some topics may have been unfamiliar to some participants. In spite of this and a very heavy schedule from 9.00 a.m. to 9.00 p.m., every session was conducted in an enjoyable and stimulating atmosphere in a rather informal style. Hopefully, this volume could convey such a spirit, containing a mixture of review and original articles which may not find its proper place in the literature of the field, thus enabling us to exchange information perhaps larger in scope than a journal article. In this sense, this book may be the first one that covers a variety of nonlinear and environmental electromagnetic waves and pulses.

In addition, the editor should note the inclusion of three special articles, among a total of twenty-seven, that have been invited in a spirit of celebrating some seventieth to eightieth anniversary of the authors' birth. They are: "Half Century Observations of Atmospherics" by Professor Kimpara, celebrating the eighty-second anniversary of his birth, "Theory of Undamped Oscillations and Electrical Instability in a Transmission System" by Professor Goto, celebrating the eightieth anniversary of his birth, and "Quasiparticle View of Wave Propagation" by Professor Marcuvitz, celebrating the seventieth anniversary of his birth. The editor has attempted to take this opportunity to make the readers aware of these outstanding studies with his deep appreciations to the authors.

Acknowledgements

The authors and the publisher have made special efforts for rapid publication of the state of the art of the selected topics. Many thanks are due to the authors for their prompt preparation and excellent presentations. Special thanks are due to Elsevier Science Publishers for their conscientious help and generous support in the course of publication.

As stated at the beggining, the Workshop was held as the first scientific meeting of the URSI Commission E before and during the next Plenary Assembly at Tel-Aviv among several meetings planned at the URSI Commission E Business Meeting in Florence. In this connection, I am particularly grateful to Professor F. L. H. M. Stumpers, Chairman of the URSI Commission E for his support and encouragement during all phases of the URSI Meetings and the Workshop and particularly for providing the "PREFACE" of the present volume, together with many suggestions and advices in the course of editorial work.

In addition, many persons contributed to the success of the Workshop. Professor R. G. Strużak, Vice-Chairman of the URSI Commission E, was most helpful, providing many suggestions and advices during the progress of the Workshop. Professor H. Ishikawa, Chairman of the National URSI Commission E, was most helpful for preparation of the technical program.* The organizer is grateful also to Professors and Drs. P. Degauque, M. Ianovici, J. Nedtwig, T. Pavlášek, R. G. Strużak, and M. W. Wik who served as session chairmen during the Workshop and contributed greatly to its success by conducting their sessions in a most engaging and stimulating manner. It is a pleasure to thank Mrs. M. Sugisawa, the director of IBF Company for her editorial assistance and Mrs. N. Suwa for her secretarial work.

Finally, on behalf of the URSI Commission E, the editor should like to extend his sincere thanks to authors, participants and supporting organizations for the success of the Workshop and for the consequent publication.

---

* In this connection, special acknowledgement is extended to Dr. Shiro Hohki, past Vice President, Sumitomo Atomic Energy Industries, Ltd. for his suggestions, support and encouragement during all phases of the Workshop, particularly in preparation for the NEMP-associated program.